LIFE

SCIENTIFIC PHILOSOPHY/PHENOMENOLOGY OF LIFE AND THE SCIENCES OF LIFE

Ontopoiesis of Life and the Human Creative Condition

BOOK I

ANALECTA HUSSERLIANA

THE YEARBOOK OF PHENOMENOLOGICAL RESEARCH

VOLUME LIX

For sequel volumes see the end of this volume.

LIFE

SCIENTIFIC PHILOSOPHY, PHENOMENOLOGY OF LIFE AND THE SCIENCES OF LIFE

Ontopoiesis of Life and the Human Creative Condition

Edited by

ANNA-TERESA TYMIENIECKA

The World Phenomenology Institute

Published under the auspices of
The World Institute for Advanced Phenomenological Research and Learning
A-T. Tymieniecka, President

KLUWER ACADEMIC PUBLISHERS

DORDRECHT / BOSTON / LONDON

Library of Congress Cataloging-in-Publication Data

Life-- scientific philosophy : phenomenology of life and the sciences
of life / edited by Anna-Teresa Tymieniecka.
v. <1-   > cm. -- (Analecta Husserliana ; <59-60>)
Based on a conference.
Includes bibliographical references and indexes.
Contents: bk. 1. Ontopoiesis of life and the human creative
condition -- bk. 2. The outburst of life in the human shpere.
ISBN 0-7923-5141-X
1. Life. 2. Phenomenology. I. Tymieniecka, Anna-Teresa.
II. Series: Analecta Husserliana ; 59, etc.
B3279.H94A129 Vol.59,etc
[BD435]
142'.7--dc21 98-8187

ISBN 0-7923-5141-X

Published by Kluwer Academic Publishers,
P.O. Box 17, 3300 AA Dordrecht, The Netherlands.

Sold and distributed in North, Central and Latin America
by Kluwer Academic Publishers,
101 Philip Drive, Norwell, MA 02061, U.S.A.

In all other countries, sold and distributed
by Kluwer Academic Publishers,
P.O. Box 322, 3300 AH Dordrecht, The Netherlands.

Prepared with the editorial assistance of Robert J. Wise.

Printed on acid-free paper.

Printed in the Netherlands.

TABLE OF CONTENTS

PART II
THE VARIATION OF ONTOPOIESIS

PART III
NATURE AS ALIVE

PART IV
TRANSITIONS: LOGOS AND EXPERIENCE

Macerata

ACKNOWLEDGEMENTS

In bringing this collection of studies – the first book of a two-volume publication, *Life Scientific Philosophy/Phenomenology of Life and the Sciences of Life* – to the international forum of readers, I want to express my warmest thanks and appreciation to those who have made it possible. First, I'd like to thank the University of Macerata, especially its Department of Philosophy and the Human Sciences under the enlightened direction of Francesco Totaro, for the most generous hospitality extended to us in the days of the conference held there on our theme, and even more for the impressive organization of that event, an effort that extended to all the Italian universities.

This was our first International Conference on Philosophy/Phenomenology and the Sciences of Life, and the exactitude and diligence put into its preparation and direction by Daniela Verducci, Secretary General of the conference, and Daria Carloni, both of the department, were invaluable.

We owe our usual thanks to Robert Wise, Jr., Assistant Editor of the *Analecta Husserliana* series, for his expert editing of this volume and the preparation of the index.

Mitchel Seagrave, my assistant, has been most helpful with the preliminaries.

Our greatest thanks, naturally, go to our dear authors.

A-T. T.

Anna-Teresa Tymieniecka, Richard Kulis, Velga Vevere, Maija Kule and Louis Houthakker

THE THEME

PHILOSOPHY/PHENOMENOLOGY IN DIALOGUE WITH THE SCIENCES OF LIFE: *Toward a Mathesis Universalis for Our Age*

With this collection, philosophy/phenomenology of life deliberately takes a radical novel turn. We have been approaching this moment step by step for the past two decades, a process that may be traced through over twenty volumes in our *Analecta Husserliana* series, especially the three volumes bearing the title *Life* (Volumes XLVIII, XLIX, and L), the last of which has the subtitle *Philosophy/Phenomenology of Life as the Starting Point of Philosophy*.

Now we have arrived at the core of our inquiry – direct dialogue between philosophy/phenomenology and the "hard sciences". It appears that the time is right for a large-scale attempt at this dialogue. Our first international conference devoted to it put life at the center. Its particular focus on phenomenology of the ontopoiesis of the life and of the human creative condition resonated with the preoccupations of many scientists and with those of scholars in other fields as well.

As shown in my early book, *Phenomenology and Science in Contemporary European Thought* (New York: Farrar, Strauss and Cudahy, 1962), the theme of which received great attention in a large number of publications, that although phenomenology has deeply influenced the human sciences – psychiatry, psychology, sociology, the theory of literature and the fine arts, etc., its forays into the "hard sciences" such as biology had been limited to questions of methodology and to related matters such as intentionality and embodiment as seen from the transcendental point of view.

Among phenomenologists, there prevailed a certain clear reluctance to deal with the empirical sciences. Behind this reluctance, priority was given to a universalizing way of conceptualizing that had no access to concrete empirical issues. Further, the early Husserlian stress on "apodictic certainty" as an indispensable criterion of truth and of the validity of cognition militated against acceptance of the empirical sciences, the theories of which are by epistemological standards merely hypothetical. We may say, that among phenomenologists there existed a natural prejudice against the natural sciences given that they held a standard of certainty that those sciences do not know. And then phenomenology was simply entrenched in the transcendental realm, to which the empiricist is not privy.

In the inaugural study of this collection, I argue that the present-day situation is a new one, the result of a profound fermentation in philosophical reflection, on the one hand, and an ever clearer manifestation of the real nature of scientific inquiry, on the other. To cut a long story short, on the side of philosophy/phenomenology the postulates of epistemological rigor, of methodological precision, have either been loosened or are being abandoned. I do not speak here of those on the fringes of philosophy who have abandoned these postulates altogether – who simply do not communicate with phenomenology. In the phenomenological mainstream intentional articulations are still held to be valid. New is that contemporary phenomenological reflection, following Merleau-Ponty and the late Husserl, has extended the transcendental into the somatic realm so that no sharp boundary separating mind and body is drawn. The empirical has entered the philosophico-phenomenological field. The last of Husserl's reflections are now being brought into prominence, and it is seen that the master himself in his private reflections was probing below the transcendental level by seeking to stretch intentionality through the entire functioning of the conscious being in the lifeworld.

In the last decades an interest in phenomenology has developed in the empirical sciences, mainly an interest in its descriptive method, its "unprejudiced" approach to things, as well as in some of the themes of transcendental inquiry such as intentionality, the mind, the body within the lifeworld.

What has been crucial, however, is that in our times, new discoveries, slowly but persistently documented, have impressed themselves on the human mind. Philosophy – and phenomenology in particular – have to take these findings into account. As I point out in my study in this collection, new inventions have established the philosophical significance of the sciences beyond cavil, given the depth to which we may now penetrate the empirical.

I affirm that the light new technologies have thrown on vital and cosmic forces has brought about an uprecedented revolution in how we view the human being and all living beings in their situation, in their world. We can no longer think of the lifeworld and the place of human beings within it, their roles and interdependencies, in the preliminary – naive, that is, prescientific fashion found in the approach to these matters in classical Husserlian phenomenology prior to Merleau-Ponty.

The pivot of philosophical reflection has now to be redefined: from epistemological certainty distilled from cognition, we move to *the vital certainty of participation in life*. The criterion of truth certainly is essential when applied to knowledge of reality, the constitution of objectivity, and the manifestation of life. It even serves as an auxiliary guide for scientific discovery and the application of it in invention, since it is an essential factor of objectivity. Nevertheless, the certainty with which we may recognize, and constitute ideal structure recedes in significance before the certainty of vital/cosmic participation, the enjoyment of which encompasses the sciences and binds in one truth complex the objective reality of the human being and that of all the universe. The technology that the natural sciences have made possible, now superimposes the criterion of practical application on real, concrete, objectively universal nature/cosmos/life, allowing the human being to not simply "cognize", but to participate actively in the laws and processes of nature to his advantage. Creating novel means for existence now takes precedence over arriving at certainty and cognizing "absolute" laws.

Despite the stance of some recent philosophical works, our culture has not given up on the need for certainty and universal validity. We have just shifted the testing of certainty away from the distillation of mental evidence into an ideal construct, an idea, an eidos, with its principles, to the "pragmatic test" that the natural sciences apply to reality. As probed by our technology, reality is not an abstract representation, but is participated in from within vitally significant perspectives. Thus, in order to cognize this reality, and philosophy has after all never renounced to do so, dialogue with the sciences is now of paramount significance. This dialogue brings us into the subterranean realm invisible to simple vision, into the becoming and radiation of life.

Developments necessitate a radical shift in philosophical reflection, away from an obsession with either the mental cognitive aspect of life and the world or with an eidetic insight into its structures and toward a full-fledged dialogue with the sciences, which not only reveal life in its generic aspects, but allow us to enter into it directly by following its tranformative responses.

This is to replace the cognitive criterion of reality and truth – recognizing its validity in its own realm – with the vitally significant creative criterion of life's becoming and transforming. We are not abandoning the traditional inheritance, we are merely shifting attention from cognition to the creativity of the scientist, the explorer, the inventor in *techné*, from marveling to experiencing, from observing to working. With this shift, we philosophers are ready to enter into direct dialogue with the sciences and seek our common goal.

But what might that common goal be? On the side of scientific research and theorizing, which in the last decades has made great advances in establishing links between the divided disciplines, the quest is for a common universal *plan* (see my study in *Analecta Husserliana* Vol. LII, *The Great Plan of Life – The Phenomenology of Life's Return to the Sources of Western Philosophy*, 1998, pp. 3–29).

I propose that in order to give form and body to such a plan, the common work of science and philosophy/phenomenology should be that of laying a common ground of a new *mathesis universalis*, a universal phenomenological science that would corroborate the main principles of the ontopoiesis of life in such a way that all the furrows that the human being makes in life by participating in and enjoying it, would find a basis and linkage, intergenerative rules and laws, constructive interdependencies and mutual prompting. Laying this groundwork for communication between all human endeavors would fulfill the strong desires of present-day scientists.

This ground can only be found in the blueprint of life, which is not a static ontological structure but a vortex of constructive becoming, enjoying, and transforming. The workings of nature, those of psyche, mind, spirit, and their enjoyment in the marvelous game of life should meet in the diffusion of the sense and enjoyment of life.

We have been laying down the principles and, as has been manifest in our publications, elaborating this point of departure for a *mathesis universalis* that may in our day fulfill the dream of Leibniz. The present collection of studies shows how a dialogue with the "hard sciences" accords with the workings of mind and spirit, and how the separations of theory and praxis, of nature and culture, vanish in the perspective of our project.

Though well advanced, it remains still a project. We expect that the grist of concrete scientific contributions will intensify our discussions giving their yield more substance.

A-T. T.

INAUGURAL ESSAY

ANNA-TERESA TYMIENIECKA

THE ONTOPOIESIS OF LIFE AS A NEW PHILOSOPHICAL PARADIGM

INTRODUCTION: IN SEARCH OF A NEW PARADIGM

1

A look at the course of the history of Western philosophy will show that it has been fixated with certain philosophical theories. The terms of their discourse have informed our culture and our whole outlook on life. Each historical period may be characterized by some notion or theory that plays a dominant role until subjected to enough scrutiny and discussion to be tested and the evidence for it found wanting. As the formulation of a theory is eroded and its appeal fades, it then under pressure yields place to an incoming idea and becomes secondary in our deliberations or recedes into the background.

In this as well as other respects, philosophy – as a historical phenomenon of human cognitive culture – resembles science in its progress. In the historical progress of human knowledge certain ideas emerge at a given point in time and direct the current of thought for an era. What philosophers and scholars in general consider to be the proper subject matter for their investigations is dictated by how the reigning model of thought defines problems.

Such a model has since Plato been called a 'paradigm.' In our day Thomas S. Kuhn put this concept to use to describe the process of scientific progress. A paradigm he describes as a theory defining a set of issues, one which seems "to guarantee the existence of a stable solution."[1]

We have seen in the historical evolution of philosophical research the repeated formation of networks of scholars who take certain outstanding core ideas as the point of departure for their research. Their effort does not aim at coming up with a novel approach or at introducing ideas that supercede the paradigm. On the contrary, they aim to analyze the core ideas in order to identify problems that are contained in them or subtend them, to articulate the core intuition. The work of the schools that formed around the thought of Plato, Aristotle, Kant,

20

24 66

A-T. Tymieniecka (ed.), Analecta Husserliana LIX, 3–28.

Hegel, Husserl and Heidegger by no means led to new theories. In a manner parallel to that found in scientific practice as described by Kuhn, these schools focus the attention of their inquirers on the unfolding of a main idea or paradigm "in detail and depth that would otherwise be unimaginable."[2] Such an investigation aims at framing problems that follow from the leading theory and finding solutions to those problems consistent with the theory. What above all attracts the attention of scholars is the articulation of the issues presented by the paradigm. The coherence of all the individual research undertaken by philosophers, scientists, scholars of any kind would be unthinkable without such unequivocal and binding model theories.

In the course of these intense investigations into nature, the cosmos, society, art, etc., insights occur that the paradigmatic theory cannot accomodate, that lead beyond it and call out for another theory to support them. This leads invariably to a disarray in viewpoints, the erosion of the received paradigm, and ultimately to a shift in paradigms.

This change itself, the disarray in thought (the "revolution," as Kuhn called it), leading on from newly discovered phenomena, intuitions, ideas to the formulation of a novel paradigm is a specific process. There is a special dynamics in this transformatory process. Upon the first impact of new facts or thoughts there is a reaction. According to Kuhn this resistance provides a safeguard from being carried away too easily by new insights, "new" with respect to the hitherto accepted dogma. But the next, significant step or move in the dynamic of a paradigm's unfolding is the necessary process by which the new insight so foreign to the ruling paradigm will "penetrate existing knowledge to the core."[3]

The first mark of a shift in paradigms in this description is the onset of a transitional period in which the leading idea is eroded and thought is in disarray, the period of what we are used to calling a "crisis," in this case a cultural crisis. The issues raised in this critical period must bear intimations of a new solution. Just enough fragmentary evidence and insight must infiltrate "settled" knowledge for the underpinnings of the old theory to undergo upheaval. The unaccounted-for facts must be impressive enough to inspire a call for a new explanation.

But the other mark of a paradigm shift is that of continuity in thought. We may say that a new theory or leading idea nourishes itself on its own insights, but it never cuts its umbilical cord, as it were, to the evidence that sustained the old model. Far from vanishing, that evidence is incorporated into the new theory.

The fact is that Heraclitean flux, Platonic ideas, Aristotelian being, Thomistic analogy, the Leibnizean monad, Kantian transcendental consciousness, Husserlian phenomenological method, and the Husserlian-Heideggerian lifeworld endure on the philosophical scene. They do not vanish, but recede into the background as the schema of thought that they dominated dissipates before new ruling ideas. The ideas that once permeated our philosophical-cultural discourse linger on, serving as points of reference in our disputations, and strangely they will reappear in the current of thought having new connotations within new discourses. This continuity in thought does not mean, however, that the old paradigms are valid or retain their past significance within new schemas of thought.

The adoption of a new paradigm marks a turn in culture. The old tissue covering the real is rent, and from the new glimpses of the real got through the opened seams there emerges a fresh insight that as it were throws a new mantel over the real. As it resolves puzzles and serves the practical needs of the time, it inspires wonder. But the labyrinth of human concerns remains, and all the past pushed into the background just will be involved in the quest and emerge from it even as present-day queries press on down new avenues of thought and enterprise.

2

There is no need to describe again the disarray in thought brought on by the contemporary discoveries that have so enlarged the scope of human powers and unsettled our social and political affairs. We are left with the broken pieces of the old worldview and desperately long for a new Promethean worldview. It is, I repeat, a Promethean worldview that we seek, because we are increasingly aware of the possibilities our powers give us for so transforming nature that we will not be subject to its vicissitudes. We are aware too, that we are in need of illumination to give us direction, significance, purpose. We need the measure of things. There is need for a key by which to grasp things at their core, to open their intricate avenues leading on to a clear perspective on all so that the very secret of existence, its orientation, its reason may be grasped.

Before I will attempt to describe a possible key to the secret spheres of existence and a new appropriate vision of All, I will first give the bearings and provide some postulates indispensable for undertaking this task.

Our key has to be a key to the common categories of reality in all its diversity: cosmos, bios, culture, human destiny. And true to the main philosophico-phenomenological principle, this key has to do justice to each of the diverse experiences in which the seeker that the human being is participates in these wonders, contributing human constitutive engagement to the task.

That implies that we have to reach the very core of all, from which everything originates and differentiates leading on to the tangle of innumerable interactions that confronts us as each entity follows its own line. We have to reach the "truth" of things below the level of human intellectualizing with its simplifying and speculative tendencies for these menace and pervert the authenticity of the real by reducing things in ways that too readily satisfy the habits of our mind.

It is the notion of *the ontopoiesis of life* situated within the philosophy of life, its center and its nerve, that accomodates these postulates and promises to fulfill the expectation of a new intellectual paradigm.

The biological and physical sciences of today also seem to take these postulates as their own, at least partly, as they seek the primary and secondary constitution of nature and enter into the operations of cosmic forces and uncover the mysteries of animal and human life. Given their explanations of phenomena and the verifying technical applications these explanations have allowed, it appears as though science may in its dizzying drive onward vanquish all the puzzlement of the human mind. There are leading scientists, such as Stephen Jay Gould, who hold this view. It bears the hope that science may expand its scope to take in all of reality.

Science, indeed, not only apprehends but also handles reality as it enters into what was previously considered an enigma. Is science capable of doing this universally? Does it not use and need philosophy in its advance?

Indeed, scientists in various fields have expressed a present need for philosophy, for ontology and metaphysics specifically. It is enough to quote from an introduction René Thom has written for a philosophical book by noted biologist, Jean Largeault. Thom states that scientists are in need of philosophy to pursue their work.

> . . . dans la mesure-assurement faible – où la science voudrait, pour se diriger elle-même, faire appel à autres critères que l'efficacité pragmatique que – par exemple – ameliorer l'intelligibilité de ses théories, alors une étude transdisciplinaire des moyens de créer une ontologie adequate à la description d'un champs phénomenal devra nécéssairement s'imposer. Il s'agira soit en heuristique d'énoncer les règles qui pour-

raient rendre efficaces les diverses hermeneutiques (découvrir le "réel voilé"), soit sur le plan d'une sociologie critique de la science, d'edifier une methodologie générale susceptible d'apporter un peu d'ordre dans la joyeuse anarchie qui caracterise actuellement – dans les disciplines peu theoretisées comme la biologie – elaborations des protocoles experimentaux.[4]

Furthermore, Thom, opposing the tendency of analytic philosophy to turn the real around without touching it in maintaining that science consists merely in formulas and predictions, stresses that at the core of scientific research dealing with changing reality there must be assumed some permanent, stable, givens "d'une certaine générativité." "Or il n'ya pas de générativité sans générateurs, entités pourvues nécessairement d'une certaine permanence et stabilité." Referring to the universal structure of algebra, he insists that "certain individualized elements (usually represented by the letters of the alphabet) must express themselves in order that the constructive and algebraic generativity of the systems may be deployed." He argues that when this generativity is efficacious and present in our experience, we cannot believe that it consists merely of symbols.[5]

Pursuing this idea further, Thom affirms that it is just a small step to affirming that "even in science, ontology is necessary; metaphysics is not dead."[6]

PART ONE: TOWARD THE ALLIANCE OF THE PHILOSOPHY/ PHENOMENOLOGY OF LIFE AND SCIENCE

1

Today disintegrating tendencies are running rampant through our culture. It is obvious that on all fronts, scientific, cultural, personal, philosophical, we are in a critical situation in which old views are evaporating and new ones are coming in like hurricanes.

As is all too well known, the latest waves of philosophical thought to burst on the scene – though long brewing – dismantle all traditional conceptual frameworks, whether realistic or idealistic, metaphysical or ethical, etc. Naturally, the epigones of earlier twentieth-century thought do continue to pursue lines of research set down by their traditions and standing personal projects. But in the main the philosophical climate is hostile or at least diffident to inherited concepts. In this climate leading ideas are abandoned, as the focus of thought turns in random directions. Philosophers seek to "update" the historic issues of the discipline with reflections on side issues and import material from the human sciences, linguistics, psychopathology, etc. in an attempt to thereby make their work salient to the interests of contemporary discourse.

The concepts of truth, the founding of cognitive data, of legitimacy seem to vanish from the scene as points of orientation for thought. Philosophy at large no longer aims at giving "final" answers to "perennial" questions. All foci, all vectors of thought are gone. Confusion reigns and is justified in a way. Contemporary philosophy is ready to forsake its vocation as even contemporary philosophical paradigms disintegrate.

We can safely conjecture that the major cause of this disarray is the enormous wealth of new scientific insight in all fields that has gone unassimilated in contemporary philosophy. With all the discoveries being made in biology, astronomy, physics, we can literally speak of the emergence of a "New Science." The disorientation in world itself that this has caused discomfits philosophy too. The new scientific reality merits our special attention.

2

We are at a stage of transition in both science and our culture at large. Humanity finds itself in a sharply delineated transition period in all spheres: cultural, social, political, and scientific, which spheres usually coincide. Thus, when the Newtonian science of the seventeenth and eighteenth centuries – which had a deterministic, mechanical model of reality and which presented the world as originating from initial conditions in a strictly mechanical fashion so that all further developments were strictly determined, each being a step in a universal mechanical process, so much so that Laplace claimed that on the basis of it we could predict the future – reverberated through the scientific world and was almost universally accepted, that was because it accorded with the social outlook of the era of the Industrial Revolution, in which society too was viewed mechanistically.

Today we are witnessing the end of the Age of the Machine. This is Alvin Toffler's thesis in his foreword to Ilya Prigogine and Isabelle Stenger's work, *Order out of Chaos. Man's New Dialogue with Nature.*[7] As Toffler sees it, the deterministic model of the world was under attack already in the nineteenth century with the discoveries in thermodynamics, Darwinian biology, and quantum physics. It could then retain partial validity as a reference point for research and the formulation of issues. But in more recent times science has undergone a truly radical transformation such that the assumption of even a basic order and rationality in nature can no longer be persuasively upheld and is losing ground in a profound reassessment as new models of reality suggest themselves.

As Prigogine, a prominent scientific researcher and interpreter, states, "Our physical world is no longer symbolized by the stable and periodic planetary notions that are at the heart of classical mechanics. It is a world of instabilities and fluctuations, which are ultimately responsible for the amazing variety and richness of forms and structures we see in nature around us."[8]

With Ilya Prigogine we may speak of a revolution in the scientific outlook, of the birth of a New Science.[9] The emphasis of classical science on the principles of stability, universality, regularity, symmetry, equilibrium recedes from the foreground and is replaced by the evidence of crucially significant states of disorder, arbitrariness, instability, irregularity, disequilibrium.

In all the sciences, not only the life sciences but those of physics, astronomy, mathematics as well, the essential role of change, transformation, evolution, event in the universe, the earth, human society has come to be appreciated. We have become aware of the birth and dispersion of elementary particles and of galaxies too, of changes in chemistry and geological upheavals that would be considered exceptional events in a mechanistic model. These are now considered to be part of a grand but hazy picture, as are the puzzling origins of living beings and the modalities of their differentiation and evolution, as are the unaccountable origins of and shifts in societal norms. The search for answers must correspondingly be undogmatic.

This new outlook has proceeded from the "discovery" of time in physics, once almost ignored and now recognized as having crucial significance. The New Science, as presented by Prigogine in his numerous books, offers a "new dialogue" between the human being and nature. At its crux is precisely a reversal in the significance attributed to the temporal aspects of becoming.

3

With the introduction of the notion of "complexity,"[10] encompassing all modes of order and disorder, we witness a bifurcation of hitherto one-sided concepts. For example, there are evident in dynamic systems contrasting processes that conserve energy and dissipate it. Similarly, we see mechanical and thermodynamic equilibria balanced by constraining non-equilibria. Moreover, Prigogine makes a sharp distinction between "closed systems" in which things originate, change, deteriorate according to fixed patterns and "open systems" in which energy maintains itself.[11] It is the open systems of becoming that are primordial; these are open to exterior forces and exchange of energy with the environment being susceptible to influences and exercising influence in turn.

The concept of open systems has emerged in response to the issues raised by Darwin's evolutionary theory and the dynamic systems observed by Prigogine. The common way of conceiving the temporality of becoming has been completely revised. When Ludwig Boltzmann set for himself the task of identifying evolutionary phenomena in the physical sciences analogous to those observed in the life sciences, he found them on the level of populations of molecules. He attempted to describe not only the equilibrium found in a population of molecules but how that equilibrium evolved. In doing so he discovered the irreversibility of the toward-equilibrium process, a time vector similar to that found in the evolution of species.[12]

Critically, Prigogine pursues the notion of irreversibility and attributes it to all open systems.[13] He shows that open systems, whether physical or biological or social, do not proceed in a reversible fashion, that the processes of constructive constitution do not go backwards. On the contrary, they follow a "vector of time." They are one-way constructions due to this irreversible vector of time.

It is precisely in such open systems having this constructive direction, interacting and exchanging energies with their environments in a random, irregular, topsy-turvy fashion that Prigogine sees the initial conditions of becoming. Biological and societal system present particularly striking open systems. Biology and genetics show us that below the recurrent scheme of life that we conventionally attribute to life's processes – what is merely an objectified, universalized surface – lie life's inner workings.[14] Under the surface innumerable sub-systems pulsate, bubble, criss-cross; instead of stability here is constant disorder and fluctuation.[15]

The very recognition by scientists of haziness, fleetingness, arbitrariness in physics and then in biology extends to all sectors of reality. In psychology too there is recognition of the turbulent life of the mind at the pre-conscious level. The pre-conscious turmoil of the psyche is also an open as well as closed system or cluster of systems, out of the interplay of which that which is irregular, crooked, fleeting, singular emerges. This is a game of chance factors, necessary dispositions, and the unforeseeable, the unpredictable.[16]

But what is most striking in all this is the rapprochement being achieved between the physical and the human, social, cultural sciences. The realization of the historicity of the human being and the course of society and culture is, following Dilthey, Unamuno, Ortega y Gasset, at the vibrant leading edge of contemporary thought. With Husserl and Heidegger's concept of the lifeworld and with Gadamer and Ricoeur's hermeneutics, this realization has entered literature, linguistics, sociology, political theory. We may safely say that it is transforming the human sciences. The goal in these disciplines is no longer the rigging of rigid, immutable models. The reality of "progress" has been discovered, the critical phases of turbulence, consolidation, and dissipation. With the recognition of the irreversible phenomena of physics and their constitutive propensities together with the vector of time, we are finding common denominators in physics and the human sciences.

The finishing touch of Prigogine's approach to becoming is his conviction that becoming is self-generative. Like Aristotle, Leibniz, Spinoza, Kant, and others, Prigogine believes that becoming emerges "from within," *sua sponte*.

Here is the gist of the phenomenology/philosophy of life. Assuming that the varieties of becoming all proceed *sua sponte* from the interplay of the regular development of forms and irregular, unpredictable conditions, Prigogine suspends the sharp dilemma of determinism and freedom, necessity and chance. Both are at work in the processes of the universe. There is a vast intermediary realm, then, in which it is interrelations that are to be investigated.

Significant in its own way within the modern intrusion of unpredictability into the abstract, mechanical order reigning in classical science is "chaos theory," that is, the study of turbulence. Altogether singular and unrepeatable is the flow of smoke out a chimney. From the same initial conditions, that flow can take innumerably different courses. Consideration of this invites a look into the chaotic disorder behind all ordering, the fleeting behind the fixed.

Furthermore, with recognition of an intrinsic mobility in all physical nature, the radical contrast between spontaneous movement in living beings and inertia in inanimate physical being vanishes. And so the mechanistic model yields to an overall organic model. The organization and finality of physics have thus come to approximate those of the life sciences.

The concept of science has undergone a radical transformation. Indeed, recognition of "chaotic systems"[17] and "catastrophe theory,"[18] has removed the backdrop of a manifest objective order of the universe, world, and life by revealing a turbulence of bubbling energies and forces running at random.

There is a new approach to scientific validity as such. The classical postulates of precision, exactitude, certainty lose their hold on the imagination. We move to viewing a hazy, imprecise, fleeting reality. In this way the "hard" sciences seem to be becoming more like the sciences of life and society.

This movement of the sciences toward each other is particularly obvious in the case of a mathematics that now treats sensitive and qualitative features as well as forms different from those of classical Euclidean geometry. This is the *fractal geometry of Nature*.[19] Although this geometry was discovered as far back as Leibniz, to whom its present inventor, Benoit Mandelbrot, refers,[20] and although it was somewhat developed at the end of the last century, it has just now gained proper acceptance and appreciation. It concerns the forms of nature, things etc. We are accustomed in life as well as in scientific inquiry to rely on forms, structures, on geometry in general as we deduce it in our constitution of reality. We seek in nature the geometrical relations so constructed. So-called "fractal geometry," however, looks past the preconceived forms usually seen in nature and the whole of the reality which science encounters, seeing that there is there a completely different composition of things.

Liberation from Euclidean geometry's circles, squares, cones – into which we have been trying to squeeze reality – shows us the structure of nature's "dislocated" irregularity, all sorts of irregular objects torn, and fluid in their relations, constructs. Here is a new mathematical approach to nature, one freeing it from the absolute rigidity of forms and structures into which classical geometry pressed reality. Going against a growing tendency of formalism in mathematics that leaves behind human intuitive representation, Mandelbrot's fractal approach to reality is all intuitive. His device is, "to see is to believe."

From the side of mathematics, then, comes a revolutionary strong affirmation of the universal significance of the concrete, unrepeatable, unique.

The infinite range of the fractal forms proceeding from mathematical algorithms effects a crucial transition in mathematics from an abstract way of conceiving nature to one which passes into the visual. Mathematics is, as it were, given senses adequate to the riches of objective experience. We move away from the classical prejudice that mathematics involves "calculability" only in a qualitative, aesthetic expansion of the discipline. The abstract science of mathematics "humanizes" itself!

At the end of this all too short survey of the revolutionary changes in science that have thrown of our hitherto cultivated worldview into disarray, recognition is due to Alexandre Kojève for his having brought out the most significant factor of the "subject," the living concrete individual who as an inquirer envisages everything around him/herself, whose role is now universally accepted in physics and the rest of science. In describing the subject's central role in scientific investigation, Kojève gave it this basic characterization: we should not identify the subject with a mathematical, abstract point, uniform and unchangeable, nor with its biological corporeity, nor as a psychological agent.[21] It remains to be seen how we must conceive of the subject, according to its function in investigations.

At this point scientific investigation encounters the Archimedean point of the philosophy/phenomenology of life. We will come to it in the second part of this study.

For the time being let us briefly look at the grand project of the "New Science" from some distance.[22]

4

To conclude our brief account, let us emphasize the striking innovative tendencies in science. 1) The physical and mathematical sciences seem to have abandoned sharp boundaries with biology as well as with the social sciences and cultural inquiries. The strict calculative nature of mathematics has taken on a qualitative aspect. These sciences seem to have become "humanized." 2) All of the disciplines have become sensitive to time and change. 3) Their theories of becoming and development seem to share some common features. 4) This sharing among the sciences without the breaching of their sharp boundaries does not allow placing them all on equal footing, nor reductively subsuming some under others, but indicates that a dynamic swing of generation, of ordering, of interactivity may well run through the entire gigantic game of existence. 5) With consideration of they the transitory dimensions, transitory trajectories of the dynamic complexes of the world, with the shift in focus of thought away from seeking closed reversible systems to apprehending open self-projecting streaks in the cosmos as well as in nature-life and social life as well as appreciating the vast territories of their attunements, interferences, gulfs of mysteries are opened for science to explore.

These are the vast intermediary, unknown areas of interlinkages, generative propensities, and seminal endowments-in-process, ever expending dynamic spheres of manifestation with their own turmoils advancing and regressing in complexity and quality, with phases of catastrophe and of regulative constructivism that have now become the fascinating areas of secretive reality. They draw our inquisitive mind wider and further. 6) But it is recognition of the central role of the subject in the process of science as such that will offer us a crucial point for the dialogue between philosophy/phenomenology of life and the New Science.

Today's science is, indeed, offering us a new vision of the universe, nature, society. In fact, the chaotic and turbulent stream, the innumerable streamlets which make up cosmos, nature, life, society and culture, in which from arbitrariness, chaos, chance there emerge segments of ordered world, such that we may acknowledge through our own existence in relatively stable societal, natural, cosmic existential conditions, opens fascinating newly to be formulated issues, views, expectations.

This preeminence given to the turbulent, fluid, accidental, irregular, disorderly in the origination and progress of All does not mean, as I have hinted at a few times, a universal "disorder" or a forsaking of order and rationality. On the contrary, it opens vistas in which we have to ask after the kinds, rules, ways of interlinking, of intermingling, molding. . . . There are no sharp divides between matter and life, nature and the cosmos, nature and human culture, but vast intermediary spheres which fascinate our imagination.

This calls for the discovery of laws of transitional phases, of coincidence, encounter, and interlinkage, of systems of spontaneous emergence, of spontaneous designs or projects, developments. This also calls for the investigation of the nature of the center point of scientific inquiry, the human subject.

With this we enter into the sphere of the philosophy of life. To a superficial glance it could appear that this new vision of the world, life, cosmos, human social life in superceding classical visions makes philosophy's traditional queries and conceptual frameworks obsolete or that science has simply replaced them. Could philosophy become obsolete, indeed subject to the penetration of scientific inquiry?

Nothing could be more hasty and erroneous. But also nothing could be more preposterous than a philosopher who believes it possible to reach reality through primary experience and the power of speculation while ignoring scientific inquiry.

The striking fact of our present situation is that philosophy needs to consult scientific data, inquiry, methods in order to be able to grapple with reality. *The natural and human sciences in turn need a philosophy that is appropriately informed by them for the more profound organization and interpretation of their findings and their own advance.*

In short the situation of our culture with all its potentials and hazards calls for *an alliance between philosophy and science.*

Remarkably enough, the radical new perspectives which science opens fall in line with those being taken by the new philosophy. The phenomenology of life and of the Human Condition emerges like the phoenix from the ashes of traditional thinking.

The project of the philosophy/phenomenology of life and of the Human Condition springs forth from the idea of this alliance. The concept of the "ontopoiesis of life" is the crucial link and vehicle of the project.

PART TWO: FOUR PIVOTAL INTUITIONS OF THE NEW ALLIANCE BETWEEN PHILOSOPHY/PHENOMENOLOGY OF LIFE AND SCIENCE

There are four pivotal intuitions and proceeding from them four critical issues which are the meeting points for the phenomenology/philosophy of life and the sciences of life and the physical sciences in general. These issues also reverberate in our time's preoccupation with order and disorder, necessity, orchestration, etc.

First of all, new awareness of temporality of *Why is There Something Rather than Nothing? Prolegomena to the Phenomenology of Cosmic Creation* (Assen: Van Gorcum, 1966), events, processes, transformations in the organic as well as in the inorganic sphere has provoked great puzzlement over the nature of "developments," that is, of the irreversible processes that carry life onwards. This is now the central issue of science. Addressing it is the grand idea of formation in which becoming may be grasped – the concept of *ontopoietic unfolding*, which constitutes the ontologico-metaphysical axis of becoming as such as well as of becoming in its lineaments. This is the fulcrum of the phenomenology-philosophy of life.

The second pivotal point of encounter between the sciences of life and philosophy of life is the whole question of the formation of "complexities" which confronts the sciences of life and of all reality "from physics to politics."[23] Whether the complex reality we are facing be a living being, a society, a political state, a work of art, etc., we intuit that there is an ultimate manifestation of "self-organization." On all levels phenomenology of life apprehends this *ontopoietic* process unfolding from within and directed by the guidelines intrinsic to the complexity-in-formation, beingness, entity.

Thirdly, philosophy of life and the sciences of life meet in the intuition of the guiding entelechial sequence of life's unfolding, the linkage between individuation and speciation, the individual and the evolution of forms.

Fourthly, and most importantly, science and philosophy of life meet in the intuition of the Archimedean point that is the ground for inquiry into all existence, that is, the *creative condition* of the investigator, whether experimenting, or observing, or speculating.

This convergence in phenomenology and the physical and life sciences of intuitions striking the same chords on the crucial issues of our culture has yielded a universal paradigm.

Therefore we will enter into our analysis of these essential correspondences by discussing the convergence between the "physical subject" of scientific experimentation and the creative human act and the more fundamental ontopoiesis of life.

The Creative Act of the Human Being as the Archimedean Point of the Encounter between Science and Philosophy of Life

1. *The "Physical Subject" in Scientific Inquiry and the Creative Mind*

It is on the point of the inquirer as "subject," as the concrete center of any investigation, a point now recognized by the New Science, that science and philosophy of life and of the human creative condition arrive at a crucial understanding. Listening to an experimental scientist talk about his experience, we enter into the heart of the matter:

It is an experience like no other experience I can describe, the best thing that can happen to a scientist, realizing that something that's happened in his or her mind exactly corresponds to something that happens in nature. It's startling every time it occurs. One is surprised that a construct of one's own mind can actually be realized in the honest to goodness world out there. A great shock, and a great joy.[24]

The experiences of the scientific discoverer are not like any other. It is an experience of the creative mind. And it is precisely the creative human mind immersed in the natural, physiological, psychic, intellectual circuits of an individual human person engaged in creative activity that calls up from its innermost core such powers as allow him or her to meet the powers of nature itself. The phenomenology of life and of the human condition proposes an evolutionary phase in which emerged the human creative condition accounting for this extraordinary synchronization of functions, energies for the constructive application of powers, for this extraordinary condensation of the entire spectrum of the universal conditions that the human creative mind emerged from, a mind that is not only capable of objectifying, differentiating, and charting the immensity of the real in which other beings are passively immersed in and participate in, but is – at the summit of its powers – capable of entering into the inner workings of that reality.

Hence it is from the point of investigation into the human creative genius that it is appropriate to enter into the exploration of reality. Here is our Archimedean fulcral point from which to probe all existence. Along these lines we pay close attention to Alexandre Kojève as he describes his views as a physical scientist on the human subject as the reference point of scientific inquiry, of all inquiry. Kojève – in his magisterial analysis of the basis upon which was founded the causal determinism of classical physics and of the principles by which it was undermined – elucidated the interpretation given by Niels Bohr to the arguments presented by Heisenberg on the essential and unavoidable imprecision of any attempt to examine the world physically, on the impossibility of speaking in physics of "exact causality in the structure of the world."[25] Bohr's interpretation is, according to Kojève, a mathematical expression of an absolutely general principle according to which no physical observation is possible without the state of whatever is observed being modified "by the very fact that it is observed."[26]

It is not that physicists were not over time aware of this "gnoseological" state of affairs, but it was Heisenberg who drew all the conclusions together. These conclusions could have been drawn already within classical physics. According to Kojève, "a necessary consequence of the classical principle of the equality of action and reaction is: if a physical entity is observed that means that it 'acts' upon the instrument of observation; this instrument has then necessarily from the outside 'acted' upon it and modified it in a certain fashion."[27] That is to say, with Heisenberg and Bohr it is the nature of experience and experimenting in physics that was brought into focus. With the theoretical assumption that physics deals with the real world and with the concepts which ultimately may be brought to experimental data, Bohr specified that physics does not deal with one world system, as it is in itself but with two systems: the system of the observed and the system of the observer.

There is no way in physics to change the fact that it moves along the borderline between these two systems, which are both opposed and inseparable. Kojève specifies: "In effect, the observed system is not accessible to experience unless insofar as it is in an interaction with the observing system, modifying it, and is in turn being modified by it."[28]

There are two consequences of this capital recognition. One of them leads Kojève to affirm that it constitutes a principle rejecting classical causal determinism in physics, effecting the passage to modern physics which holds that physics does not study the world "in itself" as idealized by the spirit but the world which is real and is given in experimentation and observation, in experience, that is, made by scientists with real, physical instruments, scientists who themselves are part of the real physical world that they study. The second conclusion that we may draw with Kojève from Bohr's analysis is the clarification, elucidation of the situation of scientific experience as such. This clarification leads to the definite acceptance of the physical subject at the center of physical inquiry, which subject belongs to the real world and simultaneously observes it and acts upon it, reaction to which in the world physics obtains in its data later.

Here comes the fascinating question of just how we should understand this subject in scientific experience. Philosophers have long since discussed these things and various of their formulations have thrown up distorting grids between the real world and the perceiving, experimenting subject. The main requirement of the new science is that the subject be seen as belonging to the same ontological region as the world and as interacting with it. In any case, I claim that given all this we cannot continue to consider cognition to be the main factor in scientific experience.

True, Heisenberg in discussing his "idealized" experiences emphasizes that he is discussing the *cognition* of the real but not the real itself. (This is also the view of Stanley Salthe, who throughout his book *Development and Evolution, Complexity and Change in Evolution*, to which we will return later, emphasizes that physics is talk about the "discourse" concerning reality and not about reality itself.[29] But in a "discourse" approach the subject is of the same significance discussed above, since he is the author of the discourse.)

However, I propose that we ask ourselves what we must understand in speaking of the subject in the experience of scientific inquiry; we have to turn our attention to the collection of scientific data, their "verification" through technology. In the perspective of this collection we find a direct interference of the subject in the real, physical nature of the world. It is not discourse about this nature that makes it possible for the inventor to apply physical principles, to put material, physical materials to use. I submit that we must keep this point in mind as we seek a more adequate description of the subject in experimental experience and that we should seek it elsewhere than solely in the cognitive faculties of the human being. These faculties have to be acknowledged to belong essentially to processes deeper than our experimenting and law formulating. We reach the workings of nature under a yet deeper jurisdiction.

To state our problem in its fullness let us call it after Bohr the problem of the relationship between *the system of the observer* and *the system of being observed*.[30]

A. It we attempt to analyze these systems, we find that the subject in the experience has to be a real physical, physiological being in order to belong to the real world. But physics, and science generally, is not interested in the variables that account for the singular features of a phenomenon. On the contrary, science is concerned with the constants. Consequently, we cannot conceive of the subject as being a singular individual with varying tastes, capacities, tendencies, etc. Inasmuch as the subject has to be concrete living being, we have to make an abstraction of its singularities and focus on its universal/concrete individuality. According to Kojève, the "physical subject" is a physical entity insofar as it is represented by a system of physical entities."[31]

B. To its system must belong the entire schema of a specifically human personality embodied within a physical, biological framework. Here is a specific type of personality which is inclined toward and endowed with the capacities for scientific inquiry and it assumes various constant forms in accord with the special scientific interests of scholars.

C. How could we conceive of the scientific subject otherwise than as one endowed not only with all the elementary sensory, emotional, and valuating faculties making it an integral participant in nature/world, but with a mature human mind with its focusing, deliberating, calculating, and speculative powers? How could any observer not endowed with these three modes of operating ever approach reality?

And yet, this is not enough. In order to complete the picture we have to acknowledge the great lights that throw it in relief. This entire system would not fulfill the expectations we commonly have of it, if it did not rotate in all its aspects around the Archimedean point that is its specific but constant axis: *the creative virtualities subtending the mind – the creative imagination inspiring it and the creative act bringing that imagination to its unique fruition.*

If we unfold the "creative system" of the human being as the scientific subject, we will understand it in the light of what is accomplished in this extraordinary interaction between the technical application of science and the workings of nature. We will also unroll and circumscribe the creative compass of all the spheres of reality/life in which the living creative subject has to participate in order to assume the role of the observer or experimenter, or discoverer, inventor, creator.

In short, I submit that only the creative mind of the human being can fulfill all the conditions set by Kojève, first, and most significantly, by legitimating its extraordinary vantage point and second by introducing us into the hidden spheres of reality itself.

2. *The Circuits of Reality Revealed Through the Creative Act of the Human Being*

The thesis of the argument we will present may be summarized as follows. Within the mental, cultural, and vital expanse of the living human being there are present peculiar vestiges of all the molds in which living beingness has progressively unfolded from the womb of the biosphere, of all the degrees of life's inward/outward directed system of unfolding. As the study of phylogeny and ontogeny shows us, none of these constructive steps can be omitted in the progression to the next level. This means that the human individual stretches vitally throughout space within the Human Condition.

But let us now begin our argument within our own context, showing that it is in the creative act that the human being retrieves the fruits of its unfolding.[32] Where physics begins with the most fundamental elements of the real, in following the creative act of the human being, we have to distinguish first the sphere of the spirit and intellect of the human being – what is most directly engaged in the intuitive, exploratory, inventive, and creatively imaginative processes. But following this thread we are led to the vast turmoil of the individual psychic life of the human person. Here, first of all, a person gathers a conundrum of habits, predilections, scales and categories of evaluation which permeate his or her functional system. All this, however, is to some or other degree conducted or inclined by the persons's will, aspirations, curiosities. We must recognize that personal factors in our psychic functional system command our feelings, emotions, wishes, aspirations, and the like and have an overall combinatorial tendency to bring the turmoil of disparate acts into some cohesive constructive composition whether merely to serve the demands of survival or at higher levels personal satisfaction, a sense of accomplishment.

We will see this psychic openness to constitutive modes much more clearly still if we will consider that it is immersed in a quite different pre-conscious turmoil, a turmoil involving the arbitrary and deformed.

And the intuition of Heraclitus comes to mind who, as interpreted by William Capelle, says: "Die Natur der Welte enthuelle sich ihm als er in die Tiefen seiner eigenen Natur hinabsteig."[33]

The idea of the human being as a cosmos in filigree is as old as Western Philosophy. Already with the Pre-Socratics Anaximander speaks of the cosmos as mirroring the human social order insofar as it indicates that its composite elements are to be kept within the confines of "justice" and "retribution."[34] Pythagoras draws a parallel between the "harmony" he conceives to be central to the order of the cosmic spheres and the human being in whom body and soul have to work together in harmony on a miniature scale.

This idea of the human being as presenting in miniature the whole of cosmos is reflected in Plato – in the *Timaeus* 35 A – when he draws a figure of the human soul and its combining opposite strivings toward the "pure" world of ideas and the "lower" world of the body as a charioteer driving two horses with great difficulty, for reason and irrational desires do not easily carry on together. The soul by partaking in both worlds plays a median role between them.

But it is in Leibniz's concept of *the monad* that we find the most striking picture of all living beings – each is animated, alive, and *reflects the entire universe*. It does so according to its own expansion and in its own perspective. Each living being is an embodiment of the universe, its living transposition in filigree, pulsating with the universe's life on its very own. Leibniz saw infinite gradations in the complexity and modes of nature, each of them reflecting the universe in its making.[35]

In his conceiving of the individual living being as a monad, Leibniz emphasized the reasons why "each created monad represents the whole universe."[36] He brings out first his general metaphysical concept that all there is is interconnected. We read earlier,

> For everything is a plenum, so that all matter is bound together, and every motion in this plenum has some effect upon distant bodies in proportion to their distance, in such a way that every body not only is affected by those which touch it and somehow feels whatever happens to them but is also, by means of them, sensitive to others which adjoin those by which it is immediately touched. It follows that this communication extends to any distance whatever. As a result, every body responds to everything which happens in the universe, so that he who sees all could read in each everything that happen anywhere, and, indeed, even what has happened and will happen, observing in the present all that is removed from it, whether in space or in time "All things are conspirant," as Hippocrates said.[37]

And then, to come back to the passage previously quoted explaining how the monad may mirror the entire universe, he writes:

> Thus, although each created monad represents the whole universe, it represents more distinctly the body which is particularly affected by it and of which it is an entelechy. And, as this body represents the whole universe by the connection between all matter in the plenum, the soul also represents the whole universe in representing the body which belongs to it in a particular way.[38]

The great question is what is the "position" of the human mind such that we may attribute to it the power to descend into the inner workings of becoming and to then lift them up from their particular irregular/regular, chaotic/leading mix to an ordering, seemingly separated from that mix and in fact involving intermediary territories. What "sight" sees into this immeasurable turmoil in which no order, no reason is visible and then distills sense from its fragments, truncated pieces, segments of ordering-in-process and by innumerable nudges provokes recognition of the wealth of rationalities which are projected by the conjunction of hazard and necessity in their constructive game?

We submit that it is precisely in the transitory phase of the Human Condition that we have within the topsy-turvy flux of constructive/destructive, advancing/receding progress within the distorted and yet constant "unity-of-everything-there-is-alive" an effervescence of the vast intermediary phase stretching from the life process getting ready for its constructive swing to the radical transition in which self-enclosed inner direction shifts toward an ever widening opening for interaction with the environment, interaction in which the soul in its "highest" swing enters into the entire spread of the "lower" bodily, organic and inorganic functioning of nature-life as well as the cosmic dynamism.

The imaginary intuitions of the Greeks, the metaphysical speculations of the moderns find an echo in the contemporary approach with its reformulations and adumbrations – its opening horizons. First, the human microcosmic realm at every moment gathers into its composition the functioning of the various preceding phases of the evolutionary process; nothing is lost, all is revaluated with respect to the new virtualities currently being activized.

We have confirmation of this in science. Paleolontologists in reconstructing the intermediary stages of the brain's development from anthropoid to full human being, have found an incremental enlargement of the brain. At the same time neuropsychologists have demonstrated that the human brain is composed of three spheres of functioning that are all the time actively adjusting to each other. That is to say, *homo sapiens* has three brain centers: the reptilian brain, the mammalian brain, and the human brain. The reptilian brain evolved first and is still maintained in the human brain. Reptiles are characterized by lack of care for their offspring. When the mammalian brain evolved millions of years later as an extension of the reptilian brain, the reptilian brain did not vanish. It remained to provide the instinctive responses needed for individual survival, while the mammalian brain extended the individual's concern to the care and survival of its offspring and its group as well, but not beyond that. We see this at work in present-day animals. Some of them, like birds, display a solidarity with their whole flock. The brain specific to humans sustains what I call "creative" activity. It allows the expansion of the social, cultural world, while relying on the instinctive and caring responses of the reptilian and mammalian brains.[39]

This "third phase" in the human brain's development was marked by the growth of the neocortex. Its development made the median position of the human being possible. Self-individualizing beingness unfolded its latent powers, virtualities, valuating capacity at this stage allowing an outburst of personal freedom by which the individual may take in hand, at least partly, its own course, forging its own identity and destiny. This is the grand transitory phase in which all that was tending precisely toward such a liberation of the latent faculties of living beings saw the dawn of the Human Condition. A measure of freedom was realized within individualizing existence. All the preceding threads of the self-individualization of life have been gathered up and reworked in the accomplishment of this transition. The individual may now employ for itself all of life's streaks of energy, forces, segmented integrations, disintegrations, powers to mold its own functioning in novel significant fashions. This is what the creative virtualities of the Human Condition offer.

Thus the human condition becomes a relatively stable station in the process of life's game, a station processing all the material coming from the "lower" circuits of existence for the establishment of a "higher" region, that of the creative mind. The novum which the human condition as a phase in the progress of life presents is precisely creative virtualities attuned to the unique conglomerate of functions gathered up in this constructive passage.

The creative act of the human being in its meanders yields insight into the "creative forge," the sphere in which our specific, singular objective oriented creative process encounters its source. The source is the human being who carries out the creative quest. In this quest the human being descends not only into the originary moment of the singular creative process but most significantly, into the networks of its existential/vital functions, which carry the creative quest as such. He discovers that the specific creative search after a shape, a form, or a substance for an object in view is carried on by a shifting schema of functions in which all of the individual's powers – the intellectual, imaginative, sentient, volitional, physiological – are involved in specific ways, employed from a center, this center being the fulcrum of force, the agency in which all the powers are gathered and from which they flow with roles being assigned to them.

In short, there is an "agency" in the performance of the creative act who plays all the strings that radiate in all directions, a "power" that gathers and distributes, directs and controls every move, a central distributor of forces and roles, a full-fledged conscious being who is obviously self-governing and self-initiating in its acts. This so ramified, versatile, imaginative, and powerful constitutive act fulgurates from its innermost. It is a simultaneous orchestration of all the faculties under the aegis of a creative imagination that projects possibilities of an intelligence that scrutinizes, compares, differentiates, etc., and of an effective will which prompts the search and the progress. All of these faculties represent the dynamic complex of the living individual, carrying the process and determining its self-promoted constructive/interpretive route. Here we gain access to the inner virtualities, freely projected from within in consistent, albeit fluctuating and changeable, directions as trial and error dictate – directions whose sequence itself knows interchangeability and mutability, is uncertain in its steps, and yet, as fragile as it may be and as unpredictable as its outcome may be, being subject to disruption and periods of stagnation, still advances with a discrete continuity/discontinuity of purpose. In all its potentialities, virtualities, advantageous situations for their actualization as well as hindrances, through progressive steps, this is a self-projecting, self-organizing system of meaning by which an entity, an object, a creation is produced by human acumen and power as it were crystallized.[40] These poietic threads reveal the lines human functional powers follow and *the poietic selfhood of the human being as a projecting and effectuating agent.*

Drawing a conclusion from the above, we may recapitulate by stating that it is due to the creative virtualities of the human condition – as a station in the evolving progress of types with all their ties to the cosmos and its laws and to the biosphere – that the human creative act may progressively penetrate into all the spheres of existence, of life, the reality in which this station is not always openly rooted, but out of which it has developed in stages maintaining permanent ties.[41] Since these developmental stages represent the becoming of the universe of life, we find here a new version of the Leibnizean monad that "reflects" the entire universe. But as we will see in our further analysis, this is a different type of monad.

The significance lies in the elucidation of in virtue of what the creative act of the human being may penetrate into the innermost workings of nature, existentially partaking of the interaction which the living being maintains with them. For this is what makes the creative human individual unique and what must be taken into account in appreciating him or her as the "subject" in scientific experimentation and experience.

To be continued in volume LX.

The World Phenomenology Institute

NOTES

[1] Thomas S. Kuhn, *The Structure of Scientific Revolutions* (Chicago: University of Chicago Press, 1970).
[2] *Ibid.*, p. 24.
[3] *Ibid.*, p. 65.
[4] René Thom, "Preface," in Jean Largeault, *Systèmes de la nature* (Paris: Vrin, 1985), p. iv; and "La creation de nouveau par le hasard; un vieux theme epicurien," *Archives de philosophie* **53** (1990).
[5] Thom, "Preface," *op. cit.*, p. iv.
[6] *Ibid.*
[7] Ilya Prigogine and Isabelle Stengers, *Order out of Chaos, Man's New Dialogue with Nature* (Boulder: New Science Library; New York: Random House, 1984).
[8] *Ibid.*, p. ix.
[9] Grégoire Nicolis and Ilya Prigogine, "Introduction," *Exploring Complexity* (New York: W.H. Freeman, 1989), p. ix.
[10] *Ibid.*, pp. 71–141.
[11] *Ibid.*, pp. 45–71.
[12] Ludwig Boltzmann, *Populäre Schriften* (1905) (Braunschweig-Wiesbaden: Vieweg, 1979).
[13] Prigogine and Stengers, *op. cit.*, pp. 257–290.
[14] Thom, *op. cit.*
[15] Largeault, *Systèmes, op. cit.*
[16] *Creation et désordre, recherches et pensées contemporaine* (Paris: L'Originel, 1987), (Interviews with Henri Atlan, Guitta Pessis-Pasternak, Gérard Ponthieu, and Michel Treguer).
[17] David Ruelle, *Hasard et chaos* (Paris: O. Jacob, 1991); Ivar Ekeland, *Le calcul, l'imprévu* (Paris: Seuil, 1984). For fascinating explanations of chaos theory geared to the layman as well as of related theories making up the "New Science," see James Gleick's, *Chaos, Making a New Science* (New York: Penguin Books, 1988).
[18] Ekeland, *op. cit.*, pp. 122–153.
[19] Benoit B. Mandelbrot, *The Fractal Geometry of Nature* (updated and augmented) (San Francisco: W.H. Freeman, 1977).
[20] *Ibid.*
[21] Alexandre Kojève, *L'Idée du déterminisme dans la physique classique et dans la physique moderne* (Paris: Librarie générale française, 1990).
[22] We may, as some scientists do, doubt the revolutionary character of the "New Science."

Several of its innovative intuitions and insights into nature had been voiced as far back as Leibniz, who discovered already the fractal forms found in nature, as Mandelbrot adverts to. The same Leibniz was bound to the mechanical model even as he was voicing the dynamism characteristics of nature-life; but in fact, he emphasized forcefully the omnipresence of living creatures and the infinite richness of their forms. We may also recall the emphasis on the dynamism of the All in the thought of Schelling, one so pervading sentient factors that he could speak of the "soul of the cosmos." It could also be said that already in the nineteenth century the discovery of thermodynamics broke the static, mechanical Newtonian model.

But today we have discoveries on all sides, in biology, genetics, evolutionary research, all dealing with the nature of "crises," with temporal, changeable, transformational phenomena being joined to these are the insights of the social sciences, economics, political theory. (See Charles P. Kindleberger, *Manias, Panics, and Crashes: A History of Financial Crises*, 3rd ed. [New York: John Wiley, 1996].) They all see at the roots of nature-life and social and cultural life the irregular, the changeable, the critical, transformation, becoming, innovation. The concomitance of all this is indeed something new.

This change of emphasis from the static to the dynamic, from the necessity of the certain to the unpredictable does not mean, however, a negation of order, an abandonment of rationality, a renunciation of solving the mysteries of life – a point to which we shall return.

23 See *Creation et désordre*, *op. cit.*, pp. 13–53.

24 Leo Kadanoff, quoted in James Gleick, *Chaos, Making a New Science* (New York: Penguin Books, 1988), p. 189.

25 Kojève, *L'Idée du Déterminisme*, *op. cit.*, p. 152.

26 *Ibid.*

27 *Ibid.*, p. 154.

28 *Ibid.*, p. 157.

29 Stanley Salthe, *Development and Evolution, Complexity and Change in Biology* (Cambridge, MA: MIT Press, 1993), p. 44: "Once more I remind the reader that what I am taking about is not the world but discourse."

30 Niels Bohr, (1928), pp. 245ff; and his *Die Naturwissenschaften* (quoted by Kojève, *op. cit.*, p. 000).

31 Kojève, *op. cit.*, p. 167.

32 For this context, see Anna-Teresa Tymieniecka, *Logos and Life*, Book 1: *Creative Experience and the Critique of Reason*, *Analecta Husserliana*, Vol. **XXIV** (Dordrecht: Kluwer Academic Publishers, 1988).

33 William Capelle, *Die Vorsokratiker, die Fragmente und Quellenberichte übersetzt und eingeleitet* (Leipzig: A. Kroner, 1935), p. 148. In my monograph *The Great Plan of Life*, in Anna-Teresa Tymieniecka (ed.), *Analecta Husserliana*, Vol. **LII** (Dordrecht: Kluwer Academic Publishers, 1997), I quote and discuss this fragment.

34 Rudolf Allers, "Microcosmos from Anaximander to Paracelsus," *Traditti* **2** (1944), pp. 319–409.

35 Gottfried Wilhelm Leibniz, *Monadology*, ed. and trans. Leroy E. Loemker, in Gottfried Wilhelm Leibniz, *Philosophical Papers and Letters* (Chicago: University of Chicago Press, 1956), 2 vols.

36 *Ibid.*, p. 1055.

[37] *Ibid.*, pp. 1054–1055.
[38] *Ibid.*, p. 1055.
[39] Steven Jay Gould, *Ontogeny and Philogeny* (Cambridge, Mass.: Bellknap Press, 1977), p. 000.
[40] For a fuller treatment of cultural creation, see Anna-Teresa Tymieniecka, *Logos and Life*, Book 3: *The Passions of the Soul and the Elements in the Ontopoiesis of Culture. The Life Significance of Literature* (Dordrecht: Kluwer Academic Publishers, 1990).
[41] See my elaboration of the Human Condition in Anna-Teresa Tymieniecka, *The First Principles of Phenomenology of Life*, *Analecta Husserliana*, Vol. **XVII** (Dordrecht: D. Reidel, 1978).

PART I

DIFFERENTIATION AND INDIVIDUALIZATION OF LIFE

At the registration: Francesco Totaro, Daniela Verducci, organizers of the congress; Louis Houthakker, assistant.

MIECZYSŁAW PAWEŁ MIGOŃ

THE ONTO-POIESIS OF LIFE IN ANNA-TERESA TYMIENIECKA'S PHENOMENOLOGY

". . . O Light Invisible, we praise Thee!
Too bright for mortal vision.
O Greater Light, we praise Thee for the less,
The eastern light our spires touch at morning
The light that stants upon our western doors at evening
The twilight over stagnant pools at batflight,
Moon light and star light, owl and moth light,
Glow-worm glowlight on a grassblade.
O Light Invisible, we worship Thee!. . ."
– Thomas Stearns Eliot

INTRODUCTION

Anna-Teresa Tymieniecka, the author of *Logos and Life*, Book 3, *The Passions of the Soul and the Elements in the Onto-Poiesis of Culture,* draws one's attention to connections between the primordial Light and the Logos of Life within the human being, the life-world and the wider setting of culture (viz., literature, art, science, etc.). In Tymieniecka's *Tractatus Brevis*, Section Two, Chapter One bears the title "The Primeval Light and The Birth of The Life-World" and Chapter Two, is titled "The Passions of Light". These chapters, especially will be the focus of our analyses.

The aim of this paper is the presentation of the thesis that Tymieniecka's concept of the onto-poiesis of life within the human being is linked with the expansion of light from the subliminal sphere into the mind's intellectual sphere. I would like to show that this phenomenologist has related the phenomenon of the translucent sphere of the human being to the Ultimate Light.

Here, the subject of the analysis is the human being's constructive advance from the Human Condition promoting of the Logos of Life inward into himself. In accordance with Tymieniecka's concept of the creative vision, the author of this paper will shows that the passions of light allow the human being to expand the virtualities of the Human Condition and realize his humanness. He will also undertake his own

A-T. Tymieniecka (ed.), Analecta Husserliana LIX, 31–48.

investigation with the purpose of grasping the existential connection between the "higher" and "lower" man. Tymieniecka's concept of the "source" can serve as the common link between both investigations.

I. TO BRING OUT THE LIGHT OF LOGOS

1. *The Schematism of the Everyday Experience and Its Overcoming*

Prof. Tymieniecka writes, that "in beginning with everyday life-experience in dealing with experience, we already deal with the human being who as such is in the lucid zone of his functioning and is a creative being; he has risen already from the brute functioning of his animal vital circuit by implementing creative imagination as the principle orchestrating his functional schema and, simultaneously, has through the operation of the newly surging faculties of memory will, and the intellect established the self-interpreting agency of his mind in the basic processing of experience. That has already transformed the brute an numb and blind moves of vital nature into complex operations through which the objective world around us and ourselves as agents in a community of others like ourselves emerge",[1] In other words, he has constituted "humanized" vital nature. Next, I may ask: How can I focus constantly upon the specifically aesthetic, moral aspects of human feelings, emotions, thought, and reflections in the situation in which one is everyday experience does not keep the subliminal soul active? For the phenomenologist "this means that our everyday life-course imposes clear-cut simplistic patterns upon the exercise of our life and confines our experience to rudimentary, simple forms".[2]

In this state of affairs, the author of *Logos and Life*, Vol. 3, seeks routes of departure for the purpose of overcoming our schematism of the everyday experience. Literature and art compensate and complete human limitations. In truth, they appear to us as crystallizations of the answers to the deepest yearnings of the subliminal soul. Thus, the human spirit is able to evoke the so-called evolutive trajectory from the primal forces of nature to its high point in a flourishing civilization.

2. *Towards the Notion of "Fabulation"*

According to Tymieniecka, light, the sea, air, and earth are constant elements that evoke or reflect a host of corresponding passions. She believes, that "fabulation, whether expressing the sequentially grasped (formulated) human predicament by enacting it in dance, ritual, mimicry, a stage play or by way of discourse in storytelling or writing, revolves around three basic principles: the tragic the comic" and the epic. Each of them performs unique role with respect to the life-significant task which it is called on to play in individual and societal life".[3]

In this way, the paradigmatic fabulations (myths, rituals, sagas) project tentacles for us by which we are able to form appreciations and invent our strategies. In fact, human dreams needs and fears release the latent propensities of the subliminal soul. That is to say, "the brute and opaque facts of life are imaginatively transfigured into elements of existence, and to confront them there emerge the elemental passions of the subliminal soul."[4] Thus, our phenomenologist treats these passions as the beacons and even the bastions of life's struggle. Furthermore, the quest for a new humanly significant equipoise is not seldom marked with "an impulse" or with "free" will. For establishing the equipoise of life it is indispensable that there be "an emotionally charged dialectic of internal forms of life's destruction and corresponding life-beneficent forms".[5] For instance,

> "Oedipus' tragedy as unfolded in the plot reveals its universal human significance when viewed in the perspective of its subliminal progression – which is carried by the passions of Elemental Light. Oedipus brings upon himself his doom by following our innermost passion to know, to understand, and to obey what the truth dictates; he is committed to justice and prompted by self-righteous pride. These are the subliminal passions of the elemental light, passions of glory, of life. Yet strangely, in pursuing them, Oedipus slowly moves toward darkness, towards condemnation. His quest for clarification, incidity, light leads to the paradoxied opposite of the darkness in which man enacts his life-course".[6]

So, good and evil govern not only the concrete moral life. They seem to be projected upon the ultimate forces of the universe, i.e., as the divine good and satanic evil.

According to the author of *Logos and Life*, Vol. 3. "myth, legend, religious mysteries, dreams all find their origin within this aboriginal context between the relentless impetus of life and the opposing individualizing constructive tendencies which struggle to capture it in coherent structured operations with directional vectors and a rationale that can process its forms in a stabilizing machinery and channel its flux".[7] We may say that out of the so-called subliminal forge within the human soul is the writer draws ciphers giving the work of art it's human life-significance. Then, that work is de-ciphered by the reader or the recipient.

To be sure, the notion of "fabulation" is treated by Tymieniecka as the primogenital factor of the creative orchestration in the initation of the human significance of the life. Myths as well as legends and mysteries express this limit situation i.e., the limit zone of human existence with its innermost struggle for survival-equipoise, the struggle to maintain oneself on the surface of the flood while in some indispensable measure controlling and dissecting it.[8]

In Tymieniecka's opinion "the fabulation that follows on wonderment intentionally transfigures life's nude facts in accordance with the propensities, virtualities, and factors of the human condition. The result is prototypical models of human character, conduct, societal organization, visions of humanity; the stage is set for the ideals which will fashion a culture's style".[9]

In my opinion, Prof. Tymieniecka has strived to reconcile "the right" of literature with "the one" of philosophy by maintaining the "substantial" difference between them.

Literature seeks to incarnate a vision of life or human existence. She writes, "the most personal vision of the writer does not appeal to he most lucid zone of the reader's thoughts but seeks out the remote regions of his/her thought, the emotional climes where in obscurity existence is carried on its way".[10]

Philosophy seeks to establish its findings in universal, objectively valid, rational language. The philosopher aims at discovering the positive truth about life, human existence and destine and so draws into a "universe" the circuit of the lines of the creative vision. In the view of the phenomenologist

> "it is philosophy's role and meaning to give man, through the exfoliaing of its vision in a systematic, rational appeal to universal and detached thinking, an understanding of reality and to enlighten him, intellectually and objectively, about the parameters of human existence, its nature and prospects, options and limitations; it offers this clarification to all men in all situations and also indicates the proper conduct for their fulfillment. Briefly, its significance is to bring to man enlightenment".[11]

In Tymieniecka's thought culture is what gives continuity to the pulsating life of the individual, the human group or society. It is transmitted as inheritance from generation to generation. At a given time, it is based where individuals are capable of retrieving ideals and corroborating their significance with their own living stream of existence. For instance, there are the rules by which Greek tragedy conducted and generated subliminal passions. It is true, that Aristotle in his *Poetics* pointed out that entire network of devices which intensified emotions (hamartia – mimesis – catharsis). It is the role of catharsis to reach into the subliminal cornucopia to evoke an emotional fullness that purifies us from all unimportant life concerns. In Aristotle's *Poetics*, fabulation involves our visual, auditory and intellectual experience.

With the phenomenologist, the Human Condition is responsible for the fullness of psychic acts and brings to them "poetic sense", e.g. the poetic Logos. However, does the writer's vision give the "light" for the intellectual understanding? Furthermore, does it give the evidence of "truth"? In accord with Tymieniecka, I believe, that it expands the human inner sphere in the sense of sentient emotive links and gives new perspectives for our thoughts, and imagination. But, it does not give the light to the intellect.

3. *The Question of "Sight" as the "One Glance"*

The human being envisages the place within the world of life in which he strives to find himself. He becomes aware of the confines of horizons. It is important for our further considerations to introduce the notion of "the sight". Because, it is exactly linked with the notion of "vision". The phenomenologist states, that

> "my 'vision' of the world does not cease at any horizon that my movement projects; on the contrary, beyond any horizon there is implied within my vision an expansion into infinite space, or space which is there. We might have learned from science that our earth is round, and we may rationally conclude from that the spaces upon which our actual horizon of the world borders are physically calculable, measurable, etc. And yet, there belongs to the experiential significance of vision an experience of the world as extending into infinite space, as commanding a feeling of repose and openness, infinity".[12]

The author of *Logos and Life*, Vol. 3, links closely the question of the "seeing" or of "vision" with infinite Being. In connection with this, the structure of sight experience is such that it synchronizes visual perspectives in such a way that one can gather them all "around" oneself "as if at one glance". Moreover, the expansion of space is linked to illumination (the primeval light, precisely, with its primogenital status).[13]

However, here appears the question; what does it mean that an expansion is implied in the vision of the world beyond any horizon, to take note that, in reality, the horizon does not exist? It merely surround is all perceived reality. But we may say that the horizon indicates any line of demarcation such as that between the natural and the spiritual. It is given as the absolute boundary, as the non-reduced one. In short, man can not eliminate the horizon. The elimination of it is the elimination of himself. Therefore, Cornelius A. van Peursen speaks of "the anthropological horizon".[14] So, the human being finds himself within a horizon when he is situated in the time.

In my mind, the resolution of the so-called contradiction between the notion of "the horizon" and that of "vision" within the thinking of Tymieniecka is to be found in the assumption that when reality passes to illumination, it vanishes. Then, the notion of "vision" becomes "current". Expansion is "the image" of the becoming of beingness.

That is why the phenomenologist makes the observation, that "the element of light as such brings together the three seemingly separate spheres of Cosmos, nature, and the human inner self".[15]

In fact, human seeing appears to be an indispensable outcome of the role that light plays in the origin of the life-world. First of all, the experience of seeing in its fullness benefits already from the orchestration of the living agency, i.e., the primeval light. In this way we approach the question of the onto-poiesis of life at the instance of the light of Logos in general.

By "onto" Prof. Tymieniecka refers to "firstness" with respect to the scale of life in its formation. This is to acho Aristotle's *Poetics*, a "poietic" process: onto-poietic.[16]

II. THE LIGHT OF THE MIND WITHIN THE HUMAN CREATIVE ORCHESTRATION

1. *From the Zone of Darkness into the Zone of Light: Imaginatio Creatrix, Memory, Will and Intellect*

According to the author of *Logos and Life*, Vol. 3, the inward vision is basic to the unity of a living beingness as the work of the subliminal forge in which our organic or vital, gregarious or psychological, and specifically human factors of significance come together within a new mold.

However, inward vision evokes the outward vision which is situated within a world. Then our inner sight reaches out into the luminosity of the "outward" life-world of our psychic vision. Hence, human life becomes as such all-luminous.

In this connection, Tymieniecka quotes from Heraclitus' *On Nature*, viz., by living is meant our awakened state in which we lead a communal existence with others; for when sleeping – on closing our eyes–we distance ourselves from active communal existence.[17] As she elaborates "we retreat from it for a while (and also from the luminous zone of the our psychic life as a field of active life-promoting involvement) into a passive sphere of our beingness, away from the world, a dark zone where in our dreams a luminosity is experienced that is not the light of primary, active, alert life functioning but, rather, a secondary memory–induced projection".[18] Here, I would draw attention to Tymieniecka's deep-rooted inspiration in Heraclitus.

But, the intellect has the faculty to spread consciousness throughout the luminous sphere of life experience that surrounds the field of self-consciousness. Simply, it gathers all experiences, their motivations, their performance, and their forces together into the unity of human beingness. It also establishes the cohesion of purposeful interlinkage and cooperative fusions through networks of rational connectedness and structural linkages.[19]

In fact, imaginatio creatrix, memory, will and intellect could not function in the zone of darkness outside the zone of light. Their functioning can emerge and progress only with the "enlightening" of the mind.

2. *Towards the Constructive Phase of the Human Condition*

The discussion of the role of light refers directly to the central point of Tymieniecka's thought, namely, to the entelechicial principle of the onto-poiesis of life in its complexity. For "it is necessary that this complexity unfolds to the point of installing its own intrinsic central agency, an agency that would centralize the automatic processes of the world performed by the separate organs by incorporating them".[20]

I question whether such a distillation will occur of the innumerable functional circuits and their interlockings so that a specific synchronizing circuit that functions "disembodied" from singular networks of operations will emerge. For Tymieniecka, the effect of this is the gathering of all into "pure rationality".[21]

Have we a synchronizing complex with so tight a hold on the reins that it really controls the entire range of organic, psychic or gregarious processes? In our phenomenologist's view, in the progress of living forms it is the Human Condition which brings about this complex and the crystallization of its virtualities within the creative orchestration leading to the emergence of the intellectual faculty. This is the so-called constructive phase of the Human Condition.

3. *The Transparency of Intellectual Consciousness*

I would like to draw attention to the fact that the light that self-awareness brings is not only the inward-outward spreading subliminal vision of the individual life. Now it is the vision of life's "roving eye" which moves at one with the network of the individual's operations.[22] These individual operations in their rational connections take innumerable experiential shapes. Thus, the network becomes transparent to the eye. It is the human beings who live within the luminous zone in their individual consciousness. Just this gathers together all human beingness. And so, the intelligibility of the rational interlockings in the phenomena of human inward as well as outward projection is exposed to the light of the human mind. According to Tymieniecka, in this light the transparency of intellectual consciousness arrives.

4. *Light and the Significant Word*

In fact, the author of *Logos and Life*, Vol. 3, differentiates various degrees of luminosity in our beingness. In the dawning of mind, it is the light of consciousness that accounts for the mind's piercing all the otherwise blind strata of individual existence. So, by this differentiation, various shapes, dimensions, color and textures of things can be grasped by our mind "in a single glance" in their "unmingledness".[23]

The phenomenologist states, that "our mind masters this differentiation, and we utter specific sounds and words to hold them before our mind: names that betoken categories and classifications".[24] In light, then, is the subliminal genesis of human reason and the word.

Tymieniecka draws attention the fact that "the sign/word or the 'significant word' is . . . the final fruit of this infinitely complex synchronization of the relevant functional links that is the emergence of the mind as a faculty synthesizing them all 'at one stroke'. The word bursts forth from the infinitely complex linkages of life's constructive processes as the decisive factor in the luminous intelligibility of life".[25] In passing from light to the word she covers the infinitesimal distance spanned in complete cycle of life's genesis.

In the situation in which the human person finds himself "on-the-brink-of-existence" the individualizing advance will encounter difficulties. On account of his loosened ties to communal life and the despoilment of this network, the human person sinks slowly into the dark circuits of the psyche. There he should overcome his critical situation and become ready to re-establish existential links with others.[26] He already knows what it means to utter the proper word. So, his effort means breaking through to the collective luminosity. Also, the word is the synthesizing outcome of all the functional circuits of life that carry the individualizing progress of the human being.[27]

III. TOWARDS THE CONSTRUCTIVE ADVANCE OF THE HUMAN BEING

1. *The Upward Growth of the Human Being in the Light*

The entire constructive progress of the human being expresses a drive toward light. Then he may attain the self-awareness of the "eye of the mind". Indeed, "man's self-interpretation-in-existence is conducted already within the zone in which the subliminal passions spring into action".[28]

Prof. Tymieniecka gives to us the image of "the new avenue" toward the light that is the mind, as the subliminal forces burst out from the specifically human womb wherein the human being is made human and culminates in the passion to create, to bring to light, to manifest an all-embracing vision of life in the form of exemplary works of art, of thought, of reflection and technique".[29]

It is important, that the individual undertake a constant effort to understand others within their own luminous private zones. But, the human being needs to look ahead beyond the circumference of the practical life. We can expect that the passions will be enlivened in the scholar, the scientist as well as in builder, inventor, etc.

The phenomenologist draws our attention to the fact that the subliminal light serves to demarcate life's zones, to mark the zones of growth, health, sanity, and darkness, etc.

So, the ascent of the human being is linked with the expansion of light

"from subliminal workshop into the mind's intelligible working-through the fulgurating images of the creative imagination, the illumination of emotions, feelings, moods, etc., by intelligible 'sentiments', the emergence from experience of a 'view', and 'insight', a 'cognition', a 'constitutive' coalescence, a decision, the realization of the intelligibility of moments and of the relevance of moments to each other is the essential feature of experience by means of which this wondrous translucent sphere of human existence emerges and maintains itself."[30]

2. *Toward the Reality of the "Higher" and the "Lower" Man*

On this level of my considerations, I would like to raise a very important question, viz., the reality of the "higher" and "lower" man and the connection between both of them. Tymieniecka's concept of the "source-experience" is linked with the crucial moment within the unfolding of individualizing life at which a transition occurs from animal action/reaction agency to the specifically human experience that simultaneously originates the human subject.[31] One should especially bring out this connection between the "higher" and "lower" man. In Tymieniecka's perspective there is first the "expanding luminosity within the human being," second, the "subliminal ground of the soul" lying at the threshold of two contrary motions –, "vital strivings towards the survival of the species" and "the creative strivings toward the human". The "moral" significance of existence offers a retreat from both of these dynamic trends.[32]

2a. *Lack of Conscience Causes an Appearance of Acts of Violence by the Human Person*

The man of conscience aims towards the "higher" man in his progress in humanity. However, the "lower" man is capable of various acts of violence against other persons. And he even becomes a "source" and "carrier" of violence. Therefore, the phenomena of violence appear in the "lower" man more because even the "higher" man does not link with them.

Thus, violence appears in the human person to be simultaneously growing in intensity and decreasing in intensity. I assume the reality of the "higher" and the "lower" man existing individual human persons by virtue of the same progress of the vital process within them. To begin with Plato or Aristotle wrote about the three aspects of the soul: the appetitive or vegetative soul, the quickened or animal soul, and the rational or human soul.

In 'Saint Thomas Aquinas' "On Being and Essence" (*De ente et essentia*), I find a characteristic thought, viz., that Socrates' understanding of humanity was the only one possible.[33] Thereby, in the realization of my "humanity" nobody can replace me. Thus, Plato's and Aristotle's theory of the soul can be regarded *here* and *now* as an example of the general division of the soul. This is confirmed by the reality of the "higher" and the "lower" man.

2b. *A Proof Through the Discovery of "Homo Absconditus" and "Deus Absconditus"*

Strongly emphasized by Saint Thomas is a particular way towards the apprehension of humanity, the discovery of "homo absconditus" and "Deus Absconditus" within the human person. On my own account I discover humanity to also be the Divine.

Even, on this level of my considerations, I want to introduce two new terms, "approach" and "distance". To my mind, the so-called "approach" of the "lower" man to the "higher" one will not mean a negation of the reality of the "lower" man. Rather, it will mean only the "turning" of the human person towards a "source". One simply links the "higher" man to the "lower" man in the course of one's life.

In this way, the spite or aggression or violence of the "lower" man in relation to the other person remains "suspended" or "restrained" by the same "source" which appears in the "lower" and "higher" man.

Already, this example conveys "something" to us about the vital tie between them. The phenomenon of "the distance" between them leads to a "vanishing" of "homo absconditus" and of "Deus absconditus" in the human person. Thus, this phenomenon causes "de-humanization". This is why the individual becomes the "carriers" of acts of violence against a "fellow creature". On the other hand, the phenomenon of the approach of the "lower" man to the "higher" one can lead to the discovery of "homo absconditus" and of "Deus Absconditus" within the human person. And the individual becomes the "carrier" of benevolent acts.

But, the "higher" man cannot do without the "lower" man in his own person, and vice versa. And so, the "higher" man "perceives" the apparent evil which is engendered by the "lower" man in the human person. But, the "lower" man is dependent on the "higher" one, so to speak; with in a common matrix there exists a "strangeness" for both of them.[34]

They are linked by the common vital tie in the human person. That is why, they cannot be separated in general.

There is a so-called "intermediary link" between them which is formed at their "base". This "link" becomes a "novum" for the vital human being.

Simone Weil was right when she said that "an intermediary link" is possible only for human persons who meet "a supernatural love", That it directs only to God.[35]

However, that philosopher "reinforces" this "link" by saying that to love "a fellow creature" equally as one loves oneself will result in itself in a reciprocation, the "fellow creatures" loving us.

Next, Anna-Teresa Tymieniecka has spoken of a common thread in the human person, of the moral sense, of the Human Condition in respect to a meaningful novum which is the "benevolent sentiment".[36] She unfolds its presence in a valuative process. This process appears to us as a transaction of the self-individualization of life for a human being.[37]

Like Weil, Tymieniecka elaborates this meaningful novum in the existence of the human person.

In my opinion, the "enlive" human being participates in the unfolding of this novum or a good conscience as the result of the discovery of "homo absconditus" and of "Deus Absconditus" within the human person.

Thus the connexion of body and soul within gathers a new significance, viz., the "corporeality" of the human body becomes more "spirtitualized" and as the human soul becomes thereby sacrified" in this progress of the "solidarity" he life of both of them.

Indeed, I desire to indicate a way to reduce this division between the "higher" and the "lower" man by the discovery as much of "homo absconditus" as of "Deus Absconditus" within the human person.

By analysis of the connexion between the "higher" and "lower" man in the human person, I have endeavoured to show their "coincidence" in the unfolding of the novum or of a good conscience. Also, I have endeavoured to show that it can be aroused by the discovery of "homo absconditus" or of "Deus Absconditus". Thus are we enabled to approach the Divine.

3. *The Sense to Surpass the Workings of "pure" Intellect into the Ultimate Light*

However, the element of light is not generated by the intellect, although, the faculty of the intellect promotes it and handles its unfolding and extension. The phenomenologist wants to distinguish strains of intelligibility other than those which enter into the workings of the intellect. Thus, her distance from the forms of the intellect as the skeleton of life's expansion grows simply, "the radiating intelligibility of human experience flows over this rational skeleton on all sides, steering the vehicle of life".[38]

Next, I must say, that the Logos of Life to which we tend radiates the constructive progress of life and becomes itself the objective of our passions to know and understand. Hence, the author of *Logos and Life*, Vol. 3, lets its luminosity expand.

To my mind, by absorbing the field of Tymieniecka's creative enterprises, we may see the connections between the "higher" and the "lower" man more clearly. The Swedish writer Pār Lagerkvist in his novel *The Dwarf* reminded us of the significant truth of ourselves that in every man there is "something" of a dwarf. The notion of the "lower" man is related to the zone of twilight or that of darkness.

This leads us also to suggest that Tymieniecka equally deeply as Lagerkvist undertakes the efforts to bring everyone out from the zone of darkness into the zone of light, that of our humanness (corresponding to the character of the prince in Lagerkvist's novel). In this way, the "higher" man is able to fructify in one's own workings, in everyone.

Earlier, Maurice Merleau-Ponty said, "Now perhaps we have a better sense of what is meant by that little verb 'to see'. Vision is not a certain mode thought or presence to self; it is the means given me for being absent from myself, for being present at the fission of Being from the inside – the fission at whose termination, and not before, I come back to myself".[39] Further, "Vision encounters, as at a crossroads, all the aspects of Being".[40]

Undoubtedly, the question of the depth of Being is already linked with the mystery of the human being. I believe, that one should discover the harmony between the "higher" and the "lower" man by seeking the depth of Being within oneself.

That is why, the phenomenologist calls for light and unencumbered flexibility of mind in a situation of anxiety, anger, fear, anguish, regret, remorse, shame, sadness, etc. For the reason, we want to grow by overcoming the misery of our being. For Tymieniecka, "sharing with others requires an interpretative effort, a disentangling of their intrinsic concatenations, and a hunt for causes and hidden motivations, and that means to bring them to the light of our clear consciousness".[41]

In accord with our phenomenologist I believe, that human beings are able to appreciate their own congenital weaknesses, incapacity, indolence at handling this or that situation, and, to recognize their limitations and their weaknesses.[42] By the way, the notion of "measure" and of "order" acquire their importance along life's course.

According to Tymieniecka, "in fact man's innermost urge toward light – to cognize, to understand, to encompass and to reach the ultimate reasons of his very existence carries him toward a light that is beyond the intelligible rationality of the life-world's horizons: it is in the guise of an 'unearthly' light that man projects his vision of the 'transnatural'".[43] In respect to this, I think, the "phenomena" of the trans-natural surge within he dialectic of light and darkness indicate the presence of divine. The human being as well, transcends the confines of his self-encircling natural luminosity. Next, he turns his gaze toward a transmundane infinity without any horizons. It appears to him the Ultimate Light, it is the unearthly Divine = God.

So, it seems to me that primeval Light/Logos enters into the world of life, that of evolutive progress. A result of this is the attainment of the luminosity of self-awareness and of knowledge of one's place, forces, limitations, and situation. The passions allow for our expanding the virtualities of the Human Condition: to realize our humanness.

Here, Prof. Tymieniecka has maintained her design that the passions of light take various particular routes in action as rays of this vast field of "enlightenment" that makes civilization and culture.[44]

4. *Man's Creative Vision and the Vision of Promethean Man*

Man conceives an unrelenting urge to surmount the limits of his existence. This is his manifestation of the urge to transcend. According to our phenomenologist, "the urge to surpass our present confines coincides with the dissection of the significance of our world of life and with the aspiration to give it a new significance".[45]

Since, awakening from complacency, contentment, etc., the human being begins to create novel forms of significance. In Tymieniecka's view, this urge to create frees the new creative vision with which he strives to go "above" his horizons as well as "below" abysmal subterranean realms to the light. Or, the rays of this vision meet within his translucent self. Hence, his striving toward the light prompts the subliminal yearnings within to understand his ultimate conditions and their reasons. In other words, the urge to transcend ignites the creative proficiencies of the human being.

The phenomenologist has said, that "the eye of the mind was followed by man's hand and was moved by his desire to not only understand but also, to make, to change, to improve, to build.[46]

Besides one's own translucent self one searches the "objective" structures, rules, and principles for "manifestation" of the microcosmos (the elementary portions, the atoms, the germinal force) as well as the cycle of the seasons punctuating the flux of life, and the natural meteorological forces, the cosmic powers of the moon, the sun, the comets, the stars and all forms of living being that make up the macrocosmos. The author of *Logos and Life*, Vol. 3, is in the right when she writes, that first great creative visions of men – the poets, playwrights, philosophers/scientists – were fruits of the purely intellectual speculation of minds that marveled at our life-world.[47] Man strove first to understand his own concrete domain, his existing way within the self-projected.

That is why this creative vision of man works deliberately on his forces and powers, upon his energies, forces, mechanisms, but also upon the mysteries of the great fluctuations in the universal schema of life. Thus is he able to penetrate and command them.

Tymieniecka needs the creative vision of a new Promethean man, one like that of the bold one who stole fire from the gods, i.e., one who would steal the secrets of the governance of the life of the cosmos. Why is this vision so important? She gives us her answer: contemporary man has freed himself from the chains of being. Simply, he feels free because his imagination has broken all the barriers of the horizon set up by past culture; he seeks no bounds.[48]

So to speak, the creative vision, viz., the "Promethean" vision, travels immeasurable distances through space. It reaches in retrospection the origins and the evolutive steps of life at the distance of millions years.

By turns, in the phenomenologist's opinion, the scientists have entered into their workings to conquer them, to expand human freewheeling in the life-network of the cosmos.[49]

5. *Toward Promoting the Logos of Life*

In her's view, "man is overpowered, and it is no wonder that the alarm is sounded over the 'death of man', the 'end of metaphysics', the 'rejection of foundations'. The analyzing or speculating contemporary philosopher, groping and lost in the pedestrian scrutiny of his expression (language) level, trails far behind what the light of the conjectural keeps uncovering, through unprecedented creative vision".[50]

In Tymieniecka's phenomenology of life as "Dionysian reason" as well as he "Apollonian reason" find their "anchorage". Clearly, our phenomenologist seeks an equipoise between the "dionysic" element and the "Apollonian" element one in her thought. However, Tymieniecka's thinking is not oppose to rationality in general.

So, our phenomenologist keeps her distance from Nietzche's philosophy, as well as structuralism, and Heidegger's philosophy. Here, there is cause to rise and assume the creative vision instead of groping on murky ground, cause "to return to the goal of determining ultimate principles".[51]

Prof. Tymieniecka does not witness the end of metaphysics but its rebirth, and does not proclaim the death of God but brings about a novel enlightenment of the mind and the heart.

I agree with Tymieniecka that the truly definitive philosophical awakening within the light of the human being is connected with the responsibility to be ready for a new era of culture beaconing at the far off horizon.[52]

In particular, the phenomenologist has offered a succinct philosophical analysis of the subject of the element of light in its primeval as well as subliminal nature. In fact, the forms of light do precedence over all other elemental forces in promoting the Logos its constructive of life.

In summary, in Tymieniecka's comprehension, literature does not give "light" for intellectual understanding. It rather expands the human inner sphere in the sense of its emotive, imaginative, reflective links. Then, philosophy endows us with the light of Logos. For Tymieniecka the Human Condition works a complex crystallization of virtualities within the creative orchestration all creation for the emergence of the light of the mind.

Hence, the individual consciousness gathers together the entire human beingness. With the light of Logos there appears the significant word. Tymieniecka relates the phenomenon of the translucent sphere of the human being to the Ultimate Light, the unearthly Divine. For my part I assume the reality of the "higher" and the "lower" can be seen in the same progress of the vital process within the person. They are linked by a common vital tie in the human person. I strive to indicate the way of accomplishing the reduction of the division between the "higher" and the "lower" man by the discovering as well of "homo absconditus" as well as "Deus Absconditus" within the human person. The phenomenologist's creative vision of man holds him responsible for his deeds and for being ready for a new era of culture beaconing at the far off horizon. The onto-poiesis of life in Tymieniecka's thought keeps its sense "in virtue of" the Light of the Logos of Life.

Technical University of Gdańsk
Poland

NOTES

1 A-T. Tymieniecka, *Logos and Life*, Book 3: *The Passions of the Soul and the Elements in the Onto-Poiesis of Culture, Analecta Husserliana*, Vol. **XXVIII** (Dordrecht Kluwer Academic Publishers, 1990), pp. 33–34. (Hereinafter, *LL* 3).
2 *Ibid.*, p. 34.
3 *Ibid.*, p. 30.
4 *Ibid.*, p. 38.
5 *Ibid.*, p. 31.
6 *Ibid.*, p. 49.
7 *Ibid.*, p. 31.
8 *Ibid.*
9 *Ibid.*, p. 39.
10 *Ibid.*, p. 20.
11 *Ibid.*, p. 19.
12 *Ibid.*, p. 80.
13 *Ibid.*, p. 79, p. 80, p. 81.
14 C.A. van Peursen, *Phenomenology and Reality*, Duquesne Studies Philosophical Series 30 (Pittsburgh: Duquesne University Press, 1972), pp. 214–238. See also: Helmut Kuhn, "The Harvard University Press, 1940), pp. 106–123; H. Pietersma, "The Concept of Horizon," in *Analecta Husserliana*, Vol. **II** (Dordrecht: D. Reidel Publishing Company, 1972), pp. 278–282. In Pietersma's opinion, the clearest statements of Husserl on the notion of "horizon" are found in the following texts: *Husserliana*, Vol. **VIII**, pp. 146–152; *Husserliana*, Vol. **XI**, pp. 3–15; *Experience and Judgement*, Par. 8.
15 Tymieniecka, *LL*, p. 77.
16 A-T. Tymieniecka, "Measure the Ontopoietic Self-Individualization of Life," in *Phenomenological Inquiry*, Vol. 19 (October, 1995), p. 40.
17 Heraclitus of Ephesus: Fragments B89, and Fragments B17, B22, B34, B72, B73, B114, (according to H. Diels, *Die Fragmente der Vorsokratiker*, 4th ed. (Berlin: 1922). See also G. S. Kirk and J. E. Raven, *The Presocratic Philosophers* (Cambridge: 1960); A-T. Tymieniecka, *LL* 3, p. 82.
18 *LL* 3, p. 82.
19 *Ibid.*, p. 82–83.
20 *Ibid.*, p. 83.
21 *Ibid.*
22 *Ibid.*, p. 84.
23 *Ibid.*
24 *Ibid.*
25 *Ibid.*
26 *Ibid.*, p. 85.

[27] *Ibid.*
[28] *Ibid.*, p. 87.
[29] *Ibid.*
[30] *Ibid.*, p. 88.
[31] A-T. Tymieniecka, *Logos and Life*, Book 1: *Creative Experience and the Critique of Reason, Analecta Husserliana*, Vol **XXIV** (Dordrecht: Kluwer Academic Publishers, 1988), p. 445.
[32] A-T. Tymieniecka, *Logos and Life*, Book 2: *The Three Movements of the Soul or the Spontaneous and Creative in Man's Self-Interpretation-in-the Sacred, Analecta Husserliana*, Vol. **XXV** (Dordrecht: Kluwer Academic Publishers, 1988), p. 13.
Note among earlier philosophers: Nicolai Hartmann (in his *Ethik* ed. 4th [Berlin: de Greuyter, 1962], especially. Part 3, Chap. 2, Chap. 2, Sec. 71: "Zur ontologischen Gestzlichkeit als Basis der Freiheit," pp. 678–685), Max Scheler (in his "The Forms of Knowledge and Culture," in *Philosophical Perspectives*, trans. Oscar A. Haac [Boston: Beacon Press, 1958], p. 14) and Cardinal Karol Wojtyła (at present, John Paul II) (in the *Acting Person, Analecta Husserliana*, Vol. **X** [Dordrecht: O. Reidel Publishing Company, 1978], p. 256; p. 315) and others paid attention on the communal connection between the "higher" and the "lower" man.
[33] Cf. Saint Thomas Aquinas, *On Being and Essence*, trans. A. Maurer (Toronto: Pontifical Institute of Medieval Studies, 1949).
[34] According to H. G. Stoker. "Im spezifischen Gewissenssinn ist das echte Gewissen/ oder: sind die echten Gewissensregungen/ nur die reelle innere Kundwerdung des Personalbösen, sei es zur Verwirklichung drangend/ wie im "warnenden Gewissen,"/ sei es prinzipiell möglich, aber reell verneint/ wie im "guten Gewissen".
H.G. Stoker, *Das Gewissen. Erscheinungsformen und Theorien* (Bonn: Verlag von Friedrich Cohen, 1925), p. 210.
[35] Cf. Simone Weil, *La Connaissance Surnaturelle*, Collection Espoir, dirigee par Albert Camus (Paris: Editions Gallimard, 1950).
[36] *LL* 1, pp. 402–404.
[37] *LL* 1, pp. 400–404.
[38] *LL* 3, p. 90.
[39] Maurice Merleau-Ponty, "Eye and Mind," in: *Phenomenology, Language and Sociology*, Selected essays ed. by John O'Neill (London: Heinemann Educational Boks Ltd., 1974), p. 307.
[40] *Ibid.*, p. 309.
[41] *LL* 3, p. 91.
[42] *Ibid.*
[43] *Ibid.*, p. 93.
[44] *Ibid.*, p. 94.
[45] *Ibid.*, p. 95.
[46] *Ibid.*, p. 96.
[47] *Ibid.*
[48] *Ibid.*, pp. 97–98.
[49] *Ibid.*, p. 97.
[50] *Ibid.*, pp. 98–99.
[51] *Ibid.*, p. 99.
[52] *Ibid.*

ANTONIO CALCAGNO

FLUCTUS, GRAVITAS ET INERTIA

A Phenomenological Reflection on the Relation Between the Human Person, the One and the Many of Life

If the philosopher had to identify a common trend or principle which "unifyingly" transcends the particularity or "many-ness" of cultures and traditions in the West while concomitantly striving to preserve what is essentially distinct or *différent* among this seemingly disparate array of cultures and traditions, one could nominate the West's insistence on the autonomy of the Ego and its inalienable right to self-determination through actualising its own will-to-power. Politically and philosophically speaking, there are two main forces which support such a thesis. First, the modern Western tradition of liberal democracy has resulted in an ethic wherein the primacy of the individual is of significant value within the wider framework of the political community, namely the state. Second, the trend in postmodern thinking and acting has tended to further rarefy the relation between the individual and the community by dismissing the possibility of a common ground or in Lyotard's words, a "grand narrative". The result of such a rarefaction is a *totalisation* (Levinas) of the individual as "difference" over the idealistic comm-unity, thereby resulting in a conception of the individual which is essentially cut off from the possibility of a common universal ground or unity – an individual so radically different that he or she can only relate to another on purely logical terms, as human beings are reduced to categorical islands of difference with no real *Übergänge* to other archipelagos of disjunct meaning. In other words, that which makes *some* islands of unity possible, and *not one unity*, is the fact that we are all Egos of difference – a basic logical fact. We are all conscious free-acting egos and our universal political goal is to preserve such an inalienable reality. Yet, we must also try to preserve the inherent difference which is concomitant with each individual ego. Postmodernity has sought to "resolve" the perennial problematic of unity and multiplicity within the structure of the Ego of *différends* or *de l'altérité*.

It will be our main contention that the West's contemporary insistence on the primacy of the Ego as the postmodern answer to resolving the question of identity and difference, unity and multiplicity, is essen-

A-T. Tymieniecka (ed.), Analecta Husserliana LIX, 49–67.

tially an empty, formal logical category which can be freely filled with any kind of content, political or otherwise, thereby resulting in a very shallow and trivialised anthropology. Proof of this misanthropy can be seen in contemporary society's desperate search for meaning amidst a lonely world of solipsism.

We propose a reexamination of the common unifying and universal ground which serves as the condition for the possibility of all difference, namely life. Life, with its three essential movements of gravity, inertia and flux, grounds the human person in a *fundamentum* which is both "comm-unifying", as opposed to isolating, and resplendent with a variety of multiple forms. It is the very fact of the being of life which allows for unity and multiplicity to coincide. Ultimately, it is life which affords the human person and his or her communities, as opposed to the isolated logical or conscious Ego, the possibility and gift of a genuine ontopoiesis.

I

Marc Guillaume and Jean Baudrillard in their work entitled, *Figures de l'altérité* remark, "Le thème de l'altérité est devenu une obsession en Europe aujourd'hui, contrepartie de la montée des courants xénophobes dans le monde occidental. Cette obsession a suscité, ces dernières années, de nombreux travaux d'un grand intérêt."[1] What Guillaume and Baudrillard note as significant is a paradoxical situation: The concomitant Western emphasis on difference or alterity and the rapid escalating trend of socio-political distance motivated by such rallying cries as race, economic equity, gender and nation (xenophobia). Philosophy too, as a human science, has profoundly influenced and been influenced by the marked postmodern phenomenon of difference.

What is so attractive to postmodern philosophers about difference or alterity is that it is viewed as a condition for the possibility of genuine and essential human characteristics or potentialities being actualised in a fruitful and non-discriminatory way. In part, this drive to recognise and preserve that which is unique about the human individual is rooted in a reaction against the modern trend of such philosophers as Kant and Hegel to systematically subordinate the dynamism of the individual to an all-encompassing human construct such as reason and the categorical imperative or the state and *Geist* coming to fully know itself. Such universal and idealised "narratives" of modernity tend to reduce the human individual to a very confined modality of being. In other words, it as very difficult to accommodate difference. For example, in Hegel's *Rechtsphilosophie* there is no space for the individual conscience to override the state. Moreover, the postmodern drive to recognise difference is rooted in historico-poltical experience. We are all aware of the devastating effects of the nihilising drive to unify a certain identifiable type under the seemingly universal banner of National Socialism or Communism. Individuals who did not conform to the universal and communal type of such movements were systematically exterminated in the camps of Auschwitz and the Gulag. Not to have an ontological space wherein the human individual was allowed to freely express herself as diverse from others had and still has stark consequences for that individual, sometimes the eruption of violence and resultant death. To recognise and create an ontological space for the difference, which is the difference or bespeaks *les différends* of the concrete human individual, is the project of postmodernity and it has as its goal the protection of the "good" which is that difference of the individual.

Postmodern philosophers, taking their lead from Nietzsche, fully recognise that the terms "good" and "bad" are relative terms, yet they do ascribe a certain value (*Wertlehre*) to the principle of difference or alterity. Even politically speaking the value of difference is recognised as essential. For example, Western democracies, like Canada, strive to promote difference in preferring to consider themselves as multicultural societies. There is a certain value that is attached to difference, and it would be fair to say that the postmodern philosophical world posits much value or weight on the notion of difference even though we do not speak of it explicitly. There is a tacit "positive" value ascribed to difference, and this tacit presupposition underlying our view of difference must not be overlooked. If there was no tacit value attached to the *discours de la différence*, then how else would we account for such a profuse amount of philosophical literature and socio-political activity devoted to this very theme?

Let us proceed to phenomenologically describe that tacit "good" or value which is difference. In order to do so we must investigate the nature of difference, namely, to ask: what constitutes difference? and, how do we recognise it as such?; and second, why is it worth promoting? what is its value? or why is it valuable? Let us begin with the latter question. We have already offered one concrete reason, namely, that it creates an ontological space wherein the human individual is free to actualise his or her own potentialities, which by nature or *ipso facto* are unique from her or his fellow individuals. It is this sacrosanct space which serves as the spatio-temporal locus wherein the individual possesses the freedom to localise himself or herself in history in a manner in which he or she sees fit. The individual's well-being is in large part dependent on this individual and differentialising space in so far as the postmodern world views such a possibility for self-individualisation as a fundamental *conditio sine qua non* for the maximisation of one's possibility of being (*Sein-können*).

The call for an individualising or differentialising space can also be translated into politico-ethical terms. Implicit in the claim of an ontological space and the positive value it entails is an ethical call for tolerance. Hobbes rightfully observed and recent history has shown that difference and the ontological space it occupies are precarious in so far as they can be easily annihilated through warring ideologies and violent weapons of destruction – a veritable *homo homini lupus*. If difference is to survive, what is required on the part of the individual is a disposition to respect the difference which is the other – a *dispositio ad alterum.* We see this ethical claim explicitly called for in the works of Levinas where he calls for a breaking through of the face of the Other. It is true that postmodern philosophers in general do not speak of ethics or even the possibility of an ethics, however, they do stress the importance of socio-political structures as fundamental to "justice" or equity. Lyotard himself, after denying the possibility of an ethics[2] in his earlier work, recognises the importance of an "ethics" in so far as he speaks of the link between aesthetics and politics in his latest analyses of Kant and in his work entitled *L'Enthousiasme*. The call for tolerance and the dispositional attitude it requires towards the other may be viewed as a starting point for dialogue or discourse. If one individual is willing to hear and respect that which is differentialised as other, then perhaps the threat of potentially violent and destructive situations may be somewhat reduced. It would be naïve to think that such an ethical attitude or *habitus* of being could prevent all crises and conflicts, but it could help temper challenging realities. This attitude will not simply come about through reason or will, but through an accord entered into by each individual to rationally will to value such a thing as tolerance. There are no guarantees that this is a possibility; however, the postmoderns, in a strange kind of way, hope that this is possible, especially given their attempts to influence the shape of various political structures. Again, it must be emphasised that such an ethico-poltical reading of postmodernity is a tacit, underlying theme of postmodern thought more than an explicit call for tolerance and a revival or incarnation of a certain ethico-political stance. A silent and benign "patior!" must be pronounced by each individual in whatever situation one finds oneself in order that this *dispositio ad alterum* may be fully or partially concretised.

There also exists an epistemological value to the postmodern stress on difference, namely, a critical and distinguishing mind. Difference challenges the critical mind to recognise that there is no one necessarily comprehensive way to view the world. Rather, there exist a plethora of foci through which to view and interact with the world. Epistemological certitude of a universal nature is simply a dream and can never be achieved. Rather, we may speak of a limited certainty which is conditioned by temporal and cultural givens. Derrida speaks of the ever transcending text – *l'itérabilité*. The very nature of meaning and interpretation is such that it can never be pinned down to an exact science, rather there is always an elusive element. Perhaps, what remains of the transcendence of meaning is a trace which may or may not be identified as a common or unifying nucleus. The *positum* of the trace can signify or point to another reality, but it is never meant to be held as being explicitly categorical. The challenge of postmodernity to philosophical thinking is twofold: Apophatically or negatively speaking, we should never be so convinced that what we believe to be true and categorical is universally valid, for the changes incumbent upon time and history usually transform our understanding of what we once held to be categorically true. Second, and in more positive or cataphatic terms, there is a challenge to constantly rethink our categories of critical evaluation. Addressing this apophasis and cataphasis of postmodernity, Lyotard aptly remarks. "Un artiste, un écrivain postmoderne est dans la situation d'un philosophe: le texte qu'il écrit, qu'il accomplit ne sont pas en principes gouvernés par des règles déjà établies, et ils ne peuvent pas être jugés au moyen d'un jugement déterminant, par l'application à ce texte, à cette oeuvre de catégories connues."[3] Up until this point we have discussed the value of postmodernity in so far as it philosophically attempts to radically define an ontological space, challenging us towards an ethical *dispositio ad alterum* and creating the condition for the possibility of a genuinely critical stance towards the world, one open to diverse cultural and historical changes. We have listed these postmodern ends as valuable in so far as they answer the question concerning the "good" of postmodernity. We now move onto a more fundamental phenomenological question, namely, that of the constitution of difference.

II

What constitutes difference? or what makes someone or something different from another? or what makes difference difference? Baudrillard and Guillaume make the distinction between *l'autre et l'autrui*:

> Pour le dire simplement, dans tout autre il y a l'autrui – ce qui n'est pas moi, ce qui est différent de moi, mais que je peux comprendre, voire assimiler – et il y a aussi une altérité radicale, inassimilable, incompréhensible et même impensable. Et la pensée occidentale ne cesse de prendre l'autre pour l'autrui, de *réduire* l'autre à l'autrui. Réduire l'autre à l'autrui est une tentation d'autant plus difficile à éviter que l'altérité radicale constitue toujours une provocation et qu'elle est ainsi vouée à la réduction et à l'oubli dans l'analyse, la mémoire, l'histoire.[4]

What is signified by the use of the pronoun *autrui* is difference which is defined and comprehended by its not being me. Radical alterity is that which is not comprehensible or capable of being "seen" as assimilated, which is even ineffable. The constitutive element or moment of difference is "me", the ego. In particular, that which renders difference comprehensible (*l'autrui*) or ineffable (*l'autre*), capable of assimilation or non-assimilation, is that which is *not* me. Baudrillard and Guillaume employ a very interesting language to describe difference – one rooted in epistemology or critical thinking as communicated through the ego. In other words, the "me" of Baudrillard and Guillame is one which determines that which is not "me" (apophasis) by means of a *cogito*[5] of sorts. Let us recall the vocablary used by both thinkers to describe this apophatic "me": "je peux comprendre, voire assimiler" and "inassimilable, incompréhensible, impensable". The vocabulary employed places emphasis on the faculty of critical judgement – a distinguishing logic, logic understood in the full and generous sense of the Greek *logòs*. The condition for the possibility of difference lies within the ego which apophatically distinguishes from itself that which is not itself.[6]

If the ego determines difference in so far as it distinguishes from itself that which is not itself, then what are the implications of such a concept as difference. First, there is the question of inter-subjectivity. How can we account for any communication or the presence of community or society? To employ Lyotard's word, wherein lie the *Übergänge* which bridge one ego to another? Postmodern philosophy in general has been reticent on this point, especially given its deconstruction of the myth of a foundation or unity. In his analyses of Kant and the sublime, Lyotard attempts to address this question by analogously linking the aesthetic with the political. The project is doomed to failure, for the ego is conceived *ab initio* as a self-contained reality. It is not dependent on the other for any reason except for its capacity to distinguish critically itself from another ego which it may or may not comprehend. Unlike Levinas' ego which must break out of the darkness which is the *il y a* in order to fulfill itself existentially, there is no need for the postmodern ego to exteriorise or go out of itself, as it is self-contained and a complete distinguishing judge-in-itself. The ego or the "me" of difference is reduced to a logical tool which facilitates the process of differentialising. In this sense, the ego becomes a logical category, for it serves to make alterity clear and distinct. The ego merely becomes an instrument of the understanding – an *ego sur-volant*, to use Merleau-Ponty's expression, which hovers above the world determining "otherness" only in so far as it is not "me". Ultimately, the *Einfühlung* which characterises intersubjectivity is rendered an impossibility, for there is no need or real possibility of the ego escaping its own confines. It always recoils unto itself by its very definition as that "me" which is capable of distinguishing that which is not me. A genuine otherness is never capable of breaking through such a self-contained and self-positing ego.

The question incumbent upon the philosopher becomes: What constitutes that "me" or ego other than its aforementioned logical or categorical function of differentialising? One cannot rely on the "narratives" of the past to guide us, as they have been proven artificial and unsatisfactory. Again, we are forced to rely on the category of difference. Each of us is different, and this categorical understanding has no formal or substantial content. Any content can occupy this formal category of difference – a freedom *für sich in absoluto*. Granted this freedom may be used to facilitate some sort of political or societal justice, however, the potential for abuse and furthering the gap (*écart*) between some already pre-existing fragile comm-unities may also come into play. The ethical ramifications of such an egologically based perspective of the world can ultimately reinforce my own view as my own. I am able to distinguish the other as "not me", but "so what?" Dare we ask, what "good" or value is consciousness for the sake of consciousness unless it is connected to something other than consciousness – that other which is not my consciousness and is not given an ontological space wherein it can appear as itself (for by its very nature it is determined by the logical category of "my" difference). In other words, I am only conscious or unconscious of an other as it is not "me". Can I truly view that other as other apart from "me"? And even if I can do this, then I am in no way obliged to respect that other, for under the category of difference I may choose to cynically ignore, reject or even thwart such otherness. The presupposition underlying the category of difference is a firm understanding of the constitution of the ego and how it relates to itself as self and as interiorly other to its own self (estrangement). The category of difference gives us no *Relationlehre*, as anything can fall under the category of difference. All egos are reduced to an identity in so far as they are differentialised by their not being "me" – every ego becomes the same in so far as each is different. In an identity there is no need or possibility of relation, as each is the same apophatically or cataphatically. We can even choose to eliminate the category of difference all together – a veritable totalitarianism. The ego simply becomes an *etwas* to be recognised and defined as simply "not me". This is the postmodern narrative – the Egology of Identity. The *etwas* which is the ego renders the whole enterprise of philosophical anthropology superfluous, as the human person is reduced to a minimalistic logical space.

In brief, what I have been arguing in the second part of my paper is that the postmodern constitutive view of difference presents the human person with grave difficulties. First, difference becomes an empty universal logical category which can be filled with any content, thereby rendering the ego or "me" an empty vacuum. The result of collapsing all egos into the grand narrative of the "Egology of Identity" is an identity crisis – which, goes without saying, has become epidemic in the West. For example, the North American "X" Generation is a perfect example of this crisis of personal identity. Second, postmodern difference gives very little account of the phenomenon of intersubjectivity and the fact that people, communities and states can relate to one another, even though they may be comprehended as different. In other words, how do we account for existing grand unities? Third, the concrete ethical implication of such a view of the world enshrined as difference is a brute relativism. Anything can be justified? Who is to say what is unacceptable or unjustified? What can make one difference more valid than another, if all falls under the logical empty category to difference? What is to prevent the human condition from being understood as a *homo homini lupus*? We see the logic of difference paving the way for a brute egotism as opposed to an healthy egoism. Moreover, there exists a grave possibility that the ethico-political channels of discursive communication may become so ambiguous and different, that no authentic communication is made possible except with myself.

As concrete evidence of the difficulties created by the postmodern emphasis on difference, we would like to offer a few remarks on the theme of loneliness. There is a paradoxical situation occurring in the Western world in so far as that on one hand the reality of the "global village" is no longer a dream. The series of communication and computer networks have brought about a greater facility in communication. Knowledge and "touching base" with people in different places and working within different fields of specialisation is virtually effortless. Technology has afforded the postmodern individual with the means of accessing a variety of different resources ranging from the most basic cooking recipe to the latest medical advances for the treatment of colorectal cancer. Computer systems such as the Internet and Virtual Reality have significantly changed the way we interact with one another. Even though an *Abstand* separates individuals communicating from different geographical settings, there is an anonymous intimacy that spans the global information highway. We use the word intimacy in that we have access to information which normally would have been privy to a certain select few.

On the other hand, we have a growing personal crisis of epidemic proportions in so far as individuals of the postmodern world are experiencing a reclusive life of greater loneliness. Loneliness is not to be confused with solitude. Solitude is a natural desire on the part of the human individual to be one and separate from others. Solitude is a temporary state of being usually expressed on the part of the individual in order to recollect or re-create oneself. It is an existential state of introspection, reflection and reshaping. By contrast, loneliness is a state of estrangement from self, other and the world. One may be lonely even while surrounded by a plethora of people and things. The postmodern individual can be characterised as dwelling in a state of loneliness, of solipsism, in so far as she or he is forced to rely on oneself for everything, as he or she is the categorical vortex of difference or alterity. The ego becomes defined in its own self-contained and self-positing reality. Nothing is given to this ego except the fact that it can comprehend or not comprehend difference – a barren reality.

We see three characterising features which typify postmodern loneliness. First, there exists a great difficulty in forming and maintaining relationships – be they friendships or deeper love interests. This is evidenced by the alarming escalating rate of divorce, family violence and counselling hours spent trying to heal the wounds of failed human relationships. Talk of relationships has dominated popular culture for the past ten years or so. The number of television programmes espousing the importance of relationships and chronicling the history of doomed relationships has come to be standard on television. There is a whole industry set up to help people meet other people, namely, anonymous chat lines, television talk shows which bring surprise crushes together and the whole computer generated meet market. These are significant indicators of a phenomenon, namely, the desire not to be alone – a desire which springs forth from a lonely reality.

A second feature of this postmodern loneliness is a general lack of imagination. We see this especially in the education of the young. Students in North America are being taught more and more functional skills necessary to meet the challenges of the corporate world. One no longer does mathematics and geometry for the sake of forming the mind, enabling it to imagine and foresee certain patterns and think through them, no matter what they may be. Rather, mathematics consists now of teaching young students how to complete basic mathematical functions in order to execute a certain corporately designated task such as the calculation of interest. There is no doubt that such corporate functions are relevant, but surely they ought not to become the sole logos through which our education system defines its purposefulness. John Ralston Saul in his *Unconscious Civilization*[7] documents the deleterious effects of the corporate culture on the individual reducing her or him to a unconscious drone. We also see this manifested in our universities where the traditional human sciences, ever reliant on the human imagination for their fulfillment, are receiving less and less funding. These arts are not seen to be imaginative but as fantastic indulgences of the spirit which do not seriously entertain pragmatic affairs of economic livelihood and practical political issues. In other words, what does a Rembrandt or an Heraclitus have to do with liquidating a national debt or solving the political strife between a sovereign Québec and a supposedly unified English-speaking Canada? Imagination is now conceived of as fantastic as opposed to re-presentational or image-making. The stress in our postmodern world is on functional thinking with a pragmatic end. The problem with this corporate functional thinking *in absoluto* is that it blunts the human imagination thereby reducing the human being to a function or capacity. My humanity becomes defined by my job or even reduced to my job. The supposed differentialising ego has become blatantly defined by a super-imposed functional thinking. The ego has been reduced to a lonely cog in a machinery called corporation. Nature abhors a vacuum and if we do not significantly choose to define the space we call difference, namely our egos, with something other than the criterion of something that is not "me", then we run the risk of having our egological space filled by a more powerful and willful ego. In this case, the ego as conceived by corporate authority. Our humanity is translated into a corporate *kosmioton*.

The third and final feature of this postmodern loneliness is cynicism. I believe we have all experienced the cynicism of our times, especially within the political world. Most Western countries have expressed this cynicism in strange ways by electing to power governments that support extreme views, not because these views are viable, but in a backlash against the ineffectuality of the previous government. More fundamental, however, is the attack mounted against the institutions and conventions of Western liberal democracy. The debate between the Straussians and the deconstructionists over the nature of the political process or the politically correct university curriculum is evidence of this cynicism. Each side distrusts the other and any attempt to genuinely resolve the problem is often met with a pre-ordained cynicism or *méfiance*. There is a general distrust of one another. For example, think of the politician who promises to solve the economic crisis of the country. Such a promise is usually met by a sight of incredulity. Each of our motivations is under suspicion and we are constantly afraid of acting lest we might be seen to be rocking the comfortable status quo. We wish to remain neutral and anyone who stands out against this status quo of mediocre neutrality is viewed with distrust or cynicism, for he or she *de natura* will never be able to solve anything. It is as if we were saying, "Why bother!!" This cynicism conditions our critical judgement in so far as we are more ready not to trust the judgement of another because we have too much trust in our own *méfiance* as a certainty. The result of such a certain trust in our own *méfiance* is a strange egological skepsis. To summarise, the postmodern condition of loneliness can be characterised by an ever increasing inability to form and sustain relationships, a lack of imagination and a growing cynicism.

III

If our investigation of postmodernity is correct, then we could best describe the postmodern *Weltanschauung* as one that has tried to preserve and encourage a discourse of difference. However, in concentrating its energies on the logic of difference, postmodernity has eclipsed the reality of the human person by reducing her to a lonely, egological existent understood within and by an all-encompassing "narrative" of difference. The discourse of difference has obscured that which lies the behind the discourse, namely, the life of the human person. In this our last section, we shall attempt to phenomenologically examine the *vita personae* not merely as "narrative" that comm-unifies the egos of difference, but as a *fundamentum* in which both unity and multiplicity/difference coincide – a veritable *coincidentia oppositorum.*

If we reflect a moment on the nature of the ego, we too like Descartes are capable of naming its various functions such as those of thinking, doubting, and cogitating. However, there is something prior which allows us to carry out such functioning both as *res extensae* and *res cogitantes.* We shall describe this as life. We all live, a basic universal fact. Moreover, there is a particular mode of our living which can also be described as universal and fundamental, namely, the fact that we are persons. What distinguishes our life as personal is the capacity of the person to creatively enact one's being – an ontopoiesis. As Augustine rightly pointed out, we simply do not exist as rocks or live like beasts.[8] Life is that force or condition for the possibility of beings and in particular, the being of the person, to actively develop and become, maximising its potentialities. Life, then, could be described as a *principium.* It is a starting point or in Edith Stein's vocabulary a zero point of orientation.

We nominate life as a fundamental starting point because of its universality – we are all living beings. It may be granted that life is common to all and could serve as a starting point for a grand narrative. However, like difference, it may be charged that life too can be conceived of as a merely formal logical category which can be filled with any content whatsoever – *a placet.* Life is not to be understood merely in logical terms, for it has a hard-core existential reality in so far as it stands as a confirmation of something's truly being as opposed to nothing being at all – the *nihil absolutum.* It is not a pure mental construct. It is physically prior to our critical capacity to comprehend the difference which is not me. How can we justify the apriori nature of life? Each of us did not think ourselves into life. Life was a given to us in so far as we were conceived by our biological parents. Moreover, we do not sustain our own lives by acts of our own volition or mind. We can choose to terminate our own lives, but the fact of suicide should not confuse us as to our sustaining our lives. Edith Stein poignantly reminds us of this existential fact, that we are kept in our being and this being held is not necessarily one of angst and fear, but one of security and sweetness. There is a feeling of assurance that we are being held in place just as a child has the feeling of freely falling while mindful of being held by his or her mother.[9] In this sense, life in its fullness (*plenitudo omnitudinis*) is a *hypostasis* in the sense that it stands under our own personal incarnation of life and sustains it.

Besides life being a given and a hypostasis, there are other universal qualities which apply to it. If we observe the fullness of life, we cannot help but be struck by the dynamic interplay of the one and the many. There is a constant relation between unity and multiplicity or difference in that we see cosmological and zoological phenomena which conform to this reality. For example, most physicists and geologists would agree that the earth had a particular beginning and will come to an end at some point. What is fascinating about their conclusion is the speculative theory surrounding the origin of the earth, indeed the awesomely differentiated universe, stemming from one source of energy. In the world of the animal kingdom one is struck at how the higher animals dwell in common yet retain some individualising qualities about them. One is struck by this dynamic when studying the life of anthropoids such as chimpanzees or gorillas. Even in human culture, especially, it would be fair to say today that human beings dwell in communal settings wherein, in optimum circumstances, their individuality and their right to union are both recognised and preserved. The Mediaevals, echoing Aristotle, thought it was by nature that humankind dwell in common. Why do these phenomena exist? What do they point to?

Husserl, in his earliest published work, *Die Philosophie der Arithmetik*, posits a fundamental insight into the nature of our psychological reflections on the nature of number, namely, that number is a unity of multiplicities and a multiplicity of unities. Husserl, would take those implications further and apply them to the *Erlebnisse* of the *Lebenswelt*. Edith Stein, in her *Untersuchuung über den Staat*, radically affirms that contained within the concept of person is a community of persons. Stein notes, "Those who confuse the concept of person as juridical person and person in so far as that person is an individual, certainly will marvel at the possibility of person being conceived of containing within herself or himself a plurality of persons."[10] What Husserl and Stein affirm as fundamental is that each individual person is not an isolated ego different from other egos only capable of sometimes seeing what is different in others. Rather, there is something which is ontologically prior, which significantly marks the life of all persons, namely, that each person has as part of his or her essence a "space" which facilitates the coincidence of unity and multiplicity. In other words, a grand narrative rooted in the life of the person is possible. It bespeaks a unity and includes both smaller communities and radical difference. It is this ontologically prior universal reality of life which is in itself the condition for the possibility of unity and multiplicity. Difference and unity are not a posteriori categories, as they are principal features of our lives.

To further elaborate the universal qualities which characterise life, we maintain that life is not to be viewed merely as a thing, static and completely observable through a microscope. Life is a dynamic drama which is composed of three essential movements which facilitate the coincidence of unity and multiplicity. These movements are: *Gravitas, fluctus et inertia*. Gravity is a particular force within life which seeks to stabilise or ground persons and objects within a livable framework. Matter, on the biological level, could be conceived of as a gravitational force which holds people and objects in place. It gives them a physical form or shape. Matter is that which makes the body livable or experiential. Gravity could also be understood in a spiritual sense in so far as it grounds the personal psyche in a certain *stabilitas loci*. For example, one's belief in the existence of a personal God may be an example of this spiritual stabilising force called gravity. One could, for example, also conceive of Love, as a spiritually grounding force. The gravity of life is essentially that which "de-fines" or sets limits around persons and objects, the world, in so far as it stabilises or localises them in particular spatio-temporal context. It is true that life as we know it has existed for over four thousand years, and this is our common legacy. However, it is the particular movement of gravity which allows particular persons to be gravitated towards or localised in a certain historico-cultural environment with a particular bodily or genetic constitution. The Scholastics used to maintain that matter (a universal principle) was he *principium individuationis* (difference) which allowed various substances to be distinguished one from the other. Gravity is that force which makes stable and experience-able the unity and multiple difference constitutive of the life of the human person.

Fluctus or flux is that particular force of life which makes active the very activity of life, for it empowers or moves the particular life of the person. The Greek *psyché* or the Latin *anima* have as root meanings motion. Here too we draw upon the motion and dynamism of life which permits all objects and persons to move and interact. This interaction is marked by a co-penetration, as objects influence and become part of the person and vice versa. Through this interactive exchange motivated by the force of flux there is a reciprocal, affective exchange or correlative onto-poiesis. As objects and persons move and interact with one another, one becomes influenced or shaped by the "doing" of the other. Not only is there an intersubjective exchange where both parties are affected, but there also exists the profound possibility of changing the nature of the grand narrative of life which unifies all of these persons and things of the world. In other words, through the force of flux, life need not become envisioned as a stagnant immovable mover. It is movable and moves simultaneously. The grand narrative of life, and the personal life which is part of this grand narrative, are not logically determined by the empty category of difference. There is a genuine ontological freedom to move and be moved depending on the nature of the interactive exchange which characterises flux. Our understanding of the force of flux permits life to be more radically open to the dynamism which is unity and multiplicity than would an egologically based understanding of difference, for life does not solely depend on an a posteriori apophatic judgement on the part of the ego to determine the nature of its being acted out. The force of flux permits the *Inhalte* of life to interact and form one another, not by difference alone, but through the ontologically prior movement of the coincidence of unity and multiplicity.

Inertia is a sickly or even a moribund force of life. It is that force which brings things to a halt while detaching them from their stabilised or gravitational ground. It paralyses flux and strips the life of the world and the person away from its *Fundament*. While this force is destructive and generally runs counteractive to life, it does present some benefits. It provides a space for those elements of life or persons which no longer can find a place in the narrative of life. For the human person, it is incarnated in the sacred space of death wherein the human person can come to rest after his or her role in the grand narrative of life comes to a biological end. It may also serve as a space for the elimination of unjust acts and tendencies, such as systematic murder or war. It may also, however, serve as a space for the breeding of violence and destruction. The force we call inertia is a mysterious force, for it can be of positive and negative value and we cannot fully grasp its movement. It is that force which makes the constitutive coincidence of life's unity and multiplicity opaque and even painful.

In general, life and its three movements along with life in its particular form as the life of the person are to be grasped as a *Fundamentum* which has as a constitutive ontological space that of the coincidence of unity and multiplicity or difference. Postmodernity in isolating difference by placing emphasis on it and deconstructing the "myth" of the grand narratives has made difference a universal category which can be easily manipulated and twisted preventing the possibility of authentic discourse and genuine respect for the life of the human person who is prior to the postmodern category of difference. We do not wish to underestimate the importance of difference. Rather, by positing life as the constitutive ground of the coincidence of unity and difference, we have sought to create an onto-poietic space wherein person and world can freely come to interact within a comm-unifying framework which houses a plethora of resplendent forms of life. We advocate no further thinking on the problematic of difference if it is to be understood simply as difference. But we do not advocate either a rethinking of the problematic of unity as it was pursued by the German Romantics and Idealists. Rather, we advocate a rethinking of the ground which is the soil for the ontological roots of the coincidence of unity and multiplicity that we find incarnated not only in the world, but in the human person. We propose a recollection and reincorporation of difference into its larger existential reality. In essence, we have tried to establish a simple *Wiederholung* back to the ground of philosophy, namely, life in general and the human person in particular. It is this ground or insight into the ground which we hope will ultimately guide the course philosophy and the other human sciences take in order to further penetrate the mystery which is, or even who is, being.

In conclusion, we must ask ourselves what this bountiful plenitude of the differentialising ground called life, can offer us. I would like to cite three possibilities: *philia, politeia*, and *eunomia*. If we recognise life as a comm-unifying ground which is concomitantly oriented towards a resplendent plethora of forms, then we must examine the strange phenomenon that permits life to continue. In other words, we must ask ourselves the question why we, who have the technological capacity to destroy life in all of its forms many times over, continue to strive to preserve life. What permits us to continue living and not kill one another? In response to such questions, we posit a tacit *philia* or *amicitia* or friendship. There exists a tacit acknowledgement, conscious or unconscious, on the part of the majority of human beings that life is an ontopoietic gift and offers each person the possibility to reap a bountiful ontological harvest. In order to preserve the possibility of such a poetico-personal realisation, there must exist a sort of love between persons. This love is not rooted in the legal jurisdiction of rights and privileges, but occurs naturally and is coexistent with life. *Philia*, understood as a natural friendship among persons, is the loving dynamic which facilitates and permits the relationship between human beings and between human beings and all other creatures, animate or inaminate. It can be understood as the natural drive for survival. As persons, we recognise that there is a tacit dependence on the other in order that we may advance our own personal life. Nature points to this reality from the moment of our conception, birth and childhood formation, as no child can completely survive on its own. The touch and care of other persons is necessary. Moreover, humankind is the only animal which remains so dependent upon others for his or her survival for so long time. Without a natural friendship which comprehends the necessity of depending upon others for personal survival and for the survival of the *plenitudo omnitudinis* which is life, then the very survival of life and our persons would be threatened. We see the reality of this threat made concrete when natural friendship breaks down and we are left with the crude insanity of war, violence and destruction. We find consolation, however, in the fact that we human beings find war and violence to be an unnatural state of being, to be something anti-life and anti-being. In other words, we find comfort in friendship rather than in hostility. It is this very natural state of friendship which permits the one and the many of life to flourish.

We now move on to a different level than that of nature, namely, that of culture and society. *Politeia* draws upon the natural givenness of *philia* and can be viewed as that element in the human spirit which attempts to draw persons into an organised societal structure wherein common similarities and differences can be maximised. The etymology of the word comes from the Greek *polis* which means a "city-state" and can best be described as the comportment both an individual and a community have towards the "good" or well-being of each other. The English "politeness" transmits a certain sense of the Greek *politeia*. However, *politeia* is not to be understood as a benign tolerance for the other. On the contrary, a politeia involves mindful persons coming together in order to maximise the conditions which would ultimately maximise the "good" of both individual and community. The individual is not a priority over the state and neither is the state opposed to the person. Rather, a discursive relationship must ensue in order to facilitate a genuine onto-poiesis of person and community. *Politeia* does not imply a panacea for the strife and struggles which we experience presently. Rather, it is a concretisation of a genuine *dispositio ad alterum* which seeks to reasonably listen and act according to the needs and desires of self and others. *Politeia* implies a listening and responsible responding to one's own personal needs and the needs of the society and other societies. It involves not only an attitudinal disposition, but a political and economic response in so far as political and economic conventions must be implemented with the realities of life, person and their onto-poeitic maximisation in mind. Life in its various culturo-historical localisations offers us the human possibility and experience of *politeia* in order to build upon the natural foundation of *philia* which is offered to all of life's participants. An example of this experience of *politeia* can be seen in the founding histories of liberal democracies over the past two hundred and fifty years where, in countries like the United States, there was a conscious attempt to mediate the exigencies of both nature and society.

Finally, life attempts to achieve a balance between nature and society. This is achieved through *eunomia*. *Eunomia* literally means good-order. Eunomia is that element in life which strives to bring a just order out of chaos. Western society has become obsessed with autonomy, literally a self-ordering, in so far as there is a drive to maximise one's own ego, a veritable will-to-power of the "me". This is facilitated and justified by the vocabulary of difference. An example of the will-to-power of the "me" is the breakdown of civil society or politeness which was so remarkable predicted and described by Eric Voegelin. In so far as I am different, I need not regard the other as relevant, as he or she merely lies opposed to me. The other becomes a *Gegenstand*. Eunomia, on the contrary, calls for a situating of the personal self within the contexts of nature, society and ultimately life. With *eunomia* the emphasis is on a broader ordering of the whole and the many while with autonomy the emphasis is on the ordering of the self. *Eunomia* seeks a good ordering of what Voegelin called the "maximum of differentiation", whereas autonomy seeks to particularise in so far as it is centred on the self. The risk of postmodern autonomy is that the self becomes too particularised, thereby paralysing the possibility of any discourse or order of being. Ultimately, we posit that *philia, politeia* and *eunomia* are contained within the "vivi-structure" of life, thereby providing the human person with the mechanism necessary to ensure the discursive plenitude of comm-unity and difference – a plenitude which is necessary for genuine onto-poeisis or self-realisation.

Catholic University of Leuven
Belgium

NOTES

[1] J. Baudrillard and Marc Guillaume, *Figures de l'altérité* (Paris: Descartes et Cie 1994), p. 9.

[2] See *Economie libidinale* where Lyotard discusses the impossibility of an ethics due to the investment of the driven libido by different individuals. He does, however, call for an *ars vitae.* Jean-François Lyotard, *Economie libidinale* (Paris: Les Editions de Minuit, 1974), p. 20.

[3] Jean-François Lyotard, *Le Postmoderne expliqué aux enfants* (Paris: Editions Galilée, 1988), pp. 9–27.

[4] Baudrillard and Guillaume, *Figures de l'altérité, op. cit.*, p. 10.

[5] I employ the word *cogito* in the root sense of literally drawing things together. This use of the Latin verb *cogito/cogitare* is meant to parallel both Baudrillard's and Guillaume's use of the term *assimiler/inassimilable.*

[6] This very much hearkens back to the Fichtean self-positing ego.

[7] John Ralston Saul, *The Unconscious Civilization* (Toronto: Anansi, 1996).

[8] Saint Augustine of Hippo, *De libero arbitrio*, Bk. II, Ch. iii, pp. 240–241 in Corpus Christianorum – Series Latina V. XXIX. For Augustine, the intellect was that which marked the life of the human being as different from the life of the beast.

[9] Edith Stein, *Endliches und ewiges Sein: Versuch eines Aufstiegs zum Sinn des Seins* (Freiburg im Breisgau: Herder, 1949), pp. 56–57.

[10] Edith Stein, "Eine Untersuchung über den Staat," in: Edmund Husserl (ed.), *Jahrbuch für Philosophie und phänomenologische Forschung* (Halle: Max Niemeyer, 1925), pp. 52–53. Translation mine.

MARIO CASULA

THE PROBLEM OF THE ORIGIN OF LIFE BY CHANCE

The present Congress is focused on the phenomenology of life in a broad sense, and thus presupposes the reality of life, and consequently its origin in a given moment in the development of the cosmos.

Judging by the most recent fossil finds, it seems certain that life appeared on planet Earth about 4.5 billion years ago. This date is very important, as will be seen below. But the real question is, why did life appear? What was the cause that made it arise?

Biologists, specifically scholars of the problem of life's origin, generally admit that, given its enormous complexity, the life-event is highly improbable but that it happened by pure chance, by a fortunate and fortuitous combination of substances and circumstances in a terrestrial biosphere which was perhaps unique and unrepeatable.

Various shades of this thesis, which in some of its motivations resembles a modern version of the antique theory of spontaneous generation, are sustained by groups of researchers such as those directed by L. E. Orgel of the famous Salk Institute for Biological Studies in San Diego, and also by numerous Nobel Prize winners, such as G. Wald, J. Monod, Manfred Eigen and Chr. de Duve (the latter has just published his views in his exhaustive book; *Vital Dust. Life as a Cosmic Imperative*).

What is the motivation for this theoretical stance? Certainly one part of the motivation came from the rapid development of microbiology, which enabled laboratory simulations of "life" phenomena such as the effect of physico-chemical reactions (beginning with famous experiments such as those by H. C. Urey and S. L. Miller in 1950–1953, inspired by A. I. Oparin and J. B. S. Haldane, and proceeding up to the more recent experiments by J. Oró, H. F. Noller, G. F. Joyce and others). I would say that above all, motivation has come from the preponderant influence of the statistical understanding of natural laws, in opposition to the classical deterministic understanding of the same laws. James Clerk Maxwell (1831–1879) introduced this understanding into physics in his formulation of various aspects of the kinetic theory of gases; Gregor Mendel (1822–1884) and especially his followers (De Vries, Tschermak, Correns) introduced it into genetics. This statistical understanding of

A-T. Tymieniecka (ed.), Analecta Husserliana LIX, 69–82.

natural laws has a physical-mathematical formulation of the first order in W. Heisenberg's so-called *Principle of Indeterminism*.

Thus, the affirmation that life arose *by pure chance* from the original "prebiotic soup" has become not only a commonplace, but a kind of dogma accepted without discussion, along with the assumption that we will eventually establish what the first life form was.

Unfortunately, philosophers have not examined the biologists' theories seriously; Monod observed that they often limit themselves to a self-sufficient smiling at them,[1] with devastating consequences not only from the religious, but also from the logical-philosophical point of view.

In fact, (and this is the deep meaning of my talk) the theory of the chance origin of life contains an implicit contradiction, since it affirms that the start of life is the most improbable event of physical nature, and at the same time says that this event happened by pure chance, billions of years after the formation of planet Earth.

In order to better evaluate the demonstration of such a thesis, it is useful to briefly review some of its fundamental concepts and principles. What is *life*? What is *chance*? What is *probability*? What is the principle of *causality*? What is the principle of *determinism*? And thus, what is the principle of *indeterminism*?

(I) *What is life*? An organism is considered living when three fundamental properties are *contemporaneously* present: growth by intussusception, species reproductive capacity and reactivity to environmental stimuli. Modern biology identifies another property of living beings, that is *communication*, the capacity to have and to transmit genetic information. Beings with this property are at times defined by the acronym citroens, that is, Complex Information-Transforming Reproducing Objects that Evolve by Natural Selection.[2]

(II) *What is chance*? As Monod already observed,[3] this term's definition is complex, dependent on the different situations in which it is used, but it can be said that the prevalent concept among biologists is that of *pure chance*, an event that happens with no definite or determinable cause, when among a multitude of *equally verifiable* possible events, suddenly one determinate event happens. In this sense, chance not only excludes a determinate cause, but, necessarily excludes any cause: for this reason one can not speak of laws of chance, as some do.[4] Very well known examples of chance events include roulette spins and lottery drawings.

(III) *What is probability*? Care must be paid to the difference between its philosophical-gnoseologic definition and its mathematical-statistical one.

(a) *Gnoseologic* probability has to do with a proposition and expresses the attitude of the conscious subject, who, uncertain as to whether the proposition is true or false, but having some motives for affirming it, calls it probable. Common terminology identifies a probable proposition with a thinkable proposition, keeping in mind, however, the fact that this thinkability must not be arbitrary, but must be sustained by more or less valid motives, which make it thus more or less probable.

(b) On the other hand, *mathematical* probability is the relationship between desired and/or expected (favorable) events and possible events, with the condition that they are all *equally possible*: this is the classic concept of probability as spelled out by Laplace. It is normally expressed as a fraction, the numerator formed by favorable events and the denominator by the possible, often much more numerous possibilities. Often the number of possible events is given by the permutations of the possible events and the number of those desired by a particular combination of the same. Thus it is clear that the result appears in decimal form, for which it has been agreed upon that an event becomes more probable as its probability nears unity, and becomes more improbable as its probability nears zero: in the first case one speaks of certainty, and in the second, of impossibility.

This clarification makes evident the philosophical meaning of the probable event: it is an *intrinsically possible* (that is, non-contradictory) and *extrinsically possible* (that is, achievable) event, in this case, by hypothesis, drawings or *repeated attempts*.

Let us look at a simplified intuitive example, one adopted from Cournot. If we place a blindfolded man in front of a box containing one million black marbles and only one white marble, and ask him to extract the white marble, if and when he extracts this white marble, we speak of chance. But if we want to calculate *a priori* his probability of extracting it, we can say that it is 1/1,000,000 or one out of a million, or one millionth, or 0,000,001.[5]

(IV) *What is the principle of causality*? It affirms in general that *any effect has a cause*. To avoid the appearance of tautology that would result if we defined an effect as the product of a cause, we define effect as any contingent entity, that is, an entity that does not exist, but that could exist (and vice versa). If it passes from possibility to existence, this necessarily happens through the action of a cause, which is thus called an *efficient cause*. From this definition it is evident that the principle of causality is the ontological specification of the principle of sufficient reason and thus of the principle of non-contradiction: that is, *ab absurdo*, an effect without a cause would be contradictory, because it would *exist*, by hypothesis, and it would *not exist*, because lacking a sufficient reason to exist, again by hypothesis, it lacks cause.

Having said this, it is evident that the principle of causality is absolutely indispensable not only for philosophers, but for the human mind in general, just as much as the principle of non-contradiction which it depends on is indispensable.[6] And thus there is no place for pure chance as the lack or negation of cause.

(V) *What is the principle of determinism*? This principle as applied to the physical order involves the principle of causality, but it is important to keep the two clearly distinct to avoid errors and false problems. According to a famous formulation by Simon-Pierre Laplace,[7] reported and translated also into mathematical terms by Louis de Broglie,[8] and lightly interpreted by us, it consists in the fact that if from a particular effect one can trace back to its particular cause, one can also do the opposite. And hence, just as by going backwards to the cause one can reconstruct the past, similarly, by going forward to the effects, one can foresee the future, presupposing naturally an intellect able to rightly distinguish cause from effect. As a classic example Laplace cites astronomy and De Broglie cites the movement of a material point in a field of known forces.

This concept of universal determinism dominated classical physics for centuries, until rebellion arose with the introduction of statistical laws (Maxwell) and their theoretical justification in Heisenberg's *principle of indeterminism*, which was initially applied to atomic and subatomic phenomena, and then extended, especially with quantum theory, to all scientific knowledge. In fact, the principle affirmed that it is never possible to contemporaneously know an atomic particle's place (coordinates) and velocity: we know either one or the other. Our instruments of observation disturb the phenomenon,[9] making impossible exact, determinate knowledge. For this reason, and also because of the imprecision of measurements, all our knowledge is necessarily, at least in part, indeterminate and thus by nature more or less statistical.

Already at this point we can make some important deductions.

First of all, the principle of indeterminism does not oppose the principle of causality, but that of determinism. This tells us practically that it is impossible for us to always determine the nature of a specific phenomenon, because we are not always able to determine its cause or causes (remember that the *cognitio rei per causas* is the highest). But this constitutes a limitation of our intellect, which, being finite, is not made for knowing everything. How many things will we never know? A banal example would be the number of the stars: is it even or odd? In and of itself, the answer is certainly even or odd – or with the birth and death of new stars it oscillates from one to the other. How can we know?

In addition, we are able to clarify the concept of *chance*. Like *pure chance* (lack of cause), it cannot be used to explain effect, because it conflicts with the principle of causality. But an admissible meaning exists which puts it in conflict with the principle of determinism, but not with that of causality. We can define chance as a *concurrence of accidental independent causes* which cross each other and/or overlap each other to produce a particular effect; its nature is unknown to us; it is difficult to define in itself or in the prevision of their meeting.[10]

This conception appears necessary in order, among other reasons, to in some way explain the overlap of man's free actions with physical events, as in Monod's classical hypothesis.[11] In such a case, however, it is necessary to keep in mind that the free action, once chosen, is always determined, just as is the physical event it crosses. However, our finite intellect can foresee neither the free action nor the encounter-clash with the phenomenon.

One thing is for certain: chance holds a strange fascination for the intellects of biologists and cosmologists. It is not by chance(!) that Midgley dedicated an entire chapter of his book, *Science as Salvation*, to *the fascination of chance*.[12] This is based not so much on an *ignava ratio* that does not want to search for causes, as it is on the fact that chance is surreptitiously *hypostatized*, as if it were an independent entity, capable of producing something. Thus people speak of the inexhaustible richness of chance, from which everything can derive in a blind game of combinations, including for example the organ of sight.[13] *Chance produces absolutely nothing*, because then it wouldn't be chance any more, but *cause*, in contrast with the hypothesis. It does not even have a *law* because it is the lack of every law. That an effect should be produced *at random* or *through chance* is not the same as affirming that it was produced *by chance*. Chance is only an idea of ours, used to mask our ignorance of cause.

Having exposed these doctrines, we can proceed to the question of *the origin of life*. We begin with an important premise: there is a very close tie between casual events and mathematical probability. In fact, theories for mathematical probability were developed to identify *a priori*, rather than through inductive experience, the relative frequency of a desired and/or awaited event among the many possible events. An example would be throwing dice.

Thus the question presents itself clearly as follows: 1) what is the probability that life would come into being by chance? 2) if this probability is almost null, how then did life happen?

It is neither possible nor necessary to examine here all authors who have written about these questions (and besides, they repeat each other), so we will limit ourselves to three of the previously cited Nobel Prize winners; they are the coryphaei of the argument and have some original thoughts on the subject.

The earliest, *George Wald* (1906–1997), affirms that the beginning of life by spontaneous generation (his terminology for chance) would be practically impossible[14] because of its very high improbability,[15] given the great complexity of the structure of the living being, which makes the most complicated electronic calculator seem like child's play compared with the simplest living organism.[16] However, life happened over the passage of time. The probability of an event increases with the number of attempts, so that something impossible within the limits of a human experience of generations becomes ever more probable in the course of geological time spans.[17] This is especially true since we are dealing with a unique event.[18] The true hero is time, by which the impossible becomes possible, the possible becomes probable and finally, the probable becomes virtually certain.[19]

Jacques Monod (1910–1976) is more brisk, perhaps also in regard to the vivacious criticisms provoked by Wald. He admits that the *a priori* probability of the unique event which is life is almost null, almost zero:[20] in fact, life was absent from the universe and the biosphere did not contain man. *We* are the fruit of a roulette spin.[21] But how, then, did life happen? Monod's response, taken up by other authors as well, is strangely circular: the universe in fact *exists*, and this demonstrates that events with "*the lowest*" probability must have happened.[22]

Strongly against the idea of life's appearance on earth by chance is *Christian de Duve* (1917–),[23] who holds that life comes from virtually living molecules of interstellar dust.[24] He is to be cited because he forwards a singular position, repeated by others, which involves a subtle sophism. The improbability of life is comparable to that of a bridge "hand" composed entirely of thirteen spades (which has never happened): it is 1 out of 635 billion, and becomes 1 out of 50 billion billion billions (5×10^{28}) if one considers all possible combinations. Well then, how did it happen? The sophist response of de Duve is the following: highly improbable events happen daily – every bridge "hand" played being very improbable – and yet we pay no attention and are only amazed by the rarer hands.[25] In other words: the reality of even extremely improbable events is a daily experience.

As can be seen by this quick overview, two common elements emerge from these theses, first, the affirmation of an enormous improbability, and second, the determining influence of time in the event's happening. The two affirmations are strongly liked. The happening of one probable event out of many possibilities requires time, since it is the necessary result of repeated attempts. Thus, the event's happening implicates time as an *a priori* parameter in the calculation. So we have the immediate necessity of specifying two quantities, *probability* and *time*. As we have seen, biologists, even Nobel Prize winners, are vague when it comes to specifying one and/or the other.

Let us begin, then, with the question of the *time* that would be necessary to transform the probability of the origin of life by chance (which for Wald is almost made impossible and for Monod is of almost zero value), into the certainty given by the fact that life exists, at least in our biosphere.

To make the result of this calculation more intuitive, we will use Stephen Hawking's famous example of "typing monkeys" from the cosmological context of his *Brief History of Time*. "It is a bit like the well-known horde of monkeys hammering away on typewriters – most of what they write will be garbage, but very occasionally by pure chance they will type out one of Shakespeare's sonnets".[26]

The Israeli physicist Gerald L. Schroeder took the trouble to calculate the value of that nonchalant "very occasionally", for the famous English poet's Sonnet 18, "Shall I compare thee to a summer's day?", which has 14 verses and 488 letters taken from the English alphabet of 26: a rapid calculation demonstrates that the possible combinations would be 26^{488} and the consequent probability would be the inverse value of $1/26^{488}$, that is $1/10^{690}$. Well, since the universe since the Big Bang is about 15 billion years old, which corresponds to about 10^{18} seconds, a tag team of monkeys working around the clock, even if they typed one letter per second, would still not have produced their sonnet: actually, just producing sentences of 16 letters, would take them two million billion years.[27]

Schroeder's calculations are not easily checked, given the size of their exponents, but it is possible to reduce them to the size of a modern pocket calculator if we use the simplest example. The importance of the argument make it worth doing just that.

We are in Macerata, and thus near Recanati, the home of the great poet Giacomo Leopardi. So then, let's take the first six verses of his famous *Ricordanze*: "Vaghe stelle dell'Orsa, io non credea solidus: tornar ancor per uso a comtemplarvi/ Sul paterno giardino scintillanti/ E ragionar con voi dalle finestre/ Di questo albergo ove abitai fanciullo,/ E delle gioie mie vidi la fine".

As can be checked easily, we have here 176 letters (82 vowels and 94 consonants). There are 21 letters in the Italian alphabet and thus the possible combinations are 21^{176}, that is, 5.13566^{232}. Given that our monkey type a letter per second, in order to know how many years a tag team of them would take, we simply divide the preceding value by the number of seconds in a year, that is, 31,536,000. The result is 1.62851^{225}, a number of years for which we have no name, except perhaps myriads.

Going on to the time necessary for the origin of life: any living molecule is composed of atoms, and thus in order to calculate the time necessary for the chance formation of the most elementary living being, such as the bacterium, we simply relate the number of its atoms to those of the universe. The universe as we know it has a volume of about 10^{78} m^3,[28] which corresponds to about 10^{100} atoms, while the non-watery part of a bacterium well-known to biologists, namely, *Escherichia coli*, contains about 20 billion atoms.[29] So the possible combinations will be given by the number of atoms in the universe raised to the power of the number of atoms in the above-mentioned bacterium. The results is that the time necessary for having life would be myriads of years. If we then change our perspective from the succession of combinations at the rhythm of seconds, and shift to atomic rhythms of millionths or billionths of a second, etc., as suggested by Morowitz,[30] the result does not change: the 15–20 billion years of the existence of the cosmos since the Big Bang would never have been enough for the chance production of even the simplest living thing like the aforementioned bacterium.

These calculations have also been done for the time necessary to produce the RNA molecule, which turns out to be 10^{15} years,[31] while the energy and thus the mass of the carbon calculated as necessary to sustain the ongoing attempts to produce even one protein is much higher than all the carbon present in the current universe.[32]

The conclusion based on the premises is evident: left to pure chance, a living molecule would still not have arisen, and if it has done so, then it was not by chance.

Let's go on now to *probability*. While biologists seemed superficial in considering the factor of time (speaking only of billions of achievable combinations in geological time without even bothering to calculate more precisely the time necessary for life to happen), when we look at probability, another curious phenomenon arises: an underestimation or misunderstanding of the meaning of the *infinitesimal probability* almost null, which has been obtained.

Morowitz, in a *Gedankenexperiment*, tried to calculate exactly the maximum probability of the rising of a life form like the *Escherichia coli* in the biosphere: his result was 1 out of 10 to the power of 100 billion, that is, $10^{-100,000,000,000}$. To give an idea of the meaning of such a figure, which in effect means nothing to us, he explains that it is a number in decimal form with 99,999,999,999 *zeros* after comma, followed by a 1.[33] Another way to illustrate the dimension of this figure, limiting ourselves to the exponent of 100 billion, would be to consider that while a group of any 4 numbers averages about 5 mm in length, the length of the number 100 billion written out would fill about 9 thousand very large books, 800 pages long each, each page containing 14 thousand numbers.

The same number can be visualized by length measurements: let's suppose, for ease in calculation, that 2 thousand numbers are a meter long and thus 2 million numbers need a kilometer-long strip of paper. Thus for 100 billion we would need a hypothetical strip of paper about 50 thousand kilometers long, which exceeds the circumference of the earth at the equator (40,000 km) and is about 5 times the distance from Lisbon to Hong Kong (10,928 km) or from London to Tokyo (9860 km). In the case of probability, this thousands of kilometers long strip of paper is full of zeros and only at the end is seen a 1, which indicates the *residual probability*, transformed now into *possibility*.

The results of all these calculations show how inconceivable it is to admit the origin of life by chance, and help us understand the profound reluctance of many biologists, such as the vitalists and the biophysicists, to accept it.

But it seems preferable not to stress these calculations too much: from the philosophical point of view they lead to an *argumentum ad hominen*, that is, a valid proof in as much as the adversary admits certain premises. So for example, the age of the universe oscillates between 15 and 20 billion years, in function of the absolute value assumed for Hubble's constant, but this supposes that the cosmos began with the Big Bang. However, what if another cosmology is correct, so that the Big Bang is a phase in a cycle of successive expansions and contractions? What if the material world is eternal (the thesis of Aristotle's *Metaphysics*)? Moreover, the world's having a beginning is a revealed datum, but it is not rationally demonstrable that the world can not be eternal (the conclusion of Saint Thomas Aquinas).[34]

So it is preferable to reject chance as contrary to the principle of causality and thus to that of sufficient reason and therefore to that of non-contradiction, in order to then resolve the difficulties raised by those who, despite everything, are still fascinated with it.

Before continuing, it is opportune to underscore a characteristic of life which for the philosopher excludes its casual happening: *finality*. It is so evident that it is underscored as such by Monod, who calls it *teleonomy*, which is, for him "entièrement logique, merveilleusement rationnel, parfaitement adapté à son projet".[35]

This characteristic is extremely important in case the first living molecule might have been the one containing the genetic code, in conformity with the research of Thomas R. Cech and others.[36] The idea that the first molecule possessed the genetic code is so evidently telic that it sets up an insuperable obstacle to the idea of the origin of life by pure chance.[37]

In fact, finality, from the philosophical point of view, requires the work of an intelligence (whether created or uncreated is another question), because in order to do something in relation to an end *to achieve*, the abstract knowledge of this end and of the means to achieve it are necessary. In fact, the object-end is not knowable by experience, since it precedes experience. The means, even though they can be present in experience, have the nature of means in as much as the knowing faculty places them in relation to the end.

Now, a cognitive faculty which anticipates experience and knows of relationships of objects as means, is necessarily an intellectual faculty, the only one which is capable of contemporaneously knowing different intelligible forms.

Now we can respond to some more common difficulties.

1. The life event, even though extremely improbable, has happened, and hence not only is it possible and probable, but it is also certain. *Response*: This is a nice example of the *petitio principii* sophism, which supposes as demonstrated that which instead must be demonstrated. From a preceding fact it follows only that the event did not happen by pure chance.

2. In life events happen continually which are also very improbable. *Response*: a) This is true, however not by pure chance, but by the concurrence of accidental causes unknown to us, among which free causality plays an important role; b) From a preceding fact one can never make the leap, "it is probable, therefore it exists". It is necessary to keep in mind that *the only existent entity is that which is real*, that which is produced by an efficient cause: the *possible* and the *probable* as such *never exist*. They reside preponderantly in our mental categories: the possible expresses the non-contradictory, while the probable expresses a possible achievable event out of many; not knowing its cause, we would like to establish the frequency of its happening. In fact, possible things and probable things happen *as they occur, one by one* from their causes, and they differ from things known with certainty because their cause is unknown, both *a priori* and *a posteriori*. In this way the production of even a very improbable event is one of many instances of a cause-effect relationship, of which daily experience is full.

3. Someone who plays the lottery often has a probability of one out of millions to win the jackpot, yet every week somebody wins. *Response*: The lottery numbers are so-called pseudo-casual numbers, for example, those generated today in a finite number by huge calculators, based on different algorithms which allow repetition only after tens of thousands or more. The decisive factor is that the lottery numbers are played more or less contemporaneously between one drawing and the other by millions of people, who themselves also produce a parallel series of casual numbers: the meeting point of the two series, one generated by the calculator and another produced by the players, mathematically implies the winning.

4. The Monte Carlo method: by generating determinate series of pseudo-casual numbers, it is possible to simulate very complex natural phenomena, very difficult or almost impossible to resolve with equations

or systems of equations. *Response*: True, but we are dealing with series of casual numbers generated for the purpose, using certain parameters in function of the known or hypothesized phenomenon to be simulated and which are changed if they do not serve the purpose.

Now, these last two difficulties impose the following question: the series of casual numbers are produced by a conscious subject using a flexible algorithm to arrive at a hypothetical result, but in the case of the production of a living organism, where is the *agent* which makes the *attempts* for thousands of millions of billions of years to try to produce even the simplest cell? We have to admit *an active immanent principle* in matter, different from the same; this reflection, carried to its end, will carry in a more or less indirect way to the fifth way of St. Thomas Aquinas: "*ergo est aliquid intelligens a quo omnes res naturales ordinantur ad finem, et hoc dicimus Deum*".[38]

University of Macerata
Italy

NOTES

1 J. Monod, *Le hasard et la nécessité, Essai sur la philosophie naturelle de la biologie moderne* (Paris, 1970), pref., pp. 12–13.
2 L. E. Orgel, *The Origins of Life* (London, 1973), p. 193.
3 Monod, *op. cit.*, pp. 127–128.
4 *Ibid.*, p. 110.
5 A. A. Cournot, *Essai sur les fondaments de nos connaissances*, 9th ed. (Paris, 1912), p. 40.
6 For the demonstration of the principle of contradiction in Aristotle, cfr. M. Casula, "La prova aristotelica del principio di contraddizione dal linguaggio," *Giornale di Metafisica* 1970: nos. 5–6, pp. 641–673.
7 "Nous devons donc envisager l'état présent de l'univers, comme l'effet de son état antérieur, et comme la cause de celui que va suivre. Une intelligence qui pour un instant donné, connaitrait toutes les forces dont la nature est animée, et la situation respective des êtres que la composent, si d'ailleurs elle était assez vaste pour soumettre ces données à l'analyse, embrasserait dans la même formule, les mouvements des plus grands corps de l'univers et ceux du plus léger atome: rien ne serait incertain pour elle, et l'avenir comme le passé, serait présent à ses yeux" (P.-S. de Laplace, *Thèorie analytique des probabilités*, III ed. (Paris, 1820), introd., p. iJ – a famous text, most difficult to find).
8 "En langage mathématique, le déterminisme s'exprime en disant que les phénomènes de la Nature sont régis par des équations differentielles dont les solutions sont entièrement determinées quand on connait leurs valeurs et celles de certaines derivées à un certain instant initial. Par exemple, le mouvement d'un point matériel dans un champ de force

donnée est entièrement connu quand on connait sa position initiale et sa vitesse initiale" (L. de Broglie, "Determinisme et causalité dans la physique contemporaine," in: *Revue de Mètaphysique et Morale* XXXVI, No. 4 (1929), pp. 433–443. The text continues citing Laplace to the letter).

[9] A very common exposition. Cfr., e.g., H. Friztsch, *Quarks* (Munich, 1981), pp. 18–22, which sees in the principle an objective limit to our knowledge, and M. Bungo, "The Interpretation of Heisenberg's Inequalities," in *Denken und Umdenken* (ed. H. Pfeiffer), (Munich, 1988), pp. 146–156. Moreover the confusion between the principle of causality and that of determinism is due to Heisenberg himself, too (cfr. his famous essay, "Über den anschaulichen Inhalt der quantentheoretischen Kinematik und Mechanik," in *Zeitschrift für Physik* 43 (1927), pp. 172–198, expl. p. 197).

[10] Cfr. Laplace, *op. cit.*, *ibid.*; Cournot, *op. cit.,* pp. 40–42, 46–48.

[11] A doctor is called for an urgent visit to a patient and enters while a worker is repairing the roof: when the doctor passes under the building the worker drops a big hammer which hits the doctor in the head and kills him (Monod, *op. cit.*, p. 128).

[12] M. Midgley, *Science as Salvation*, (London, 1992), pp. 43–50.

[13] Just so Monod still –. *op. cit.*, pp. 140, 112.

[14] G. Wald, The Origin of Life, in *Scientific American* 19 (August 1954), p. 46c (letters after the page number indicate the column).

[15] *Ibid.*, p. 47b.

[16] *Ibid.*, p. 46c.

[17] *Ibid.*, p. 47a.

[18] *Ibid.*, p. 47c.

[19] "Time is in fact the hero of the plot. The time which we have to deal is of the order of two billion years. What we regard as impossible on the basis of human experience is meaningless here. Given so much time, the "impossible" becomes possible, the possible probable, and the probable virtually certain. One has only to wait: time itself performs the miracles" (*ibid.*, p. 48a).

[20] Monod, *op. cit.*, p. 160.

[21] "L'Univers n'était pas gros de la vie, ni la biosphère de l'homme. Notre numéro est sorti au jeu de Monte-Carlo" (*ibid.*, p. 161).

[22] *Ibid.*, p. 161.

[23] Chr. de Duve, *Aus der Staub geboren*, German translation (Heidelberg, 1995), p. 36, (English original: *Vital Dust. Life as Cosmic Imperative* [New York, 1995]).

[24] *Ibid.*, e.g., p. 50 (given, not conceded: whence comes life in the other world?).

[25] *Ibid.*, pp. 35–36

[26] S. Hawking, *A Brief History of Time: from the Big Bang to Black Holes* (New York, 1988), p. 123.

[27] G. Schroeder, *Schöpfung und Urknall*, German translation (Munich, 1993), pp. 244–245 (English original: *Genesis and the Big Bang* [New York, 1990]).

[28] K. Stierstadt, *Physik der Materie* (Weinheim, 1989), p. 453: this value concerns the universe the light which has reached us and would be about 1/10 of the real volume.

[29] H. J. Morowitz, *Energy Flow in Biology* (Woodbridge, Connecticut, 1979), pp. 12, 65. This is one of the best biophysical studies on the subject, much stimulated, as the preface affirms, by E. Schrödinger's *What Is Life?* (Cambridge, 1944).

[30] *Ibid.*, p. 12.

[31] J. Guitton, *Gott und die Wissenschaft*, German translation (Munich, 1992), p. 61 (French original: *Dieu et la science* [Paris, 1991]).
[32] Schroeder, *op. cit.*, p. 153.
[33] See Morowitz, *op. cit.*, pp. 5ff. for the *Gedankenexperiment* (this very well-known German term is his), in which the value of the probability indicated is on p. 7 and the considerations on the number that render it "*vanishingly small*" on pp. 11–12. Cfr. also *ibid.*, pp. 64ff.
[34] *De aeternitate mundi (contra murmurantes)*, in St. Thomas Aquinas, *Opera omnia* (ed. Leonina), Vol. XLIII, (Rome, 1976), pp. 88–89. As for Aristotle, cfr. *Metaphysics*, e.g., XIII, 6, 1072a, and *ibid.* 8, 1073a.
[35] Monod, *op. cit.*, p. 32.
[36] J. Rennie "In the Beginning: Evidence Grows the RNA Was the First Self-made Molecule," *Scientific American* 261 (Sept. 1989), pp. 13–16; L. E. Orgel, "The Origin of Life on the Earth," in *Scientific American* 271 (Oct. 1994), pp. 53–61; Th. R. Cech, "RNA as an Enzyme," in *Scientific American* 255 (Nov. 1986), pp. 76–84.
[37] Schroeder, *op. cit.*, p. 152.
[38] St. Thomas Aquinas, Summa Theologica, I, q.2.a.3.c.

ROBERTO VEROLINI AND FABIO PETRELLI

A NEW CREATIVE PARADIGM: CHAOS AND FREEDOM

THE SUBJECT

"In earlier times there was only one field of scholarship and this was philosophy. Nowadays, we are often told – mainly by practitioners of the humanities – that only in recent time has there been a split, a schism in scholarship between natural science and the humanities. This is however a fallacy stemming from a basic misunderstanding of the essence of natural science" (Lorenz Konrad, 1993).

With these world Konrad Lorenz introduces the first chapter of *The Natural Science of the Human Species. An Introduction to Comparative Behavioural Research. The Russian Manuscript.* A work in which the great ethologist in a scholarly mixture of scientific rigour and language accessible to the layman explains his theory of a realistic conception of nature as opposed to the idealistic formulations of classical philosophy. In particular, Lorenz puts his finger on the fundamental, epistemological problem of philosophy and modern science: the possibility of bringing about a new union between *natural science* and *the humanities*.

This fundamental aim eschews the circumspection with which this union is treated by our contemporaries, both scientists and philosophers, who these days have to compete with an ever growing number of doctrines, paradigms and philosophies in a situation which is increasingly fragmented and relativist. A situation which is totally different from the profound philosophical synthesis typical of the great thinkers of the past. Yet it is fundamental to consider that never before has man had such interpretative tools at his disposal with which to be able to outline new philosophical and scientific paradigms, tools which are reliably objective. This seems to be the aspiration and preoccupation of K. Lorenz and this is the challenge which awaits us on the threshold of the second millennium.

Some of Lorenz's ideas, in particular those on the *evolutionist theory of knowledge*, have been accepted by one of the great names of Western philosophy, Sir Karl Raimund Popper who died on 17th September, 1994 (Popper 1975, 1994). One of the remarkable contributions made by this

A-T. Tymieniecka (ed.), Analecta Husserliana LIX, 83–114.

epistemologist is that the showed the falsification criterion of suppositions that form the basis of the theoretical progress of science. Popper develops the principle of a progressive *objectivization* of man's knowledge following an ideal path in search of an *objective* truth towards which man may strive but at which he will never consciously arrive. A *given* truth then, irrespective of the fact that it can never be grasped by any conscious being. Current scientific theories are like composite *fragments* of human knowledge which have been proved during the course of numerous verifications. Following in the wake of Tarski, epistemology and modern logic teach that these *fragments* are inevitably marked by an apodictic definition of several of their basic axioms.

Even scientific theories are *composed* of undeniably metaphysical conceptions, a fact we must always remember, especially when discussing the most important gnosiological subjects. Scientific research justifies its presence in the realm of a metaphysical theory which affirms the real existence of an extrasubjective physical universe existing irrespective of whether it is observed or not. Man is a part of this universe and we can perceive it through our senses and therefore describe it using our intellect. It is interesting to not how, within this metaphysical theory, there has been an attempt to direct scientific research along two *polarizations*, which embody the great existential themes of the history of man: atheism and theism. According to the atheistic polarization, scientific theories form a conception of reality in which the physical universe represents absolute being, the entirety of being. The theist polarization tries to form scientific theories within a metaphysical framework in which the physical universe does not represent absolute being, but only *one* being which ontologically derives from an external being, which represents the prime, absolute being: God.

In this context it is necessary to consider whether the formal contents of the two metaphysical theories, round which rage heated controversies, represent the best framework in which to evaluate scientific theories, taking into account the features peculiar to Western culture. Or if, on the other hand, it may be possible to extract further systems of metaphysico-epistemological co-ordinates more suitable for the results of scientific observation.

A new realistic synthesis: this is the goal, the direction in which modern science can lead us with a contingent, finite contribution, unique in its objectivity, placing its results at the service of a revision of the fundamental assumptions of philosophic speculation. Science, in perhaps an irreverent way, *must* these days compete with theology, by introducing new conceptual categories and its formal rigour. But, above all, it must drag philosophers and theologians into using, where possible, experimental evidence and empirical data. Only in this way will it be possible to go back and consider fundamental philosophical concepts in a new light. Concepts like necessity, contingence, reality, time, space, matter, essence, *epoché*, spirit, subjectivity, creation, God, ethics, man and freedom.

A new metaphysical theory then, which is *transversal* original, realistic, scientific, clear-cut, autonomous and mature, characterised by a profound analysis of observation, empirical data and evidence. Just as from the verification and evidence of a cosmology, of an anthropological theory, and of a study of man, the great Greek philosophy was born. This was the first step. Then, finally, we will be able to reach a new comparison between these and other metaphysical theories: a new, open, constructive and objective comparison which will certainly make us . . . *discover things new and things old*. This is our ambitious goal.

The theories on which modern science hinge are quantum mechanics, the cosmological theory of the *Big Bang* and the very recent chaos theories. The latter are a follow up to the conceptual revolution in the field of theoretical mathematics which originated, in particular, from the works of Gödel and Poincaré. The heuristic horizons proposed by these theories have raised questions that Western philosophy, since its origins, has never before been called upon to answer. It is possible, then, in the field of realistic metaphysics, to propose evaluating modern scientific theories within a scientific paradigm and a new anthropological theory. We can hope for unusual results if we put the present scientific paradigm into a metaphysico-theological context given the close, indestructible relationship which exists between cosmology and theology.

THE QUANTA

The theory of relativity and still more quantum mechanics, more than any other theories have challenged human logic and intellect. Quantum mechanics results from a need to interpret the phenomena of the sub microscopic world which clearly eludes the logic of classical physics and as Bohr aptly said anyone who is not surprised and confused by quantum theory has not understood it. This theory has definitely challenged the thinking on the meaning of reality, on philosophico-logical induction and deduction, the foundations of physics and the philosophy of the past. Many controversies caused by the introduction of this physical theory revolve around the epistemological problem.

The orthodox formulation of quantum theory, known as *the Copenhagen interpretation*, uses a mathematical function, Schrödinger's wave function, to represent the *complete* state of a microscopic system, for example, an electron that orbits round the nucleus of an atom. This wave function has a particular characteristic. If it were observed the phenomenon would collapse into single, mutually incompatible solutions. The obvious unpredictability of these physical phenomena which result from this *collapse* of even the most deterministic wave function would explain the indeterministic, objective nature of physical reality. This unpredictability means that it is impossible to know determinate pairs of physical parameters of a quantum process at the same time, like the speed and position of an electron. The *uncertainty* confirmed by the Heisenberg principle, in fact, results from this. This fundamental principle does not however cover exhaustively every aspect of the phenomenon of the complementary nature of the quantum processes. This phenomenon, which is even more profound and essential in quantum theory, is also found in other forms and modalities which are not necessarily connected with the possible measurement by an observer, and the *uncertainty* does not only stem from the fact that it is impossible for an observer to make exact measurements but it is implicit in the wave function itself and is seen as a *complete* description of the phenomenon.

The Copenhagen interpretation states that there is an objective indeterminacy in subatomic events. It denies, in particular, the existence of underlying relations with laws and *hidden* variables. The quantum nature of these phenomena are expressed then in a *superimposition* of separate quantum states implicit in their own fundamental nature, which are resolved following a possible observation. This is opposed by other theories which refute the indeterminacy introduced by quantum mechanics. They consider the concept of the *collapse* of the wave function equivocal.

One of the alternative theories is that of the physician David Joseph Bohm. His theory avoids the *collapse* of the wave function. He uses a mathematical description of subatomic systems in which every particle is held up by a pilot wave which is not seen as meaning of a mere descriptive, mathematical object but rather of a real physical entity in the same way as, for example, a magnetic field. The theory, out of respect for the classical principle of causality and determinism, which underlie classical physics, denies all hints of quantum uncertainly. For example, it explains eccentricities like the wave corpuscle duality which emerge from experiments with diffraction through slits, by means of a physical interaction which takes place between the subatomic particle and the respective pilot wave. We should point out that this mathematically valid theory has been criticised from an epistemological viewpoint because what it proposes, i.e. the physical existence of a pilot wave which is completely imperceptible to any form of observation, decidedly clouds this metaphysical hypothesis.

Paradoxical experiments have been made concerning the fundamental assumptions of quantum mechanics. Recent experiments like those conducted by Professor A. Aspect's team in Paris and by other researchers like Professor Leonard Mandel and his collaborators at Rochester University, would seem to confirm, at least for the present, the Copenhagen interpretation as against others, thus endorsing the situation feared by Einstein who was an indomitable opponent of quantum uncertainty (D'Espagnat, 1980; Horgan, 1992).

Some interesting experiments have been have been made where it is possible to observe the reversibility of the collapse of the wave function even after observation of this phenomenon. In one experiment parametric converters are used. These are devices which make it possible to divide a photon into two with halved energy and direct them along different paths without, however, altering the quantum characteristic. The photons emitted by a laser are beamed onto a parametric converter. The photons that emerge from the converter do not meet again as they are guided along separate paths. In the end the photons are made to converge contemporaneously onto photon detectors which show interference patterns proving that photons follow both the alternative paths, just as waves do. Now, it is known that photons emitted by a laser are distinguished by the same direction of polarization. A polarizer is successively placed along the exit paths of the converter and positioned so that it is possible to determine the path taken. This is achieved by means of a differential polarization which marks the photons arriving on the photon detector. And so the interference pattern immediately disappears. But it re-appears if two polarizers are placed in front of the photon detector, both identically oriented so as to cancel out the different polarization produced previously in order to show the path followed by the single photons!

It is clear that quantum theory interprets these phenomena as a close correlation between the real *resolution* of the phenomena and observation. Perhaps it is this acquisition of information observed through the *collapse* of the wave function which gives a concrete, determinate meaning to one of the states inherent in it. Some researchers have managed to put off even further choosing the type of observation to use. The polarizers in front of the mirrors are replaced by beam splitter polarizers and the data relative to the single photons emitted plus registration of arrival times and planned polarizations are stored in a computer on separate files. One file contains all the arrival times at the photon detector while the other the planned polarizations which have been set up.

The reconstruction of the events made starting from the uncorrelated data, in other words using only the file with the arrival times, without any reference to polarization, shows that the particles have a corpuscular nature. The photons seem to pass alternately from one path to another. It is possible to determine the paths and polarizations of the single photons from the correlation analysis of the observations memorised on both files. These observations cause interference patterns which is a sign that the photons follow both routes and thus behave in the same way as waves. It is thus possible to restore the wave corpuscle duality notwithstanding observation. It is as if nature regained the indeterminacy lost by the collapse of the wave function in periods of time following observation without bringing to the phenomenon any *invasive* physical action. This proves the effectiveness of the Copenhagen interpretation and of the fact that the wave function represents only an exhaustive description of the process.

These results raise serious question about theories on reality and also about Bohm's theory since it does not seem difficult to see how it has any effect on how the pilot wave, seen as a physical entity, can be called upon to interpret these experiences, in which falsified time acquisitions of the values observed are put into effect. But, as these experiments have not yet led to sufficiently exhaustive conclusions and leave the subject open, we must be both wary and prudent in giving an exact interpretation of these eccentric phenomena. However, the impression remains that some peculiarities of the sub-atomic world represent a quality which is intrinsic to this dimension of reality; for example, the absence of locality, the quantum correlation, the perfect chaos of the wave reduction process. As many physicists and epistemologists hope, further in depth theorisations might substitute the present meaning of *collapse inherent in conscious observation* for a possible physical which is anyway impossible to compute.

CHAOS, GOD AND FREEDOM

Well, we would like to make some remarks on this point.
It seems justifiable to fear a fundamental conceptual distortion in the intentions of some scholars especially when it comes to the epistemological valuation of the debate on determinism and indeterminism which derives from these phenomena. We would like to show how this possibility seems actually to be a real obstacle to reaching a meaning of reality which is harbinger of both interesting scientific and philosophic significance. Although we fully agree about the opportunity to verify contradictory, experimental theories of paradoxical natural phenomena, we would like to point out that the attention given by scholars to certain interpretative paradigms of reality can be excessive and that this may cause us to risk missing the opportunity of grasping a peculiar aspect of nature. A quality which we may deduce from a pluridisciplinary vision of reality, but which perhaps the paradoxes inherent in quantum theory make us apprehend in a more immediate way. We can clarify this by quoting Einstein's remark that "God does not play dice!" which expressed his *uneasiness* about the indeterministic results of quantum mechanics which since the beginning of the century has stood side by side with his revolutionary theory of relativity. And although we admit the relevance of Einstein's criticism of quantum theory, we would like to emphasise the inappropriateness of intellectual *uneasiness* implicit in this theory. Planck's attempt to claim the existence of an *ideal spirit* free from the gaze of the human observer is a good example of this *uneasiness*. According to Planck this spirit, could act causally on microscopic phenomena beneath the threshold of uncertainly of the subatomic world.

Why does this indeterminacy of nature induce *uneasiness*? In the light of what principles or metaphysical expectations? With regard to this, we should remember the continual attempts to describe bio evolutionist dynamics as ways of determining the emergence of a particular living species, man, in the context of a theological, creationist vision. Is this search for deterministic explanations a sign of *healthy* intellectual curiosity and consistent scientific work or is it only a contingent need, attributable to paradigmatic conditioning? It is clear that the dividing line between these two approaches is very elusive and the question asked may be controversially seen as not relevant from the epistemological point of view. Yet an analysis of this possibility is very necessary.

In classical physics the deterministic principle to explicit in its law. The deterministic principle of cause and effect, also applied in realistic metaphysics, is usually represented as an uninterrupted sequence of events which are linked by the laws of nature. These evens permeate the entire physical universe: $E_1 \Rightarrow E_2 \Rightarrow E_3 \Rightarrow E_4 \Rightarrow \ldots \Rightarrow E_k \Rightarrow E_{k+1} \Rightarrow E_{k+2} \Rightarrow \ldots \Rightarrow E_\infty$.

This sequence is valid as a generalisation in which one can trace a universal synchrony of *continuous* events seen as *discrete*.

Quantum indeterminacy seems epistemologically to create an incomprehensible break in this uniformity. Yet there is nothing irrational or epistemologically unacceptable in indeterminism, whether it is that of quantum theory or any other. Actually quantum indeterminism, when correctly interpreted expresses the unexpected significance of an extremely positive meaning of nature with regard to a whole series of profound epistemological and philosophical reflections.

The supporters of determinism are trying to reintroduce a view of nature in which they regain, even in these phenomena, the *healthy* ability to predict which has been destroyed by quantum mechanics in the infinitely small. In this way all levels of reality would have a single guiding principle.

We should point out however, that this correctness or interpretative symmetry, introduces the possibility of an synthetic conception of nature that raises many questions. The contribution made by the current theories on chaos and complexity and on the dynamics of complex systems strongly conditions, in that it is more difficult to accept these **pan-deterministic** theories, some determining ontological factors which we are going on to discuss.

Both from quantum and mathematical theories on chaos and complexity there emerges a view of certain aspects peculiar to the world of physics that is only apparently violated by the profound *opposition* represented by that basic indeterminacy which makes thinkers and *philo-determinists* suffer such *uneasiness*. This can be more easily understood if we consider the epistemological meaning of the recent theories on the dynamics of complex systems, known as *sensitive to initial conditions.*

The inadequacy of deterministic mathematical models in representing complex phenomena like atmospheric fluctuations, the turbulence of fluid flow and bio-evolutionist dynamics have led to the adoption of models based on theoretico-mathematical methods which are intrinsically indeterministic like the calculation of probability, chaotic attractors, fractal algorithms and topology (Crutchfield et al., 1987; Gutzwiller, 1992; Ruelle, 1992; Stewart, 1993).

The verification of the non-computability and the non-linearity of the very great majority of natural dynamics throws new light on the horizons of both physics and mathematics, both analytical teachings par excellence, thus pushing Laplace's dream of the perfect and total determinism of the physical world irretrievably into a nightmare of inconceivable depths (Popper, 1975). What epistemological meaning could lie in the fact that beneath the macroscopic world, there may or may not be an intrinsic, objective source of indeterminacy. A macroscopic world permeated by this particular characteristic of close and reciprocal physical relations between infinite factors. On the contrary, what could be the implication of the possibility that right at the extreme lowest limits of the real an indefatigable, rigid deterministic principle could exists?

Let's try and analyse this second hypothesis in depth aware, however, that this will obviously appear epistemologically and scientifically outdated. The situation that appears before us in this case, strong in its elegant and absolute, deterministic symmetry, is perfectly plausible. In it there seems to rise the spirit of seventeenth century investigation which led Laplace to write in his *Philosophical Essay on Probability*, in 1814:

> An intelligence which, for a given moment, might know all the forces by which nature is animated and the respective situation of the beings which are part of it, if it were vast enough to analyse these data, it would embrace in the same formula, the movements of the largest bodies in the universe and of the lightest atom: there would be nothing uncertain for it, and the future, like the past would be before its eyes.

If, for the sake of paradigmatic *symmetry* or perfect determinism, we defend a deterministic vision that ventures into the depths of the infinitely small, of the sub nuclear, then this epistemological exigency must, also, take into consideration the above mentioned theoretical principles of chaos and complex systems.

But in the light of these theories is not the picture which emerges, with unheard of violence and impact, simply the reconfirmation of Laplacian determinism? The influences which pass between systems which are extremely sensitive to initial conditions weave very subtle relations between infinitesimal events, even if they are situated at incommensurable distances from each other. If we extend this determinism to cover the whole of reality is not what remains nothing but the image of a gigantic and unchangeable mechanism in which, in a much more profound way than it was possible to imagine until the last century, every object and natural dynamic turns out to be an inescapable *effect*?

Let's say, for example, that Bohm's theory holds true. On the same lines as the pilot wave of every infinitesimal particle we can go as far as admitting the existence of a deterministic wave function of the entire universe which is able to comprise in itself every complex or macroscopic object which is part of reality. Let's take the earlier representation of the causal sequence of events $E_1 \Rightarrow E_2 \Rightarrow E_3 \Rightarrow E_4 \Rightarrow \ldots \Rightarrow E_k \Rightarrow E_{k+1} \Rightarrow E_{k+2} \Rightarrow \ldots \Rightarrow E_\infty$. Now we will adapt this expression to the concept of the wave function of the whole universe. To do this we will have to set some formal restraints. We will suppose then, even if it is absurd, that all the elementary physical entities E are constantly given and immutable on levels k of the causal sequence. Present information theories show us that in the observable universe there does not exist a spatial-temporal dimension physically capable of containing "the intelligence, that in a given moment would know all the forces which animate nature and the respective situation of all the beings which form a part of it . . .". This however, does not stop us from postulating the existence of a deterministic solution, in the light of which "there would be nothing uncertain, and the future like the past would be present". This solution would exist irrespective of the fact that some intelligence could understand it or not. We will write it as follows ψ_k here k is an index which indicates the causal succession of elementary events E.

This evidence is decisive. The theories on the dynamics of complex systems which are sensitive to initial conditions, affirm that a given event $E_{x(k)}$ can, in principle, be conditioned by all the n elementary physical entities E on the level $(k - 1)$ of the sequence pertinent to it, even if only on account of gravitational effects. However these effects cannot be ignored in the complete evaluation of every physical dynamic. Every event $E_{x(k)}$ is then a function of the summation of all the n events $E_{n(k-1)}$. This is expressed by the equality $E_{x(k)} = f(\Sigma E_{n(k-1)})$. And in its turn $\Sigma E_{x(k)}$.

This implies that the causal sequence $\psi_1 \Rightarrow \psi_2 \Rightarrow \psi_3 \Rightarrow \psi_4 \Rightarrow \ldots \Rightarrow \psi_k \Rightarrow \ldots \Rightarrow \psi_{k+1} \Rightarrow \psi_{k+2} \Rightarrow \ldots \Rightarrow \psi_\infty$, which describes the interaction between the physical n components expressed by the wave function ψ of the universe in each of the k terms of the same, is correctly determined. Each term $\psi_{(k)}$ is therefore equivalent to the others. This means that, in the event of being able to define an initial term, ψ_0, this term comprises in itself all the successive terms. It should be pointed out, incidentally, that this hypothesis implicitly takes us back to a realistic, theistic metaphysical theory, but that this idea can also be applied to an atheistic view. This however, obviously does not add anything to the Laplacian idea.

The picture changes drastically, however, when we consider the way in which natural dynamics follow a prevalently non-linear evolution, that is they are extremely sensitive to initial conditions. This is contrary to the expectations of seventeenth century philosophers and mathematicians. This means, as a result that the sequence $\psi_1 \Rightarrow \psi_2 \Rightarrow \psi_3 \Rightarrow \psi_4 \Rightarrow \ldots \Rightarrow \psi_k \Rightarrow \ldots \Rightarrow \psi_{k+1} \Rightarrow \psi_{k+2} \Rightarrow \ldots \Rightarrow \psi_\infty$, cannot be reconstructed and understood prevalently in a *linear* way, that is, counting on a relative causal independence of events, events which are sufficiently separated from each other.

Both from the thermodynamic point of view and from the physical and biological viewpoint, natural events, irrespective of their absolute relevance, cannot be seen and analysed as local, closed systems. This is true whether they are microscopic, subatomic, macroscopic or galactic. It is impossible then to postulate such reciprocal *causal* independence, especially on spatial-temporal dimensions typical of the most important verifiable events of the currently observable universe such as the formation of planetary systems, organic evolution etc. Actually, in reality in every term of the sequence we have a predominant effect in the reciprocal interaction of all the physical components n of ψ. Every event is, in fact, a function of all the events on the preceding level and is a part of the *causal* terms which will influence all the events on the successive level. This they will do in the same way and irrespective of their relevance. The history of every event $E_{x(k)}$ is given then by a *causal* succession in which a number of interactions is expressed which increases enormously along the sequence in physical reality a n^k factors or dimensions. Every event place in the sequence is a *holistic* expression of a network of inconceivably complex relationships. All these relationships are from the *causal* point of view identically essential for determining the physical coming into being of every event. Now let's try and place these observations in a metaphysical framework in which a teleological aim expressed in physical reality is affirmed, identifiable in event $E_{z(p)}$, and a creative event of the universe. We suppose then the existence of an event ψ_0, the original expression of the plan which is traceable back to the absolute being: God.

In other words, we consider the hypothesis of deterministic, theistic metaphysics, like that advocated by the present schools of theistic, creationist thought and by the present Western religions, in which the earthly intervention of a divinity in the determination of natural dynamics is postulated.

What happens to our earlier considerations in the fact of the idea that physical reality may be the expression of a given teleonomy founded on the actuation of event $E_{z(p)}$? Well, all the *n* event E of all the preceding $p-1$ levels are equally essential because of event $E_{z(p)}$ and they must be contained in ψ_0. the sequence expresses, then, an inflexible *causal rigidity*. Now, the fundamental philosophic fact is that this mind boggling, deterministic conception irreparably destroys all hypotheses on even the weakest free will. In fact there develops a whole series of insoluble problems in the light of the epistemological principles centred on the emergence of conscious forms and their free will. Every entity existing in reality, every dynamic, event or living being turn our to be stages in an inflexible pan-deterministic sequence. They are mere *physical automatons* which would explain the initial state of reality in deterministic terms. No natural dynamic, however chaotic or complex, can escape this fundamental, background determinism. Well, it would certainly be impossible for any physical intelligence to grasp this reality, but at the same time it would be *ineluctably* given. This is the case since the existence of a ψ_0 in a pan-deterministic sequence cannot comprise within it any degree of freedom. Information theories, the assumptions of indeterminacy which are mathematically computable have shown that the extension of natural events and the complexity of physical dynamics are really such that they make it impossible for any finite intelligence to have a deterministic understanding of the entire universe. But at the same time the deterministic hypothesis supposes ineluctably the existence of a ψ_0 in which is contained the contingent reality of the universe. This is, or has been, *advanced* and from these initial conditions emerges the inflexible *causal* chain of events, which in this hypothesis, would determine the real oneness of creation. The existence of the ψ_0 precludes then all allusions to any hypothesis on free will for any finite, conscious being.

It is obvious that a real, finite, conscious being, like man, for example, finds himself in a position where it is objectively necessary to make decisions and choices founded on a partial and imperfect evaluation of the causal elements of every infinitesimal event. As he is not able to understand the infinitude either of the physical factors involved or the exact numerical sequence of the parameters and variables at stake in any natural dynamic, a being like this could not but deceive himself into supposing a *voluntariness* or conscious *circumspection* in making his choices. This voluntariness and circumspection would turn out to be only a perfect illusion of *ontological liberty* which is absolutely inexistent. With this in mind, we will try to define what *ontological freedom* means for a conscious being, We can say he is ontologically free when he is "straightforward, autonomous and therefore objectively responsible for making ethical choices whether operative or not, following his finite perception and evaluation of a range of independent, possible alternatives. These are counter factual, in which a plurality of degrees of freedom is expressed." This means that, if we define an area of the counterfactual events which are inherent in the theoretical opportunities for a given self aware living creature to make decisions – this we can intuitively represent by means of a tri-dimensional volume, every event P_a, identifiable in one point of this volume – it must be able to be connected to a successive event P_b by means of a causal evolution which is not eminently computable or deterministic. A causal relation then which is not derivable from the knowledge of the factors and laws relative to physical levels (macroscopic and microscopic, biotic and abiotic) underlying this event, but an event which is willed and has been decided on, possibly ethical and ontologically attributable to the ego of the conscious subject.

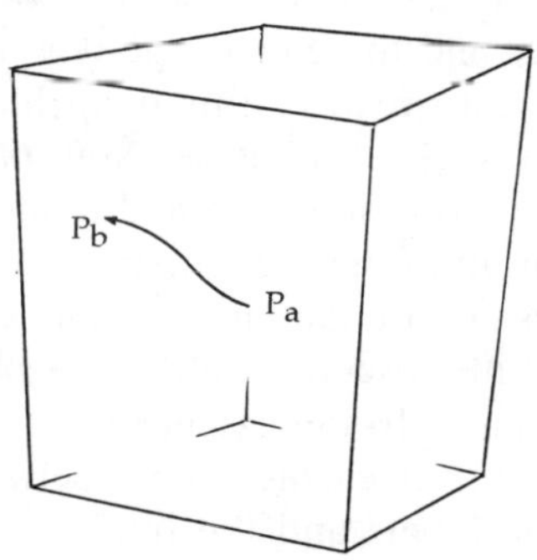

In a pan deterministic reality we would have only the illusion of making choices, gestures, following principles and feeling emotions which would be subjectively seen as conscious and *free*, while being absolutely unaware of the deeper reality which governs every event. These entities are the expression of obscure dynamics which underlie all possible perception of individuality and free choice. But this is not all. The frustrating and irremediable thing is that this happens now in the awareness that the fundamental causes of every infinitesimal, natural dynamic reside, in the final analysis, beyond the *wall* established by Heisenberg's uncertainty limit.

So every answer to our mind boggling questions is irreparably hidden beyond that $m\sigma\chi * \sigma\nu \geq h/4\pi$ (where $\pi = 3.14159\ldots$, h = Planck's constant, $m\sigma\chi * \sigma\nu$ the product of the indeterminacy of the position and velocity of the mass particle m), which shuts out our observation of the world of the infinitely small.

In the case of relations underlying he microscopic events described indeterministically by quantum mechanics, we would place Heisenberg's uncertainty wall along the terms of a totally deterministic sequence, precluding the determination by any physical, macroscopic observer of all the lower levels.

⇐ Heisenberg's uncertainty threshold

$$\psi_1 \Rightarrow \psi_2 \Rightarrow \psi_3 \Rightarrow \psi_4 \Rightarrow \ldots \Rightarrow \psi_k \Rightarrow \ldots \Rightarrow \psi_{k+1} \Rightarrow \psi_{k+2} \Rightarrow \ldots \Rightarrow \psi_{p-1} \Rightarrow E_{z(p)}$$

This eventuality would definitively confirm both the perfect and total predestination of reality and the equally perfect and total impossibility for any conscious being to resolve this disturbing unawareness. It is not without foundation, then, to consider how *this* paradoxical reality seems to create a much more objective and immediate *uneasiness* than that feared by Einstein and other scientists who refuse the *paradoxes* posed by quantum physics. If quantum mechanics were to represent the correct interpretation of microscopic reality we can immediately observe how radically thing change. To the extent that Heisenberg's uncertainty threshold represents an objective causal break between the classical determinism of macroscopic levels and the intrinsic indeterminacy which derives from the collapse of the wave function relevant to quantum levels, we obtain a generalised representation of reality in which we have no unfathomable *breaking down* of this threshold:

	$\Leftarrow$ Heisenberg's uncertainty threshold
chaos	
quantum	$\Rightarrow \psi_1 \Rightarrow \psi_2 \Rightarrow \psi_3 \Rightarrow \psi_4 \Rightarrow \ldots \Rightarrow \psi_k \Rightarrow \ldots \Rightarrow \psi_{k+1} \Rightarrow \psi_{k+2} \Rightarrow \ldots$
	$\Rightarrow \psi_{p-1} \Rightarrow E_{z(p)}$

this implies some very strange considerations from the metaphysical, epistemological point of view. First of all the determinism of reality reaches the threshold of the quantum world and this would permit us to observe the way in which nature expresses intrinsic indeterminacy, at the very beginning of the levels accessible to observation by conscious, physical beings. One simple consideration: is it really unfounded to see this fact as a very meaningful and relevant *message* from nature?

Every deterministic chain *breaks* by continuously introducing something new to every quantum event, an indeterminable component of physical chaos which acts as an inexhaustible, indeterministic lung for all the successive dynamics, in both the spatial and temporal dimensions, especially regarding the dynamics of complex systems and chaos. This phenomenon, which keeps introducing at every moment and in every place perturbing elements into the causal sequence $\psi_1 \Rightarrow \psi_2 \Rightarrow \psi_3 \Rightarrow \psi_4 \Rightarrow \ldots \Rightarrow \psi_k \Rightarrow \ldots \Rightarrow \psi_{k+1} \Rightarrow \psi_{k+2} \Rightarrow \ldots \psi_\infty$ makes even the most elementary natural dynamic intrinsically unpredictable.

On every level infinite quantum events I, causally separated by the sequence in progress and absolutely not attributable to any underlying deterministic law, unexpectedly *enter* it, infusing reality with newness in a continuous, infinite *creation ex novo* of further causal events which are absolutely unforeseen in ψ_0.

This situation could be generalised in the following way:

$$\begin{array}{ccccccccccccccccc} nI_1 & & nI_2 & & nI_3 & & nI_4 & & \ldots & & nI_k & & \ldots & & nI_{k+1} & & nI_{k+2} & & \ldots & & nI_\infty \\ \Downarrow & & \Downarrow & & \Downarrow & & \Downarrow & & & & \Downarrow & & & & \Downarrow & & \Downarrow & & & & \Downarrow \\ \psi_1 & \Rightarrow & \psi_2 & \Rightarrow & \psi_3 & \Rightarrow & \psi_4 & \Rightarrow & \ldots & \Rightarrow & \psi_k & \Rightarrow & \ldots & \Rightarrow & \psi_{k+1} & \Rightarrow & \psi_{k+2} & \Rightarrow & \ldots & \Rightarrow & \psi_\infty \end{array}$$

The inextricable complexity of nature, as already pointed out, is peremptorily confirmed by the joint action between the dynamic evolution of physical systems and quantum mechanics. The coming into being of the universe, in each infinitesimal expression, is not then contained in the number of *initial conditions* from which it has originated.

The knowledge of the laws of nature and the initial conditions can no longer *completely* represent the dynamic evolving of physical systems.

The *non linearity* of the laws of physics, which are often defined by complex relations between numerous variables that describe these transformations, exponentially extend the influence exercised on any event by even the most infinitesimal component of the entire universe. The result of this is that the progress of almost all physical systems, right from the most simple, is intrinsically unpredictable, in other words perfectly chaotic. Most of the evolution of physical systems, from the movements of atmospheric strata to biological dynamics, show a behaviour pattern in which the deterministic element of the laws of physics intersects with a fair portion of chance, thus causing the prodigious complexity and unpredictability of reality. What are the most significant questions raised by this conception of reality? What does this conception imply as regards the idealistic existence of a superior intelligence, or to paraphrase Einstein, God playing dice?

As to these questions it is possible to make important metaphisico-epistemological considerations. The superior intelligence if it wants to organise the evolving of any simple natural dynamic in a teleological plan, has to assume the extenuating duty of continually analysing the enormous number of elementary physical events going on every single level of reality. In terms dear to theological metaphysics referred to more or less unconsciously by Laplace, Einstein and Planck (authors already mentioned in this work) the teleology of creation, *divine providence* or the influence of the divine on the physical universe, in the earthly sphere, cannot possible get out of being totally and continually dedicated to knowledge and putting things into effect, a fact which divests its creatures of any hint of free will. If this were the case God would have to supervise every *single throw* in an exhaustingly taxing *game of dice* with the human species as the stake in the game. A game which has been going on for at least fifteen billion years all the time and in every place in the universe! All radio-active decay, every energetic transformation going on in an atomic orbital, every quantum collision, every natural dynamic, from the fluctuations of an invisible particle to the violent cosmic tides, even the most insignificant genetic mutation, every beating of a butterfly's wings, all times and events in reality must be subjected to the careful and obsessive control of this intelligence. All the evolutive events, from the prebiotic synthesis of the first proto cells to the genetic-metabolic mutations that led to the first eukaryotic cells, from the impact of the asteroid which, it would seem, decreed the extinction of the dinosaurs to the acquisition of prehensile hands by protohuman hominids, must all have been carefully supervised by every divinity which had had in mind *to create* the human species on the third planet within the solar system.

The only conceivable alternative to this careful and wearisome divine undertaking, which is anyway rather unconvincing and incredible, is that in which a divinity could be postulated as the origin, the *causa causarum*, of a universe . . . left to itself. A universe then, in which one does not try a priori to effect deductively any specific natural dynamic. The immediate objection which comes to mind is: "Yes, but what sense would the teleological design, the concept of creation have then?"

Surprisingly we can find the answer in a holistic consideration of the mathematical and physical theories stated above and in other conclusive scientific theories of today. We find it in the very indeterminacy brought about by the *quantum caesura* and it develops a metaphysical theory which is completely consistent with the present cosmological and evolutionist theories.

THE BIG BANG AND EVOLUTION: A WAGER MADE BY GOD?

Modern cosmology has pushed much of philosophic and scientific thought on the concept of the universe, reality and creation towards heuristic horizons. Modern theories simply add a further, conclusive contribution to the meaning of intrinsically indeterministic reality.

Many theoretical aspects of today's cosmology rely on quantum mechanics and on QCD (quantum chrome dynamic) which derives from the former and so on. This is true for the theoretical extrapolations of the physical states of the exact initial moment, the Big Bang, to the problem of entropy of black holes, from the breaks in symmetry from which derive the separation of the fundamental forces of nature to the *ex novo* formation of particles due to quantum fluctuations in the vacuum, to the calculation of the intensity of background fossil radiation to theories on inflation (Brush, 1992: 5; Davies, 1984, 1993; Guth and Steinhardt, 1984; Hawking, 1988). Today those pioneering the research on the origins of the universe are working increasingly closely with those involved in experimental atomic physics on account of quantum theories on the infinitely small.

Another disquieting example of this could be taken starting from the fascinating and still open question which arises from the implications of the quantum approach to the oneness of the universes raised by some theories on cosmology which follow the implications of an eminently quantum approach. Some theories of cosmology postulate the existence of *parallet* universes, independent of each other, in order to resolve the eccentricities of quantum *collapse*. With regard to this we must make reference to the Soviet astrophysicist, Andrei Linde's original theory on the inflationary, self-reproducing universe (Linde, 1995). According to this scholar, the universe in which we live is just a single *process of cosmological evolution* which exists within the framework of a complex *fractal tree* of universes which originate from one another in a perfect chaotic sequence through simple *Quantum budding*. Now if we were able to put a *beginning* even to this immense, chaotic cluster of universes it is clear that it would be very difficult to given a possible deterministic option, a super wave function $\zeta_{0.}$ common to all these cosmological events. The action of a possible divine creator would be further removed by the subtle realization of a given physical reality and this imply makes both the conception of pandeterminism and the intrinsic uncertainty of the alternative indeterministic definition of evolution more absurd.

Charles Darwin's theory of evolution, put forward in 1859 in his pioneering work *The Origin of the Species*, represents the fundamental axiomatic basis of all teaching on biology (Darwin, 1973). The advanced stage reached by the modern, cosmological, evolutive paradigm seems then to indicate that the time is ripe for it to spread its influence *beyond* its most typical field of application. Starting from this paradigm we can derive a realistic, theistic metaphysical theory essentially consistent with it leading to unknown metaphysical hypotheses which can only be fully understood by redefining the ideal figure of the *divine creator*. So, it can be demonstrated that these features do not represent an *elementary* set of characteristics but rather composite coacervates, due to the apposition of metaphysical assumptions which are unconnected.

Theoretically theologico-cosmological systems can be imagined in which there are entities which express only some of these qualities. We should remember that the way orthodox theology expresses the conception of divinity, creation and other formal aspects is founded on a fixed, anthropocentric and geocentric vision. All Scholasticism, all the thinking of authors like St. Thomas, St. Augustine, St. Bonaventura,

Spinoza, Kant and many others, are essentially understandable if we consider the problem of the existence of God, the relationship God – man, the antithesis between good and evil and so on, from a theological perspective based on ideas on cosmology and anthropology which are extremely different from those deducible from the modern theory of today. Modern science has questioned these metaphysical conceptions and these two visions cannot possibly be *reconciled*. This is why every *reconciliatory* attempt has failed. Only an unknown, radical theologico-scientific synthesis can do justice to these questions. And this is what we are proposing.

A different perspective on the problem leads to a fundamental revision of theological thinking on the idea of God, meanings of creation, teleology, eschatology, revelation, sin, evil, salvation and so on. For example, it is possible not to invoke a divinity in a metaphysical system which contemplates life after death, or rather it can be conceived in such a way that it does not posses any of the qualities usually attributed to it merely out of cultural habit. These hypotheses are perfectly valid and consistent from the philosophical point of view. We have no reason to believe that our metaphysical theory can express all the facets of the divinity any more than we can objectively maintain that our theory can represent the *best* unless we go back to obscurantist, sectarian, ethnocentric explanations. Only using an *objective* method of comparison can we solve the problem. That is why one appeals to scientific judgement to back up given theologico-cosmological metaphysics.

A consideration, centred on the verification of a cosmologico-anthropological idea which indirectly leads us to be able to opt for one or other metaphysical theory. And the potentiality of this application of scientific knowledge is just as valid and unexplored. Science today, for example, has been driven to place serious vetoes on the possibility of postulating certain *finalities* in the putting into practice of natural evolutionist dynamics.

The impossibility of grasping the phenomenon of evolution in its single forms as an *instrument* of an original plan centred on the emergence of the human species in creation, has represented until now an insurmountable obstacle to the understanding of natural dynamics as compliant *effectors* in a magnificent divine plan organised around man (Greene, 1971; Hall, Hall and Boas, 1991; Monod, 1970). Evolutionary processes, left to themselves, are not capable of working towards the emergence of a predetermined objective, unless they call upon continuous *corrective intervention* in earthly events by means of *divine providential supervision* of every infinitesimal natural dynamic.

As we have seen the latter eventuality irreparably undermines the principle of human free will. On the contrary, in the context of a chaotic, indeterministic, self-organising universe, we must underline a fundamental fact about the possibility of affirming any teleological content. Every creature that emerges in creation through the evolutionary process has absolutely unpredictable characteristics given the intrinsic fortuity present in the interminable series of events which have led to its origin. The spatial temporal context is likewise absolutely unpredictable.

The conception of a universe in which, in the end, some predetermined living creatures must originate and on which the respect for supernatural, ethical principles is imposed is totally incompatible with the universe described by current scientific theories, despite the orthodox teleological beliefs held in our culture. This is the case, in particular, as regards the metaphysical canons of orthodox theologies in our culture (Dawkins, 1980, 1988, 1996; Wilson, 1980). How can we expect that a given line of evolution will lead to that particular living species and even more to those peculiar biological characteristics which will determine, for example its sexuality? From the phenomenological and causal point of view, how can we establish before the creative event itself, *above* the level ψ_0, absolute moral principles, in this case sexual taboos, if the anatomico-ethological characteristics of a living being – event $E_{s(k)}$ where $k \gg 0$ – are absolutely unpredictable in the course of millions and millions of years of evolution?

These entities or contingent qualities cannot possibly be seen as *goals* in an original aim entrusted to the spontaneous, natural realisation of the cosmo-biological process of evolution. This impossibility is then peremptorily confirmed by the modality of the dynamic evolution of physical systems, by quantum mechanics and the very recent studies on *chaos* which have become very important in the bio-evolutionary context.

The coming into being of the universe, and in particular living forms, in their every infinitesimal expression, is not contained in the number of *initial conditions* from which natural reality has originated. Nor can the knowledge of the laws of nature and initial conditions *completely* represent the dynamic evolving of physical systems. But is it scientifically impossible then to support any teleonomic content at all in natural dynamics? It does not seem to be. We can give bio-evolutionary processes a universal meaning taking into account that these phenomena are *statistically recurrent in the universe* but at the same time unique, unrepeatable and unpredictable in their particular characteristics. The process of natural selection can be seen as a phenomenon to which we can attribute the possibility of giving rise to progressively more complex biological situations. These occur in time and space yet in forms and contexts which a priori are absolutely unpredictable. In particular it is possible to postulate the origin of bio-evolutionary frameworks characterised by the presence of organisms which have progressively more refined psychic structures, capable in the end of sustaining the emergence of intelligence and a reflective conscience. The latter observation, if appropriately evaluated, becomes extremely important from a theological point of view.

The undeniable indeterminacy of a given bio-evolutionary path *disappears* in the *universal* value of evolutionary phenomenon. The *finalistic* contents associated with this emerge clearly if we understand this process as beyond the contingency of the events which have marked, for example, the success of life on earth or the characteristics of individual living beings.

Studies on the origins of planetary systems, on the abiogenetic origin and the capacity for selective auto-organisation of complex biochemical components, lead us to have a fairly clear picture of the cosmos (Kastling, and Toon, 1988; Black, 1991; Gould, 1994).

In the attempt to support a finalistic interpretation in the evolutionary processes, these results can lead us to favour a framework which postulates a repeated emergence (thus statistically reliable) of self-aware, living forms in the universe. Besides, we can also recall the question of the so-called *fundamental physical constants* of nature. These magnitudes which are defined precisely by scientific research, take on a harmony of extremely critical and interdependent values which do not seem to originate from any underlying natural law or relationship. Why is this agreement of values decisive in making the emergence of life possible? Why *this* universe? Up to a certain point we go along with the opinion of some authors who see these facts as representing a clue to a teleological finality but we do not believe in any ethical or genetic predestination (Von Ditfurt Hoiman, 1981).

As we have seen, it is possible to provide a model which is both scientifically coherent and theologically valid. This we can do by omitting theological theories which are irrationally anchored in an untenable *anthropocentric* view and favouring a finalism less focused on man. In this model, although man is *ousted* from his unfounded role as the *apex of evolution*, he can maintain the same satisfactory relationship with the divine as he had before.

It is clear that in all this even the divinity undergoes a formal *reformulation*. The ontological relationship between God and man is presented differently thus emphasising the new ideas. In an indeterministic, evolutionist context the teleology of creation is not centred either on a particular living species or on the existential where there are superior ethical values, which intervene, watch over, veto and sanction. The *creative gesture* would be a plan in which we do not predetermine the emergence of any ethical or genetic dimension which have to be submitted to the observation of ties with the supernatural. This is something which a divinity is neither *used to* nor *desirous* of. In this *plan* creation would be left to evolve, *to be*, completely freely as an intrinsic, spontaneous manifestation of natural laws. In the same way, man would be totally free to be and become, both as regards his prerogatives and his natural limits. From this would arise the ontological condition of a being which is ethically *independent* of the divinity. These beings are, in the final analysis, ontologically *free*. Free to be. Responsibly and autonomously free even before their own creator. *This* is the only possible teleological result of a process of biological evolution that pertains in the ontopoiesis of the reflective conscience and in the emergence of thought in creation. *This* would represent the *teleological aim* of creation by a divinity: natural and total freedom for its creatures.

GENESIS: A NEW METAPHYSICAL THEORY, A NEW THEOLOGY

We can imagine the way in which the explicative and philosophical potentialities of this model defines new principles which make a definite break with the canons of current philosophy but which does not detract from theistic theory. No theology had ever been able to coherently originate from a similar alliance between indeterminism, chaos and freedom. No theology has ever been able to place the ontological relationship between God and man, between creator and created into a framework like this. This represents a new milestone in human thought, quite undreamed of in all our past. A philosophic step of incommensurable importance. This model is opposed to all those ideologies that have formed the most sneaky instruments of ideological and political oppression, of the forced conditioning and control of the conscience and of the masses which have ever been put into effect in the entire history of mankind. Ideologies which are a crystallisation and sublimation of the fundamental, sexually repressive authoritarian principle of the socio-economic and cultural structure of all historically known societies. These ideologies have been criticised by writers like Freud, Neitzsche, Fromm, Marx and many others. So, for the first time in the story of modern scientific thought we finally have at our disposal a new metaphysical theory which is able to provide an alternative to all these inconsistent, pernicious doctrines. But man reaches this new horizon with helped by the results of further, conclusive philosphico-scientific research.

A recent research project of ours, printed in a Camerino University publication in 1994, entitled "Metamorphosis of Reason. Evolutionist Psycho-sociological Exegesis of Genesis 1,3 and Bio Ethical Implications" has demonstrated the way in which this original metaphysical conception, which has interesting theological implications, can be actually inferred by a scientifically valid exegesis of the book of Genesis. This is backed up by an imposing amount of experimental data (Verolini and Petrelli, 1994). It is possible to trace in the theology of Genesis, in a philosophically concise way, a definite verification of our ideas. The God of Genesis, unlike that sustained by the orthodox, millenarian exegesis, seems to share this metaphysical theory. A hermeneutic issue that has caused a great stir.

The almost binding assertion of the current scientific conception of nature as evolutionary and indeterministic has stimulated an enormous amount of theological research. The great commitment of well known writers like Maritain, Guitton, Balthasar, Rahner and Alszeghy, the attention and ostracism with which the Church has treated these problems, show how important this aim and these themes are.

Teilhard de Chardin is the scholar who more than anyone else, evokes the effort made by today's theologians to find a synthesis between evolutionism and traditional theology. However there is great dissent over his interpretations while still maintaining respect for the strength of his intellect. These attempts are, in fact, only creationist conjectures, which are not easy to verify scientifically and with which scholars try to force the orthodox interpretations of the first three chapters of the book of Genesis into a evolutionist vision of nature. This is the result of the *classical* custom of approaching these themes in an very philosophical way, without feeling it necessary to make any experimental, empirical verifications of the assumptions and postulates. This *praxis* ruins the chance of reaching scientifically verifiable hypotheses which are in line with the principle of confutability advocated by Popper's theories.

In this particular instance, we see the problem as a need to represent exhaustively the themes of divine creation of the universe, living forms and in particular man and more important the *fall* or *original sin*. This elusive event sets off the whole salvation dynamic which is fulfilled in the figure and redemptive mission of Christ recounted in the New Testament. Up to and including today, no interpretation seems to possess the energy and truth necessary to resolve the contradictions caused by an in depth evaluation of the evolutionary model as regards these aspects of faith. The official position of Catholic Doctrine, what's more, still revolves around a literal, historical interpretation of the facts recounted in Genesis which is absolutely incompatible with the modern scientific paradigm.

It seems that since a *literal* of the texts is no longer possible, modern theologians are not able to give a scientifically intelligible interpretation of the allegorical stories found in them. There have been many hypotheses formulated until now and some are even interesting and valuable. But all, without exception, lead to inconsistent although permissible scientific interpretations which invoke explicative assumptions and principles which cannot usually be verified in a concrete, scientific way. Therefore a revolution in this field of research is even more to be wished for, causing modern science to contribute its findings even in this area.

Some people might object that it is an obvious contradiction to attempt an interpretation of a text which should carry an absolute, supernatural *truth* by relying on the scientific method which is based on the category of contingency, on the intrinsic limitations of empirical data, of a never accomplished experience and on ever growing knowledge. We are convinced, however, that implementing current scientific ideas in this research is the only valid way of making a new, objective, non-ideological evaluation of these questions. In other words ". . . new wine in new bottles".

Our work develops a theologico-cosmological model which is perfectly compatible with today's evolutionist-indeterministic vision of nature. It makes assumptions which are fundamentally antithetical to known orthodox doctrines. This makes it possible, for the first time in the history of modern thought, to submit to an objective verification, the various metaphysical interpretations which attempt to interpret Genesis. And it is carried out fully respecting the traditional scientific method without inevitably having to exclude the theistic hypothesis. The theory is characterised by a philosophico-theological conception in which the divinity functions exclusively as the creator of reality and admits the existence of an pre-existent being which is distinct from the world which is seen as its work, as a direct emanation. But no further ethical privileges over its creatures are attributed to it. This model may seem reminiscent of the enlightened idea of *deism* but it is distinguished from this by its particular interpretation of Genesis, by its being placed in an evolutionist framework and by the way it affirms a special relationship between God and man. We demonstrate how this model is intrinsically compatible with a wide range of anthropological, sociological and psychoanalytical evidence.

A vast amount of literature on ethnology shows how similar theologico-cosmological ideas are typical of the oldest known societies and of those at a socio-economic stage of development before the setting up of political and legal structures. This permits the model to be justifiably considered as illustrating the most primitive religious belief developed in the inhabited world. The interpretation has a mainstream historico-evolutionary vision of religion shared by several disciplines from cultural anthropology to the history of religion, but is clearly distinguished in its approach to the usual ideas of religious evolution (Eliade, 1984; Eibl-Eibesfeldt, 1984; Harris, 1990; Leroi-Gourhan, 1970, 1977). From this paradigm there derives a theology where the divinity has new qualities with respect to the models in which it has characteristics centred on its ethical all seeing nature. Thus this model assumes a conclusive meaning. This fact, obviously, leads us to postulate, contrary to what has been done until now, the impossibility of an evolutionary *continuum* between the different religious paradigms.

The distinction is centred around a dichotic classification of the *theistic* universe, of religious beliefs, into two fundamental types: that of real *religions* (systems which do not have moral divinities) and that of *theo-eto-tomies* (a neologism meaning models where the divinity takes on all seeing and moral qualities, leading to a division – *tomy* – in the ethical sphere), and the division is determined by the observation of how these models, each radically different from the other in their metaphysical assumptions, originate from completely different existential needs and are linked with socio-cultural situations which are antithetical to each other. A fact like this would show their distinct raison d'etre and their different formal, historical, ethnological and sociological position.

The importance of this aspect, which has been undervalued until now, is that it leads a new evaluation of the correlation between a theologico-cosmological model, the cultural system and socio-economic conditions. We can demonstrate that the intrinsic quality of the *religious* models perfectly with the physical and cosmological evolutionist indeterministic conception, in particular with the evolutionist theories on the origin of man. First of all a divinity which is eminently *a creator*, like that in religious models, would easily find a solution to its expulsion from any *active* role in natural dynamics, by simply being *antecedent* to creation as *causa causarum*. This divinity, without losing anything theologically, is completely indifferent to the possible modalities with which the universe is created.

The teleological value hinges then on a universal connotation of bio-evolutionist processes so that these phenomena are *statistically recurrent in the universe*, but at the same time always unique, possessing unrepeatable and unpredictable characteristics. This fact, as we have seen, does not go against current scientific trends. A religious divinity, one not imbued with moral significance, co-exists without any problem or theological degeneration within the limits imposed by the present view of nature. It does not need to determine the ontological reality of creation especially with regard to its characteristic function of being morally all-seeing.

But the essence of *Metamorphosis of Reason* consists in the identification of *the fall* narrated in Genesis with the transition due to the passage from socio-cultural models of the kind we have called religious to those that are theoetotomist. The theory plainly demonstrates how the biblical texts, unlike the way they have been understood until now, support the superiority of the religious model over the theoetotomistic one, and the attributes to theoetotomistic cultures that undergo a thorough philosophico-ontological degeneration with regard to both the community and the individual. This means we can given an in depth interpretation of the facts which appear in a metaphorical form in Genesis 2,3 which are relative to a mysterious *degeneration* in the human species, the so called *original sin* or *fall*. Thus our theory puts the contrast between evolutionism and faith into a new framework (Harris, 1984).

Until today this controversy has resulted from the fact that it has been impossible for theologians to find a way of making the orthodox reading of Genesis 1,3 and the evolution theory co-exist in a concise, scientifically acceptable way (Harris, 1990; Eliade, 1984; Eibl-Eibesfeldt, 1984). The evolution theory denies the particular origin of man, showing that man descended from pre-human primates in the course of millions of years of natural, phyletic evolution. This irrevocably contradicts the argument in favour of the original, genetic monogenism of the human species (Binford, 1990; Camps, 1985; Holloway, 1984; Leakey and Roger, 1979). All this, moreover, refutes the possibility of considering the so called *original sin* in the same way as the Church. In other words it cannot be seen as an event from which has stemmed the successive ontological corruptibility of mankind, as directly traceable to our protoparents, Adam and Eve.

Until now no probable hypothesis, which is scientifically acceptable, has been put forward to explain this event which is so crucial to the theological interpretation of biblical texts. Where should this event be placed in the evolution of the human species? Who was responsible? Homo sapiens, Homo erectus or even Homo habilis? Was this event caused by one individual or more than one? How could it be immune to the influence of the instinctual *genetic* component, attributable then to human nature and thus *directly begotten by the Creator himself* and of the determination of human behaviour? Where did this ontological cataclysm take place? Why was sexuality involved in all this? As it is impossible to uphold man's direct genetic descendence from one foundation couple, how is it possible for this state of man's corruption to have spread in such a thorough and absolute way? Would this event represent a *local* accident which involves planet earth only or does it extend to the whole universe?

Until now these questions have had no other plausible answer than faith which is deaf to all the promptings of modern science. Now if we demonstrate that in the theistic universe the divinity does not necessarily take on ethical and authoritarian connotations we can see that things change drastically.

In the metaphysical universe of theism the individual is not *inevitably* involved with the divinity in a relationship based on ethical subjection and moral obedience, nor is he the object of external, divine conditioning when it comes to making choices.

A similar paradigm, even thought it is placed within the theistic ideal, is characterised as follows:

1. It succeeds in overcoming the problem of genetic monogenism – both of the couple and the group – which results from an orthodox exegesis and which is unequivocally refuted by genetic polygenism because it no longer sees this as a determining factor for a positive reading of Genesis 2,3.
2. Without appealing to self-styled supernatural causes it identifies the historical event to which we associate the legendary *original fall*, with a socio-cultural fact: the beginnings of theocratic society typified by political and legal state systems, class and theoetotomy.
3. It brings together historical-geographical documentation and the texts. It does this by placing the verification of a process of the cultural diffusion of the theoetotomistic model, which is associated with states run along political and legal lines, within the evolution of mankind (Childe, 1978). This model spread all over the earth on account of the great political, military and economic efficiency of these societies with the result that the religious type of culture was eclipsed and the transition, according to historical evidence, first affected the Neolithic peoples of the *fertile half-moon* of the Middle East.
4. Without appealing to the supernatural, as the various theological schools of thought have done, it bases its argument for the transmission of this state not on genetics but on a natural transmission or socio-cultural diffusion.
5. It has a definite effect on philosophico-theological speculations on theodicy as it refutes the personal meaning of the supernatural principle of *evil* without limited the theological and eschatological value of Genesis 2,3.

These observations would mean that the socio-economic metamorphosis and even more so the psycho-social metamorphosis are attributable to sociological and psychological mechanisms which are dealt with the detail in this theory. Other authors have examined these mechanisms in an attempt to understand this transformation but without success.

This theory differs from previous attempts to explain the origin of our historically class-ridden, theocratic, *authoritarian repressive* society by starting from *egalitarian*, early cultures, and relying exclusively on contingent, economic ecological, technological and demographic aspects, etc.

In conclusion, the indeterministic religious model is characterised by the establishment of a relationship between the individual and the divinity as a result of which:

1. There is a radical alteration in the theological and teleological meanings of creation and eschatology.
2. There is no conception of the ontological corruption of creation exclusively linked to a *non monistic* theologico-cosmological definition.
3. There is a complete release from a fixed vision, implicit in the theoetotomistic paradigm, and consequently from the all seeing and active role of the divinity which is present there.
4. We can resolve lay criticism of theism in a different way, (criticism which I share to a certain extent) when it accuses this faith of causing the degeneration and alienation of man and society.
5. The meaning of Marxist criticism is reconsidered in that we demonstrate that theism should not be accused *tout court* but rather his accusations should be directed against the theoetotomies.
6. It removes the patriarchal and *authoritarian sexually repressive* aspects from the figure of the divinity as they are the theoetotomistic meaning of divinity (Clastres, 1984).

At this point it is interesting to note another significant link: these aspects of theology have led Freud to criticise theism (Freud, 1969). We can demonstrate the existence of a positive correlation between the more or less hypertrophic and long lasting formation of the Super Ego, pregenital regressions connected with Oedipal dynamics and the affirmation of the theoetotomistic ideal (Dacquino, 1981; Freud, 1971, 1982; Fromm, 1975, 1981). We can draw an analogy between the effecting of repressive ethical attitudes regarding sexuality and the expression of the degenerate *basic personality* (modal aspects of the individual in a given social context) verified in the psychologico-ethnological research into political, and legal based states and class ridden cultures. Even from the point of view of psychoanalysis, this helps to put into a new perspective, the character *mutation* due to an affirmation of a theoetotomistic ideal of the individual thus overthrowing many of Freud's theories.

In conclusion, this research proposes a conception of divinity, of God and of the human being, capable of putting forward a theological, anthropological and cosmological theory which is finally free from heavy ideological content which until now has always formed our ideas of God and divinity. This new approach is both objective and intelligible. This point of departure could also be carried usefully into interpretations of the New Testament, in particular the Gospels, in which the figure and redemptive mission of Christ is re-evaluated in the light of orthodox theology. It identifies facts which are relevant but absolutely explosive and contrary to traditional interpretations. We can imagine the way in which this different ontological reality will enrich our daily routine and the self-knowledge of the individual and it will profoundly modify the epistemological debate relative to the basic themes of human experience: faith and the relationship with God.

Camerino University
Italy

REFERENCES

AA. VV., Articles by Gould, Stephan Jay and Weinberg, Steven and Sagan, Carl and Calvin, William H. end other, *Le scienze. Versione italiana di Scientific American* 316, special number "La vita sulla terra". November 1994.

AA. VV., *Catechism of the Catholic Church* (City of Vatican: Libreria Editrice Vaticana, 1992).

AA. VV., *Homo. Viaggio alle origini della storia. Testimonianze e reperti per 4 milioni di anni* (Venezia: Cataloghi Marsilio, 1985).

AA. VV., "Il Paleolitico," *Le scienze*. Quaderni (June 1986).

AA. VV., Letture da Le Scienze. "L'universo, problemi ed incognite," 1976.

AA. VV., *L'evoluzione dei primati* (Milan: Jaca Book, 1986).

Binford, Lewis R., *Preistoria dell'uomo* (Milan: Rusconi, 1990).

Black, David C., "I pianeti delle altre stelle," *Le scienze. Versione italiana di Scientific American* 271 (March 1991).

Brush, Stephen G., "Come la cosmologia divenne una scienza," *Le scienze. Versione italiana di Scientific American* 290 (October 1992).

Camps, Gabriels, *La preistoria. Alla ricerca del paradiso perduto* (Milan: Bompiani, 1985).

Chagas, Carlos, *L'evoluzione dei primati* (Milan: Jaca Book, 1986).

Childe, G., *L'alba della civiltà in Europa* (Turin: Einaudi, 1978).

Clastres, Pierre, *La società contro lo stato. Ricerche di antropologia politica* (Milan: Feltrinelli, 1984).

Crutchfield, James P., Doyne J. Farmer, H. Norman, and Robert S. Shaw, "Il Caos," *Le scienze. Versione italiana di Scientific American* 222 (February 1987).

Kastling, J. F. and J. B. Toon, "L'evoluzione del clima sui pianeti terrestri," *Le scienze. Versione italiana di Scientific American* 236 (April 1988).

Dacquino, Giacomo, *Religiosità e psicanalisi* (Turin: Saggi S.E.I., 1981).

Darwin, Charles, *L'origine delle specie per selezione naturale o la preservazione delle razze privilegiate nella lotta per la vita* (Rome: Newton Compton, 1973).

Davies, Paul, *Dio e la nuova fisica* (Milan: Arnoldo Mondadori, 1984).

Davies, Paul, *La mente di Dio* (Milan: Arnoldo Mondadori, 1993).

Dawkins, Richard, *Il gene egoista* (Bologna: Zanichelli, 1980).

Dawkins, Richard, "La natura: un universo di indifferenza," *Le scienze. Versione italiana di Scientific American* 329 (January 1996).

Dawkins, Richard, *L'orologiaio cieco* (Milan: Rizzoli, 1988).
D'Espagnat, Bernrad, "La teoria dei quanti e la realtà," *Le scienze. Versione italiana di Scientific American* 137 (January 1980).
Eliade, Mircea, *Il sacro e il profano* (Turin: Boringhieri, 1984).
Eibl-Eibesfeldt, Irenäus, *Etologia della guerra* (Turin: Boringhieri, 1984).
Evans-Pritchard, Edward E., *Teorie sulla religione primitiva* (Florence: Sansoni, 1978).
Flich, M. and Z. Alszeghy, *Il peccato originale* (Brescia: Queriniana, 1971).
Freud, Sigmund, *Il disagio della civiltà ed altri saggi* (Turin: Boringhieri, 1971).
Freud, Sigmund, *L'Io e l'Es* (Turin: Boringhieri, 1982).
Freud, Sigmund, *Totem e tabù* (Turin: Boringhieri, 1969).
Fromm, Erich, *Anatomia dell'aggressività umana* (Milan: Arnoldo Mondadori, 1975).
Fromm, Erich, *Fuga dalla libertà* (Milan: La comunità, 1981).
Gale, George, "Il principio antropico," *Le scienze. Versione italiana di Scientific American* 162 (February 1982).
Greene, John C., *La morte di Adamo. L'evoluzionismo e la sua influenza sul pensiero occidentale* (Milan: Feltrinelli, 1971).
Guth, Alan H. and Paul J. Steinhardt, "L'universo inflazionario," *Le scienze. Versione italiana di Scientific American* 191 (July 1984).
Gutzwiller, Martin C., "Il caos quantistico," *Le scienze. Versione italiana di Scientific American* 283 (March 1992).
Hall, Rupert, A. Hall and M. Boas, *Storia della scienza* (Milan: Il Mulino, 1991).
Harris, Marvin, *Antropologia culturale* (Bologna: Zanichelli, 1990).
Harris, Marvin, *Cannibali e re. Le origini delle culture* (Milan: Feltrinelli, 1984).
Hawking, Stephen, *Dal big bang ai buchi neri* (Milan: Rizzoli, 1988).
Holloway, Ralph L. "Gli antenati dell'uomo. I cervelli degli ominidi fossili" *Le scienze. Quaderni* (October 1984).
Horgan, John, "La filosofia dei quanti," *Le scienze. Versione italiana di Scientific American* 289, Genuary 1992.
Kastling, J. F. and J. B. Toon, "L'evoluzione del clima sui pianeti terrestri," *Le scienze. Versione italiana di Scientific American* 236 (April 1988).
Korner, Melvin, *L'ala impigliata. I condizionamenti biologici dello spirito umano* (Milan: Feltrinelli, 1984).
Kung, Hans, *Dio esiste?* (Milan: Arnoldo Mondadori, 1979).
Leakey, F. E. and Richard Lewin Roger, *Origini. Nascita e possibile futuro dell'uomo* (Bari: La terza, 1979).
Lenneberg, H. Eric, *Fondamenti biologici del linguaggio* (Turin: Boringhieri, 1982).
Leroi-Gourhan, André, *Le religioni della preistoria* (Milan: Rizzoli, 1970).
Leroi-Gourhan, André, *Il gesto e la parola. Tecnica e linguaggio* (Turin: Einaudi, 1977).
Lieberman, Philip, *L'origine delle parole* (Turin: Boringhieri, 1982).
Linde, Andrei, "Un universo inflazionario che si autoriproduce," *Le scienze. Versione italiana di Scientific American* 317, January 1995.
Lorenz, Konrad, *L'aggressività* (Milan: Il Saggiatore, 1969).
Lorenz, Konrad, *La scienza naturale dell'uomo. Il manoscritto russo* (Milan: Arnoldo Mondadori, 1993).
Lorenz, Konrad, *L'altra faccia dello specchio. Per una storia naturale della conoscenza* (Milan: Adelphi, 1974).

Molari, Carol, *Darwinismo e teologia cattolica* (Rome: Borla, 1984).
Monod, Jacques, *Il caso e la necessita* (Milan: Arnoldo Mondadori, 1970).
Morin, Edgard, *L'uomo e la morte* (Rome: Newton Compton, 1980).
Muller, A. Richard, "La radiazione cosmica di fondo e la nuova deriva dell'etere," *Le scienze. Versione italiana di Scientific American* 119, July 1978.
Nietzche, Friedrich, *L'Anticristo. Maledizione del Cristianesimo* (Rome: Newton Compton, 1977).
Nietzsche, Friedrich, *La gaia scienza* (Milan: Adelphi, 1989).
Pettazzoni, Raffaele, *L'essere supremo nelle religioni primitive* (Turin: Einaudi, 1965).
Popper, R. Karl, *Conoscenza oggettiva* (Rome: Armando Armando, 1975).
Popper, R. Karl, *Verso una teoria evoluzionistica della conoscenza* (Rome: Armando Armando, 1994).
Ruelle, David, *Caso e Caos* (Turin: Boringhieri, 1992).
Stewart, Ian, *Dio gioca a dadi*? (Turin: Boringhieri, 1993).
Verolini, Roberto and Petrelli Fabio, *Metamorfosi della Ragione. Esegesi evoluzionistico psicosociologica di Gn 1,3 ed implicazioni bioetiche* (Dipartimento di Scienze Igienistiche e Sanitario Ambientali, University of Study of Camerino: Italy, 1994).
Von Ditfurt, Hoiman, *Non siamo solo di questo mondo* (Milan: Longanesi & C., 1981).
Wilson, O. Edward, *Sulla natura umana* (Bologna: Zanichelli, 1980).

MARTIN NKAFU NKEMNKIA

THE NOTION OF "ELENG", THAT IS OF TIME, IN AFRICAN ORIGINAL EXPERIENCE: THE VITALOGICAL DIMENSION

1. CHARACTERISTICS OF AFRICAN CULTURE

Before starting the reflection on the notion "*eleng*", that is of time in the African way of thinking, I would first of all like to highlight some characteristics of African culture. By so doing, I hope to illustrate the originality of the African way of thinking, which in itself finds its maximum expression in the neologism "*vitalogy*". Vitalogy is a term which I have found extremely appropriate in identifying, and distinguishing, the African way of thinking from all the other forms of thought world-wide.

Life in itself is the very limit within which the African mind operates. Thus, my reflection on the originality of the African way of thinking, and above all on the notion of time in the latter, should take place in a vitalogical manner. In fact, reality in all its manifestations introduces us to the vital principle by means of which every living being defines him or herself within the complex of the universe.

It will therefore be within the framework of the original vision of the world, of the living being (human being), and of God, in a comunitarian perspective, that we shall investigate the so-called ontological problems, including that of *time*, that of *eleng*.

African thought is born from life and persists throughout its duration. Everything is therefore *life* in itself. Life is what remains identical in every moment of its perception. Having said that, one could say that, every demonstration of life, which in its turn is taken as an expression of divine creativity, takes place within a so-called "vital force". This "force" that unites everything, ensures that no one can think without being and, at the same time, precludes that a being without life be conceived.

Therefore, most Africans consider life to be the first transcendental reality expressed by "*eleng-in time*" through its multiple manifestations.

For this reason, some realities of ontological nature, such as the world itself in all its complexity, like the mystery of the human being and the reality of God in His transcendental and immanent nature, are not considered to be abstract objects, but as ontological ones without

A-T. Tymieniecka (ed.), Analecta Husserliana LIX, 115–129.

which, nothing has a meaning. All these realities are inter-dependent, in a hierarchy of values that range from God through human life to inanimate things. In this context, the reality of "*eleng-time*" assumes a qualitative dimension, being in itself a pre-conceptual, pre-logical and transcultural truth such as God, human beings themselves, and the world in its totality. All these ontological realities together with others (which will not be discussed in this context) are called "universal truths" and, can be simply said to be "universal" and therefore "necessary".

I. *Reality and Time*

In all this process, the human being is seen as a gregarious being. He forms relationships between himself and his peers, himself and the world, himself and God. Such a vision implies the notion of "eleng-time". This forms the basis of the ontology of the collectivity which is innate in the African. That is, *"I" is equal to "you", that is to say, that "I" is equal to "the other", and that "the other" is equal to "us". One can thus say that "us" simply means "the living"*. Thus, the community is considered to be a real social subject.

Well, all this necessarily implies certain extractions, and therefore some "eleng-time" that would facilitate the operation between the *"I", "you" and "us"*.

One of the characteristics of African culture, like that of thought itself, is unity in diversity. In fact, in the way of thinking of the individual, the presence of the "Totality" to which the individual belongs, can be discerned. In the African way of life, everyone always presents oneself as the synthesis of one's culture through all time. All that we are saying always implies the existence of "eleng-time".

The world as a representation of the objective reality in its multiple forms, would not remain new and beautiful if all this were not constantly changing. Furthermore, this change requires that there be a *"before" and an "after"*, it requires "eleng-time" within which the passages from one form to another are realised. The African world (*made of the material world, that of the ancestors, that of the spirits and that of the stars and planets*) is not simple to understand, properly because all this worlds are not considered simultaneously and also because of the multiple ways and aspects that present themselves to the mind in the length of "eleng-time", in order to enable the conscious subject to come into full knowledge of these different worlds.

Regarding the truth that is God (*Who is always thought of as Creator, He who lives, Existence itself and the "Vital Force" which gives life to every being*), one cannot think of such a being without implying a certain form of "eleng-time". Africans usually do not think of *"God"*, for the simple reason that this reality *"God"* is essentially involved in their lives. For this reason, everyone relates to God not only by means of thought but also by means of words, prayers, rites and different cults. All this takes place in the normal way in which one would relate to a real subject *"You"*. A certain dimension as "eleng-time" is therefore required within which and through which this knowledge and this relationship with God, both by individuals and by group of people, is realised.

All this confirms once more that the reality of time is always present; otherwise man, and therefore the community as a whole, could not experience successive sensations of liberty and goodness brought about by the meeting with the Divine or with God as He manifests Himself firstly within and then with us.

II. *African Vitalogy*

In the African way of thinking concerning Vitalogy (*that is, the logic of life and the science of the reflected life, logic of life*) every concept and value is found inside an environment in which everything is possible. The *"Whole" or the "Totality"* consist in the unification of the parts (*the totality is what is made known immediately to the mind of the thinker*). *This "oneness of the totality"* is what never imposes itself on the parts (*suffocating or eliminating them*), but always shows itself as an absolute novelty in which, every part which is annulled can find itself once more in a new shape, thus realising itself in a different way within the "*totality*" itself. This enables us to each time, find ourselves as new beings full of vitality and of beauty and urge each other to desire the other, without which one is nothing, is senseless, is meaningless.

It would be contradictory, and impossible, to think of a "*Totality as a unique reality*" deprived of one of its parts. In the same way, it would be just as difficult to think of the parts without having an idea of the *"One", the "Whole", the "totality"* to which each part belongs. In such a global vision of reality, it would be spontaneous to affirm that to all that shows itself as a reality should be added all that *should* be and *could* become of it. Therefore, means and ends, life and death, sensitive and spiritual reality, before and after, go hand-in-hand and are therefore invisibly thought of, or conceived as really existing. Thus, one should always be able to say something again, something more, about all that happens and will happen. Therefore, every appearance or phenomenon presents itself in the global vision of *"One"*. We can therefore say that visible realities in their various forms correspond to invisible *"realities"* which simultaneously constitute the source (*and the truth*) of what is seen, as well as its principle. In this case, the reality of "eleng-time" is considered in its ontological dimension.

In African Vitalogy, one has therefore a global vision of reality in which there is no space for dichotomies between matter and spirit, body and soul, religious tension and everyday life, the world of the living and the living-dead or decreased (*the reality of the ancestors*). This statement still pertains to the sphere of the state of mind. What matters is not *"when" but "being here and then"*. No one bothers about yesterday because it is gone. Neither do they worry about tomorrow because it has not yet come. Everything is concentrated in the present moment.

In fact, one is not born and does not die in the past or in the future, but in the present moment. In the contemplation of reality there is unity between *"before" and "after"*, so that no one feels the need to delay his or her joy and happiness to another moment, so that no one finds it possible to live the joy of a time gone by. According to our recent reflection, this statement may appear, at least initially, to be a contradiction, but it seems to me that things are not quite like that. The vision of the world includes both material and spiritual things. Both of them are real in thought and therefore in the thinking subject. What enables everyone to have this global vision is precisely the dimension of "eleng-time".

In the African way of thinking, there has never been a moment that *time* was not linked to the relationship between man and the rest of the universe, and to life in its multiple forms. Therefore, it is impossible to speak of *time* as of an entity by itself that can condition and make someone lose something absolutely. The notion of *time* finds its reason for being in a new way each time. Since African *time* is the present moment, we cannot but say that it is a predicate of reality itself. Reality is then, the subject. There are therefore different ways, or dimensions, in which to understand the reality of *time*. Let us now look at some of these ways and dimensions.

2. DIFFERENT WAYS AND DIMENSIONS OF CONSIDERING TIME

III. *Time as a Relational Dimension*

What we have been discussing so far allows us to say that in the African way of thinking the first dimension in which to consider time is of an ontological nature. It is a subject which presents itself as a way and a quality of reality, as its invisible predicate. As things thus stand, the reality of *time* all along presents itself under a double aspect: "*real time*", which can also be considered as physical time, for the fact that, it is clearly present in our minds in its ontological dimensions, as we have shown previously. In fact, it is precisely all these that enable us to have a global dimension of life itself. It thus presents as a cloak of reality and that which is in real – which is ontological, what can simply be thought of as a dimension and a state of mind.

We have affirmed before that time presented itself as a reality that is perceived at its fullest in the present moment. We shall now add that time in its relational dimension is conceived in function of the subject who thinks and perceives it, firstly as an immanent reality and then as a reality in itself by means of the relationship between diverse peers (*thinking human individuals*) and the environment that surrounds them.

In fact, the individual as a subject fits itself into a complex reality where diversity is not immediately perceived, and it is evident that such an intellectual operation, such a relationship between other subjects requires a "*certain lapse of time*". In this case, time exist only when there is someone or something, or any how when the individual relates to another individual or to things. It follows that the reality of time thus becomes the condition of real experience of individuals amongst themselves and within the nature. *Time thus turns out to be the instrument of reciprocity, and as such it becomes a "criterion of evaluation". Time is really born when there is "Something Else" or "Someone Else".*

In the company of others, the reality of time assumes a social character when, for example, welcoming the new born, participating at the village feast, being present at the agony of an elderly person awaiting death, marvelling at the growth of a plant – all experiences lived in *time* together with others.

IV. *Social Time*

Social time in its vitality and dynamism is realised within the entire period of a lifetime that runs from the birth of the individual in the community to his growth, maturity and death. In the social time interwoven with relationships, one is aware of time without being able to grasp it and without being conditioned by it. Therefore, it would be out of place to say that one never has enough time to do a certain thing or to listen to someone etc. In the social experience one is aware of time. Time is experienced in such a way that one knows of its existence but does not manage to give it a real definition, because its value is always so new and is always linked to a relationship with others, so that no one would like to think of possessing it or of not having enough of it. One cannot have *time* at one's disposal in the same way as one possesses little or no money, nor can one say that one has lost or spent too much *time*, thus referring to *time* as something that can be found at a later date or at a cheap or expensive price.

Let us once more state therefore that the reality of "*time* is a fundamental element of reality itself, so that *one lives with time in its eternity in the present moment*. In fact, in the span of one's life one often says that they have seen time passing, but in reality one feels and lives in or within *time*. Therefore, *social time is what can be expressed as states of minds or of souls*.

V. *Memorial Time*

"*Memorial time*" follows "*social time*. In fact, the ways in which the past is remembered varies from one place to another. When one thinks of the future one remembers the past. When one talks about the death of a King, of a Queen, or of a person dear to them, one simply says *that "s/he has gone just in the same way as s/he came*". We often hear people saying: "*who knows what waits us tomorrow . . .*" All the experiences we relate are the contents of our memory and, therefore, of our past. We ask ourselves if it is possible to live in the past in a coherent manner, or if the future can occupy a lot of space in our minds. Concerning the question of "*later or after*", I would like to present the thoughts of John Mbiti, an African thinker, as regards to "*time as sasa and zanani*". This approach is particularly interesting for meaning which refers to religion.

VI. *Time According to John S. Mbiti*

According to Mbiti,[1] the whole life of Africans can be explained through religion. Religion, he claims, is an integral part of everyday life. "*Time and space*" are all ways in which faith is lived. One of the values of African Traditional Religions, as Mbiti sees it, is that of associating the present with the future. Life in the present is the most important concern of activities and beliefs, so that the real reason that one believes is not a view of a certain future or of a better future life. Belief does not consist of a hope in the future. Physical and spiritual life go hand in hand, there are no distinctions between *body and soul*, between *now and after*. Even life after death is conceived in material and physical terms. Thus, heaven must be a concrete place, a place that can be represented with images, or, at least, one should be able to experience it in the world of the living, otherwise it would be impossible to imagine its real existence.

> . . . Africans have their own ontology, but it is a religious ontology, but to understand their religions we must penetrate that ontology . . . To see how this ontology fits into the religious system, I propose to discuss the African concept of time as the key to the understanding of the basic religious and philosophical concepts. The concept of time may help to explain the beliefs, attitudes, practices, and general way of life of African peoples, not only in the traditional set up but also in the modern situation (whether political, economic, educational or Church life). . . .[2]

In this statement, we find the importance of the reality of "*sasa and zamani time*" as an essential element for the study of African ways of thinking and religions.

Since all peoples are intimately linked to their religious concepts, it follows that their history is also the history of their religions. Furthermore, every history implies the notion of "sasa and zamani – time". The topic of time in the approach of Mbiti refers above all to the question of existence.

Mbiti claims that for African peoples, the notion of sasa and zamani, of time has no scholarly interest. "Time" as such is perceived within the composition of events which took place, which will take place shortly, or which will inevitably take place, or will possibly take place in the near future. In the traditional African way of seeing things, *time* is a bi-dimensional phenomenon with a long past, a present, and virtually no future. What has not taken place and will not take place now, cannot enter into the constitution of the notion of *time*, because it would not be a cause for worry. What has not taken place, or what has no likelihood of an immediate occurrence falls in the category of no/time. What is certain to occur, or what falls within the rhythm of natural phenomena, is in the category of inevitable or potential time.[3]

The author claims therefore that the reality of time in its *potential* dimension consists of future events destined to take place for sure and which form the inevitable rhythm of nature. With regards to "*real time*", it is what is present or what is past. This time orientation, governed as it is by the principle dimensions of the present and the past, dominates the understanding of the African individual, collectivity, or of the African world itself.

If, however, future events are certain to occur, or if they fall within the inevitable rhythm of nature, they at best constitute only potential time, not actual time. What is taking place now no doubt unfolds the future, but once an event has taken place, it is no longer in the future but in the present and in the past. Actual time is therefore what is present and what is past. It moves backward rather than forward; and people set their minds not on future things, but chiefly on what has taken place."[4]

VII. *From the Numeric Calendar to the Calendar of Phenomena*

According to Mbiti, the computation of time is made in view of concrete purposes in connection with events, and not in a purely mathematical spirit. For this reason, there are no numerical calendars in the African context. In their place, there are calendars of phenomena. A typical example is that of pregnancy or of the moon, be it full or half. In practice, it is events that give a place and a significance to time.

Instead of numerical calendars, there are what one would call phenomenon calendars, in which the events or phenomena which constitute time are reckoned or considered in their relation with one another and as they take place, i.e., as they constitute time. For example, an expectant mother counts the lunar months of her pregnancy; a traveller counts the number of days it took him to walk (in former years) from one part of the country to another. The day, the month, the year, one's life time or human history, are all divided or reckoned according to their specific events, for it is these that make them meaningful.[5]

Regarding the concept of the past, the present and the future, the author suggests two Kiswahili words,[6] "*sasa*" and "*zamani*",[7] to explain these dimensions of time in their conceptual shadings.

"*Sasa or near time*", gives the idea of immediacy, of vicinity and of the momentary. "Sasa" is the period that immediately touches the person.

It therefore indicates that which is about to happen, which is happening. For this reason, in many languages "sasa" would correspond to the concept of the near future. Thus, the word "sasa" is used to express all events that have already taken place (*for the fact that they are already part of the present*), or else to indicate those that are about to happen. For this reason, "sasa" is the most significant period for the individual. Even though "sasa" involves extension expressing direct happenings, it is not a mathematical constant. "Sasa" projects itself both into the near future and above all into the past, that is, into "*zamani or unlimited time*" (*unlimited past and unlimited near future*). In fact, "zamani" is the uncertain future, distant and, at times, conceived of even as the unlimited past. While "sasa" has its near future, a dynamic present and a past lived directly (taken in the sphere of microtime), "zamani", on the other hand, is not limited to the past, present or future. "Zamani" is realised on a much bigger scale. "Zamani" is the sphere of macrotime.

Sasa is really an experiential extension of the now/moment stretched into the short future and into the unlimited past (or Zamani). . . . Sasa is not mathematically or numerically constant. The older a person is, the longer is his sasa period. The community also has his own Sasa, which is greater than that of the individual. . . . Sasa is the time region in which people are conscious of their existence, and within which they project themselves both into the near future and mainly into the past (zamani). . . .[8]

"Zamani – unlimited time" overlaps with "sasa", even though they cannot be separated. In fact, "sasa" flows from and disappears into "zamani". "Zamani" thus becomes the period beyond which one cannot proceed, thus becoming the graveyard of time. These two terms can be referred to as comparative adjectives, similar to big, small, long and so on; however, all this must be taken in reference to a particular event or phenomenon. Zamani is the graveyard of time, the period of termination, the dimension in which everything finds its halting points. It is the final storehouse for all phenomena and events, the ocean of time in which everything becomes absorbed into a reality that is neither after nor before.[9]

"Zamani", in the final analysis, connects individuals to their environment, which in its turn is the period of myth that confers a sense of base or security to "sasa" time. "Zamani" connects all created things, so that they are all contained in the macrotime.

Mbiti claims that the history of all the African peoples should be read in the light of this conception, in terms of "sasa" and "zamani". They form part of the "*whole reality*" beyond which nothing can proceed (zamani).

In response to the question of whether or not there exists a future towards which everything could tend, Mbiti claims that in the African way of thinking there is no concept of history that moves towards a culminating climax in the future, or towards a certain ultimate aim. His statements confirm the assumption that the future as such does not exist beyond the span of a few months. Therefore one should not expect any golden age, any better world different from the one that already exists in "sasa" and in "zamani".

For the author, both history and prehistoric periods belong to mythology. There are, however, many myths about the creation of the world, the clan, and the tribe. If one tries to place these tales and these traditions (passed down through generations both by spoken and written word) on a mathematical scale of time, it would at first sight seem that, they embrace only a few centuries, while in reality they go back much further. These traditions appear to be under the form of a myth, they elude all efforts and attempts to describe them in a temporal mathematical manner.

We can therefore conclude that the African vision of history (taking into consideration the terms "sasa" and "zamani") does not foresee that the world will have an end, and consequently that, in one case, may one think that history itself might have an end. What in African Vitalogy is understood as "the end", or as "eternity', is only God, to whom even the reality of the beginning is applied in fullness.

In African Vitalogy, with regards to experience, life has an eternal cycle. Everything starts at birth, continues through the initiation during puberty, through matrimony, procreation, old age up till death (*which is to be considered as the deceased's entering into the community of the ancestors in the company of the spirits, and as the birth in a new form of life*). This is what makes up the ontological rhythm of the life of each individual, and of the collectivity as a whole. In fact, at the communitarian level, there is always a cycle of seasons with diverse activities, such as sowing, cultivating, reaping and hunting, as we have previously illustrated.

Human life does knows some unusual events like eclipses, drought, the birth of twins and so on. These events always have an explanation, as well as a meaning, of religious nature. All these so called anomalies, which are inserted in human life in relationship with "time", are seen as an *invasion of the ontology harmony.*

In spite of the complexity of the problem of "*time*", we have to admit is existence, no matter how virtual it might be. This is because any temporal dimension implies the idea of a future, even if only the near future. In fact, even the calendar of events would not be such if the future was not conceived of as existing in some way.

However, we could say that the traditional concept of time is linked to the logic of life and of Vitaloty, because people do not project their daily thoughts into a distant and uncertain future. They are conscious of the fact that the future is potentially present, and that it is inevitable that it take place in a certain way, as, for example, the seasons of the year and the rhythms of birth-growth-procreation and death in individual's. This consciousness of the future of time is expressed in the reality of everyday life. When a tree is planted, it is known that it will bear fruit within four to five years' time, or that a ten-year-old offspring will one day get married (even though it is not important whether this will take place in ten or fifteen years' time).

What follows from Mbiti's exposition is that for Africans the telos of life is living, because in African culture no catastrophic end is foreseen. Hours are marked by the movements of the sun and events are given a place on that basis. For the African, says Mbiti, a year does not exist, one, two, five, or one hundred do not exist. Man exists. What really exist is life.[10]

VIII. *Cosmic Time, The Cycle of The Seasons*

Cyclic time is a reality perceived through the two main seasons of the year, the summer time and the rainy period. These are characterised even by the sowing of provisions and their harvest; from the moon is appearance to when it becomes full, to its new appearance. Under the African sun, nature lives and moves in continuation. Flowers always bloom at the same time of year, rain always comes during the same period, the day always starts in the same way "*eleng yoh = in the same solar hour*", with the sun rising, and it ends when it sets.

This type of "eleng-time" is very important for Africans, because it allows them to desire in turn both the dawn of a new day and the sunset of a particularly tiring day worked in the fields. This experience of the real in its complexity leads to the desire for eternity. We affirm that African eternity should not be thought of as the extreme limit of temporality for, as previously stated, eternity belongs to the reality of God.

In fact, everyone think of the reality of the ancestors as taking in not only the deceased of the past, but also the living of the present before eternity, especially before God. Such a relationship with the reality of the ancestors awakens in all the desire for immortality, for everlasting life.

IX. *Time as A Myth and as History*

In African Vitalogy, the concept of history belongs to the reality of Amieuteh, that is, of the memory. All the myths about the origins of the world, of human beings and of the tribe itself derive from a tale that takes memory further and further back in "eleng-time", with a strong bearing on the present. "*Eleng ghih-in those days*", in the days of King so and so . . . who conquered the world . . . these are all expressions which refer to a past at times distant and at times very near, but which can not all be given a specific date.

In fact, none of these expressions present reality in its certainty. Other expressions can once more help us to understand what we are reflecting upon: "*Ela eleng mohg = at a certain time, it happened that* . . . , once upon a time . . . , when the dry season was about to finish or to start . . . , when the young one was growing . . ., after the time that followed his initiation . . . , when he had committed himself in his turn . . ." etc. Well, very many of these expressions do not allow Africans to think of history only in terms of dates and facts, of a succession of events but rather as "*eleng lequiet*", that is, historic time, the time of history or of tales, a vital time. We can say that the reality of time in each of these dimensions can be considered to be "*essential measures of the living reality*".

X. *Time as a Property of Living Being*

Every culture has its own vision of time. For Africans, *time is the living being that which the human being has at his/her disposal.* "*Time* or *non-time*" that which conditions his/her activities. One can understand that "*time is expensive*", but would dispute the fact that there is an urgency to finish a certain job just because it was foreseen. When an African makes an appointment for 2 p.m., one must bear in mind that all the afternoon could be taken up with that appointment. This does not mean that the African does not have a sense of time. The main motive that we can imagine is found in the relationship that is lived with the other. It is finally in *being at the disposal of others* that the reality of time can be defined. It is expensive because it enables itself to be consumed by others. We shall conclude this reflection with only one statement: "*time belongs to life and life belongs to God*".

We cannot conclude this argument so dear to our minds without referring to the reality of *Aliuh*, that is, to the reality of *space*, which, together with time, allows human beings to conceive movement towards others and towards everything, which allows us to experiment, with and at the same time to admire the beauty of life and the goodness of its creator. Thus, what has been said of time can be applied totally to *Aliuh*, to Space. *Space is in fact a dimension of time.*

NOTES

1 John Mbiti is the first African thinker who, in his works, researched and talked about everything and presented in an original way the reality of time, inserting it, however, in the sphere of religion. See *African Religion and Philosophy* (Nairobi: Heinemann Kenya Ltd., 1969).

2 *Ibid.*, pp. 15–16.

3 *Ibid.*, p. 17.

4 *Ibid.*

5 *Ibid.*, p. 19.

6 "Swahili" is a language spoken in numerous countries in East Africa. It is one of the Bantu languages spoken on the continent.

7 "Sasa" and "zamani" are two Kiswahili words that respectively qualify time in its immediacy and in its future virtual reality.

8 Mbiti, *op. cit.*, p. 22.

9 *I bid.*, p. 23.

10 *Ibid.*, pp. 24–28.

REFERENCES

De Postioma, Adalberto, *Filosofia Africana* (Luanda: Seminario Arquiepiscopa, 1968), 56 pp.

Agblemagnon, N., "The Existence and Nature of African Philosophy in The Past and in The Present," *Journal of African Religion and Philosophy* 1/2 (1990).

Agblemagnon, F. M., "L'Afrique Noire: La Métaphysique, l'Ethique, l'Evolution actuelle," *Comprendre* 21–22 (1960), pp. 74–82.

Alassane, Ndaw., *La pensée négro africaine* (Dakar: Nouvelle ed. africines, 1983), 284 pp.

Anyanwu, K. C., *Nature of Black Cultural Reality* (Washington, 1976).

Arthur, J. K. M., "African Philosophers and African Philiosophy," *A Journal of Religion and Philosophy In Africa, The African Mind* 1 (1989), pp. 26–35, Makerere University.

Bahoken, Calvin J., *Cahiers métaphysiques africaines. Essai sur la philosophie et la Religion chez les Bantu du sud-Cameroun* (Paris: Présence Africaine, 1967), 124 pp.

Bodunin, P. O., ed., *Philosophy in Africa. Trends and Perspectives* (Nigeria: University of I.F.I. Press, 1985), 279 pp.

Brink, A., *Kennis van die Aand* (Cape Town, 1973); English version: *Looking on Darkness* (London, 1976); Version française: *Au plus noir de la nuit* (Paris, 1976).

Bwanga Wa, Mbenga, *La philosophie du language de A. Kagame, contribution à la problèmatique sur la philosophie africaine* (Louvain-la-Neuve: Ottingnies, 1988), 165 pp.

Cesaore, A., *Cahier d'un retour au pays natal* (Paris, 1947); English version: *Return to My Native Land* (Harmondsworth, 1969).

Devaux, V., "La philosophie Bantoue," *Bulletin des Juridictions Indignes et du Droit Coutumier Congolais* 8 (Mars-Avril, 1946), 225–230.

Dougali, James W. C., "Philosophie Africaine, Characteristics of African Thought," *International African Institute Memorandum* X (Oxford: University Press, 1932).

Eboussi, Boulaga F., "Le Bantoue Problèmatique, critique de la philosophie Bantoue de P. L. Tempels," *Présence Africaine* 66 (1968), pp. 3–40, Paris.

Estermann, Ch., "Manifestaçao tardia do monoteismo na evoluçao da humanidade?," *Portugal em Africa* (1946), pp. 135–148.

Griaule, M., "Philosophie et Religion des Noirs," *Présence Africaine* (1950), 307–322.

Hodkin, Thomas L., *Nationalism in Colonial Africa* (London, 1956; New York, 1957).

Ittmann, P., "Volkskundliche und religiose Begriffe," in *Nordlichen Waldland von Kamerun* (Berlin, 1953).

Kagame, A., *La Philosophie Bantu-Rwandaise de L'Etre* (Brusselles, 1955).

Kagame, A., *La philosophie Bantoue Comparée* (Paris, 1976).

Kagame, A., "La place de Dieu et de l'homme dans la religion des Bantu," *Cahier des Religions Africaines*" II, 213–233, III, 5–11 (1968/69), Kinshasa.

Kamptedza, G. J. de D., "O signifcado filosofico dos costumes bantos," *Portugal em Africa* (1965), pp. 17–30.

Kaunda, K., *A Humanist in Africa* (Nashville, New York, London, 1966).

Kiambi, B., "L'être chez les Bantous," *Revue du clergé Africain* (1966), pp. 28–35.

Laburthe, Tolra, P., "Y'aura-t-il une pensée africaine spécifique en philosophie et théologie?," *Cultures et développement* 13/3 (1981), 447–468.

Laleye, I. P., "La problèmatique de la philosophie en Afrique: effort d'élucidation et de thématisation," *Cahiers Philosophiques Africains* 6 (1974), (1976), pp. 7–31.

Lufuluabo, F. M., *La notion Luba-Bantoue de l'être* (Louvain, 1964).

Marwick, M. G., "How real is the Charmed circle in African and Western Thoughts?," *Africa* 43 (1973), pp. 59–70.

Mbiti, J. S., *African Religions and Philosophy* (Penguin, 1969).

Mulago, V., *L'Union vitale bantu chez les Bashi, les Banyarwanda, les Burundi face à l'unité vitale ecclesiale* (Rome, 1954, unpublished).

Neugebauer, C., "München, Kinshasha, Afrikanische Houchschulschriften," *Einführung in die Afrikanische Philosophie* (1989), 312 pp.

Nkafu Nkemnkia, Martin, *Il Pensare Africano come "Vitalogia"* (Rome: Città Nuova editrice, 1995), 215 pp.

Nyerere, J., *Ujamaa: The Basis of African Socialism* (Dar es Salaam, 1968); Version française: "Les fondaments du Socialisme africaine," *Présence Africaine* 47 (1963), pp. 8–17.

Odera Oruka, H., *Trends in Contemporary African Philosophy* (Nairobi: Typeset by Ananda Printers, 1990).

Odera Oruka, H. and D. A. Masolo, "Philosophy and Cultures," in *Proceedings of 2nd Afro-Asian Philosophy Conference* (Nairobi, October/November, 1981).

Odera Oruka, H. and D. A. Masolo, *Sage Philsophy. Indigenous Thinkers and Modern Debate on African Philosophy* (Nairobi: African Centre for Technology Studies, 1991), 272 pp.

Ruch, E. A., "Is There an African Philosophy," *Second Order* III, IFE (Nigeria), (1974), 3–21.

Ruch, E. A., "African Attitude to Knowledge," *Mohlomi: Journal of African Historical Studies* 1 (1976), pp. 15–31, Rome (Lesotho). .

Ruch, E. A. and Anyanwu, K. C., *African Philosophy* (Rome: Catholic Book Agency – Officium Libri Catholici, 1984), 412 pp.

Tempels, P., *La Philosophie Bantoue* (Paris: Présence Africaine, 1949). 2nd ed., *Tradution de Néerlandais par A. Rubbens* (Paris, 1961).

Tshiamalenga, N., "La vision "Ntu" de l'homme: essai de Philosophie linguistique et anthropologique," *Cahiers des religions africaines* VII, 14 (1973) (1975), pp. 175–197.

Tshiamalenga, N., "Qu'est-ce-que la philosophie Africaine?," *Recherches Philsophiques Africaines* 1. (1977), pp. 33–46, Kinshasha.

Tshibangu Wa, M., "Métaphysique, cette philosophie qui nous vient d'ailleur," *Cahiers philosophiques africain* 3–4 (1973), pp. 41–49.

Van Parys, J. M., "Philosophie en Afrique, Analyse du Séminaire sur la philosophie africaine d'Addis-Abeba," *Recherches philosophiques Africaines* 3 (1976), pp. 33–70.

Wiredu, K., *Philosophy and an African Culture* (Cambridge, 1980); Mambo, M. "L'étrangeté d'un socialisme Africain," *Présence Africaine* 50 (1964), pp. 20–37, Paris.

Wright, Richard A., *African Philosophy. An introduction, Bibliography* (Washington: University Press of America, 1977), 277 pp.

CHARLES W. HARVEY

TRANSCENDENTAL PHILOSOPHY, PLURALITY AND RESPECT FOR THE REAL

Transcendental philosophy has always been humble before the Real. It has always bowed down before what showed itself as beyond human making. To expose, describe, or even account for the conditions for the possibility of *X*, where *X* might be anything at all, is first of all to accept *X* as *X*, to accept the thing as the thing that it is: this is so whether the thing is a thing, whether it is an idea, a way of life or a culture; it might even be, as in Kant, the world itself (not "in itself," mind you, just "itself").

Perhaps paradoxically, the philosophy that would be most radical of all, that would provide epistemic security once and for all, presupposes what it would secure – ordinary life amidst ordinary things – the ordinary world in its many modes of appearing. What transcendental philosophy would ground, in fact, grounds it; what it finds it apologizes for as if having lost it. Transcendental philosophy is philosophical piety expressing itself in a complex, uncommon chant, an ode to what is, a philosophical mantra in respect of what is, a voice that can sing its object only because it has attained a properly respectful distance from it.

This essay is a mantra about the mantra-making of philosophy; it suggests that that which is amazed at and respects *what is*, is itself amazing and should be respected for the respect it promotes – even when, oddly, it seems so disrespectful to so many. I will say what the "transcendental" of "transcendental philosophy" amounts to, and suggest why transcendental awareness is significant for life in a culturally diverse world. I also want to argue that the word "transcendental" is in some sense always redundant when attached to the word "philosophy." By doing this I do not mean to exclude from philosophy all those ways of thinking that do not declare themselves "transcendental." I rather understand myself as deflating, thinning, expanding and perhaps even weakening what has been called 'transcendental," and claiming that at the center of this idea lies also the center of philosophy, that the two are coextensive, and that each is both much more pervasive and much less unusual to human being than normally believed. By making its suppressed pervasiveness explicit, I hope to reveal that the philosoph-

A-T. Tymieniecka (ed.), Analecta Husserliana LIX, 131–143.

ical attitude is of fundamental significance for life in a culturally diverse world.

"Transcendental awareness" is awareness that is conscious of itself as separate from the object of which it is aware. Awareness, explicitly aware of itself as awareness of objects is "transcendental." Not all awareness is like this – most is simply aware of objects and only covertly aware of itself being aware. As Sarte pointed out, consciousness usually loses itself in its object. Like a great empty wind, consciousness constantly blows towards what is and almost, but not quite, exhausts itself there.[1] The specification "not quite" is very important. It hints at the transcendental tinge that colors awareness, what Aron Gurwitsch called "marginal consciousness."[2] What is left over from first-order awareness, somehow, is us ("me" in my case, "you" in yours) – but, again, not quite. It is "me" without empirical content (so in one sense, not really "me"), "you" without empirical content (so in the same sense, not really "you"); it is what is aware of me, intimately, as a collection of contents, and of you, intimately, for you, in the same way. Transcendental awareness is the empty "I," the translucent placeholder of simple awareness, the part of us that even we can't know, except to know that we're it, it's us, somehow. It is pure reflexivity, what Rimbaud sensed when he said "I is someone else."[3] This awareness that each of us has or *is*, is what Husserl called the transcendental ego,[4] what the Advaita Vedanta calls "Atman-Brahman"[5] and what Thomas Nagel calls "centerless selfhood."[6]

Let me clarify this. A person can be amazed that he or she is a particular someone.[7] "How odd," we think, "that that person, with these and these traits, there, amongst all the others, is me." This occurs when one gets disconnected from the "me," disconnected from the empirical, role playing self that he or she is in everyday life. When this occurs one sees oneself as one more person-thing amidst seven billion others who are also simply there, each in its own way, quite a particular person. When one has this experience one is being transcendental. In Nagel's terms, one has become an objective, centerless self. One is outside what one is normally inside – oneself, in fact; and the transcendental seer is amazed that "it," somehow, is also that wholly particular, deeply attached being of which it is now disconnectedly aware.

In his poems, E. E. Cummings often celebrated, and suffered, this experience of doubled selfhood. In 1925 he wrote:

> my mind is
> a big hunk of irrevocable nothing which touch and
> taste and smell and hearing and sight keep hitting and
> chipping with sharp fatal tools
> . . . i
> feel that i cleverly am being altered that i slightly am
> becoming something a little different, in fact
> myself
> Hereupon helpless i utter lilac shrieks and scarlet
> bellowings.[8]

To see, as did cummings, that one is becoming oneself, and to even see how, is to be outside of oneself, to be separate from the self that one is (and momentarily isn't). It is to recognize that, on the one hand, one is a passive recipient of the world and one is built by it, while on the other hand, one comprehends the building process and can thereby, to some extent, join in the building. One can, within limits, selectively determine the sharp, fatal tools that will hit and chip and gradually make one what one is. And one can recognize, however unclearly, that each and indeed all of the others have been and are being hit and chipped upon, too.

Philosophy and this centerless seeing have been a long time connected. I want now to say how common transcendental experience is, to show that it is always there, the inner-side of mundane experience, and how, thereby, the philosophical attitude lies waiting to emerge, crouched ever at the center of everyday life. I will then say why the cultivation of this trait, why occasionally being a transcendental, centerless no one, is a good way to be.

The birth of philosophy, unsurprisingly, gives us a clue to the origins of the philosophical attitude which I, as did Husserl and as does Nagel, will roughly equate with the transcendental, centerless viewpoint. Note first that philosophy has never not been in a pluralistic world. It had its start there. When the Ionian philosophers felt the need for a single, nonculturally embedded explanation of the cosmos, they were responding to the diversity of explanations that had been spawned by the *Lebenswelten* of their time. These philosophical pioneers set philosophy on the track of objectivity – a fast track that in short time led Plato right beyond the world's borders.

When the first philosophers were nudged out of their world by the entrance of alternative worlds into their own, they saw the world in a new way. Suddenly there were options, what had been actual became merely possible: ethical values, aesthetic values, and even truth values went up for grabs. Which ones were true? How could they say? At the very least, the need to say indicated that the self was distant from its world even while living within it. An array of worlds appeared to the thinker, and at least one of them included herself. This marked the advent of what Michel Foucault, twenty-five centuries later, would call that "strange empirico-transcendental doublet":[9] there was one self ever world-framed, ever becoming itself, being hit and chipped upon always by the sharp, fatal tools of worldliness; while the other self, "a big hunk of irrevocable nothing," stood witness to the contingencies and vicissitudes of selfhood and world. Occasionally, in wonder and *Angst*, this transcendental self tried to speak: it grunted, pointed with words at what could not be said, uttered poetico-philosophical sounds, all, Merleau-Ponty proposed, a cognitively disguised panegyric to the mute presence of wild being.[10] A very strange wedding indeed: two types of self, each the perpetual shadow of the other, each chained to the other with links forged from nothing but *differance*.

Philosophy has always existed in a pluralistic world, but it has rarely been happy about this. This, too, tells us something about the transcendental experience. When Plato posited his transcendent *Eidea*, he wanted to reconcile *differance*, resolve the diverse happenings of daily life into one grand account of all of them. He even would have liked, I believe, to subsume them all, to freeze-frame the differences, to make static what moved, and to let us rest with Plotinus (whom he didn't know, but I'm sure would have secretly liked) in contented contemplation of *the One*.

But why did philosophy seek to "cure" this condition of personal, cultural and metaphysical *differance*? From its inception, philosophy seems to involve a condition where the self is driven out of itself in an attempt to locate itself in a confusing, multifaceted world. Philosophy, even in Thales' attempt to find the single *ousia*, seems driven to objectivity, to a position outside of a merely personal perspective, in order to find a frame that would unify difference, locate the self. I don't wish to claim that particular acts of philosophy are causally determined, but the act of philosophy writ large does seem demanded, insisted upon, by the inconsistent, unreconciled differences of the world. Too many stories, too many explanations, too many things, behaviors and cultures, shove us out of ourselves in an attempt to find the "real" story that would let us locate ourselves by placing ourselves in the midst of the Real.

Think about situations where philosophy happens. The self knows itself in terms of its world. Its identity is a placed affair, a located, situated selfhood, that sees itself through the world more or less darkly. The more layers of meaning coughed up by the world, the more layers the self must take on. And as goes the world, so goes the self. The self blurs to the extent that the borders of the world become hazy. We become what we behold: too much to behold, too much to become. The self is driven out of itself. If this occurs, so too may philosophy – the drive towards objectivity and the correlative experience of transcendentality. This happens when the self is forced out of itself and attempts to see where it is in order to get back into itself.

Plurality, then, helped produce transcendental, centerless selfhood. This helped produce philosophy, which gradually, along with a host of other factors, deepened the sense of centerlessness in the attempt to get back to a world and self-center. The philosophical attitude intensifies this situation because the attempt to get back to the unified self demands a more objective look at the self's situation, but this objectifying process is itself precisely a distancing from that self-world relation. The "me" of everyday life, the empirical self, becomes an object for the "I" of centerless, transcendental selfhood. To use the phenomenologist's language, the empirical "I" living within the everyday world appears to the transcendental "I" as a correlate to the experience of the world as object. But to experience the self and world as objects, as thing within thing, is not always pleasant. It is, perhaps, this uneasy experience of transcendental disconnection, of not-being-at-home in the world, that accounts for philosophy's long-term desire to assuage this condition, as well as for the suppression of philosophy by the status quo. But if my claims have been cogent, this condition is itself the condition for the possibility of philosophy. A philosophy that would freeze-frame all differences, that would "fix" the condition of doubled selfhood, would simultaneously eliminate the conditions for the possibility of itself. Hence, I now want to say why "successful" philosophical therapy would be bad to achieve, and why its opposite – maintaining and even encouraging the philosophical attitude – is a good thing.

These days, the "centerlessness" of centerless selfhood is being felt all over the place – and it is rarely appreciated, it is often loudly lamented.

This second-order awareness, this experience of transcendental disconnection that originally kicked-off philosophy, lies also, I think, behind most forms of contemporary reactionary fundamentalism – in religion, philosophy, and politics.[11] Once one becomes centerless, the cozy center of previous life may look pretty good; it proffers warm fuzzies in place of the cold pricklies of the centerless view, but it may also confine and oppress and make one blind to all the good forms of Otherness. Philosophy frightens; it frightens because it takes us far out, to the Twilight Zone or Outer Limits as it were, it leaves us uneasy, unsure of the world.

But while defamiliarization with the ordinary may be initially disconcerting, it can also expand and enrich what is Real, and make it more vivid. Seeing things strangely may help one to realize that there really are no "ordinary" things, people or events, that things, people and events become reified and mundane only when we do. As Robert Nozick has said, "The most ordinary things yield surprises to attentive awareness. Chairs, tables, cars, houses, torn papers, strewn objects, all stand in their place, waiting, patiently."[12] When focused on fully, each thing seems to stand forth, almost as if it is defiantly pushing itself from the Nothing, against all metaphysical odds – exhibiting thinghood. Transcendental disconnection may be the prompt for the ordinary's appearing exotic, and it may help us to reflectively witness the unlikely event of Being. The amazement or wonder in which children and poets often live, and in which Plato and Aristotle thought philosophy started, may indicate that one has slid, at least for a moment, from the empirical to the transcendental side of that strange and mysterious doublet.

It was, I think, a big and important event, and one too often ignored, when philosophy shifted its starting point from wonder to doubt. The words "wonder" and "doubt" indicate significantly different self-world relations. Think of their synonyms: For "wonder" we think of "astonishment," "awe" and "amazement." For "doubt" we conjure "suspicion," "worry," "qualm," "apprehension," perhaps even "distrust" and "misgiving." These various words have very different emotive undertones. "Wonder" and its cognates have a positive valence, "doubt" and its cognates have a negative one. Wonder, astonishment, awe, amazement, all suggest that there is something to stand before humbled, something to revere, something greater than ourselves and our problems, and something deep and mysterious that is quite beyond our correcting. Doubt, suspicion, worry, on the other hand, all suggest an uneasy feeling (to use the pragmatist's language), something to be fixed, gotten past and corrected, it suggests a puzzle-solving mentality that will lead us from doubt, give us some place other, and better to go. But it seems misguided even to suggest a "solution" to wonder; in wonder there is nothing to be fixed or corrected, no place better to go. It is a satisfied, simple state, a state of recognition in strangeness, where the complexity of the simple appears, as does the simplicity of the complex. Since this doesn't make good logical sense, I guess it is mystical: hence, an embarrassment for the modern – so metaphysically mature, so prematurely unmetaphysical. "If we can't say it, we can't whistle it either," F. R. Ramsey purportedly quipped about Wittgenstein's domain of the shown.[13] And so, presumably, we ought not to try.

But rather than say with Descartes, "I think, therefore I am. And *from this* I can deduce the whole world," the transcendental seer says "The world is, therefore I am. And *for this* I am eternally grateful – and at least slightly amazed." This is not to say that doubt, or calculative thought – philosophy out to solve problems – shouldn't be done, or even that it isn't philosophy; it is only to suggest that non-calculative thought, perhaps call it "*panegyric philosophy*," marks an alternative center for philosophical thinking. If, to use Heidegger's phrase, "questioning is the piety of thought,"[14] then wonder, which lingers in the domain of the question, not doubt or suspicion, which departs in search of an answer, marks thinking's optimal site. If, from Cartesian doubt comes our modern hermeneutics of suspicion, then from wonder comes the hermeneutics of respect. And though many will feel that since the world wars and the holocaust we are beyond such innocent beginnings, anyone who has children, I think, will hope this untrue. Given our world, we want our children to be calculatively clever – to know how to spot shysters when they meet them – but we also hope that their cynical wisdom is tempered by wonder, that their capacity to think calculatively is nested in a natural trust that tells them they needn't always use that capability. We want them to feel that sometimes it's okay, it is safe, and it is good, simply to stare, to linger, astonished, in the presence of what is.

From these reflections we can note another good that accrues to transcendental awareness, to philosophical life. While the unexamined life is surely worth living, the examined one has extra goods added to it; a reflective life provides an existential bonus. Reflection is not an ontologically neutral, non-efficacious act; it adds to experience, makes it richer and deeper. Transcendental disconnection that allows us to look back on taken-for-granted everyday happenings flows back into (later) everyday life and makes it more full of wonder, more wonder-full; it makes life more of a good thing just in terms of itself. There is, I think, a cumulative effect to transcendental awareness, a reflective residue that remains within the ordinary, making it vivid and perhaps even profound. Though I resist becoming a bliss bunny, I call this experience "transcendental joy" – the condition wherein philosophical wonder infiltrates everyday, mundane experience and produces an enraptured joy-at-Being – a joy at feeling the-world-in-oneself and oneself-in-the-world, the joy, perhaps, of proprioceptive contact witnessed and felt by one disconnected. In such a condition, one stands as receiver and witness to what Heidegger (1972) called "the giving,"[15] one sees as if for the first time, and with an odd sense of reverence one realizes that, *in spite of all explanations*, the world's appearing is indeed *a magical presencing*.

In my mind, exemplary instances of this type of witnessing can be found in the writings of Nikos Kazantzakis, in most of E. E. Cumming's poetry, and perhaps in the perceptions of the Kogi priests from the mountains of Colombia. For the first two, the world's perpetual producings seemed ever miraculous – love, war, flowers and children – fairy tales occurring. I don't want to "reduce" the world to a fairy tale – anymore than did Kazantzakis or Cummings – nor do I wish to aestheticize, nor anesthetize us to, all the world's horrors. But anyone who has read Kazantzakis or Cummings has encountered souls awed by the world, by both its wonders and horrors. For these two, the world was enchanted, in Cumming's words, a place perpetually dividing and multiplying "thises times thats," and "whens times wheres,"[16] astonishment with which, he insisted, would always be much more important than the day "some one-eyed son of a bitch" invents an instrument to measure Spring with.[17] For these wonderstruck writers, the world was an irreal event, much like a fairy tale – a fairy tale in the center of which they had somehow been thrown, a gift suddenly granted, giver unknown.

Having a spirit like Kazantzakis or Cummings – being in a world enchanted – is, I think, a good way to be. It is a way to be humbled before the great cornucopia of worlds that the world gives, a way to genuflect without resistance or fright or servility, and a way that philosophy, with all its unanswers, can help us to be.

When I think of the emotive tenor of the transcendental witness, I think, too, of the Kogi priests from the mountains of Colombia. The Kogi are an ancient people who, until recently, have maintained centuries of deliberate isolation, and who, after speaking to us – their "younger brothers" – have now returned to their solitude.[18] On the first day of birth, a child destined to be a Kogi priest is placed in a dim room and kept there nine years. He is given only enough light to prevent blindness, only the blandest of foods in the minimal amounts required to survive. There are no things in the room. He is visited daily by adult priests of the tribe, he is spoken to and taught about the world outside. And everyday, the child, the priest-to-be, is asked to imagine the world about which he has heard – imagine it hard. Then, after nine years, the young priest is taken outdoors. He sees, tastes, smells, hears and touches the things which he had only imagined. He sees ants and butterflies and clouds and flowers and streams and rocks and colors and sky, he sees people at work, and for the first time, he sees women. He tastes new things, and smells and hears and touches them all for the very first time. And then he is asked: "Did you imagine *this*? Even with all our descriptions, *could you have*? See how great is the Mother to have made such a world and to constantly give it?"

The priest, I think, will have become "transcendental." A true transcendental priest he'll be, showing himself humble before the Real, bowing down before all that is beyond human making. Though, perhaps like the rest of us, he'll sometimes get calloused and forget how "mud-luscious" and "puddle-wonderful" the world really is,[19] it will be harder for him. He'll always recall that second first-day of his life, the day he was born, nine years old, because he was given the world. For him, the event of the gifting will not be forgotten.

How does all of this poetry, anthropology and high metaphysics relate to the role of philosophy in a pluralistic world? Do I recommend nine years isolation for all of our newborn? Should we aim for a culture of phenomenologically frazzled priests who will always worship the world? My response, I think, need not be so radical. It has, I hope, been suggested already, murmuring between the lines all along, waiting to be said. Let me now try.

The experience of transcendentality, of doubled selfhood and reflective disconnection, is triggered by the encounter with multiple worlds, with plurality. One experiences oneself as a transcendental self when an array of worlds spread out before one – when big, dense sets of encompassing contraries, contraries that encompass large portions of people's lives, appear in the life of a single person. The person is suddenly outside and inside, and this causes discomfort. This initial bad feeling is important to note. The response of the natural attitude – the attitude that knows what's what, that lives secure under the sacred canopy of local knowledge – is to grab onto a regional truth and deny the newest reality. The tendency is to shut-out the new world, to say it is wrong, immoral, not even pretty. But it *is there* – and *that* can't be denied. It is existentially real, it has ontological status, and it won't go away (unless we make it, of course – lock it up, banish it, bomb it, poison it, wreck it somehow).

This "natural" response is exhibited in philosophy, too. This is only to be expected since, if my claims are correct, philosophy begins when the world starts to divide and then multiply, giving birth to itself in various guises. Indeed, philosophy, along with religion, is sometimes the spiritual vanguard of the naturalistic defense against too many meanings, too many worlds worlding, the "one world" unraveling. So Plato's attempt to freeze-frame the differences, and Husserl's early hope for essential structures by which to define all appearing, run the gamut of philosophy's history – and both are, if I am right, attempts to tame and domesticate the too-many-Reals the world tends to throw at us. This might account even for more personal-philosophical acts like Heidegger's romance with National Socialism. This unfortunate tryst was motivated, I think, by his desire for an idyllic, simple life, a life focussed and resolute, and made most probable in a world without so many Reals. The nostalgic romanticism found in his later critiques of modern technology reveals the same kind of attitude. Theodor Adorno, a Jew in exile from Nazi Germany, saw this temptation to totalize as philosophy's most dangerous inclination – an inclination which, when actualized, (other) philosophy must be vigilant to resist, to deconstruct, to reconstruct negatively.[20] Adorno suggested that whole social movements such as National Socialism have used, and would continue to use, this fear of too-many-meanings and too-many-Reals to galvanize whole peoples towards a new Eden.[21] More recently, I think, communitarians have not been free of this tendency – based (not without reason) on the same kinds of fears.[22]

But one can get past this natural resistance to too many meanings, too many Reals. This "getting beyond" can be found in some types of current political life, and in philosophy, too.

The ironic, postmodern, bourgeois liberal, as described by Richard

Rorty[23] (a liberal not really so different from the one described, 100 years earlier, by John Stuart Mill[24]), has something like the right attitude – it is a "philosophical" attitude in the popular sense of the word, it is, I think, an inchoate form of the "transcendental" attitude in the terms I have been using. The liberal minded person is the one whose views have been displaced often enough that she's able to back away from totalizing viewpoints, able to loosen her grip on one claim to reality, and able not only to tolerate, but perhaps even appreciate, the many worlds and ways of world-making around her. This is the person who has had to mix and match multiple worlds in order to get along in the world. She is able to "do business" with all sorts of people, and she has even probably come to take joy at the diversity of everyday life. Like liking more than one genre of music or film or art or cuisine or literature, she is able to enjoy and appreciate and perhaps even imbibe many world-views.

If the self is a web of contingently generated beliefs, as Rorty believes,[25] then perhaps it is right also to say that it lacks a central anchoring strand. Or perhaps if there is one, it should be described as a situated openness, *das Lichtung* in Heidegger's language, a passivity of sorts, a readiness-to-accept what shows itself real – where "real" is real enough simply because it's the object pole of the Other's intentional acts, of his or her propositional attitudes. So, perhaps paradoxically, being disconnected connects us with more. We are most in the world only when we are finally not engulfed by it. We find the world Real and respect is as such only by losing it first, by acknowledging its birthing of multiple worlds and its subsequent blurring, and then brightening, by them.

John Stuart Mill famously asked of a suffering Socrates, if Socrates were to compare his lot to that of the satisfied dunce, "Would he trade places?"[26] The answer seems invariably "no". So, likewise, with the liberal suffering from displaced ontology. Would she return to life under a single, comfortable canopy? Choose to see less in order not to know more? Again, except in cases of extreme suffering, the answer is invariably "no". She has become comfortable with the "unnatural" attitude. In order to be able to help make her own world, she has chosen to cease trying to unmake the others – after all, she understands, amongst all those others just might be the "right" one. Or perhaps better, she might understand that *the* right one just doesn't much matter. What matters is making each one a little bit better, a bit more humane. How is this done? By the slow process of cross-cultural comparison and contrast: by the slow fusion of horizons that have appeared to a decentered seer.

Adopting the centerless viewpoint might help more people see why Mill and Rorty request that in matters cultural and political we hold in abeyance worries about metaphysical and ontological truth: it might help them to appreciate cultural diversity and to experience it as a good thing. Bertrand Russell thought that because philosophy showed us how much there was to know and how little we did, it helped to alleviate that "impertinent insolence towards the universe"[27] which he thought was promoted by science, theology and daily parochial belief. If he was right, then philosophy might equally reduce that impertinent insolence in matters cultural and subcultural, too.

By encouraging us to see a little too much, and at a deep, ontological level, discouraging us from believing too much, philosophy broadens, heightens and enriches experience. It opens up minds, it makes minds transcendental. And while there are many tough problems that arise from cultural diversity – problems that cannot be resolved by theory alone – appreciating diversity, perhaps even responding to it with fascination and glee, is the best first-response we can make to it. Enough philosophy may help us to do so; in my mind, doing so, for most folks, is philosophy enough.

University of Central Arkansas
Department of Philosophy and Religion
USA

NOTES

[1] Jean-Paul Sartre, *The Transcendence of the Ego*, trans. Forest Williams and Robert Kirkpatrick (New York: Noonday Press, 1966).

[2] Aron Gurwitsch, *Marginal Consciousness*, trans. Lester Embree (Athens, OH: Ohio University Press, 1985).

[3] Jean-Nicolas-Arthur Rimbaud, *Rimbaud: Complete Works, Selected Letters*, trans. Wallace Fowlie (Chicago: The University of Chicago Press, 1966), pp. 303–304.

[4] Edmund Husserl, *Ideas Pertaining to a Pure Phenomenology and to a Phenomenological Philosophy*, trans. F. Kersten (The Hague: Martinus Nijhoff Publishers, 1982).

[5] J. N. Mohanty, *The Possibility of Transcendental Philosophy* (Dordrecht: Martinus Nijhoff Publishers, 1985), pp. xxix–xxxi.

[6] Thomas Nagel, *The View From Nowhere* (New York: Oxford University Press, 1986), pp. 60–66.

[7] *Ibid.*, p. 65.

[8] E. E. Cummings, *Complete Poems* (New York: Harcourt Brace Javanovich, Inc., 1968), p. 199.
[9] Michel Foucault, *The Order of Things* (New York: Vintage Books, 1973), p. 318.
[10] Elearnor M. Goodway, "Wild being, the prepredicative and expression: How Merleau-Ponty uses phenomenology to develop an ontology," *Man and World* 26 (1993).
[11] Charles W. Harvey, "Paradise Well Lost: Communitarian Nostalgia and the Lonely Logic of the Liberal Self", *Philosophy in the Contemporary World*, v. 1, No. 1 (Spring 1994), pp. 9–14.
[12] Robert Nozick, *The Examined Life: Philosophical Meditations* (New York: Simon & Schuster, 1989), p. 59.
[13] A. J. Ayer, *Part of My Life: The Memoirs of a Philosopher* (New York: Harcourt Brace Jovanovich, 1977), p. 115.
[14] Martin Heidegger, *The Question Concerning Technology and Other Essays*, trans. William Lovitt (New York: Harper & Row, 1977), p. 35.
[15] Martin Heidegger, "Time and Being," in: *On Time and Being*, trans. Joan Stambaugh (New York: Harper & Row, 1977).
[16] Cummings, *op. cit.*, 664.
[17] *Ibid.*, p. 264.
[18] Arkansas Educational Television Network, "From the Heart of the World: The Elder Brother's Warning" (May 22, 1994).
[19] Cummings, *op. cit.*, p. 24.
[20] Theodor Adorno, *Negative Dialectics*, trans. E. B. Ashton (New York: The Seabury Press, 1973).
[21] Theodor Adorno *et al.*, *The Autoritarian Personality* (New York: Harper and Row, 1950), pp. 658 ff.
[22] Harvey, *op. cit.*, "Communitarian Nostalgia."
[23] Richard Rorty, *Contingency, Irony, and Solidarity* (Cambridge: Cambride University Press, 1989). "Postmodern Bourgeois Liberalism," in: *Objectivity, Relativisim, and Truth: Philosophical Papers*, v. 1 (Cambridge: Cambridge University Press, 1991).
[24] John Stuart Mill, *On Liberty* (Indianapolis, IN: Bobbs-Merrill, 1985).
[25] Rorty (1991), *op. cit.*, p. 199.
[26] John Stuart Mill, *Utilitarianism* (Indianapolis, IN: Bobbs-Merrill, 1979), p. 14.
[27] Bertrand Russell, *A History of Western Philosophy* (New York: Simon and Schuster, 1945), p. xiv.

OSVALDO ROSSI

LIFE, GENIUS AND THE WORK OF ART

Before the theme of life become a central object of interest to the philosophy of science which aspired to assert itself as the science of life, it had constantly been central to the philosophy of art which, after Kant, developed into the romantic theory of the genius and "Erlebniskunst".

As a matter of fact life has had more importance in modern aesthetics than in the philosophy of science.

In the present contribution we will illustrate the paramount importance of life in some moments of nineteenth-century aesthetics. We shall try to organize this paper by developing the themes of rationality and life.[1]

AN APPROACH

It was with Kant that modern aesthetics reached a philosophical formulation in which emotions and feelings were granted a dignity of their own, similar to that acknowledged reason and will.

Kant distinguished the sphere of feeling from "Pure Reason" and founded his theory as a *Kritik der Urteilskraft*. This distinction, on its positive side, contributed to the definition of a precise field in which the concept could not be instrumentally exploited although it was open to further developments. The Kantian distinction might also be expressed, from the Husserlian point of view, as a separation of mathematical and physical objectivity, given in the natural world, from "the world of life" which may be contemplated phenomenologically.

Kant's research aimed at re-establishing the unity of the natural world and the world of artistic production through the subjective faculty of taste, considered as the ability to reflect the inner vital force which made man a part of nature. Although he used it with a typically human attitude of his own in order to strengthen and expand that taste of his own through which he "judged" beauty in nature. Taste was nothing other than an aesthetic dimension of life in which nature and thought converged as a transcendental condition of that particular emotion felt for Beauty.

Owing to this double convergence Kant was obliged to apply the

A-T. Tymieniecka (ed.), Analecta Husserliana LIX, 145–157.

idea of taste both to beauty in nature and in art by introducing the concept of genius. He was thus able to unite the field of artistic production to the finalism and the harmony of the natural world. Yet the idea of genius was necessary to support the extension of the "judgement of taste" to the field of art, which happened to run the risk of being excluded from the more general sphere of human judgement. If that approach justified the general position of kant, it involved a substantial misunderstanding of the vital force of feeling which was subordinated to the requirements of the transcendental "I".

The notion of genius in the arts, introduced by Kant, has the same qualities of the transcendental "I" in the world of phenomena: he justifies the relationship between beauty and rationality through his creativity. The "I" builds the world by using the "defining judgement" which "reflects" that legal order conferred on nature by the transcendental "I". The "thinking I" and the genius are not realities which can be considered to be different beyond their function: the genius is the aesthetic equivalent of the "thinking I" which is the legal aspect of the genius.

Here follows the union which Kant makes between the many forms of *Kritik*: the common denominator is the same subjectivity which operates in three precise modalities: determination, action, creation, although they share a common methodology.

German idealism operated the first attempt to go beyond the Kantian perspective by taking two new directions: a) by abandoning the constraints in which art had been confined; b) by overcoming the critical formalism in the direction of its dialectic transformation into "an aesthetic theory of content" as in Hegel.

The notion of genius therefore contributed to the development of thought beyond the sphere of art, beyond nature, thus reaching the philosophy of Idea of *Geist*. Genius and Nature appeared to be closely related in the unitary dynamic of the activity of the absolute Spirit, of which the man of genius was the protagonist. More generally speaking, against the background of a philosophy of identity, the Kantian dualism between phenomenon/noumenon, nature/reason was no longer valid.

This was the first time in the history of the romantic-idealistic philosophy that it was possible to make taste and genius coincide not only in order to explain the genesis of the work of art, but also to grasp its meaning and consequently to interpret it.

The extension of that idea, associated as it was, with creativity, found a "true apotheosis" in the romantic notion of unconscious production which was spread a broad through Schopenhauer, to the philosophy of the unconscious.

The theme of genius lead down the path to *Erlebnis*, which, by taking motivation from the concept of "reinforcing the vital feeling" of aesthetic pleasure in Kant, would assert itself as a true and authentic fact of consciousness.

The theory of genius, widely supported by lovers of the arts, readers and the public, who looked at the increasing perfection of work of art, was thought to give an answer to the problems which Kant had raised about the contemplation of beauty in nature, problems which appeared to emerge from the content of the work of art. Consequently the man of genius was no more thought of as the aesthetic guarantor of the internal laws of nature, as well as of the content of the work of art; instead, he was considered to be the unconscious creator of the perfection of his work.

Although that theory was no longer popular among the thinkers of the beginning of our century owing to its clearly irrational fallout, nonetheless the idea that the genius was responsible for the perfection of his masterpiece continued to be shared by the vast masses of bourgeois society.

As a consequence, a philosopher such as Gadamer, for instance, has pointed out that the aporiae, the difficulties in Kant were substantially disregarded by the aesthetic thought of the nineteenth century.

It seems apparent, at present, that the aporiae, promoted by the Kantian partition of art from nature and the idealistic separation of reality from appearance, are to be reconsidered, whether they are seen in the context of the Kantian independence of art or from the point of view of idealistic totality or, almost paradoxically, from those crypto-idealistic perspectives which we still find in our own times.

There is one more consideration which I would like to put in the form of a question: in what sense are the Kantian aporiae preserved in German idealism?

Kant separated art from nature, as well as the scientific approach to nature from the aesthetic one, even if they were actually associated in the same formalistic method. In the romantic-idealistic atmosphere an attempt was made at reconciling art and nature through an aesthetic interpretation which saw nature not as an outward "appearance" ruled by indefectible laws which impose themselves on reason, but as the interpretation and creation of the vitalism which the genius bestowed on his “sleepwalking” creation. In this sense Goethe’s panism, Hölderlin’s pantragism, Novalis’s magic idealism are already interpretations which reduce nature to an aesthetic dimension, that is, to art itself.

Consequently, the romantic genius in his unruly and productive excitement does not live in nature but in the world of art of which nature is only a glittering glare, a product of his “poiesis”; therefore nature becomes an aesthetic creation-illusion which finds in beauty her own accomplishment.

In this scenario who is genius? He claims for himself the function that Kant had reserved to the intuitive intellect of nature: the creating power and the power of imposing a finalism.

Art is the phenomenic accomplishment of that creative and teleologic process that the man of genius unfolds from time to time.

If we analyze this relationship in its specific terms, we cannot avoid noticing a kind of circularity. If art is the fundamental moment of the creating genius, yet art is also the vital dimension which suggests and strengthens the genius towards new and untrodden aesthetic paths. This accounts for the reduction of life to literature, while at the same time the real flows into the aesthetic ideal.

In the sphere of romantic idealism, in comparison with Kant, it is possible to notice a further development in the fall of one of the poles of the dialectic triad (subject, art, nature): nature.

But what is the relationship between the unconscious activity of the genius and the absoluteness of the Spirit in the limits of their bipolar dialectic?

This relationship is not only to be considered as originating from nature reduced to an aesthetic function but also from that absoluteness which is *Geist*.

Up to now we have emphasized the relationships between genius and creation-interpretation, we can now add that from the point of view of the *Geist*, the genius is its historical interpreter. The interpretation assumes here a dual value, a double meaning closely connected to the term “create”. If, on one side, the genius is the historical interpreter of the *Geist*, as he interprets its creativity, on the other, he is the mediator and the *medium* between the absolute in its originary dimension and the aesthetic products of his interpretation.

It is therefore possible to ask whether the romantic-idealistic conception has truly overcome Kantianism or if it is only its simple aesthetic interpretation.

The Kant's *Kritik der Reinen Vernunft* there is a triple dimension: an original, postulatory, necessary, exigent sphere, the "noumenon"; a creating principle in the rational sense of a legislator and unifier (*Ich Denke*); the result of this creating act is nature or the world of phenomena.

In German idealism we have a corresponding triple dimension: the noumenic sphere is the *Geist*, a type of immanent transcendence which cannot be defined; a creating principle which assumes the "sentimental" connotations of the genius: the product of the creating principle is art or the world of phenomena.

Apart from the metaphysical language and a different methodology of approach in both philosophies, the genius is his own and unique referent in the creative primacy to which he is entitled, that is, his creating power, as both "noumenon" and *Geist* almost give consistency and objectivity to his "poiesis". Therefore they are to be thought of as external points which prove the legitimacy of his creative powers.

The value that the work of art assumes here is the one of being a piece of the genius's life, in the sense that he puts to work his vital essence.

But out of this context the work of art does not possess any other value. The autonomy of art is absolutely non-existent: it is the autonomy of the Genius. On the other hand the phenomenic world itself is the rational experience of the *Ich Denke* without which *it* could not exist. Paradoxically the autonomy of nature is resolved, here, in the autonomy of the *Ich Denke*.

But it is in this context that the contradictions of the romantic aesthetics become apparent. They may be reduced to: the non-significance of the work of art and, as a consequence, the arbitrary formulation of the idea of genius and, simultaneously, the speculation on his activity. In those paradoxes there is an inevitable and clearly dissolving issue.

Romanticism, which asserted itself as the coryphaeus of aesthetics, considered on the side of the genius quality of life, ended up denying and dissolving genius into an idealistic nihilism deprived of any externally-objective reference, while, in Kant, that issue had been avoided to some extent, in virtue of an autonomous consistency of nature, even though in a formalistic context: here, the transformation of this "vestige" in the genius's life prepared the way for the preliminary statement of the Hegelian "death of art" in the sense of its philosophical overcoming (acquisition of truth).

Romanticism, in this sense, surprisingly represented the primacy of idealism, of which philosophy and aesthetics were a two-pole interpretation of the same reality: the "Ego" generated the genius or/and the Idea. Consequently, art and philosophy were operating projections of a common ideology: the reduction of the world to the fruition and creation of the subject. A typically burgeois ideology, as it turns out.

In the post-romantic resumption of Kantianism the thesis of "the development of the vital feeling" from which the romantic idea of genius had had its origin, favoured the establishment, in a unilateral manner, of "Erlebnis" and from here of "Erlebniskunst", a concept which, owing to its original source was charged with previous idealistic elements.

TALKING ABOUT "ERLEBNIS"

The term *Er-lebnis*, as far as we know was first used in a meaningful way by Dilthey and presents relevant traces of the immediacy and irrationality of life, besides being considered and a cognitive criterion of life itself and its products.

The prefix *-er-* in fact, indicates the intensive and repetitive character of life (*Leben*) considered as an incessant going back to our inner self, a truly experienced *Er-leben*, the coming back of a kind of reality which cannot be described since it is indefinable and quite original. For this peculiar character, the notion of "life" has always coincided with an interior experience, which may be religious and/or poetic.

Goethe's poetry, which acquires new light just because it is the result of his own experiences, has transferred his experience of life even into his mythological poetic fiction. In *Prometheus*, for example, the hero speaks to Zeus (vv. 48–59) and declares his attachment to life, presenting himself as the symbol of the artist and creator.

In the lyrical production of the English poet, John Keats, we can see the same feature. In *Ode to a Grecian Urn* he wrote these lines

> Heard melodies are sweet, but those unheard,
> Are sweeter, therefore, ye soft pipes, play on:
> Not to the sensual ear, but, more endeared,
> Pipe to the spirit ditties of no tone. . .
>
> When old age shall this generation waste,
> Thou shalt remain, in midst of other woe
> Than ours, a friend to man, to whom thou say'st,
> 'Beauty is truth, truth beauty' – that is all
> Ye know on earth, and all ye need to know.

In this context the work of art expresses the only valid truth. A beautiful work of art represents the continuity, the permanence of life which is, on the other hand, precarious and transient. Art is superior to life since it can seize the fleeting moments when everything is perfect and make them immortal.

Dilthey too, in his well-known essay on Goethe, a critical work that was elaborated over a long period of time, correctly defined poetry as a "representation and expression of life".

In the pre-idealistic thought of Fichte, the philosopher stresses the immediate, rewarding and subjective character of a way of life not yet filtered by reason. Life formed the metaphysical background of the whole of German idealism and was strictly connected to the notion of *totality and infinity*.

Up to that time both life and the concept of living through experience (*Erleben*) had not been submitted in their respective fields, to a rational approach.

The formulation of the concept of *Erlebnis*, a technical concept referring to a precise philosophical thought, an expression of life and representation – according to Dilthey – of the reality of life, one claimed to be the principal requirement of the transcendental knowledge of our experience of life. The term has, in fact, found its first meaning and justification in the field of aesthetics, as we can see from Dilthey's essays. From that moment on, *Erlebnis* became prevalent as an alternative criterion in art.

Because of this gnoseological pretension the concept of "Erlebnis" was to reveal its full ambiguity. The ambiguity was not inherent in its origin, but in the assumption it had to satisfy: it should, in fact, reconcile immediacy with objectivity, irreducibility with knowledge. To satisfy this methodological assumption the concept of "Erleben" turned into a positive fact ("datum") and acquired a thematic content.

The ambiguity of the term has not been commonly perceived and for this reason a rather simple interpretation of romantic aesthetic values has prevailed and was to last till our days, as Gadamer pointed out.

As a matter of fact, the difficulty of reconciling immediacy with rationality is still felt. To reconcile these terms means to make one of them prevail over the other. In this specific case, as the mediation between the terms is a rational act, immediacy actually disappears only to reappear in the form of "immediate" rationality.

This result is not caused by the cold rationality of enlightened ideas prevailing over life, but is already present in the idealizing and romantic character of life. The Diltheyan concept of "Erlebnis" has only brought to the extreme consequences an option that was also present in the Romantic aesthetics of Goethe and Keats: for them art is life.

A paradox is, in some ways, hidden in this conception. Art can make life immortal and even replace it only if art succeeds in overcoming its limits, its fragmentary nature, its vitality and contingency. In this way, both in Goethe and in Keats, art has been turned into a "perfect crime" committed against life.

Dilthey has accepted the invitation of the Romantic poets and has transformed it into a philosophical approach to life and to history through the new category of "Erlebnis".

Owing to the potential loss of the "vital" aspect, *Erlebnis* can be reduced to a form of dialectics which is implied in rationality itself, for which life loses the wealth of its immediacy and the experience of life coincides with what rationality tells us about life.

Once we have reduced life to a form of dialectics between mediation and immediacy, "Erlebnis" comes to be a work of art, that is, a world in itself which, in the act of subordinating that world to itself, claims the right to define and reinterpret the world of life.

In the field of aesthetics, the concept of "Erlebnis" includes, under the form of a gnoseological criterion, the somnambulistic immediacy of a gifted consciousness.

It follows that *Erlebnis* not only appears to be rooted in experience but claims the pretension to found itself in the artistic consciousness as a guarantee of the work of art: such pretension is to be traced in the theory of *Erlebniskunst* which is evidence of this ambiguous and twofold aspect of the concept.

As someone has pointed out, on the one hand, the term means that this kind of art derives from "Erlebnis", or better, that it is its natural expression. On the other hand, the term is also used from the point of view of the interpreter who considers the work of art as the expression of "Erlebnis". The two meanings are closely connected to each other: what is the expression of "Erlebnis" can really be understood in its full meaning only through an "Erlebnis".

As we can easily realize, art has now been reduced to a form or affectation of an aesthetic way of life in its twofold valence, including production and fruition. In other words, we have here both the Kantian idea of a "vital sentiment" and that of a gifted way of living of the genius, which ideas are typical conceptions of Romanticism. They have here been elevated to an aesthetic value and considered as both the content of art and the modality for approaching it.

Even though Gadamer expressed some doubts about *Erlebniskunst*, we can recognize in it a first hermeneutic step leading to a kind of dialogue between the work of art and the person who will enjoy it.

According to idealism the admirer was called to reproduce in himself the conflict and the enigmatic quality of the genius, therefore living a passive role. Now, on the contrary, *Erlebniskunst* having been turned into a direct living experience of life, it involves a more intimate meeting between the experience of life of the genius and that of the reader and admirer of a work of art. *Erlebniskunst* induces the reader to look for a personal relation involving a dialogue with the work of art.

While in idealist thought the experience of the reader comes to be assimilated into that of the genius, giving way to a relationship of imitation/identity, now, on the contrary, even though in an elementary and temporary way, the reader/appreciator tries to find agreement based on a common denominator: the experience of life (though always respecting the various points of view). Here is a dawning presentation of a early modern form of interpretation.

The reader, the one who enjoys the work of art, who feels he is involved in the work, tries to reconstruct the world of the work of art. This must not be considered as a fall into relativism as it can imply the attempt to find a way of coming into contact with art.

Gadamer sees the fall of the interpretation of the work of art into the relativism of experience as a negative development actually, in this historical phase, it may represent a first step toward the claim to a sort of diversity in the identity of a common experience.

The exaggeration with which some have insisted on submitting the work of art to the control of the contemplator and critic, has only had this result: the work of art is subject to the approval, or better to the criteria of interpretation, of the critic. *Erlebnis* can be seen as an attempt to put the work under the control of the reader.

If we consider the path that has been followed so far, we realize that there has been a shift from a conception of art as *Erlebnis-Geist* to one that sees it as an *Erlebnis* of the reader/enjoyer in the sense that it is only his history and life that can give a meaningful value to the work of art; he expresses the aesthetic quality of the work of art.

On this basis one tries to identify, to recognize the work of art through its symbolic character which, however, must be found on the level of *Erlebnis*. In this withdrawal of the symbolic into life aesthetically interpreted, the unique character of the symbol itself has obviously been misplaced.

In contrast, in the Greek-archaic world the artist did not assign any personal meaning to his work, he was not a creator in the true sense of the word: the community made his work meaningful by interpreting it.

It is only in the Greek polis that the artist becomes a "poietes", a creator of meanings through his work which was charged with the public, ritual function of strengthening the citizens' consensus on law and order.

But the work of art did not exhaust itself in this public function: a new dimension emerged which alluded to the sense the artist gave to his work, a personal interpretation of a public performance (as happened with tragedy).

In the Greek "polis" a complex and nuanced dialectic of art was announced, one bound to become an influential stream in Western aesthetics.

THE RECOVERY OF THE WORK OF ART

A radical form of criticism of the aestheticism of consciousness and of nineteenth-century culture is represented by Nietzsche; to him art and *Erlebnis* in general are only elaborations and reflections of that daily horizon of life in which our will plans life itself.

In Nietzsche we see an extension of the concept of "Erlebnis" which is not confined to the limited sphere of those who have an interest in art, or have an interest in genius (men of letters, artists, musicians . . .) in its special sense. They are not the only ones entitled to live aesthetically and transform the futility of daily life into a work of art. The same concept concerns all those who plan their existence as a different way of being which Nietzsche expresses through the notion of "dance", which is the joy of life, a way of being grounded and deeply rooted in the earth.

According to Nietzsche art is "the world of life" (*Lebens-welt*) in which man plans to exceed his limits.

Nietzsche's anti-bourgeois and anti-Philistine radicalism has been timidly adulterated by Husserl's phenomenology for which life represents the subjective consciousness in its constituting act in the world of phenomena.

In this perspective there is already a separation between consciousness, as the original dimension of the vital force in a pre/ante-predicative direction, and its phenomenalism which is predicated in some specific intentional acts. *Erlebnis* is the name given by Husserl to any intentional act of consciousness.

Through the intentional and vital acts of consciousness, Husserl eventually tried to reach the pre or ante-predicative horizon of the "World of Life", using these acts as its phenomenological evidence. The methodological essence of *Erlebnis* achieves its maturity here, as *meth-odos* (a way to/or of remaining along the same path) and as the horizon of the vital paradigm of consciousness. The former was only the guiding aspect of a search of which the latter founded the gnoseological aspect.

These modalities of *Erlebnis* underline a twofold intention: on the one hand it tends to be the path towards the "world of life", while on the other, it tends to be the proof of its existence. The consequence of this is the basic difficulty of explaining something that is at the same time homogenous and different as to its original derivation.

This difficulty points out a positive aspect concerning the aesthetics of Romanticism and a reductive aspect that concerns Nietzsche.

Phenomenology has formulated a new way of considering the work of art and the artist, a conception which can be seen as a break with the idealising point of view of the Romantics. Romanticism saw them as the product of a genius and his experience of life, and in Romanticism we have the frequent use of terms such as imitation, appearance, derealization, illusion, magic, dream.

Phenomenology does not accept an aesthetic sphere distinct from the concreteness of the work of art, which is *one*, because in a work of art the real content is always related to the unity of form and meaning. It is possible to find an access to the work of art and to the artist through visual perception which is not something abstract, but basically art is a way of grasping connections and signs together with a meaning.

To insert a work into the world to which it belongs, means to grasp its relational and dialogical character; it means to remove it from the ideal of perfection which is at the basis of the conception of the genius and from the subjective relativity of *Erlebnis* in order to stress those characters that the romantic conscience generally considered to be extra-aesthetic.

If we start from phenomenological considerations, the romantic idea of "genius" appears completely inadequate owing to the lack of critical discernment before the work of art which is a concrete reality given in the relation to which it belongs. The romantic theory tends to steal it from the phenomenologic perception and its demands for meaning. The Romantic theory, in fact, prevents us from perceiving the real meaning of the work of art, a matter left to considerations based on its free and absolute perfection.

We face other difficulties from a phenomenological point of view when we start our analysis from the Romantic concept of *Erlebnis*, a concept that we have examined before.

The problematic and dialectical character of the production and enjoyment of the work of art in the Romantic age was therefore bound to fail in the debate over its completeness and its fragmentary nature, between its perfection and dispersion.

Phenomenology itself presents several limits which are caused by the paradoxical character of conscious experience. As we can see from the latest notes written by Husserl about phenomenological evolution, he was torn between a reasserted transcendental exigency of "ego" and the "world of life".

The problem was that of reconciling a transcendental subjectivity that claimed the right to intentionally "constitute" the world, and simply revealing its very essence out of a kind of subjectivity which derived from the *Lebens-welt* through a reduction or phenomenological *epochè*. The two different approaches, a methodological one and a founding one, were to be unified in the general theme of the "world of life".

As a matter of fact, the difficulties that progressively emerged in Husserl's analysis brought out with even greater evidence the irreconcilableness between transcendental subjectivity, which had a methodological-contemplative nature and a more primitive form of subjectivity that was trying again to take possession of its vital horizon.

Husserl was moving inside a vicious and paradoxical circle: how could transcendental consciousness, intentionally oriented found that "world of life" from which it had originally emerged? At the same time, subjective consciousness claimed the right to be considered both as a product of the *Lebens-welt* and its transcendental foundation.

This difficulty seems to have led Husserl to privilege the methodological aspect to such extent that this aspect acquired greater importance and became the constitutive dimension of that temporality into which the "world of life" was to develop. Georg Gadamer is right when he considers the theory of *Lebens-welt* only as an expedient to give support to the transcendental reduction of consciousness.

Nietzsche's radicalism was simply neglected, misunderstood, but not surpassed.

What was misunderstood was the productive character of *Lebens-welt* which was reduced to an epiphenomenon of consciousness. The loss of a praxis oriented, enthusiastic and antimethodological dimension of life was also noticed in the scarce interest in the work of art that Husserl showed; it can be seen also in the attempts of other phenomenologists to give the work of art more importance, in the direction of a consciousness to be "filled up" through experience.

In the revenge of vitalism on transcendentalism (Kant-Dilthey-Husscrl), in which vitalism was victorious, we can see the shift from criticism to Romantic aestheticism and to their decline. Phenomenology, standing between reason and life, reopened the Hegelian dilemma between art and philosophy, but offered at the same time, through the Heideggerian secession, an incentive to reconsider the theories between Hegel and Nietzsche, to reflect on the decline of art and on a new beginning of art itself.

San Benedetto del Tronto

NOTE

[1] Cfr. Hans Gegor Gadamer, *Wahrheit und Methode Hermeneutik I*, Gesammelte Werke, 1 (Tübingen: 1986), pp. 48–106: Anna-Teresa Tymieniecka, "Phenomenology of Life and the New Critique Life and of the Human Condition", *Analecta Husserliana*, Vol. **XXXIV**, pp. 3–15.

AGNIESZKA NOGAL

THE FEMININITY AND THE MASCULINITY, AN ATTEMPT AT PHILOSOPHICAL ANTHROPOLOGY

Henryk Elzenberg is one of the most important figures in contemporary Polish philosophy. Thirty years after his death, his work is becoming increasingly popular. His work comprised a wide of interests. He dealt with axiology, ethics, esthetics, history of culture, history of philosophy and religion. In all of them Elzenberg sought to forge his own independent stance making it impossible to classify him within any of the currents in contemporary philosophy. Interestingly enough, in his work he cast a critical eye on contemporary philosophy, which he viewed critically, preferring to look for inspiration in the remote past: ancient rationalism, Buddhism as well as Stoicism. His attempt at rationalism, understood here as a postulate for the search for the proper measure of things is also connected with seventeenth-century rational thought. This kind of rationalism is open to intuition which is rooted in common knowledge, while permiting a deeper reach into the sphere of reality. Philosophical anthropology is part of Elzenberg's work. His approach to this question, apart from dealing "the human condition", also include a "feminist" part. Distinguished in his work are two kinds of human nature: (1) the male nature and (20 the female nature.

The feminist aspect of Elzenberg's philosophical thought, also connected to "rational plot", is less known. His thoughts on the nature of human sexuality have been underresearched and deserve great attention. This problem is, in essence, anthropological. In writing about human nature Elzenberg accents sexuality, by distinguishing a human being as a women, and a human being as a man and the bonds existing between them. This part of Elzember's philosophical investigations has not been published – only short notes could be found in *Kłopot z istnieniem* (The Trouble with Existence).[1] Most of his work on the subject still remain in the form of manuscripts left in the Archives of the Polish Academy of Sciences in Warsaw.[2] They are part of what is titled: *Woman and Love, Erotic.*[3] These notes differ: from the quickly scribbled notes, to deep critical analyses. However, there are some plots, out of which a solid and coherent picture of human nature evolves, on highlighting the basic differences in sexuality. Let us try to show and analyze this picture,

A-T. Tymieniecka (ed.), Analecta Husserliana LIX, 159–164.

considering too Elzenberg's concept of love as seen in his polemics with Schopenhauer: "Schopenhauer, Metaphysick der Geschleshtsliebe"[4] and in his work: "Love".[5] To construct the definition of masculinity and faminity the following two texts also were used: "Psychology of women"[6] and "The Ideal of Woman and Man".[7]

This focus on the duality of human nature brings Elzenberg closer to those who are regarded today as feminists. Feminism emphasizes that human beings manifest themselves in two forms: man and woman. The relationship between those two varieties of human nature can be interpreted or explained in different ways. Part of the feminist movement claims that the nature of man and woman are identical and that all the differences are only biological or sociological but not psychological. We might refer to it as *the thesis of the identity of the two sexes*. The other school of feminism claims that the human nature of a woman is different from that of a man. Let us call this *the thesis of the difference in nature of the two sexes*. This second position is advanced in two different forms: (1) moderate and (2) extreme. The moderate form asserts that despite the differences the two forms of human nature – man's and woman's are anthropologiclaly equivalent; this is *the thesis of the equivalence in the nature of the two sexes*. The extreme position holds that the woman's expression of human nature is not only equal to the man's but even higher and better; this is *the thesis of the superiority of woman's nature*. At times this view adopts the belief that the world will be saved by women.

Elzenberg does not accept the thesis that man's and woman's nature are the same and leans towards the thesis claiming equivalence. In his opinion the constitution of a woman's soul is different from that of a man's. Elzenberg thus attempts to describe the "soul" of sex. According to this thesis the human beeing can be viewed on two levels: first – the more external – the level of humanity: here every man and woman is the same and they are evaluated on the same basis. Everybody who is subject to law is a human being in this sense. There is nothing wrong with this kind of thinking. But on a deeper level both kinds of human beings are completely different. And that difference manifests itself in love, and love shows the power of sexuality. It is possible to love a woman and hold her in contempt as a human being and there is no contradiction here. It is possible because we are only referring to the different sphere of anthropological nature.

The natural state of a woman's soul is calm, while the natural state of a man's soul is anxiety and connected to this anxiety, disturbance. Those two states of the soul – calm endurence and increased anxiety – are complementary: they do not exclude each other, but they cannot both be a part of one soul either. In that Elzenberg sees the essence of the attraction of one to the other. Man needs the calm and balance of feminity, while a woman craves for some anxiety, which saves peace and calm from the element of monotony.

Men and women are different from each other not only in the state of their souls but in their having different ways of perceiving the world also. A man perceives the world objectively and also evaluates it constantly. In accord with his nature, he sees things as they are, independent of his personal attitude or stance. Objective thinking, typical of a man, and a natural disposition towards making value judgments generate a conflict between representation and valuation of the world. It tears a man's soul apart in that he sees and evaluates himself, as a part of the world, objectively, but an egoistic state of superiority in evaluation of the world from that position is also present. Such an inner contradiction in the way of evaluating is called a "disruption," which is a source of constant disturbance and anxiety. To overcome this disruption brings forth an effort to liquidate this discord between being and existence. Therefore, a man modifies his world, arranges and organizes it in agreement with the objective exigencies of the mind.

A woman, on the other hand, according to Elzenberg, perceives the world subjectively, i.e. she sees it through the prism of personal reference. There is no conflict between the spontaneous subjectivity of every soul and the objective way of seeing and evaluating the world. The disturbing element, that is, the objective look at herself "from outside", is not present in a woman's thinking. She perceives reality only from her own perspective, because of her dominant position. The subjective thinking of a woman, inspired by intuition, necessarily leads to the modification of the world so that it accord with her personal perception. She modifies her world, arranges and organizes it in agreement with the subjective exigencies of her soul.

Culture, according to Elzenberg, is a sum of realized values. The two varieties of subjectivity found in feminity and masculinity display differing attitudes toward the world of culture. A human being thrown into the real world can chose two totally different ways of thinking and behaving: either he or she can accommodate reality (here we are not dealing with culture) or attempt to change it. A man modifies his world, arranges and organizes it in agreement with the objective exigencies of the mind. He tries to rise above his individual existence. On the other hand, a woman, using her subjective thinking, inspired by intuition, necessarily modifies the world in accordance with her personal existence. A man creates the culture of objective values, a woman creates the culture of subjective and utilitarian values: what will be important for her is what brings profit or pleasure. In confronting the world a man and a woman by nature behave differently: feminity, strongly connected with individual reality, balances a masculinity inclined to escape into the sphere of abstraction. In this way, a woman is necessary link between a man and the world.

Two kinds of human nature and two kinds of culture are postulated in Elzenberg's research. Feminity and masculinity are both incomplete. It is only in love that each complements the other and in being complemented becomes a full human being. The power of sexes, important in every sphere, is particularly manifested in love. Let us reiterate what we said before: it is possible to love a woman and hold her in contempt as a human being and there is no discrepancy between the two facts. We are only referring to the different spheres of anthropological nature of a woman. Only a woman, as a representative of the opposite sex, can be loved by a man. That plot evolves as the main part of the Elzenberg's polemics with Schopenhauer. Both, Schopenhauer and Elzenberg treat love and its power to influence human life very seriously. Schopenhauser beholds the source of this power in the biological will to survival of humankind. According to him love reduces to sensuality, which serves only to lead a human being towards the act of procreation and is, as a matter of fact, the voice of animalistic side of human nature. Elzenberg does not accept this view. He claims a human being can sacrifice life for love, thus acting against its own biological imperative. Instead of giving life, via procreation, he may first deprive himself a life.

According to Elzenberg, love does not end in the sensual. It goes deeper – even below the psychological level of desires and needs. It is obvious that a human being desires closeness, safety, and love sometimes does fulfill these desires. Sometimes, however, it does not fulfill anything. According to Elzenberg love is something more: it gives sensual pleasure, it can comfort and answer needs, but there is also something deeper to it – *the mysterious attraction between the two sexes.* It is a primary inclination, rooted in the nature of humankind, always changeable depending on sexuality. This conception of love we can fall androgynous. Femininity and masculinity are both seen as incomplete. It is only through love that man and woman complement other and each becomes a full human being.

It is interesting to see how Elzenberg's "woman's nature" is connected to the different theses advanced in today's feminist movement as formulated above. As was mentioned before, Elzenberg believed that both natures are equivalent. According to him, despite the differences, neither of the human sexes is better or worse than the other. They are complementary and only together do they create the full human being, only together do they have all the possibilities of developing destination of humankind. Although equivalent to a man, a woman, however, is totally different from man and a distant pole of humanity. It is not possible to achieve complete communication between man and woman. According to Elzenberg, manliness starts where womanhood ends.

The above-described point of view has a specific character, i.e., it betrays a typical masculine way of perceiving. Woman is seen through a man's eye. And she is described only in terms of the interest of man. Trying to describe womanhood, a mystery from a man's point of view, Elzenberg stresses the differences: in ways of seeing the world and in attitudes towards reality. It is hard to say whether this enquiry was supposed to be self-standing or was made only during his analysis of love and so but partially concerned with femininity. Perhaps "feminine thought" as a theme was only part of a larger enquiry into the phenomenon of love. The typically masculine philosophical-anthropological approach taken in Elzenberg's research shows the elementary dilemma in all discussions regarding woman. Elzenberg, like everybody else, is at once a judge and an advocate of a certain position. Maybe this argument, detached from the service of interests, would eliminate the gap – as with any description this one also requires some perspective. But in this case the boundary between a man's feelings (where a woman is the source) and the description of a woman's nature often gets blurred or erased. Let us try to explain this: a man feels anxiety and only a beloved woman can soothe this anxiety. Elzenberg concludes too much too easily about a woman's soul on the basis of men's experiences and feelings. A woman can bring about calm and peace, but this does not mean that she herself and in-herself is an epitome of calm.

Elzenberg's analysis of the sensual, psychological and spiritual aspects of love may inspire present day discussions about love. Elzenberg's thoughts on human nature are still alive and valid, despite the fact that they are from the beginning of the century. The two kinds of subjectivity that are femininity and masculinity display different attitudes toward the world of culture. Man creates a culture of objective values; woman creates a culture of subjective and utilitarian values. But femininity and masculinity are both incomplete. They complement each other, and so eventually a man and a woman can become full human beings. Perhaps this way of thinking can also inspire present-day feminism, even if the description of a "woman's nature" here is problematic for the representatives of that movement, for the level of feminist interpretation at the end of twentieth century very often falls short of Enzenberg's analysis.[8]

Political Philosophy Department at Institute of Political Studies
Polish Academy of Sciences
Warsaw

NOTES

1 H. Elzenberg, *Kłopot z istnieniem* (The Trouble with Existence) (Krakow: Znak, 1963).
2 Elzenberg papers, *Archives of the Polish Academy of Sciences* in Warsaw, number III/181, position 44. We are thankful to Mr. Bogusław Wolniewicz and Mr. Jan Zubelewicz. It would have been impossible to find and work up some of the materials without their help.
3 Elzenberg Papers.
4 *Ibid.*
5 *Ibid.*
6 *Ibid.*
7 *Ibid.*
8 This subject is also covered in my article, "Antropologia płci Henryka Elzenberga (Henryk Elzenberg's Anthropology of Sex)", *Teksty Drugie* 3/4 (1995), pp. 113–132.

PART II

THE VARIATION OF ONTOPOIESIS

A group of participants.

GABRIELLA VALACCA

AUTOPOIESIS IN THE ORGANIZATION OF LIVING PHENOMENA

A Comparison Between Heteronomous and Autonomous Systems in Cognition

The theory of autopoiesis[1] was developed by the Chilean biologists Humberto R. Maturana and Francisco J. Varela. Starting from their research on the organization of living phenomena, they endeavoured to extend their investigation to situations lying beyond the boundaries of classical biological disciplines with regard both to methodological issues and the problemic aspects of theoresis.

The present paper deals with the essential aspects of autopoiesis as a systemic theory which investigates the basic characteristics of living systems. In particular, we aim to clarify the extent to which the theory in question provides insights into our understanding of organization and behaviour in psychology. In this sense we shall develop several observations on the theory and the method adopted by Maturana and Varela to analyse living systems and to extrapolate from certain aspects rather than others.

SEVERAL ESSENTIAL ASPECTS OF THE THEORY

The concept of autopoiesis is part of the search for a general theory of systems and makes use of several definite reference to cybernetic research aimed at characterizing complex systems in non reductionistic terms with regard to the problem of autonomy. As their general premise the Authors focus principally on "understanding the organization of living systems with regard to their unitary character".[2] In this sense the general theory of systems allows research to be performed in *metasystemic* terms, in which the living system is viewed as a definable set of components linked by relations (interactions) that characterize their constituent topological properties. These properties are therefore qualitative and, as such, are independent of the measurement of separate characteristics of living phenomena.

A living system is described as an *autopoietic machine*, i.e.,

A-T. Tymieniecka (ed.), Analecta Husserliana LIX, 167–172.

a machine (defined as a unit) organized as a network of processes for the production (transformation and destruction) of components . . . which . . . through their constant interactions and transformations regenerate and realize the network of processes (relations) that produce them.[3]

The Authors focus on the following essential aspects of the system.

(a) The first aspect is related to the very concept of autopietic system, defined as a self-referential system that produces and reproduces by itself the elements of which it is composed. An autopoietic system possesses the property of autonomy and cannot be reduced to any model, however complex, intended to represent it.[4]

(b) "Living systems are cognitive systems, and living, insofar as it is a process, is a process of cognition",[5] in that it is the computation of the behaviours that allow the organism to survive in its environment.

Furthermore, survival "is not mere conservation. The system may undergo changes as a result of its interactions: the important thing is that these changes should not detract from its capacity to enter into further interaction".[6]

(c) The autopietic system is characterized by an *organization*, i.e., by relations which define the system as a unit and determine the dynamics of interaction and transformation which the unit as such may undergo.[7] The interactions with the environment are characterized by computation mechanisms in keeping with the self-referential method, which involves the "closure" of an organization defined as the application of an organization to itself. This means that the result of an operation is again subjected to the same operation (recursivity).

(d) "The actual components (including all the properties) and actual relations . . . that materially realize a system . . . go to make up its *structure*".[8] This means that "the structure is the place in which the living system implements its autopoietic organization in the physical space".[9] While the notion of organization indicates the invariant relations that necessarily characterize all living systems of a certain type, the notion of structure indicates the relations characteristic of a specific living system at a given time.

(e) The way in which a living system adapts to its environment (*medium*), thus, avoiding disintegration (that is, maintaining its organization) represents a *structural coupling* between unit (system) and medium. "The notion of structural coupling thus captures the fact that each living system is basically conservative in its interactions with its medium and therefore also with the other living systems with which it comes into contact".[10]

OBSERVATIONS ON THE THEORY

One essential feature of autopoiesis regards the relations that take place between the organization that specifies the general characteristics of the living system and the structure denoting its particular characteristics in a given situation in time and space. The authors give priority to the characteristics of organization wherever the particular features of implementation in reality are valid only on the strength of those possibilities of interaction with the medium within which the organization characteristics remain *unchanged*. In this sense the autopoietic system is a closed system in that it reproduces itself and cannot be described in terms of input-output relations vis-à-vis the medium in a general sense. The concept of constraint characterizes the organization of the system insofar as it denotes its *invariant* relations. While the structure of the system is related to the actual material medium in which, and only in which, situations of system *uncertainty* are envisaged.

The Authors do not explicitly describe the limits imposed by the autopoietic organization on the interactions with the medium. Nevertheless, one is aware of the impossibility of any general theoretical approach to the problem of possible limiting interactions jeopardizing the very organization of the system. Consequently, the description of a system structured in a given medium the behaviour of which is directed towards its own survival is reduced to a form of tautology that, although, on the one hand, justifies the autopoiesis at the level of a general definition, on the other, is not concerned with the phenomenology of the system, defined as a particular unit located in a given situation in time and space.

THE CONTRIBUTION OF COGNITIVE PSYCHOLOGY: DIFFERENT INFORMATION PROCESSING MODELS

With regard to cognitivism, autopoiesis, insofar as it is based on Maturana's idea that cognition is a biological phenomenon, reframes in radical terms the conception of living systems as systems that process information and that can be described on the basis of processes involving computations, transformations, storage, retrieval, recognition and other forms of sensory input.

In particular, in the nineties, cognitive psychology developed the research on this in two basic directions: a) the study of internal states and of the representation of knowledge, and b) the architecture of self-organizing systems in natural and artificial intelligence (AI).[11]

In brief, it may be asserted that both in AI and in cognitive psychology, a given field of knowledge may be coded into a program by means of representations (facts, declarations, analogic images, etc.) fixed in the memory or else coded in the form of a set of procedures or actions to be carried out. The main difference between the two types of coding lies in the fact that coding by declarative representations involves the storage of information (knowledge) which is memorized in suitable registers and recalled when needed. Coding by procedures, on the other hand, does not require any information retrieval from the higher levels of the system and implies a bottom up (i.e., data driven) functioning of the system. We are thus looking at two different models of knowledge organization. The first consists of *knowledge by representations*, which involves the so-called central processing of information and implies computations, categorization and symbolization of the information as well as comparisons between experience stored in memory and current experience. This type of mind architecture is based essentially on a logical-symbolic paradigm. The second model consists of *knowledge by procedures*, which utilizes the immediate processing of input. The conceptual referent of this second model is a sub-symbolic paradigm.

The mind architectures based on logical-symbolic paradigms are in many instances comparable to autopoietic systems and, like the latter, refer to a *more or less abstract knowledge* (involving concepts or categories), and in any case to knowledge that is always *constructed* by means of information handling processes, and require the input to be ordered in suitable registers according to criteria specific to the system.

While these competencies, on the one hand, justify the representations (classification, categorization, comparisons), on the other hand, they point to a possible distinct separation between the real world and the mental world, to the extent that it becomes practically impossible to recover the phenomenology of the living system through the contribution of these processes alone.

The logical-symbolic paradigm thus implies having to accept at the theoretical level an inevitable dualism between abstract (general) knowledge and empirical (specific) knowledge. Any resolution of this dilemma certainly does not go in the direction of reducing, let alone cancelling out, the meaning and the contributions of the abstract knowledge, but rather in that of providing some more precise indications of the (genetic and factual) modalities of generalization itself.[12]

Furthermore, it must be pointed out that

meaning is not significant for information. In computational terms, information contains a message that is precoded within the system. . . . Information processing cannot handle anything beyond a set of data input in a well-defined and arbitrary fashion. . . . Such a system cannot cope with indetermination, polysemy, and links of a metaphorical and connotative nature.[13]

Conversely, since the theories of the mind founded on the sub-symbolic paradigm enhance immediate input processing, they presuppose an *empirical and practical type of knowledge*,[14] namely, a knowledge directed towards the realization of a given type of action by the perceiver of the stimulus. In this regard, therefore,

one accepts from the outside the existence of particular constraints among processors and certain stimuli. In this sense, the functional validity of the processing at first sight lies in a behavioural context that can be represented by means of a systemic model for the study of interactions between the organism and the environment and the states of dynamic equilibrium of the system itself.[15]

Such a systemic model is not autopoietic, as it is not self-reproducing. Even though it possesses self-organizing properties its conduct depends on the factual and instrumental context of its genesis and its evolution.

Rome

NOTES

1 The term was used for the first time by Maturana and Varela, in H. R. Maturana and F. J. Varela, *Autopoiesi e cognizione. La realizzazione del vivente* (Venice: 1985 [1st ed. in Spanish, 1980]).
2 Maturana and Varela, *cit.*, p. 127.
3 Maturana and Varela, *cit.*, p. 131.
4 Cf. Maturana and Varela, *cit.*, p. 133.
5 Maturana and Varela, *cit.*, p. 59.
6 A. Preite, Autopoiesi: elementi per un'analisiv. Comunicazioni Scientifiche di Psicologia Generale 5 (new series), (Naples: Edizione Scientifiche Italiane, 1991), ("Sistemi ad auto-organizzazione"), p. 11.

[7] Cf. Maturana and Varela, *cit.*, p. 201.
[8] Maturana and Varela, *cit.*, p. 33.
[9] A. Preite, *cit.*, p. 12.
[10] G. De Michelis, "Prefazione" to Maturana and Varela, *cit.*, p. 14.
[11] Cf., "Sistemi ad auto-organizzazione," Comunicazioni Scientifiche di Psicologia Generale 5 (u. 1), (Naples: Edizione Scientifiche Italiane, 1991).
[12] G. Valacca, *I processi modulari: un confronto fra teorie ed alcune evidenze sperimentali*, Comunicazioni Scientifiche di Psicologia Generale 8 (new series), (Naples: Edizioni Scientifiche Italiane, 1992), p. 72.
[13] J. Bruner, *La ricerca del significato* (Turin: Bollati Boringhieri, 1992), (1st ed., 1990), p. 22 (cited in the Italian edition).
[14] In this connection the remarks by J. J. Gibson concerning the ecological approach to perception are particularly significant. See J. J. Gibson, *The Ecological Approach to Visual Perception* (Boston: Houston Mifflin, 1979).
[15] G. Valacca, *cit.*, p. 71. For an exhaustive treatment of the systems approach based on von Bertalanffy (1968) and its applications to psychology, see M. Olivetti Belardinelli (*La costruzione della realtà come problema psicologico*, Turin: Boringhieri, 1986) who has provided the most complete formulation of a "systemic" psychology so far available. Cf. also E. Pessa, *Intelligenza artificiale: teoria e sistemi* (Turin: Bollati-Boringhieri, 1992), pp. 190–203.

HORST MATTHAI

LIFE AS EONTOPOIESIS[1] AND SELF-INDIVIDUALIZATION

The inversion in the title above of the third general theme for this conference given in the Call for Papers is intentional and has as its purpose the destruction of age old prejudices. It is in the nature of man to inquire about the world in which he finds himself, as well as about his own presence therein. Aristotle expressed it in a clear manner at the beginning of his so-called *Metaphysics* or *First Philosophy*. Παμτες ανθρωποι ειδεναι ορεγονται φυσει[2] is the sentence at the beginning of its first book. Few translators perceived the underlying meaning of these words, translating "All men by nature desire to know."[3] However, το ειδεναι does not express, at least not exclusively, the idea of *to know*, but, as we see in Passow, also that of *to wise*.[4] Now, the English language not being very helpful in our case, nor the German dictionaries which treat *wissen* and *kennen* as synonyms, we find in Heidegger a clear distinction between these verbs: "Thus is distinguished the never surpassable difference between common knowledge and observation on one hand, and true wisdom on the other."[5] Hegel, though not using identical terms, expresses the same idea: "When we speak of thought, we must distinguish between finite thought, which is mere understanding, and infinite thought that is rational."[6] We hold that two thinkers of such renown arrive at this distinction only after the most profound reflection, not doubting that an Aristotle did likewise, the moreso if we take into account his famous separation of sciences that explain the *what* of reality from the science of its *why* or *causes*. Our translation, consequently, should be: "All men by nature tend towards wisdom."

Now we all know that Aristotle, following the path of causes, arrives at the impass of infinite regress, which leads him to the assertion of a primal, but also final cause, which is God. Plato, his predecessor, expressed, at least implicitly, a similar view, when he held that the creator, wanting to build a world as similar as possible to himself, had created only one world. Here we see that both philosophers, although belonging to a tradition of polytheism, had in fact developed a theory of monotheism, which made them most welcome guests to later Christianity, one to Saint Augustine, the other to Saint Thomas. What had been lost,

A-T. Tymieniecka (ed.), Analecta Husserliana LIX, 173–188.

however, was the polytheism of the archaic Greek thinkers, ignored in its deeper meaning even in their own time. Of course here we do not refer to the rather crude popular polytheism of ancient Greece, criticized by, for instance, Heraclitus when he says in B 5 that praying to statues is like conversing with buildings. What the Ephesian seems to expose here is what we observe in many cultures, where the most sublime creations of the human spirit are eventually reduced to metaphors accessible to the less brilliant minds of the general population. But, as we know from history, the pristine conceptions of reality were kept alive and transmitted to future generation through initiations kept secret from the unworthy.

Godhead, as something immanent in all things, was one of many truths held by all ancient thinkers. There is an infinite number of gods in the world, Thales of Miletus already sustained. Anaximander, his disciple and successor, amplified this by affirming an infinite number of heavens, each of which contained an infinite number of worlds. He called both of them gods. That such godhead also belonged to humans, was held by most of the ancient cultures, as in the case of India where the Brahmans attributed divinity to transcendent individuals who attained higher levels of spirituality,[7] or even Plato, who in his *Symposium* developed his theory of love as the primal mover, but allowing for the presence of such love only in individuals in what he called a state of enthusiasm.[8] Now enthusiasm derives from εν and θεος, that is the god within us, obviously not understood by Plato to mean a separate god in each of us. However, that is precisely how we should understand it, in the sense of presocratic philosophy. We must not forget, that Plato – as well as Aristotle later on – did not grasp the depth of the archaic thinkers, as he himself admits.[9] Therefore, let us not follow a tradition that has shown its incapacity to solve the human problem, as Heidegger clearly saw,[10] but reexamine the basic tenets as they appear among the archaic thinkers.

Life was expressed in ancient Greece by the terms ζωη and βιος. Of these βιος generally denoted the duration and quality of life, while ζωη signified life as physical existence.[11] Of these we are interested for our subject only in the latter. Plato seems to have been the first to identify life, ζωη, and movement,[12] a conception that was strengthened by Aristotle to mean continuity and eternity.[13] It is true, that the Stagirite assigns these aspects to God, but, as already explained, by then human consciousness of the existence of an infinite number of gods, that is polytheism in its deeper sense, had already been lost. For us, therefore, the qualities assigned to the one supreme God by Aristotle, have to be allotted to each individual. Now, individual, or its Latin counterpart, *individuum*, equals the Greek expression of ατομος, *atom*, what cannot be separated, what is indivisible. Indivisibility implies infinity, which in turn excludes any notion of magnitude,[14] so that an individual or atom is as well great and small, as Anaxágoras already said.[15]

We saw before that for the ancient thinkers life is movement. However, the real nature of movement has been lost since then. The Latin *movimentum* disappeared and was substituted by *momentum*,[16] our present moment. However, the term *momentum* still signified movement, as an immanent force,[17] while our present-day moment is limited to *a minute portion of time, an instant*.[18] It is true that in mechanical engineering *momentum* is still in use as a technical term, meaning "... the property of a moving body which determines the length of time required to bring it to rest when under the action of a constant force or moment."[19] This definition – leaving aside the *bringing to rest*, which from a philosophical standpoint is, of course, impossible, there being no absolute rest – still keeps alive the identity of force and moment. Now, according to Georges, *momentum* carries two connotations; one expressing the length of movement in space, the other the duration of movement, *generally with the conciseness of brevity*.[20] The German *Begriffsprägnanz* or conciseness, however, derives from *prae* and the ancient *gnascor*, *what has been born before*, implying the priority of time over space. But time, once conceived as brevity, is not any arbitrary brevity, such as a day, a minute or the fraction of a second, but absolute brevity, which would be *nothing*, or *punctual time*, as Bielawka sustains.[21] Applying the principle, already mentioned, of each individual or atom being simultaneously great and small, we can then hold their mutual inclusion, in the sense already suggested by Leibniz.[22]

Marx, in his doctoral dissertation of 1840–41, quotes from Pierre Bayle:

> Avant lui, (c.-à.-d Epicure) on n'avait admis dans les atomes que le mouvement de pesanteur, et celui de réflexion. (. . .) Epicure supposa que même au milieu du vide, les atomes déclinaient un peu de la ligne droite, et de là venait la liberté, disait-il. . . .[23]

We accept Bayle's interpretation of Epircurus' famous saying, which concurs with Plato's identification of movement and life. As far as Marx is concerned, however, it seems strange to us for him not to have exploited the obvious implication in the saying of Epicurus, in the sense that atoms have a will of their own – or liberty in the words of Pierre Bayle – the more so since Marx himself sustained that man is the creator of his world[24] and nature his inorganic body.[25] Of course it is quite possible that Marx – who tended towards materialism, such as he eventually developed after his encounter and collaboration with Engels – wanted to avoid the potential indeterminism implied in Epicurus' assertion, but not so others, as for example Leibniz, who at a certain occasion remarked that some monads, present in the coffee he had just sipped, might some day become humans,[26] or Teilhard de Chardin, who considered the initial quantity of consciousness, contained in our mundane world, to be present in what he calls *infinitesimal centers*,[27] in other words: *atoms* or *monads*.

It must be kept in mind, however, that Teilhard de Chardin, being in a way what we would call a prisoner of conventional science and, of course, his theological commitment, even after admitting the presence of consciousness in the particles (*Teilchen*) of the Universe, still does not identify consciousness and life, calling the passage from molecule to cell a leap *towards* life – whatever such an expression might mean. Where conventional science, or theology, is not the determining factor – as should be recognized by all responsible investigators[28] – the recognition of the presence of life in all matter is simple logic. Maybe in our days theosophy does not merit much credence with its affirmation of the presence of life in minerals, or even postulate what they call elementary life;[29] nor does Chinese taoism, wherein we find the original *One* as life,[30] life that is one with the forces of heaven and earth.[31] However, Aristotle's *living God*[32] – that eventually found entrance in Christian theology – was, for him at least, the primary cause of all that exists, though he assigns to him not only life and eternity, but also *continuous* always-ness. The text is as follows:

Φαμεν δη τον θεον ειναι ζωιον αιδιον αριστον, ωστε ζωη και αιων συνεχης και αιδιος υπαρχει τωι θεωι· τουτο γαρ ο θεος.

Translation: Truly we hold the god to be a living being, eternal, most good, so that life and continuous always-being and eternity are foundation and beginning for him: For this is the god.

Given the importance of this passage, we shall analyze it carefully. We start with the term ζωιον, the termination of which suggests the gerundial denotation of the participle of ειναι, to be. Now, we all know that the Greek language does not have a gerund, properly speaking; still Kühner-Gerth[33] hold that the participle with regard to its *meaning* corresponds to the gerund of other languages, which, as in the case of Latin, expresses action,[34] but an action of the verb, as Webster notes, ". . . as generalized or in continuance (*ars vivendi*, the art of living)."[35] ζωιον, then, is not only a *living being*, but is such in continuance, that is, as shall be explained later, in *separateness*.

αιδιον, according to Passow, derives from αει, always, which for the ancients meant *separate* repetition, as we can show in Homer who uses αιδιον in this sense.[36] Taking into account that αιδιον also denotes eternity, we now find that eternity or αιδιον is not, as presently surmised,[37] an infinite duration or endless time, but, as we shall show later, a simultaneous presence of an infinite number of *nows* or moments.

αιων, a noun, equally derives from αει, but directly referred to life, as was apparent in the ancient Sanskrit *âyu*, life,[38] or in Homer who uses it to express life's force.[39] Let us remember in this context that the term moment, as said earlier, implied an immanent force, so that we can now connect αιων's second connotation, eternity, with the idea of *separate* moments,[40] since life's force is fully present in each one of such moments, it being inconceivable that such force were distributed among the totality of moments lived through a lifetime, devolving only a tiny fraction to each one of them.

Now, the Greek συνεχης – the last term we want to discuss – from συνεχω is homologous to the latin *continuus* from *contineo*, both verbs meaning *to hold together*.[41] To hold together, however presupposes the *separateness* of whatever is being held together, which in our case is time. This is so much more important, in as much as time and life are identical, as we saw in the ancient thinkers' life being the same as movement, and movement or moment being the pristine *presence* of time. Now, *present* derives from *prae-sens*,[42] the being *before [in time] something or in front of something*, which in our case employing Parmenides' language[43] – would be *being now, simultaneous, all, one and continuous*. Many translate ομου with *together*, implying a spatial nearness of separate entities forming a totality, παν. In that case, however, ομου would be synonymous to συνεχες, that is *continuous*, violating the *laconic* style, which Plato found to be the common denominator of all the archaic thinkers.[44] We, on the other hand, prefer ομου's temporal connotation of *simultaneity*, which would mean that there are a multiplicity of nows, as a matter of fact, an infinite number of them, and all of them, being at the same time, not only form the all, (το) παν, but being is, thus conceived, also *the one*, εν, and continuous, συνεχες.

With the help of Parmenides and Aristotle – the latter most probably would not approve of our interpretation – we have now established the identity of time and life; however, time being simultaneity – that is, something radically different from our ordinary concept of time, i.e., past, present and future –, we are compelled to use a different linguistic symbol for its designation. The term that seems to us most appropriate is that of *primordial time* or *Urzeit*, as suggested by van der Leeuw.[45]

It is strange that the idea of simultaneity, applied to time itself, were lost since ancient times. The presocratic philosophers incorporated it in their writings, although, as already pointed out by the ancients,[46] they employed an enigmatic way of expressing themselves. Where we find an explicit mention of primordial time in the old days is in Plotinus in his reference to the ancient god Kronos who, according to mythological tradition, swallowed his progeny as soon as it emerged,[47] that is, Time swallows its *moments* as soon as they emerge, not allowing for a past or a future.

A most fascinating example of primordial time in modern physics, we find in de Gortari's reference to subatomic physics, where he says:

> . . . each elemental particle fills by itself alone the whole system to which it belongs and, consequently, has simultaneously a multitude of special and different locations. Such property of occupying many positions at the same time is the *polytopy* of the elemental particle.[48]

Further on his text says:

> On the other hand, since polytopy is a common property of each and all of the elemental particles, it follows that they also have the property of *homotopy* with regard to any location contained within the system to which they belong, since each location is simultaneously occupied by all of them.[49]

Extending de Gortari's idea to reality in general, we can then assert not only *polytopy* and *homotopy* with regard to all entities or beings, but also their *polychrony* and *homochrony*. When de Gortari introduces the first two of these terms, he clearly notes that their assertion places us *outside common sense* and *classical physics*.[50] But he never explains what he means by *simultaneity*, except referring to it as *at the same time*. Time however, is perceived here as *ordinary* time, since otherwise he would have given a detailed account of it. In other words he takes it for granted that the occupation of a single location by all beings, or the occupation of all locations by any single being, happens in one and the same time for all, for which reason he did not have to include it as something that is *beyond common sense and classical physics*. Now, physics have been conditioned along the lines of the Einsteinian notion of the space/time continuum, the very term clearly suggesting the primacy of space with respect to time. As a matter of fact this is a problem that has plagued many modern thinkers, such as, for instance, Heidegger who admitted in his own words that his *Being and Time* failed, due to the inadequacy of the language employed. We all remember that Heidegger tried to access time through being, admitting later, however that what he called the *Kehre*, the return, would be the return of our thinking from *Sein und Zeit, Being and Time*, towards that of *Zeit und Sein, Time and Being*.[51] Although Heidegger does not explicitly say so, we interpret this to mean that for him we cannot access time through being, but only being through time which, taking into account that he speaks of a *return*, implies the return to something more pristine than the previous starting point of his philosophy. However, Heidegger never achieved this return, nor could he, in their famous Heraclitus Seminar, follow in the steps of Fink,[52] for whom the heraclitian fire was the time that produced the *was*, the *is*, and the *shall be*, that is, a punctual time from which brusquely emerge or break out – in Fink's words – the three dimensions of empirical time.

Now being, as noted above, is not an adequate linguistic symbol with which to represent both metaphysical and ontic reality. The word is used to express what we might call *be-ness* – (το ειναι, *l'Etre, das Sein* or *el Ser* in other languages), which belongs to the realm of metaphysics, and *being* (το ον, *das Seiende, el ente*), which is the expression of ontic reality. Unfortunately, many translaters render both, το ειναι and το ον, as *being*, as we can appreciate in reviewing selections from Aristotle's *Metaphysica*:

1. 993b31 . . . ως εχει του ειναι, ουτω και της αληθειας
   . . . so that as each thing is in respect of being, so is it in respect of truth
2. 1018a35 . . . επει δε το εν και το ον πολλαχως λεγεται
   . . . since "one" and "being" have many senses[53]

But to translate both terms indiscriminately as *being* seems to us to be a gross inaccuracy, the moreso in the absence of explanatory footnotes. However, our problem becomes even more complicated, once we notice the difference between the Ionic εον and the Attic ον. This remarkable divergence is brushed aside by most commentators as just another one of the many phonetic discrepances that exist in the various dialects of a language. But this case is vastly different from such minor divergencies in as much as the short ε in εον shows the pristine presence of be-ness in this *substantivated* participle. It was Heidegger who contrasted the ον and οντα in Plato and Aristotle with the εον and εοντα in Parmenides and Heraclitus, seeing in the ε of these forms the root of the verb *to be*, ειναι, of the εσ in εστιν, as well as in the Latin *esse* and *est*. It is Heidegger's conjecture[54] that in Plato's or Aristotle's ον and οντα we deal with corroded forms of the pristine words of εον and εοντα. Moreover, where Parmenides[55] and Heraclitus,[56] as their Milesian predecessors had done, still thought an infinite number of worlds, and consequently, at least explicitly in the case of Parmenides (B 3), identified thought and be-ness[57] – concretely possible only in the individual, *qua* incarnation of his own and private world – Plato held the ον to be a single world, because for him there was a creator, ο ποιων, who wished to make a world as similar to himself as possible, so that he, being only one, also created only one world.[58]

Not feeling bound by the Platonist and Aristotelian tradition (perpetuated until today through Christian theology, as already observed by Heidegger), however, we can proceed to view reality in its proper dimension which is infinity. There is an infinite number of worlds, each corresponding to an entity that exhibits two basic qualities, i.e., individuality and movement as manifestation of life. Of these two, individuality, as we saw above, implies infinity as well as totality, either one of which is self-individualization. It is self-individualization, because each entity discovers in all the others its own being, since all the others are but an infinite variety of the former. Leibniz exposes this in his *Metaphysical Treatise*:

> That each singular substance expresses in its own manner the whole totality and that in its concept are contained all events with their circumstances and the whole sequence of the exterior things.[59]
>
> . . .
>
> And since all the other substances express this [singular] one in its own manner and adapt themselves to it, one can say, that it extends its power over all the others in imitation of the omnipotence of the creator.[60]

Now, when Leibniz speaks of substances he refers to what he calls monads in others of his works, and the extension of the power of a substance or monad over all the others implies the use of its free will (cf. n. 23), that is, that each monad is the creator of its own world. This was also clearly seen by Schopenhauer – for whom *will* was the supreme principle – as we can infer from the following quotation:

But this multiplicity (of individuals) does not affect it, the will as thing in itself, but only its appearances: it (the will) is totally and undividedly present in each one of these and views in its surroundings the innumerably repeated image of its own essence. This essence in itself, however, that is, the truly real, it finds immediately only in its interior.[61]

Even Marx, whom we cannot easily classfy among the idealists, conceived man – for Marx collectively speaking, of course – as one who views himself in a world of his own creation.[62]

The very idea of each entity, atom or monad creating its own world, not in ordinary time, but in primordial time, in the *now* already previously discussed, requires a radical change in our way of thinking. As Spengler already noticed,[63] the Germanic peoples tend to think in a vastness of infinity, and we hold that the *leap* – to use a Hegelian expression[64] – from the dimension of the finite towards infinity is *the* condition for a cosmovision that allows us to conceive of life as something without beginning and without end. Just as anthropologists find what some call the *horror of incest* in many of the primitive societies, so does man in general abhor the notion of absolute *lability* – that is complete unstableness. Having lost faith in himself – depositing it in a transcendent being devoid of any rational foundation, his cowardice impels him to cling to anything that promises a feeling of security. Even a thinker like Archimedes looked for a fixed point in order to give meaning to his action, said point being, however, spatial.[65] Had he only known that the crucial point is the point of *time*, the Parmenidean *now*, the καιρος[66] of the ancient thinkers, maybe mankind would have avoided the postulate of a *giver of life*, that eventually was to become the scourge of mankind. Ancient Tibetan wisdom already held that man is his own ancestor,[67] who gives life to himself anew in every instance. There is nothing outside of life, outside the *now* we are living in each instant. There is no past and no future,[68] clinging to them is to deprive the present of its infinite potential.

Now, when speaking of ontopoiesis we must first address the problem, already mentioned, of the difference between the archaic εον and what Heidegger called the corroded later form of ον, both of them participles of the verb ειναι, to be. In our time it appears to us appropriate to speak of this verb as an auxiliary verb, raising the question, however, of whether this was always its function. When we say, for instance, that somebody is a teacher, it does not occur to us anymore to ask ourselves whether he really *is* a teacher. The same may be said of any other subject like whether a McDonald's Hamburger *is* food fit for human consumption, that is, devoid of detrimental ingredients and, consequently, of nutritional value to further our physical and mental well-being. In other words, do we use the verb *to be* as an expression of the *be-ness* of something, or only as that of an accidental attribute of a subject, as in saying: a teacher is a member of a bowling club, or a McDonald's Hamburger is a sign of modern life in some countries?

When speaking of ontopoiesis we should, in view of the aforesaid, replace this concept with that of *eontopoiesis*. The archaic being, the εον of the presocratic philosophers, was identified with thought by Parmenides and later named *mind* by Anaxágoras, who considered it (B 12) as something separate from all things and without limit, though dominating all of them. We shall refer later to the subject of limit, while in the meantime we can explore the relation between εον and εοντα, that is the use of this term in the singular or plural. For Parmenides εον is that something which can only, in its strictest sense, be designed by the expression εστιν or *is* – emphasized through the acentuated first syllable (B 2.3) – even though elsewhere (B 3) he refers to it as ειναι, that is *be-ness*, in the sense of the German *Sein* or the French *Être*. As already stated in our essay on *Parmenides' theory of thought*,[69] most scholars amend the original text by inserting a subject implied, according to them, in the εστιν: Diels opts for *IS is*, Burnet for *IT is*, others, like García-Bacca, for *of Being is be-ness* or, in the case of the Academy of Sciences of the USSR, *Be-ness is*. However, the original Parmenidean *is* is not the *is* of this or that subject, but *the* all-inclusive *is-ness* devoid of the need of any specification whatsoever. Once this premiss is accepted, all the other *is-ses*, as Schopenhauer had already said, are but its own infinitely repeated image, that is, the εοντα of the ancient thinkers. Still, each εον, that is, each one of this infinite number of εοντα or be-nesses, also contemplates in all the others *its* own image, so that each one creates its own world, the eontopoiesis mentioned before, while simultaneously establishing its own and exclusive individuality. It may be difficult for contemporary man to accept that, as already said, each time that an *is* or εον discovers in another its own image, this implies the giving of life to same, but that precisely is the meaning of an ancient Indian saying:

From the substance of this *supreme soul* there escape, *like sparks of fire*, innumerable vital principles, that incessantly communicate movement to the creatures of diverse order.[70]

Be it noted that the expression *creatures of diverse order* refers to *all* creation, i.e., organic *and* inorganic, reconfirming Plato's identity of life and movement, as well as the presence of life in all things as held, among others, by the theosophists.

Finally, and extending to mankind as a whole Heidegger's admonition that the fate of the Occident depends on the correct translation of the archaic εον,[71] we can now return to the problem of limit, referred to above. Let us imagine that a circle is traced on a piece of paper which we then cut out carefully with a pair of scissors. As a result we obtain two circles, one forming the outer edge of the cutting, the other forming the inner edge of the left-over paper. Leaving aside the question – which should be answered by the mathematicians – whether the circle being one only, now belongs to the cut-out or to the left-over paper, *or to both*, we see the circle as a limit. In the case of the cut-out we obtain an *inclusive* circle, but in the case of the left-over paper an exclusive one. Now, when Parmenides speaks of the *is* as something similar to a well rounded sphere (B 8.42–44), we have but the tridimensional extension of a circle. If we were to carve out of a solid block such a sphere without hurting the rest, we would have, in an analogy to our example of the circle, an inclusive sphere and an exclusive one. But Parmenides' sphere has an *ultimate*[72] limit, for which reason it is all-inclusive, comparable to a sphere of infinite extension, from which nothing could be excluded. However, this all inclusive sphere *is*, in the words of this ancient thinker (B 8.5–6), in the[73] *now simultaneously all, one, con tinuous*. For Parmenides, therefore, *only in the now*, that is in punctual time, do the separate *is-ses* – creations of the original *is* in the form of an infinite number of self-images – conjoin as *the all* and also as *the one*.

Given the identity of thought and be-ness, as held by the Eleatian (B 3), our infinite all-inclusive sphere then is the εον, accessible only through infinite thinking, the realm of wisdom. It also is eontopoiesis, life's creation – thought being the supreme form of movement – and, given the sphere's infinity, life's self-individualization. The exclusive sphere, on the other hand, limited towards its interior and, therefore, finite, holds ontic reality in the form of experience – ex-perience signifying what we perceive *outside of a limit* – and belongs to knowledge, the area of finite thinking, an unsurmountable obstacle to individualization.

Universidad Autónoma de Baja California
Mexico

NOTES

[1] The meaning of this term will be explained *infra*.
[2] 980a21.
[3] The Oxford edition, W. D. Ross, transl.
[4] We are quite conscious of the non-existence of this verb, which would correspond to *wisdom as to know* corresponds to *knowledge*. Webster* includes the verb as obsolete, but registers the *slang* expression *to wise up*. It is suggestive that the English language has abandoned this philosophically most important symbol of expression. Should this be concomitant with Marx' famous assertion** that "materialism is the native son of Great Britain" and the elimination in general by the English of words with an idealistic or metaphysical background?

 * Webster's *New Collegiate Dictionary* 1961.
 ** Karl Marx and Friedrich Engels, *Die heilige Familie*, *Marx Engels Werke*, I, p. 823.

[5] Martin Heidegger, *Nietzsche*, I (Pfullingen: Keste, 1961), p. 310: "*Damit wird der wesentliche, niemals überbrückbare Unterschied zwischen dem gewöhnlichen Kennen und Zuschauen auf der einen Seite und dem eigentlichen Wissen auf der anderen Seite angedeutet*".
[6] Hegel, Glockner edition, VIII *System*, § 28: "Wenn vom Denken die Rede ist, so muß man das endliche, blos verständige Denken, vom unendlichen, vernünftigen, unterscheiden".
[7] *Devas* = Los seres angélicos que primeramente fueron homres, pero que adelantaron en la escala espiritual hacia el estado semidivino, los seres brillantes.*

 * Ramacharaka, *Filosofías y Religiones de la India*, F. Climent Ferrer, transl. (Buenos Aires: Kier, 1984).

[8] Cf. Hofmann: "ενθουσιασμος *göttliche Begeisterung: von* εν–θεος *Gott in sich habend, von Gott inspiriert*"* (Enthusiasm, divine spirituality: from εν–θεος, having god within one's self, inspired by god).

 * J. B. Hofmann, *Etymologisches Wörterbuch des Griechischen* (Darmstadt: Wissenschaftliche Buchgesellschaft, 1966).

[9] *Theaet.* 183E–184A: "It seemed as there were a not at all common profoundness in his discourses; I fear that we do not comprehend his words and that we do not properly penetrate his thoughts."
[10] M. Heidegger "Der Spruch des Anaximander," in *Holzwege* (Frankfurt a/M: Vittorio

Klostermann, 1963), p. 297: "Das unausgesprochene Richtmaß für die Deutung und Beurteilung der frühen Denker ist die Philosophie von Platon und Aristotles. Beide gelten als die nach vorwärts und rückwärts maßgebenden Philosophen der Griechen. Diese Anschauung hat sich auf dem Wege über die Theologie des Christentums zu einer allgemeinen und bis heute nicht erschütterten Überzeugung verfestigt". (The not explicit measure for the interpretation and judgement of the early thinkers is the philosophy of Plato and Aristotle. Both are held to be the determining Greek philosophers for all that was previous or subsequent. This notion – on the road through the theology of cristianity – has coaleseed into a general conviction, not shaken till the present time.)

11 Franz Passow, *Handwörterbuch der griechischen Sprache*, 4 vols., 5ª ed. (Darmstadt: WB, 1983): ". . . das Leben, die physische Existenz."

12 *Laws*, 895C7–8 and *Phaedrus*, 245C6–7.

13 Aristóteles, *Met.*, 1072b29–30.

14 *Ibid., Met.*, 1073a10–11: ουκ εστιν ουθεν απειρον μεγεθος (No magnitude is infinite); cf. Hegel, *op. cit.*, p. 139: "Eine unendlich kleine oder unendlich große Größe ist keine Größe mehr (An infinitely small or infinitely great magnitude is a magnitude no more.)

15 B 3: προς εαυτο δε εκαστον εστι και μεγα και σμικρον (. . . for itself, however, each [thing] is as well great and small).

16 According to Karl E. Georges: *Ausführliches Lateinish-deutsches Handwörterbuch: momentum* from *movimentum – ad moveo.

17 Georges, *op. cit.*: "momentum, die Bewegung als innewohnende Kraft." (. . . movement as immanent force.)

18 Webster's New Collegiate Dictionary, 1961.

19 *Ibid.*

20 Georges, *op. cit.*, "Momentum . . . (aus *movimentum, zu moveo), die Bewegung als innewohnende Kraft, . . . übtr.: die Bewegunslänge im Raume, . . . die Bewegungsdauer, in der Zeit, gew. mit der Begriffsprägnanz der Kürze." (. . . movement as immanent force, fig. the length of movement in space, . . . the duration of movement, in time, usually with the conciseness of brevity.)

21 Maria Bielawka, "The Mystery of Time in Roman Ingarden's Philosophy," International Conference, Jagiellonian University and the Cracow Branch of the Polish Philosophical Society, October 25–27, 1985: "The present is conceived here as being punctual in nature . . ."

22 Leibniz: *Monadología*, § 60, como mónada *sin ventanas*, cada una incluye las demás mónadas y *viceversa*.

23 Cf. Cicero, *de nat. deor.* I, 69 and 73, where he has Epicurus sustain the *declination* of atoms.

24 Karl Marx, *MEW*, *Okonomisich – philosophische Manuskripte 1844*, Erganzungsband I, p. 517: ". . . und sich selbst daher in einer von ihm geschaffnen Welt anschaut".

25 *Ibid.*, p. 516: "Die Natur ist der inorganische Leib des Menschen, nämlich die Natur, soweit ste nicht selbst menschlicher Körper ist".

26 Peter Shalimov, "Leibniz and the Russian Philosophy in the 19th and 20th Century," in *Leibniz und Europa*, VI. Internationaler Leibniz-Kongreß, Vorträge. (Hannover: Gottfried-Wilhelm-Leibniz-Gesellschaft, 1994, 1995), I. Teil, pp. 748–749: "Eines Tages

trank Leibniz in einer Gesellschaft Kaffee und sagte, daß möglicherweise in dem von ihm heruntergeschlürften Kaffee einige Monaden sind, die mit der Zeit Menschen werden könnten".*

* Source: N. O. Lossky: "Personalistischer Idealrealismus," in *Kant-Studien* 51 (1959–1960), p. 406.

[27] Pierre Teilard de Chardin, *Der Mensch im Kosmos* (Münich: Beck, 1959), p. 50: "Das Anfangsquantum von Bewußtsein . . . stellt vielmehr eine Masse von infinitesimalen Zentren dar, die durch ihre Ursprungsbedingungen und ihre Entwicklung strukturell miteinander verbunden sind".

[28] Wild, Wolfgang, "Was William Shakespeare und die Thermodynamik gemcinsam haben; Naturwissenschaften und Geisteswissenschaften – immer noch zwei getrennte Kulturen?," *Die Welt*, Nr. 219, Sec. Geistige Welt, Samstag, 20. September, 1968, p. 1: ". . . the majority of scientists in natural sciences have distanced themselves from this conviction (that of obtaining objectively valid knowledge). They recognize that objectively valid knowledge cannot inductively be derived from experience."

[29] C. W. Leadbeater, *Los espiritus de la naturaleza* (Buenos Aires: Federación Teosófica Interamericana, 1973), p. 25: "Vida elmental, Vida mineral".

[30] Dschuang Dsi, *Nan Hua Dschen Ging*, Richard Wilhelm, transl. (Düsseldorf, Diederichs, 1951), p. 88: "Im Uranfang war das Nicht-Sein des Nicht-Seins, war das Unnennbare. Daraus erhob sich das Eine. Dieses Eine – in seinem Dasein, aber ohne Form, das die Dinge bekommen müssen, um erzeugt werden zu können – heißt LEBEN".

[31] *Ibid., loc. cit.*: ". . . wird es eins mit den Kräften des Himmels und der Erde".

[32] *Met.*, 1072b28–29: φαμεν δη τον θεον ειναι ζωιον αιδιον . . . (Transl. Ross: We say therefore, that God is a living being, eternal. . .).

[33] Raphael Kühner and Bernhard Gerth, *Ausführliche Grammatik der griechischen Sprache, Zweiter Teil*, 2 vols. (Darmstadt: WB, 1966), § 485.2.

[34] Kühner and Friedrich Holzweissig, *Ausführliche Grammatik der lateinischen Sprache, Erster Teil* (Darmstadt: WB, 1989), § 155.b.

[35] Webster, *op. cit.*

[36] αει ον αιει, *Il.* 1.107, 12, 211, 23.648; *Od.* 15.379, 9.386.

[37] Webster, *op. cit.*

[38] Arthur Anthony Macdonell, *A Practical Sanskrit Dictionary* (London: Oxford University Press, 1958).

[39] *Il.* 24.725.

[40] Each moment being in itself eternity.

[41] συν and εχω, *con* and *teneo*.

[42] Kühner and Holzweissig, *op. cit.*, § 194, 7: *Praesens* derives from *prae*, in front of, *ens*, being, and the *s* of the arcaic *-sens*.

[43] See Hermann Diels, *Die Fragmente der Vorsokratiken* (Berlin: Weidmann, 1964), B 8.5–6: ". . . [εον] νυν εστιν ομου παν, εν, συνεχες."

[44] *Prot.*, 342 D-E.

[45] G. van der Leeuw, *La réligion dans son essence et ses manifestations* (Paris: Payot, 1955), p. 165: ". . . certaines tribus australiennes ont des noms spéciaux pour «le temps primordial» (Urzeit), inaccessible dans le temps". (Certain Australian tribes have special names for «the primordial time» (archaic time), inaccessable within time.)

[46] Theophrastus, as quoted by Kirk: G. H. Kirk, *Heraclitus. The Cosmic Fragments* (Cambridge: At the University Press, 1962), pp. 308–309.

[47] Brehier translates (Emile Brehier, transl., Plotin, *Ennéades*, 5 vols. [Paris: Les Belles Lettres, 1954–1956]), V, 1, 7: "Cronos le dieu très sage qui reprends toujours en lui les êtres qu'il engendre".
[48] Eli de Gortari, *Dialéctica de la Fisica* (México City. Ediciones Océano, 1986), pp. 54–55: ". . . cada partícular elemental Ilena por si sola todo el sistema al cual pertenece y, por ende, tiene simultáneamente una multitud de localizaciones especiales diferentes. Esta propiedad de ocupar muchos lugares al mismo tiempo es la politopla de la partícular elemental".
[49] "Por otra parte, como la politopla es una propiedad común a todas y cada una de las partículas elementales, resulta, que éstas tienen también la propiedad de la homotopla respecto a cualquier lugar comprendido dentro del sistema al que pertenecen, puesto que cada lugar es ocupado simultáneamente por todas ellas".
[50] *Loc. cit.*: ". . . (quedamos) fuera del 'sentido común' y de la física clásica".
[51] Martin Heidegger, *Platons Lehre von der Wahrheit* (Bern: Francke, 1947), p. 72.
[52] Martin Heidegger and Eugen Fink, *Heralklit Seminar Wintersernesten 1966/67* (Frankfrut a.M.: Klostermann, 1970), p. 57.
[53] English version from Ross' edition, the Greek version from Jaeger's Oxford edition.
[54] Martin Heideggger, "Der Sprunch des Anaximander," in *Holzwege,* p. 317.
[55] Cf. Horst Matthai, *PENSAR Y SER IV*: *La teoría parmenídea del pensar* (Mexicali: UABC, 1990), p. 63.
[56] Cf. *idem, PENSAR Y SER III: Heráclito el obscuro* (Mexicali: UABC, 1997), Chapter IX.
[57] Later the Cartesian *Cogito, ergo sum.*
[58] *Tim.* 31B.
[59] Gottfried Wilhelm Leibniz, *Kleine Schriften zur Metaphysik, Metaphysische Abhandlung* (Darmstadt: Wissenschattliche Buchgesellschaft, 1965), pp. 77–78: "Daß jede einzelne Substanz das ganze All auf ihre Weise ausdrückt und daß in ihrem Begriff alle Ereignisse mit allen ihren Umständen und die ganze Folge der äußeren Dinge enthalten sind. (. . .) Und da alle anderen Substanzen diese eine auf ihre Weise ausdrücken und sich ihr anpassen, kann man sagen, daß sie ihre Macht auf alle anderen in Nachahmung der Allmacht des Schöpfers ausdehnt".
[60] Of course, here we have to eliminate Leibniz's reference to the idea of a creator, once we admit his having catered, as is well known, to the religious traditions of his time.
[61] Arthur Schopenhauer, *Sämtliche Werke*, 2nd ed., *Die Welt als Wille und Vorstellung* (Wiesbaden: Brockhaus, 1949) II, p. 391: "Aber diese Vielheit [von Individuen] trifft nicht ihn, den Willen als Ding an sich, sondern nur seine Erscheinungen: er ist in jeder von diesen ganz und ungeteilt vorhanden und erblickt um sich herum das zahllos wiederholte Bild seines eigenen Wesens. Dieses selbst aber, also das wirklich reale, findet er unmittelbar nur in seinem Innern".
[62] *Marx Engels Werke* (Berlin: Dietz, 1968), Ergänzungsband I, p. 517: ". . . und sich selbst daher in einer von ihm geschaffnen Welt anschaut".
[63] Oswald Spengler, *El hombre y la técnica*, 2ª ed. (Buenos Aires: Espasa-Calpe, 1947), p. 77: ". . . la inclinación germánica a lo infinito".
[64] Glockner edition, I, pp. 535–536, IV, p. 459.
[65] Δος μοι που στω, και κινησει την γην, was his famous saying: Give me where I stand, and I shall move the Earth.

[66] No less than five of the Seven Sages of ancient Greece mention this most important concept in their sayings.

[67] *Tibetan Yoga and Secret Doctrines*, W. Y. Evans-Wentz, ed., 2nd ed. (Oxford: Oxford University Press, 1967), p. 47: ". . . man has been and will be, in a way as yet unsuspected by our biologists, his own ancestor."

[68] *Jaina Sutras*, 2 vols. (New York: Dover, 1968), I, p. 32: ". . . there is no past thing, nor is there a future one."

[69] *La teorla parmenidea del pensar*, p. 26.

[70] MANAVA-DHARMA-SASTRA, *Leyes de Manú*, Juan España, transl., 2nd. ed. (Madrid: Lorenzo Hijos, 1936), 12.15: "De la sustancia de esta *alma suprema*, se escapan, *como las chispas del fuego*, innumerables principios vitales que comunican incesantemente el movimiento a las criaturas de diverso orden".

[71] "Der Spruch des Anaximander," in *Holzwege,* p. 318: "Wir können mit Übertreibung, aber mit ebenso viel Gewicht an Wahrheit behaupten: das Geschick des Abend-Landes hängt an der Übersetzung des Wortes εον, vorausgesetzt, daß die Über*setzung* in der *Über*setzung zur Wahrheit dessen beruht, was im εον zur Sprache gekommen". (We can sustain with exaggeration, but with just as much weight of truth, that the fate of the Occident depends on the translation of the word εον, provided, that the trans-*lation* is based on the *trans*lation towards the truth of that what has been spoken in the εον.)

[72] Passow: "πυματος der äusserste, hinterste, letzle, sowohl vom Raume als von der Zeit". (. . . the most extreme, ulterior, last, of space as well of time.)

[73] Only Diels, contrary to all other commentators, renders the Greek νυν as a noun, giving emphasis to this most important word.

MARCELO SÁNCHEZ SORONDO

HEGEL: LIFE BETWEEN DEATH AND THOUGHT

If we take a preliminary look at the second major work of Hegel's maturity, namely the *Science of Logic*, which he published after the *Phenomenology*, we note that in Hegel's usage the word life has three different meanings, each of which is related to one of the three parts into which he subdivided his masterpiece, the *Encyclopaedia*. We thus have a "logical life" that can be distinguished from "natural life" and "life inasmuch as it is united with the spirit".[1]

The least problematical of these three notions of life is probably the second, natural life. This life indicates the living organism, vegetal or animal, and the crystallized form of life that is represented by the earth, which we can still call nature, as did the Greeks. All this is the object of a very precise philosophical discipline that Hegel, following Aristotle, calls *Physics* and which concludes and crowns the Philosophy of Nature. But the relationship of life with the spirit, i.e., with freedom and thought, is rather more complex. Indeed, since life is the highest level at which nature can arrive in its process of self-formation, natural life is precisely that from which the spirit gradually differentiates itself. Hegel notes that such expressions as "life of the spirit" or "nature of the spirit" are devoid of sense, because they pretend to explain the spirit in terms of something to which it cannot be reduced, almost as if one could speak of "spiritual matter".

But there is yet another dimension in which life is placed not in opposition to but in unity with the spirit. I am referring to the passage-overcoming from Nature to the Spirit. Here life is the beginning of interiority, and therefore the link that joins the exteriority of nature to the absolute interiority of the spirit, that is to say, the thought of thought.

Ever since Kant, philosophers have sought to delve into the contraposition according to which, face to face with the world of nature that Greece tells us about, there is this other world – the world that Hegel, thematizing for the first time, calls Spirit. If we wanted to use a formula to pinpoint what differentiates Nature from the Spirit, we would find, together with Hegel, a central notion. Nature is that which *is there*. And the spirit is that which is the '*I*'. Nature is therefore being there,

A-T. Tymieniecka (ed.), Analecta Husserliana LIX, 189–203.

At the folkloric meal: Louis Houthakker, Anna-Teresa Tymieniecka, Marcello Sanchez Sorondo.

exteriority. For Hegel, more precisely, Nature is "that-which-is-in-itself", *dynamis*; Spirit, on the other hand, is "that-which-is-by-itself", being with oneself (*Beisichselbstsein*),[2] "selfness" or pure *energeia*. This is the starting point of Hegel's proposal.

There can be no doubt that in prior centuries philosophy had been familiar with, above all, the νοῦς of Aristotle and the freedom of Kant, moving in this element – as Hegel would say – of the spirituality of the spirit. And it is also certain that the spirit in Kant is not a second world placed by the side of the first, nature, but rather claims to have a rank superior to nature inasmuch as the latter is subject to necessity, while the spirit is spontaneousness, liberty. Hegel however holds that the radical sense in which one can speak of Nature and Spirit had not been understood before him.

Attributing to the word universe a wider sense than was the case in Greece, we can say that our universe, unlike that of the Greeks, is made up of nature and of spirit, of objectivity and subjectivity, of world and I, of exteriority and interiority, of necessity and freedom. The significance of life, therefore, depends on the type of relationship that is assigned to these terms.

Schelling kept on repeating that this relationship does not consist of anything other than a kind of basic identity in which Nature ceases to be nature and the Spirit ceases to be spirit: an identity, and therefore a simple indifference, that, by virtue of the very fact of its being indifference, is – so Hegel would have it – something similar to the darkness of the night "wherein all the cows are black".[3] For Hegel, therefore, the identity of nature and spirit does reside in a simple formal identity, a vacuity, but one which has the following very concrete meaning: since there exists an essential relationship between nature and spirit, there must be something that is positively the common foundation of this relationship. Understanding nature and spirit is to see how this foundation underlies both of them: how this foundation, inexorably, *becomes* Nature and Spirit.

Now, how does Hegel qualify this common foundation? Within it there move or seek support both Nature and Spirit, that is to say, *all* that exists. Thus, inasmuch as this foundation encloses within itself everything that exists – not only the nature of the Greeks, but also the free and self-realizing spirit of Christianity and modernity – this foundation from which everything emerges is in an eminent and fundamental manner the authentic and true All. It is the foundation of all the rest; as Hegel tells us, it is the absolute in itself and for (by) itself.

For Hegel, therefore, "everything depends on understanding and expressing the true (*das Wahre*) not only as substance (*Substanz*), but also as subject (*Subjekt*)".[4] Hegel could also say that everything depends on understanding the true not only as Nature but also as Spirit. In the *Encyclopaedia*, as is well known, the Philosophy of the Spirit comes after and crowns Logic and the Philosophy of Nature. This does not mean that the Spirit appears only after nature; it was spirit already before being the Spirit, this at least in the sense that it was the dialectical road towards which, by which and also contrary to which the Spirit self-realizes itself as Spirit. "The true is the whole. But the whole is only being that completes itself by means of its development".[5] If one wants to understand this development, one has to distinguish, as it were, two different states. The first is what is generally indicated as aptitude, capacity, being that-which-is-in-itself, *potentia, dynamis*. The second, on the other hand, is that-which-is-by-itself, fully actuality, the *actus purus*, as Hegel puts it.[6] In this way Nature is counterposed to the Spirit.

For Hegel, nevertheless, Nature makes us understand the Spirit as that-which-is-in-itself, possesses a finality, precisely by means of the notion of life. But what exactly does Hegel mean when he says this? In trying to answer this question we have to bear in mind not only the challenge that Hegel takes up, but also the sources on which he relies to give a positive answer to the challenge. The challenge is that thrown down for him by modern thought and, more particularly, by Kant; the sources that Hegel reconsiders and reinterprets are those of Greek thought in general and Aristotle in particular. Faithful to his conception of speculative thought, in fact, Kant had denied that man's intellect was capable of grasping the noumenon of the "formative force", though its efficacy was right under the very eyes of anybody who reflected about the explosive phenomenon of life. The ideas of Aristotle about the *soul* as an act that possesses its end, on the other hand, is what makes it possible for Hegel to glimpse a certain intelligibility peculiar to life. Here we are not concerned with an imperfect act where the end is outside the process, but with a perfect act in which the end is the process, that is to say, an ἐν-τέλος-ἔχειν, where ἐν means unit, τέλος end, and ἔχειν possession.[7]

In fact, Aristotle observes that the act of movement is an "imperfect act" (ἐνέργεια ἀτελὴς),[8] because its end (τέλος) is outside the movement as such, that is to say, in this movement the act is not the end, that is, οὐ γὰρ τέλος.[9] But there does exist a type of act that is decidedly dissociated from the act of movement (motion) and is associated with that of action in the sense of praxis and theory. What makes this act different from movement is the fact that it has its end within itself; for example, at one and the same time one sees and has seen, knows and has known, thinks and has thought, whereas one cannot teach and have learned, nor heal and be healed. In these perfect acts that possess their end within themselves, the dissociation between the act and the movement is sustained by a grammatical criterion concerning the operation of the tenses of the verb, namely, the possibility we have of saying "at one and the same time", "together" (ἅμα): he has seen and he sees, he has thought and he thinks, he has lived well and he lives well, he has been happy and he is still happy. On the basis of this disjunction, therefore, Aristotle can say "there is not only an act of movement, but there is also an act without movement".[10] Decentering this immediate experience of the perfect acts among our actions in the direction of a corresponding ontological foundation in the interior of our identity (sameness), Aristotle – for the first time in the history of philosophy – thinks of the soul as the first act, the ἐντελέχεια ἡ πρώτη.[11] Whereas in the days when he was an adept of Plato's *Dialogues* he spoke of the soul's being ἐνδελέχεια or "continuous and everlasting movement", as Cicero translated the term in passing it down to us,[12] in the days of his maturity, Aristotle coined the term ἐντελέχεια to signify the perfect act that possesses its end.[13] This change is due to the new notion of act arrived at by Aristotle, one that can not only explain movement, but also found a new ontology of living beings. "Not all things", as Aristotle puts it, "are called act in the same way, but only by analogy: as this is to this or with respect to this, so also is this other to that or with respect to that. Some things, in fact, are called act as movement with respect to power, others as substance with respect to some matter".[14] Whereas ousia and eidos are mere universals for Plato and therefore unable to act and without life, for Aristotle the substance that is a real subject is principally acting and life in a living being. The polemics of the later Aristotle against Plato consist precisely in defining Plato's Idea as simple δύναμις; to this concept he opposes the affirmation that the idea, recognized by both to be the principal part of the individual, has to be considered rather as ἐνέργεια, that is to say, as the immanent form that confers unity and definition upon a reality.[15] In the soul as energy there is thus already present an active principle that is neither totally separate from temporal becoming nor completely subjected to time, but which, rather, proposes itself as being in becoming and becoming in being, as a principle that traverses and subjects to itself the necessity of time, impressing the form of the end upon it. We turn now to see in a more analytic manner how Hegel arrives at the speculative determination of life.

* * *

In his maturity Hegel admonished Kant for having crushed the truth in a dualist system of abstract intellectualistic oppositions (subject-object, intellect-reason, matter-form, thing in itself-thought, nature-God, freedom-necessity, etc.) incapable of rising to a superior point of view.[16] Nevertheless, Hegel showers unreserved praise on the *Critique of Judgement*, wherein Kant, expounding on the realm of ends, presents finality as life and recalls the position of Aristotle. Hegel increasingly recognized that Kant, with his distinction between relative or external finality and internal finality, had the signal merit of having reopened for modernity the great highway to the concept of life and of the idea in general. The most decisive formulation of this judgement is to be found in the *Encyclopaedia* of 1827–30: "What is notable in the *Critique of Reason* is that with it Kant has expressed the representation or, rather, the thought of the Idea. . . . Many, especially Schiller, have found in the idea of *artistic beauty*, or of the concrete unity of thought and sensible representation, the way out from the abstractions of the intellect; others above all in the intuition and consciousness of *living vitality* (*Lebendigkeit*) in general, be it natural or intellectual.[17] It is only natural to think that when Hegel recalls others who accepted Kant's teaching on teleological judgement with respect to living beings, he had in mind not only the road covered by Hölderlin and Schelling, but also his own itinerary in leaving behind intellectualistic abstractions incapable of theoretically reaffirming the reality of life.

Kant speaks of the "technique of nature" when he refers to the manner in which organized beings operate. But what restrains Kant from assigning to the principle of finality of nature some cognitive power with respect to the noumenon of objects is the fact that, between the causality of nature in its products (i.e., organisms) and the causality that manifests itself in the technical operation of the human being, there exists only the remote analogy that is proportional correspondence, and therefore a purely extrinsic relationship. Indeed, the difference between the technique of nature and human technique resides in the fact that man's technique always presupposes the intentional nature of a project and choice made in freedom, while in nature finality is something that is spontaneous and neither free nor intentional. Thus, nature's operation shows that it needs no instruction from without to realize itself and therefore it does not give evidence of anything within that in any way comes close to the qualification of the intentionality and freedom found in the causality of technique. Experience certainly shows nature to be everywhere impregnated with ends, but it can never show that these are at the same time intentions.

Nevertheless, one may note that qualifying the activity of organic beings as the "technique of nature",[18] be it even unintentional, not only betrays a Cartesian mechanistic residue but, above all, manifests an element of anthropomorphic ambiguity in Kant's teleology. In a human technical operation, in fact, the project or the form, as an ideal anticipation of the work to come, constitutes something derived from without with the aim of configuring matter that was previously unformed that therefore serves only as a paradigm to be kept in mind and nothing more. In human technique, therefore, the form does not by itself possess the capacity (δύναμις) to bring to completion that of which it is the form. For this very reason, a human operation stands in need of another cause, other than form, to carry out the indications contained in the form or project. In other words, human technique stands in need of an efficient activity (ἐνέργεια) that realizes the work or product (ἔργον) in external matter that is susceptible to being formed in accord with this paradigmatic form. In an organic being of nature, on the other hand, the form of the end is already there in the efficient principle and foundation of the realization of the work had within itself. This means that in organic beings in general, with the exception of thought only, the form is not introduced into the process from without, but, on the contrary, the organic process is in some way the very activity of the end that is present in the form, without there being any extrinsic perspective imparting instructions or transmitting information. When molecular biologists today talk of codes, instructions, information, etc., one has to note that these notions, which derive their origin from man's own technical operations and are used to explain nature's living organisms, end up betraying the complexity of the life of organic beings, because one cannot but recognize that the code of a natural being codifies itself, i.e., that the instructions are received by exactly the same subject that transmits the information. All said and done, therefore, if we limit ourselves to analytically reconstructing the process of a living being in the succession of its various phases in an analogy using the model of human technique, we run the risk of overlooking the very thing that is essential, namely, the being's actual proceeding from within itself, its ceaseless self-configuring, that which constitutes the dominant characteristic of life. It is precisely this sequential performance of the living subject that Hegel claims to grasp with more adequate conceptual tools, and it is for this reason that we have in him a rehabilitation of Aristotle.

In fact, though Hegel, as we have seen, recognizes that Kant had the merit of introducing the concept of internal finality and, thus, of opening the road to the idea of life, he immediately adds that, in so doing, he did nothing other than to reawaken "the solid definition that Aristotle has given of living being, namely that it is to be considered as operating in accordance with its end".[19] In his *Phenomology*, Hegel already affirms that the Stagirite had taught him this notion of activity in conformity with an end (*Zweckmässiges Tun*). "Aristotle," – Hegel tells us, "defines nature as operating in accordance with an end, this being immediate at rest, the immobile that is also the motor".[20] In the *History of Philosophy*, again, Hegel sings the praise of Kant, because "in modernity, it is only in his philosophy that this Aristotelian concept of intrinsic finality is rehabilitated".[21]

Now, to rediscover the foundation of this "solid definition" of Aristotle one has to go back to a point upstream of the inadequate analogy between human technical operation (τέχνη) and the productions of nature (γένεσις) that one meets so frequently in the thought of the Stagirite, especially in the classical statement of Chapter VIII of the second book of his *Physics*.

Here Hegel finds, first and foremost, the stimulus to detach the concept of end from that of intentionality or of conscious and deliberate thought, though without ends for that reason ceasing to have reality in nature. In fact, Hegel recalls that for Aristotle "it is absurd to hold that there is no generation with a view to an end if one does not realize that the principle of movement is capable of deliberation (βουλευσάμενον)"[22] "As regards the end," Hegel affirms, "one need not think, or need not think only, of the form in which it occurs in consciousness as a determination that is present in a representation".[23] Thus, if there is no resolution of Kant's antithesis, by virtue of which an end was either real as a concept – and therefore could only be the principle of conscious human operation – or was a non-intentional attribute of the technique of nature, then an end can only be a subjective design of reflective judgement; but Hegel, following in the footsteps of the Stagirite, recognizes that even a non-intentional or non-deliberate end could have objective reality in nature in the way in which it manifests itself, as, for example, in a seed or an instinct.

Hegel's first example of such a teleological being in se of life is the seed (*der Keim*). The seed contains "the essentials of development"; this unity of seed and development is "a speculative concept".[24] Moreover, when in the *Berlin Lectures* he again takes up this notion of the teleological dynamic of the seed, it is almost as if he were reciting Aristotle's idea. Hegel says: "The structure of evolution is such that something that is at first hidden takes place. Thus, for example, in the seed there is contained the entire tree with its differentiation into branches, leaves and flowers, etc. That which is simple, which contains this diversity within itself, the *dynamis*, while in a seed that has not yet developed, has not yet passed from the form of possibility to that of existence".[25] Aristotle, for his part, says in *Metaphysics*: "The seed has the potentiality of the eidos, and that which proceeds from the seed is in a certain way homonymous", that is, (τὸ σπέρμα) ἔχει γὰρ δυνάμει τὸ εἶδος, καὶ ἀφ' οὗ τὸ σπέρμα, ἐστὶ πως ὁμώνυμον.[26] And, again, in *De Anima* he says: "The seed and the fruit constitute that which is potentially a body of that species", that is, τὸ δὲ σπέρμα καὶ ὁ καρπός τὸ δυνάμει τοιονδὶ σῶμα.[27] Therefore, the seed and the fruit in which there is conserved the seed of the tree are potential with respect to the living body "enacted", but it is by virtue of an active teleological power that the living body unfolds. For Hegel, the reality of the seed resides in its being "act and its activity consists in producing the tree",[28] in realizing the end, in finalizing self-activity.

A second and more immediate example is offered by the teleological nature of instinct (*Instinkt*). An instinct is in effect a demonstration of the reality of finality that works in an unconscious manner. Since it is an originary teleological drive (*Trieb*) in a living being, its activity appears to have determination, that is to say a form or plan, which consists in being first and foremost something subjective. Now, for Hegel "since the content of the determination is originary and is maintained in the activity, and by means of this activity only is made evident, that content is end, purpose".[29] The instinct that precipitates activity, this originary drive, is therefore the most immediate example of operation ordered to an immanent interior end of which the living subject is not conscious. Thus, "the instinctive drive is not conscious of its end, the animal does not know its end as an end; and Aristotle gives the name φύσις to unconscious activity ordered to an end".[30]

Nevertheless, in this selfsame Aristotelian source Hegel also finds the clear identification of the limit beyond which the analogy between the operation of nature and the operations of art cannot be pushed. What ultimately prevents art from operating like nature is the fact that nature has a principle of activity, an active *dynamis*, within itself. In contrast to nature, in fact, human technique does not out of its own ordering deploy an energetic principle, but must always procure it by having recourse to something else; on the other hand, it could ever come into being as technique if it were not ultimately to root itself in the teleological self-acting unfolding of nature. Even today, this seems to be the insuperable difference between the latest generation of informatics products and the life of nature, which is finalized self-acting.

Hegel unconditionally accepts this Aristotelian entelechial principle: "in what Aristotle here affirms we have the whole of the true and profound concept of what is living, which has to be considered as an autonomous end in itself: here we have something that is identical with itself, that independently impels itself and, in manifesting itself, remains identical with its concept – in short, an idea that realizes itself".[31] In conceiving of nature as that which is the cause and the principle of activity, Aristotle – so Hegel would have it – succeeded in going straight to the essential distinction between internal and external finality. In this way, in fact, "he understood nature as life, that is to say, as that which has its end and purpose in itself, is oneness with itself, which does not pass into anything else, but thanks to this principle of activity determines changes in conformity with its own content and thus conserves itself in them".[32] "Nature signifies precisely that a thing becomes something that it already was already in it (nature) right from the beginning: it is this internal universality and finality that realizes itself; thus, cause and effect are identical, because all the individual parts are tied to this at oneness with the end".[33]

* * *

Thus, the entelechial principle that manifests itself in life as self-activity operating as ordered to an immanent end, as seen, for example, in dialectics of seed and fruit and need and instinct, serves to indicate in an analogical manner the appearance of the spirit on the basis of its real-acting potentiality with a dynamism for teleological development. Hegel nevertheless establishes an insuperable distinction between the life of nature and what improperly, may be called spiritual life. In fact, unlike the spirit, natural life is prisoner to a dialectic that condemns it to finiteness and the seed to death. "The seed develops in itself, emerges as itself. The realization of this emergence goes only as far as what was implicit in itself. Its movement has an end, has a configuration, a finality, but a predetermined, non-spontaneous finality: the fruit".[34] What happens, therefore, is that the initial principle and the final term – the seed and the fruit – are two separate entities; this doubling results in a scission into two different individuals. "In animal life, in just the same way, parent and offspring are different individuals, but of identical nature".[35] In the final turn, therefore, the radical impotency of natural life "comes to the fore in the fact that what is beginning and what is result become dissolved".[36] In fact: "In nature, life – which springs from death – is in its turn individual life; and if the species is considered that which is substantial in this succession, then the disappearance of the individual is a falling back of the species into singularity. The conservation of the species thus subsists only as a uniform repetition of the same mode of existence. . . . In nature the species dos not progress".[37] In the vital process of nature, therefore, the living individual, who is the only entity that is really alive, is in itself transient and survives only by generating another individual; and the species, which is what is permanent in this process of the generation and death of single individuals, has in actual fact concrete life only in the mortal individual. This disparity, therefore, "is its originary illness and its innate germ of death".[38] In nature, the need and the quest for infinity thus have as a response only an infinite reproposition of finiteness in accord with a process of generation and decay within the species.

But what happens in the act of the spirit is altogether different. The spirit is "pure act";[39] "it is this infinite motion, ἐνέργεια ἐντελέχεια (energy, activity) that never comes to rest, something that has abandoned its first stage and becomes drawn towards a new state, that elaborates it and has found itself in its work".[40] Thus, "Activity (*Tätigkeit*) is its being".[41] "Therefore", Hegel underscores, – "its first formation, as real being, is self-activity only (*Selbst-tätigkeit*). Its being resides in activity (*Aktuosität*)".[42] We are not here concerned with an activity that causes an effect different from itself, for "the spirit is essentially the result of its activity".[43] Thus, "that which the spirit produces is its object, is the spirit itself",[44] so that, inasmuch as the spirit is "characterized by having produced itself, by having become of its own accord, having made itself",[45] "its beginning is also its end".[46]

Thus, while life in nature is an active *dynamis* or first act, like the Aristotelian soul, the spirit is the pure actuality of being-with-itself. With life conserves itself only for an eternal return of the same species in accord with the process of generation and corruption, the spirit progresses *ad infinitum* towards itself and its own act. While life, which is segmented into beginning and end (seed and fruit) and split into individuality (parent and offspring), does not arrive at becoming a being-by-itself, the spirit, where the end returns to its beginning, "succeeds by virtue of one and the same principle, and not any other, in being a being-by-itself".[47]

In reality, however, according to the model of the Aristotelian νοῦς, "the spirit converts its being-in-itself into an object-by-itself and thus becomes one with the object"[48] in the unity of the act. In this way, when the subject comes back to itself, it finds itself to be not with something else, but rather with itself. It therefore follows that, in the last and definitive analysis, the entire generation of the universe, the entire history of the universe is nothing other than the entry of the spirit into itself, the realization of the Idea. In each of the steps that the absolute spirit has taken, not only has it not gone outside itself, but what it has really done is to rediscover itself, meet with itself, as it were. This permanent rediscovering itself with itself is what traditionally has been called eternity. For this reason, history (in the Hegelian sense of the term) is not a succession of things that happen in time, but rather the being of this succession. In a certain sense, the being of history is eternity. History is the concretized reality of the Idea. And life has neither historicity nor eternity unless it be through the spirit.

Grasping the absolute spirit, absolute knowledge, in its eternity, but doing it in an absolute manner – this and nothing else, as far as Hegel is concerned, is what *philosophy* consists of. Philosophy is not thinking about the absolute, but rather the explicit form of the absolute itself. It follows from this that to history there belongs essentially its own history. "The idea of philosophy," Hegel tells us at the end of the *Encyclopaedia* "has as its middle term self-knowing reason, the absolutely universal: this middle term which 'dualizes' itself into spirit and nature, turning the former into the precondition of reasoning, and the latter into its universal extreme inasmuch as it is process in itself and objectively so. . . . The self-knowing concept, eternal in itself and per se, actuates itself, produces itself and enjoys itself eternally as absolute spirit" (§ 477).

As is well known, the final seal of the *Encyclopaedia* is constituted by a transcription of the theological text of Aristotle's *Metaphysics* in Greek. In the translation and interpretation that Hegel gives of this text in the *Berlin Lectures on the History of Philosophy* (according to K. L. Michelet's first edition) he says: "the fundamental moment of Aristotelian philosophy is therefore this: that thinking and what is thought are one, that the object of thought and the energy of thought are one and the same thing. 'For it is that which receives the objects perceived and the existent'. Thinking is the thought of thought. As far as thinking is concerned, Aristotle says '[It] operates to the extent possesses [i.e., ἐνεργεῖ δὲ ἔχων]' (in other words: its is but a single whole with its action); 'so that this' [activity which it thinks] 'is more divine than that which thinking reason (νοῦς) believes itself to have of the divine' (the νοητόν). The most excellent is not what is thought, but the energy of thought itself. 'Speculation (ἡ θεωρία) is thus that which most partakes of the delectable' (beatitude of the highest degree, ἥδιστον) 'and the most excellent (supreme)'".[49]

To conclude: Hegel, making explicit reference to Aristotelian origins, especially in his later writings, conceives philosophy as the self-process of the Spirit within the active potentiality-act, that is, first act-perfect act plexus, after the likeness of science is gown become science in act, that is, contemplation. Thus is the stagirite's notion of active potentiality or first act, as "progression within itself and in entelechy",[50] interpreted now taking due account of the hint of the Christian Trinity, as the absolute potentiality of the subject as the principle of life. But Hegel, in this rereading of Aristotle, seems to reserve Aristotle's notion of act as being, i.e., the thought of thought, for explaining the spirit. The first overcoming of the external is thus life, which in this way realizes the first form of interiority, that is to say, the soul, the first entelechy, though subject to growth and decay, to generation and corruption, while the ultimate point of arrival is the spirit as the perfect and absolute act, Aristotle's thought of thought.

Nevertheless, Hegel deprives the theological proposal of the Stagirite (act as the being of the thought of thought) of life, which occupies as central position in the doctrine of Aristotle, where God is either attributed eternal life or God is referred to as the Living One. But a position very different from Hegel's was assumed by another great Aristotelian, Thomas Aquinas, who explained the Trinitarian processions on the basis of God's generative potential (power), i.e., life that "non dividitur contra actum, sed fundatur in eo".[51] All the same, the difference resides in the fact that Thomas conceives God on the basis of being, as the act of being by essence, and not in the uniqueness of His substance as Spinoza has it, who considers life to be bound up with *res extensa*, and therefore with matter and death. But that is a topic I take the liberty of reserving for another occasion.

Libera Università Maria Santissima Assunta
Rome

Translated by Herbert Garrett

NOTES

[1] G. W. F. Hegel, *Wiss. d. Logik*, Collected Works, Vol. 12, ed. F. Hogemann and W. Jaeschke (Hamburg, 1981), pp. 179–181.
[2] G. W. F. Hegel, *Vorl. ü. d. Philos. d. Weltgesch., Die Vernunft in der Geschichte*, ed. J. Hoffmeister (Hamburg, 1955), p. 55.
[3] G. W. F. Hegel, *Phän. d. Geistes*, ed. J. Hoffmeister (Leipzig, 1937), p. 19.
[4] *Loc. cit.*
[5] Hegel, "Das Wehre ist das Ganze" (*Phän d. Geistes, op. cit.*, p. 21).
[6] Hegel, *Vorl. ü. d. Gesch. d. Philos*, ed. H. Glockner (Stuttgart, 1965), Vol. 17, p. 49. A parallel Text in *Einleitung in die Gesch. d. Philos.*, ed. F. Nicolin (Hamburg: 1959), p. 101.
[7] Cfr. *Aristotle De Anima libri tres*, ed. F. A. Trendelenburg (Graz, 1957), especially pp. 243–262.
[8] *Phys.*, III, 2, 201 b 31f.
[9] *Metaph.*, IX, 6, 1048 b 22.
[10] *Ethic. Nic.*, VII, 14, 1154 b 26f.
[11] *De Anima*, II, 1, 412 a 27.
[12] *Tusc. Disput.*, I, 10, 22.
[13] Cfr. E. Bignone, *L'Aristotele perduto e la formazione filosofica di Epicuro*, Vol. 1 (Florence, 1973), pp. 227–272. For a definition of the question and overcoming (resolving) it on the basis of various philological approaches, see the important article by E. Berti, "Il concetto di atto nella Metafisica di Aristotele", in *L'atto aristotelico e le sue ermeneutiche*, Marcelo Sánchez Sorondo, ed. (Rome, 1990), pp. 43–61.
[14] *Metaph.*, IX, 6, 1048 b 6–9.
[15] "If there are some realities and substances of the type that dialecticians call Ideas, then there must be something that is more knowing than knowledge-in-se, and there will be something much more mobile than movement-in-se, indeed, these would have a greater

claim to being acts, while the Ideas would be powers of these" (*Metaph.*, IX, 9, 1050 b 35–1051 a 3).

16 G. W. F. Hegel, *Enz. d. phil. Wiss.* (1830), especially §§ 56–60.

17 *Ibid.*, § 55.

18 See the special emphasis, particularly in the *Erste Einleitung* to the *Critique of Judgment.*

19 Hegel, *Enz. d. phil. Wiss.* (1830), § 360; cfr. also *Enz. d. phil. Wiss.* (1817), § 154.

20 Hegel, *Phän. d. Geistes, op. cit.*, p. 22.

21 Hegel, *Vorl. ü. d. Gesch. d. Philos., op. cit.*, Vol. 18, p. 341. As regards Aristotle, cfr. *Phys.*, II, 8 passim.

22 *Phys.*, II, 8, 199 b 26–28.

23 Hegel, *Enz. d. phil. Wiss.* (1830), § 204.

24 Hegel, *Einleitung in die Gesch. d. Philos., op. cit.*, pp. 106f; cfr. pp. 101f, 108, 127, 348f.

25 *Ibid.*, pp. 82f.

26 *Metaph.*, VII, 10, 1034 a 34–b 1.

27 *De Anima*, II, I 412 b 26f.

28 Hegel, *Einleitung in die Gesch. d. Philos., op. cit.*, pp. 102f.

29 Hegel, *Enz. d. phil. Wiss.* (1830), § 360. Cfr. also *ibid.*, § 204.

30 *Ibid.*, § 360, Addendum, ed. H. Glockner (Stuttgart, 1965), Vol. 9, pp. 633f.

31 G. W. F. Hegel, *Vorl. ü. d. Gesch. d. Philos.*, ed. C. L. Michelet, *Werke*, Vol. 15 (Berlin, 1842), p. 305.

32 *Ibid.*, p. 303.

33 *Ibid.*, p. 305.

34 Hegel, *Einleitung in die Gesch. d. Philos., op. cit.*, p. 108.

35 *Ibid.*, p. 109.

36 Hegel, *Vorl. ü. d. Philos. d. Geschichte, op. cit.* (Stuttgart, 1971), Vol. 17, p. 119.

37 Hegel, *Vorl. ü. d. Philos. d. Weltgesch., Die Vernunft in der Geschichte, op. cit.*, p. 153.

38 G. W. F. Hegel, *Enz. d. phil. Wiss.* (1817), § 282.

39 G. W. F. Hegel, *System d. Philos.*, § 378, Addendum, ed. H. Glockner (Stuttgart, 1965), Vol. 10, p. 12.

40 Hegel, *Vorl. ü. d. Philos. d. Weltgesch. Die Vernunft in der Geschichte, op. cit.*, p. 161.

41 *Ibid.*, p. 55.

42 *Ibid.*, p. 74.

43 *Ibid.*, p. 72f.

44 Hegel, *Einleitung in die Gesch. d. Philos., op. cit.*, 109.

45 Hegel, *Vorl. ü. d. Philos. d. Weltgesch., Die Vernunft in der Geschichte, op. cit.*, p. 74.

46 *Ibid.*, p. 55.

47 Hegel, *Einleitung in die Gesch. d. Philos., op. cit.*, p. 110.

48 *Loc. cit.*

49 Hegel, *Vorl. ü. d. Gesch. d. Philos., op. cit.*, Vol. 18, pp. 330f.

50 *De Anima*, II, 5, 417 b 6f.

51 ST, I, q. 25, a. 2 to 1.

GRAZIELLA MORSELLI

THE LOGOS BETWEEN CRITIQUE AND GENETICS

The Greek word *logos* has found its way into the technico-scientific culture of our day via the derived term *logic* that Nietzsche, linking it with the will for truth, defined as the attempt to determine the world in accordance with a pattern of Being and rendering it calculable.[1]

Nietzsche also wrote that logic is derived from "a dominant impulse to treat similar things as alike":[2] face to face with things that are numerous and differentiated, the dominant impulse is to stop their procession and reduce them to satisfy a need for reassurance. Indeed, identifying things is to fix their likenesses in such a way that all difference is compressed into a few and constant "identities" and to establish units of measure by means of the logico-mathematical method, applying them to the largest possible number of cases: in either way, things are thus reduced to variants of the ever repeated identical.

But the term *logos* involves also the person who produces ordered knowledge, rather, it unites him with these "products" of his in such a way that – seeing that logic holds supreme sway – he himself is *logos*. Personified in this manner, the *logos* tends to take the place of the man, hides him from sight, makes him incapable of recognizing himself in the "things" he has brought into being, rendering the world equal to himself. Man no longer feels the need for questioning himself about his life and his disturbing progress towards the death he calls time.

In fact, time itself has been determined and rendered calculable in the objective views of scientized history (among which one must include also the evolutionist view) and seems foreseeable in accordance with the laws of historicist reason, so much so as to have become functional for the sciences that concern themselves "impersonally" with man and yet elude the ultimate questions. Logos-man, producer of objective knowledge, wants to dominate time to take his distance from the fecund flow of life, so that he may observe it and explain (or "unfold") it, i.e., make it flow in an orderly manner, starting from the "folds" he has impressed on it by superimposing his own hypostasis. Nevertheless, since he himself – as a living being – finds himself immersed in this flow, he can neither transcend time nor really determine it. A living being,

A-T. Tymieniecka (ed.), Analecta Husserliana LIX, 205–211.

in fact, experiences not only immanent time, but also the time constituted among the living (who are like him), and so-called objective determination is but a part, and not even the greater part, of this intersubjective constitution.

Husserlian phenomenology has clarified this temporal being of ours, has taken away from the sciences and restored to philosophy the questions connected with it: here, therefore, philosophy has to bring its critical force into play to unmask the dominion of logic and to pose once again the decisive questions abut life and death. Before phenomenological ontology there has thus opened the great field of the senses of being, the field of the new *logos* of life that is the consciousness of internal time.

Husserl set himself the difficult first task of analyzing this consciousness, but was well aware that from this there would inevitably derive another closely connected task: to inquire into the genesis of the psychophysical "I". Indeed, it is precisely this empirical "I", coincident with every subject in the world, with every living "I", that experiences time as an immanent flow of memorizations, presentifications and expectations, and therefore as a total unity of his lived experience. And it is for this living "I" that the idea of a subjectively constituted sense of time is valid as his personal consciousness of time flows within his experience.

As a result of his genetic analyses, nevertheless, Husserl came to see in this further task of phenomenology a "superior dimension" of problems concerning coming-into-the world and generative connections: "Does one not have to see in all this the more serious and essential problems of a phenomenology constitutive as transcendental philosophy?". But, as he did not fail to add, these problems were very premature at the initial stage in which phenomenological research found itself at the time.[3]

And yet, as one can note in this selfsame quotation, this brief indication already contained a philosophical question regarding everything that in principle precedes and renders possible – through the "generative connections" – the constitutive consciousness of the empirical "I", with the perspective sensitivity that is inherent in it and the corporeal complex to which it is bound in the form of physical dependence, and also with one's own "empathy" that links it to the extraneous "I" in the world.

As regards the problem of empathy in particular, the answer to this question could have opened a different road, providing the inquiry into the copresence of temporal community and entropathic experience that Husserl carried out in his *Crisis of the European Sciences* and *Cartesian Meditations* with more valid bases than those of consciousness' own transcendentalism. Husserl knew that what he had up to then achieved in this field was still in danger of remaining nothing other than the construction of a theoretical consciousness that presented the other to itself by means of abstract analogies based on the vision of a body with a likeneness to its own. But Husserl aimed at a more direct access to otherness when he proposed to obtain also a theory of empathy from his analysis of the genesis of the "I".

In two fragments dedicated, respectively, to procreation and to the child we can find some precious indications as to how he would have continued this analysis, which unfortunately remained nothing but an intention. In the brief space of time of the present paper I cannot give a full account of my attempt, already published elsewhere,[4] to work along this road, but I shall try to summarize the results.

Husserl would certainly have had recourse to the concept of corporeal intentionality that he had already developed very extensively in *Ideas II* and *Experience and Judgment*, as also in his many analyses dedicated to the specular genesis of the connected senses of estrangement and objectivity, to further develop his notes on the intersubjective relations that precede and accompany the birth and the growth of the child. From the act of copulation to gestation and care of the child, following the phases of an articulated complex of impulses, perceptions and consciousnesses, he would gradually have been able to grasp the originary constitution of the sense of otherness that he had called "first empathy".

The copulation of the parents, taken in its optimal acceptation as an encounter of two persons aware of their responsibilities, is defined by Husserl as an "intersubjective act of procreation" from which there derive "motivated natural processes in the other life, different processes of self-temporalization".[5] From this there follows, I would add, a subjective and intersubjective constitution of time in which there are involved two intentionalities directed towards a new process in their own time of existence, but this involvement not only transforms both poles of the union, but also produces the premises for a new constitution of time that will be realized by a future intentional subject.

This can be explained better by assuming that the two empirical subjects live a twofold process of natural causalities and motivations of their own, which interact and give rise to many variables, ultimately connected in a manner unforeseeable in the intentionality.

Irrespective of whether the coitus has occurred by pure impulse or has been either preceded by a desire for procreation or followed by acceptance of an unforeseen offspring, the beginning of the new life must therefore be placed in the lived time of the two parents before it is placed in the objective time of biology or perception. At the precise moment of their lifetime when a couple decides to become parents, they also decide for the time of their children before these effectively exist.

In fact, inasmuch as they intentionally (as consciousnesses, but also in their sensuality and sensitivity, that is, as intentional corporeities) constitute the time of a new life, what the parents effectively do is placed in a pre-time with respect to the pre-I of the child in which they invest their feelings and their raison d'être. The pre-I thus already has a non-actual time and world that later, in real existence, will become for the child a pre-time and pre-world that have already been; but in this way there is also determined an interweaving of times, by virtue of which he will have had his part in the past that for others was present, while others have had their part in the future that has now become his present.

Apart from this temporal union, one also has to consider (always on the basis of Husserl's suggestions) the spatial union of the "maternal corporeal link", which has two moments for the child: the intrauterine phase and the enveloping and nutrient contact of the postnatal phase.[6] Continuing this analysis, there comes to the fore a vision of a dialectic organic development governed by a rhythm of transforming and generating exchanges that – inasmuch as it involves some psychic processes of different degrees of complexity and significance – is not just an organic and functional symbiosis.

One may therefore speak of a merging of senses and a compenetration of acts, but also of an intersection of experiences and a transposition of intentionalities. Nor does it seem excessive to postulate that the current idea of subjectivity, and even the philosophical representation of an individual essence, may well go back to the exclusive and direct contact peculiar to the psychophysical interaction relation between mother and child, directed as it is towards the peculiarity of a new being, for which it provides, as it were, reassurance and confirmation of the continuity of one's existence, eventually conferring upon one a self of one's own.

In this sense, therefore, the relation would acquire the significance of a "trans-subjective" dynamic.

In this connection one cannot but note that such advanced and so-called genetic sectors of the sciences as neurology, psychoanalysis and biology have arrived at some interesting findings regarding the child's brain formation in the fetal and neonatal phases, placing special emphasis on placental cell exchange processes that imply even a transformation of the maternal organism and neuropsychic effects of this exchange.

Even when due account is taken of significant contributions of the empirical sciences, it still remains for phenomenology to inquire into the corporeal and consciousness tied aspects of the procreative relation in its entirety. There thus becomes delineated an originary complex of which the dynamics seem to be articulated in a cognitive potential that exalts a number of phenomenological constructions and throws them into a new light, especially those concerning the subject-object, real-possible, immanent-transcendental oppositions and intersubjective relationships.

This complex can therefore be referred to by the name of the "maternity model", and this not only in the descriptive sense of a synthesis of concepts, but also in the methodological sense, because it has the function of bestowing on the multiplicity of human relations in a unified origin. In this way it also becomes possible to bring out the historical obscuring of the sensing of this origin, which has hitherto remained relegated by ancient prejudices to the pre-cultural sphere of facts of an instinctual and libidinal nature.

But this model does not indicate a grateful nest in which we can regress to infancy; quite the contrary, it is placed under the banner of the hard labor of rationality. And this has to be said not only of the far from easy inquiry it calls for, taking us well beyond the schematicism of the empirical sciences and the prejudices of the philosophies of the Identical, but also of the opening to the possible that it heralds. In fact, it is a model that belongs to the *logos* of life, that is to say, to a reason committed to research and the constitution of a truly intersubjective sense of life.

Since it is strictly rooted in the primary experience of each one of us, moreover, the model can furnish new foundations for ethics in accord with the a priori significance of the community, i.e., with what Husserl called the "genetic self-constitution of intersubjectivity". In fact it represents in an efficient and on the level of being also real manner (that is, before it does so on the level of having to be) the sense of indistinctness and compenetration with the other interiorized before birth, so that it represents something that can become reactivated in the adult who – in his turn – is capable of generating, something like a condition waiting to be recognized to give rise to a new ethic.

The community a priori, on the other hand, is linked with different contents of the same experience, according to the positions of the child or the parent that each one assumes there, bearing in mind also the fact that the procreative relation attains its maximum degree of intentional (and not merely physical and emotive) participation in the position of the mother. This specific content can nevertheless be communicated both in the direct manner of transmission of the gesture, the attitude and the voice, and symbolically also in the signification of words and other messages. What is therefore offered for our comprehension and learning is the special generative relation of union and exchange, in which each partner was previously in the originary passive position, as a person come into the world, and by virtue of which it is given to him to advance to the active position as someone who can put into the world.

For the purposes of the ethical significance we have to deduce from this, obviously, the active position has to be understood as a generative disposition in relation to others, that is to say, as capacity predisposed to others' living a lives worthy of being lived, a capacity for "putting-into-the-world" also in the wider sense of re-giving life to someone who is already living, renewing, confirming, replenishing the vivifying funness of a life of human relations. In this sense, therefore, every relation founded on active respect for the existence and freedom of the other is equivalent to putting a new human being into the world.

Procreation, moreover, is manifestly conditioned by the will of the person who generates, in other words, it is an asymmetric relation, a power relation, whereas in the maternity model, which is based on the optimal condition of a childbearing aimed at otherness, the asymmetry assumes the significance of a power that affirms itself by denying itself for the other.

Inasmuch as the mother's consciousness fore-sees the other as the bearer of a right to be, it recognizes the independence of the other and "at one and the same time" (while it performs the "temporalization" of this other consciousness) creates the conditions for the return of power through recognition by the child.

The model's significance for a priori community thus seems to be fully realized in the circular movement of recognition that tends to bind all of us in Husserl's "intermonadic temporal community" through the links of generative continuity.

University "La Sapienza" – Rome
Italy

English translation by Herbert Garrett.

NOTES

[1] Cfr.: F. W. Nietzsche, *Der Wille zur Macht* (posthumous work) (Leipzig, 1906).
[2] *Id., Die Frölige Wissenschaft 1882*, in *Werke* III (München, 1956).
[3] E. Husserl, *Cartesianische Meditationen*, 61, in *Husserliana*, Vol. **I** (The Hague, 1950).
[4] G. Morselli, *Ri-cognizione dell'origine* (Manduria-Bari-Roma, Piero Lacaitz, ed., 1994).
[5] E. Husserl, Ms E III 5 ("Universale Teleologie", 1933), in *Husserliana*, Vol. **XV** (The Hague, 1973).
[6] *Ibid.*, Ms K III 11 ("Das Kind, die erste Einfühlung", 1935).

FABRIZIO FORNARI

LA QUESTIONE FENOMENOLOGICA DELL'ENTE-UOMO

1. FENOMENOLOGIA E PSICOLOGIA IN HUSSERL

Come è noto, la filosofia di Husserl è connotata da un ampio e pregnante interesse per le scienze, la logica e la matematica. E' altrettanto noto come questo interesse trovi le sue radici nel matematico Karl Theodor Weierstrass, con il quale Husserl collaborò in veste di assistente, e soprattutto in Franz Brentano, che lo indirizzò verso l'analisi psicologistica dei fondamenti della matematica, culminata con la pubblicazione della *Philosophie der Arithmetik* avvenuta nel 1891 (cfr. Husserl, 1970). I temi legati alla ricerca di una fondazione della scienza erano, del resto, ampiamente dibattuti nell'ambiente intellettuale *mitteleuropäisch* e non deve pertanto stupire il fatto che Husserl li ritenesse essenziali (cfr. Riverso, 1991). Tuttavia, proprio l'interesse per la scienza e soprattutto per la logica lo costringerà ad abbandonare presto – anche in forza della critica di Gottlob Frege – l'orizzonte della *Philosophie der Arithmetik*, per far posto a quel nuovo ambito di studi che confluirà nelle *Logische Untersuchungen* (1900–1901). In esse si ribalta la posizione psicologistica contenuta nell'opera del 1891. Contro il relativismo scettico cui condurrebbe lo psicologismo, Husserl delinea qui una complessa configurazione della coscienza, intendenola come trama dei vissuti intenzionali, come *status* fenomenologico dell'io e come percezione adeguata.

Tali argomenti segnano profondamente il cammino teorico husserliano e ritornano in tutte le opere successive, costituendo essi la base su cui la fenomenologia trascendentale istituisce una serrata critica della scienza moderna, rea di essersi separata dal soggetto effettivamente fungente e dal suo *telos* razionale. Alla cieca volontà di una ragione priva di soggetto bisogna fenomenologicamente sostituire la trascendentalità della coscienza, quale orizzonte insuperabile di ogni autentica operazione conoscitiva. E'questa la svolta delle *Ideen* – da un io come semplice insieme dei vissuti intenzionali a un *ego trascendentale*, orizzonte e fondamento del mondo – che spaccherà al suo interno la scuola fenomenologica, sorta subito dopo la pubblicazione delle *Logische Untersuchungen* nel tentativo di applicare ad altri campi del sapere e della

A-T. Tymieniecka (ed.), Analecta Husserliana LIX, 213–232.

cultura il metodo e lo stile, non più deduttivo, della stessa ricerca fenomenologica (cfr. Landgrebe, 1963). Nel suo isolato cammino, Husserl appare subito lontano non solo da Kant e dall'idealismo speculativo, ma soprattutto dal metodologismo dei neokantiani, mentre sembra approssimarsi alle tendenze empiriste, rispetto alle quali però introduce importanti novità come l'intuizione categoriale e l'*apriori* materiale. Alla decisa confutazione del metodo deduttivo, egli fa seguire anche il rifiuto del metodo induttivo della logica classica, che tenterebbe di risalire dal fenomeno ad una causa esplicativa (cf. Pöggeler, 1991). In quest'ottica, il problema della realtà – in che misura le cose esistano realmente – viene elaborato nella direzione di un atteggiamento problematico, segnato da una sospensione radicale del giudizio (*epoché*) e proprio per questo in grado di neutralizzare tanto il dubbio metodico intorno ad essa, quanto l'accettazione scontata e assoluta della sua evidenza (nella convinzione che il fatto, per se stesso, staccato dal suo orizzonte di apparizione è appunto fenomenologicamente insignificante). Anche rispetto al campo delle scienze umane, la fenomenologia ed in particolare Husserl assegnano un'assoluta priorità all'indagine che lo coglie in quanto ordine di fenomeni e quindi internamente ad una sua rilevanza eidetica (cfr. Tilliette, 1983).

Come si evince da quanto detto, Husserl rifiuta sia l'obiettivismo delle scienze (l'epistemologia dei fatti), sia il particolarismo che consegue alla negazione umanistica dell'obiettivismo stesso (l'ermeneutica storicista dei valori): egli cerca una dimensione dell'oggettività che non sia mera assunzione naturalistica del dato, né accesso ad un *multiversum* di significanti sempre eccedenti la possibilità del significato medesimo di dispiegarsi in sé. Il lavoro sistematico di Husserl risulta, piuttosto, un'indagine interamente protesa alla elaborazione di un orizzonte trascendentale, al cui interno si renda conto del *soggetto* e delle sue operazioni costitutive. E' precisamente tale soggettività ineludibile che non trova luogo o ragion d'essere nelle scienze della natura e si riduce a cosa descrivibile empiricamente nelle scienze dell'uomo. Da un punto di vista husserliano, la crisi del sapere scientifico nasce precisamente dall'oblio, in esso operante, della soggettività umana, oblio che concerne anche le *Geistwissenschaften*, se è vero che queste ritengono che l'identità dell'uomo sia un dato tra gli altri, una cosa tra le cose. Per Husserl, l'uomo non è qualcosa di cosale (*kein Ding*); esso è invece unità attuale di atti intenzionali, inoggettivabile come cosa, il cui senso, come sosterrà poi Max Scheler, sta nel compiersi di essi.

Si tratta pertanto di far emergere la dimensione necessaria della soggettività, imprescindibile tanto per il naturalismo obiettivistico, quanto per l'ermeneutica storicista e ciò non contro l'interesse per le scienze, la logica e la matematica, dal quale Husserl stesso prende le mosse, ma mediante esso e lungo la linea di sviluppo del pensiero fenomenologico.

Punto di partenza è la critica al concetto di soggettività presupposto e usato dallo psicologismo. Si è già accennato alla confutazione husserliana di questa dottrina, un tema capitale della fenomenologia. L'argomentazione svolta in merito da Husserl è per molti aspetti analoga alla critica cui oggi Putnam sottopone, richiamandosi a Wittgenstein, il relativismo del solipsista metodologico (cfr. Putnam, 1989). Per Putnam, le tesi del relativista e del solipsista si negano da sole, si autoconfutano; a sua detta, infatti, il relativista non può capire la differenza tra il credere di aver ragione e l'aver ragione stesso, dal momento che egli afferma proprio ciò che viene negando (afferma come vero il fatto che non esiste alcuna verità) e non può dunque seriamente pensare, ma solo credere di farlo. Analogamente il filosofo che intenda essere psicologista tenderà a relativizzare i problemi di pensiero, riconducendoli al condizionamento storico, sociale, psicologico che li genera, privandoli perciò di un valore intrinseco. Ma così, secondo Husserl, i conti non tornano più, giacché nulla vieta di ritenere che in futuro possa prevalere l'ipotesi opposta (se il relativismo psicologistico lo negasse, la sua tesi sarebbe assoluta; il che è appunto autocontraddittorio).

Come afferma Merleau-Ponty anche il sociologismo – simmetrico sviluppo della tendenza psicologistica – si espone alle stesse critiche. Se tutto il nostro pensiero fosse soltanto l'espressione di una situazione sociale i cui limiti imprigionassero il senso d'essere delle cose (secondo un'impostazione che fu già di Marx), allora si finirebbe con l'impossibilità di distinguere i tratti di una società razionale, i cui valori siano controllabili, da quelli di una società ingestibile perché incontrollabile, finendo con il ritenere per vero, ma solo a parole, tutto e il contrario di tutto (cfr. Merleau-Ponty, 1982). Ne seguirebbe un irrazionalismo politico nel quale nessuno avrebbe argomenti per contrastare le scelte del più forte.

La ricerca husserliana di un *telos* razionale intende del resto mettere al riparo da questi rischi, valorizzando la *logicità* del pensare ed evitando, al contempo, di chiudersi reattivamente in un assolutismo logico del tutto presupposto. La logica è da valutare nell'ambito di quella idealità che non si lascia ridurre alla struttura formale di un pensare vuoto, né ai meccanismi psicologici della coscienza.

Come è stato ricordato, nelle *Logische Untersuchungen*, Husserl osserva

che nessuno dubita che la fisiologia del pensiero sia un fatto materiale, ma che su tale base si giustifica altrettanto bene tanto un ragionamento giusto quanto uno errato. La differenza specifica tra un ragionamento corretto e uno scorretto non si può misurare in kilocalorie, dicendo per esempio che il pensiero corretto costa più sforzo e quindi consuma di più, o invocando una misteriosa attività psicofisica come "scatola nera" di cui non sappiamo nulla (Melandri, 1989, p. 27).

La differenza

non può consistere che nel fatto di ragionare come si deve e ciò vuol dire usare l'apparecchio fisiologico *in conformità* con le leggi, non a causa di esse. Così una calcolatrice aritmetica (Husserl pensava ad una di quelle macchinette di fine secolo che funzionavano con un giro di manovella) funziona bene non perché sia in sé logica, ma perché strutturalmente è stata costruita in modo da produrre i risultati desiderati; se fosse guasta o costruita male produrrebbe altrettanto bene risultati errati (*ivi*, p. 28).

La logicità del pensiero – il *logos* effettivamente vivente – non si subordina ad alcun ordine psicologico; essa condiziona la psicologia e limita, al contempo, un'indebita assolutizzazione della logica formale, dal momento che, per Husserl, una logica senza mondo, una logica che non si confronti con un mondo che logico non è, non ha alcuna rilevanza fenomenologica. In altre parole, egli cerca di non cancellare i pregiudizi stabiliti in noi dai condizionamenti esterni, ma di renderli presenti alla coscienza (senza negarli). La critica antinaturalistica vale cioè come negazione di una estraneità incontrollata (quella del dato). Si tratta, in breve, di verificare ogni obiettività, trasformando l'irrazionalità del dato stesso in un processo plastico di coscienza, nel quale l'io si dia come "unità di una storia" (cfr. Husserl, 1959, p. 109, trad. it. p. 124).

Scopo della riflessione fenomenologica, non solo husserliana, è pertanto quello di penetrare nell'ambito della nostra esperienza temporale, giacché la filosofia deve *sich einstromen*. Ciò significa, detto altrimenti, che il fenomenologo non può permettersi di saltare logicamente il problema del tempo, stante che la logica è strutturalmente tempo e il pensiero sarebbe nulla se non discendesse da un *flusso temporale* (cfr. Husserl, 1966). Da un punto di vista fenomenologico, l'uomo non vive in un mondo di fatti – i fatti sono interpretazioni, aveva già detto Nietzsche (cfr. Vattimo, 1994) – ma in un orizzonte di senso (in un aver già interpretato, dirà Heidegger in *Sein und Zeit*). Compito della "riduzione fenomenologica", espresso nelle *Ideen* come *epoché*, è di mettere allo scoperto, manifestare, far vedere, tale orizzonte. Essa non nega, ma porta alla coscienza e porta alla coscienza anche l'esistenza come noi la viviamo prima della riflessione, in quell'operare anonimo e fungente che è il mondo-della-vita (*Lebenswelt*). Tale risulterà appunto il guadagno della *Krisis* (cfr. Husserl, 1954).

In questa prospettiva, la riduzione fenomenologica non ha alcuna pretesa di immettere il soggetto in una radicale oggettivazione del tempo, quasi fosse possibile superare quest'ultimo nella direzione di un eterno logico. Il tempo è piuttosto l'orizzonte insuperabile del mondo e dell'essere (cfr. Husserl, 1966). Husserl stesso affiderà ad Heidegger nel 1928 una serie di lezioni e di manoscritti sul tempo affinché ne curasse la pubblicazione. Come è stato giustamente affermato, "questo è un episodio capitale per comprendere il cammino del pensiero del Novecento". Infatti,

> proprio muovendo dallo studio di questo manoscritto del maestro (studio iniziato ben prima del 1928) Heidegger perverrà alla convinzione che l'orizzonte del tempo è il tratto decisivo dell'essere, donde il tema del capolavoro di Heidegger stesso (*Essere e tempo*, 1927) che aprirà la stragione dell'esistenzialismo, e che però si annuncia per ammissione esplicita dell'autore, come uno sviluppo delle intuizioni fenomenologiche husserliane relative alla temporalità della coscienza (Sini, 1992, p. 16).

Articolando l'interpretazione del tempo, Husserl giunge a dire che non vi è un'unica modalità di apprensione della temporalità, essendovi piuttosto una molteplicità di modi di apprensione temporale: vi è, ad esempio, il modo passivo in cui il tempo è soltanto subìto (in proposito Husserl usa l'espressione *innerzeitigkeit*) e vi è un modo di ricezione temporale che vede il tempo come incremento della coscienza e del proprio vissuto intenzionale (un vissuto che resta sostanzialmente estraneo all'analoga tematica proposta da Dilthey, nella quale la storicità del vissuto resta strutturalmente passiva).

E' altresì vero che Husserl parla della fenomenologia come se fosse una scienza delle verità eterne. Tali verità sembrerebbero però specificarsi, come egli stesso segnalerà negli ultimi anni della sua vita, non già in quanto verità situate oltre l'ordine del tempo, bensì in quanto verità onnitemporali, valide in ogni tempo (cfr. Merleau-Ponty, 1982). Si tratta cioè di approfondire non di oltrepassare la temporalità e in questo senso la logica diventa correlato concettuale del nostro avere il mondo alla mano (heideggerianamente del nostro essere-nel-mondo). In tale direzione, si spiega l'affermazione husserliana delle *Logische Unter-*

suchungen per la quale le leggi del nostro pensiero sono per noi leggi dell'essere: non nel senso hegliano di una perfetta coincidenza di reale e razionale, bensì in quello di una *coestensività* del nostro pensare a tutto ciò che è affermabile. Tant'è che se si intendesse immaginare un altro pensiero (non umano) in grado di contraddire quanto pensiamo e come lo pensiamo, esso sarebbe *nulla*, per noi. Se è qualcosa lo è in conformità alle leggi del nostro pensare: se esulasse da queste leggi non sarebbe per noi pensabile. Per questo Husserl sostiene che Dio stesso non potrebbe avere una esperienza del mondo che non fosse prospettica, per profili, e con tutti gli adombramenti conseguenti. Qualora Dio pensasse, dovrebbe pensare seguendo le leggi della logica o non penserebbe affatto. Quando Husserl afferma che soltanto lui e Heidegger sono i veri positivisti, egli intende sostenere che la razionalità non si basa su *un diritto anteriore al fatto*, in grado di condizionare la nostra recezione del fatto stesso, ma che ogni conoscenza razionale deve incontrare direttamente le cose stesse (punto questo che sarà sempre ribadito anche da Heidegger fino a *Zeit und Sein*). Tale conoscenza non potrà mai presentarsi come sistema assoluto, secondo gli assunti della onto-teologia, come discorso in grado di chiudere una volta per tutte la questione dei fondamenti. Husserl considera la fenomenologia, ossia la filosofia in quanto scienza rigorosa, un metodo in divenire, un campo di conoscenze in cui trova riscontro la forza emancipatrice della ragione, nel suo distendersi entro un progressivo accrescimento (Husserl è, in fondo, un evoluzionista, come lo era un altro grande teorico del progresso infinito del pensiero, Charles Sanders Peirce). Certo questa idea sospinge la fenomenologia husserliana ai confini estremi della *modernità*, se è vero che la categoria di moderno si caratterizza, secondo la lettura di origine heideggeriana, per l'idea di progresso, sulla cui base il vero si identifica con il nuovo (cfr. Vattimo, 1985) e il nuovo è visto come *pars construens* della storicità: la mera ripetizione non dà storia, sebbene anche ciò che è solo ripetuto, per il fatto stesso di essere ripetuto, non è quello che fu (cfr. Vitiello, 1991).

Il problema della temporalità della coscienza (e del pensare) porta con sé l'esigenza di esplicitare la questione del trascendentale. Smontando l'obiettivismo galileiano e aprendo così la stessa possibilità dell'ermeneutica odierna (cfr. Ricoeur, 1975), Husserl mostra come intenibile la pretesa di dedurre lo scibile da un unico principio. Da un punto di vista fenomenologico, difatti, il mondo non può venire sistematicamente dedotto dall'io, come ritenevano Cartesio e Fichte (cfr. Gadamer, 1963).

Egli mira ad un'oggettività indubbiamente diversa da quella metafisica e scientista, non può certo e tuttavia fare a meno di un riferimento in ogni caso oggettivo e se si preferise *intersoggettivo*, dal quale poi si renda accessible lo stesso precategoriale mondo-della-vita (che è un altro aspetto che assume nell'ultimo Husserl la critica all'obiettivismo).

Come si evince dalle *Ideen* (cfr. Husserl, 1952), la soggettività ultima, radicale, fenomenologica – quale polo egologico trascendentale – è in realtà una Intersoggettività (cfr. Held, 1973; Schütz, 1957). Quando si parla di trascendentale, ha osservato Enzo Paci, il pensiero va a Kant, per il quale trascendentale è ogni conoscenza che, pur non avendo origine dall'esperienza, non può oltrepassare i limiti dell'esperienza stessa (cfr. Paci, 1987). Così in Kant, ad esempio,l'intuizione dello spazio non ha origine dalle esperienze spaziali dell'uomo: infatti io non vedo lo spazio, ma lo spazio è nel vedere e per questa ragione esso è *a priori*. Ma se dicessi che Dio è spaziale oltrepasserei indebitamente il campo di applicazione dell'intuizione dello spazio che rende possibile la geometrica (cfr. Derrida, 1987). Nel caso di Husserl, però, il riferimento a Kant non coglie nel segno (cfr. Kern, 1964). Husserl raggiunge infatti un maggiore livello di radicalità (cfr. Paci, 1987): l'intuizione dello spazio è una forma di esperienza del soggetto umano, il quale è esso stesso trascendentale e lo è perché ha esperienze dirette e non perché, come in Kant, condiziona *formalmente* la possibilità dell'esperienza medesima (e non esperisce nulla direttamente, come sottolinea ancora Paci). La percezione trascendentale kantiana, in quanto risponde ad esigenze *puramente formali*, non tiene conto che il soggetto sono io in carne ed ossa e che gli altri soggetti sono altri io come me.

In quest'ottica, il soggetto fa sempre esperienza del mondo, ma lo trascende in una tipicità del significato. Ciò che percepisco direttamente come "rosso" è sì quel rosso particolare che si dà nella percezione e tuttavia non è solo questo: esso è *il tipo*, l'*eidos* di qualsiasi colore rosso (e anche Max Weber, come si sa, sia pure in altro contesto, parla di 'tipi ideali').

Questo modo di conoscere, questo emergere di un orizzonte irriducibile al bruto dato di fatto, mediante l'intenzionalità, consente di precisare il trascendentale alla luce del concetto di intuizione. L'individuazione del tipico – quale orizzonte trascendentale: io sono trascendentale perché mi oltrepasso, perché scopro tipi ideali – è infatti, per Husserl, "intuizione" (grande tema dibattuto in quegli anni ed approfondito soprattutto da Henri Bergson). L'intuizione, a sua volta, è essenzialmente intuizione delle essenze. In forza della sua duplice articolazione, concreta e universale, fattuale e ideale, la percezione, in senso fenomenologico, coincide con l'apertura intuitiva nella quale il dato diviene "essenza". Così intesa, la riflessione sull'intuizone può offrirsi quale mezzo per una radicale trasformazione del campo d'indagine delle stesse scienze umane. Queste ultime dovrebbero, in ultima istanza, comprendere come il nostro vissuto, *Erlebnis*, possa essere determinato da fattori assolutamente contingenti e individuali e tuttavia esso possa anche strutturarsi intersoggettivamente, almeno per la parte in cui si comprenda il nostro continuo essere trascesi nel tipico, il nostro essere sempre in contatto con una *intuizione eidetica*. Per certi fatti del tutto casuali, scrive Merleau-Ponty in merito al rapporto tra scienze umane e fenomenologia, si può scoprire l'esistenza della Nona sinfonia; scoprendola però si svela anche qualcosa che non dipende da quei fatti casuali per i quali si è venuti a sapere che la Nona Sinfonia esiste. Essa non esiste solo nel momento in cui qualcuno la scopre (e in questo senso l'intuizione eidetica è profondamente anti-idealistica); piuttosto essa appare nelle *diverse* esecuzioni o *interpretazioni* che se ne danno. La Nona Sinfonia è cioè un oggetto culturale (simbolico) che si manifesta nel concerto che ora possiamo ascoltare o stiamo ascoltando, ma che non è completamente riconducibile all'interpretazione che di essa si sta offrendo. Questo stare nella particolarità dell'interpretazione infatti non elude, secondo Husserl, e anzi implica il riferimento all'*essenza* di ciò che l'interpretazione medesima coglie solo nel particolare. Tenendo presente questo, se si riuscisse ad elaborare la nostra esperienza sulla base di ciò che essa importa, tematizzando ciò che si offre in quell'irripetibile nomento, noi potremmo arrivare a cogliere ciò che non è singolare, né contingente e che è la Nona sinfonia (cfr. Merleau-Ponty, 1982).

L'uomo e il suo vissuto intenzionale sono dunque sempre conchiusi in una prospettiva e allo stesso tempo sono sempre protesi verso l'*orlo* di essa, sicché non vi sarebbe limite (una particolare interpretazione della Nona Sinfonia), né l'oggetto limitato (la sinfonia stessa) se non si ripresentasse nuovamente la gestualità strutturalmente intenzionale della coscienza (intenzionalità e trascendenza sono in Husserl indissolubilmente legate).

Ma va poi anche aggiunto, e questo per Husserl è un punto decisivo, che il "tipo" percepito (l'universale) *non è* slegato dalla serie potenziale delle mie altre percezioni ed anzi vi è necessariamente connesso. Se penetro chiaramente il delinearsi di una percezione (visiva, sonora e così via), accanto ad essa riconoscerò le altre, stante che i tipi sono tra loro in rapporto e l'uno è *indice* degli altri (considerazione preziosa anche per la sottesa teoria del linguaggio husserliana). Nella *Krisis*, Husserl scrive:

Da qualsiasi punto di vista consideriamo la cosa, risulta che qualsiasi essente, che valga in quanto realmente essente per me e per qualsiasi altro soggetto pensabile, è in una correlazione e, per una necessità essenziale, è indice della sua molteplicità sistematica (Husserl, 1954, trad. it. p. 193).

In questo senso, l'io non abita alcun luogo puro (cfr. Brand, 1955; Carr, 1987), ma è sempre interno all'esperienza (e come centro delle operazioni logiche tale io trascendentale mi impedisce un accesso alla verità senza che si sia verificato alcun contatto con l'esperienza contingente, come invece vorrebbe la moderna logica formale). Siccome ogni mio atto percettivo è *mio* e al tempo stesso è *tipico*, l'uomo si trascende fenomenologicamente restando *presso se stesso* (cfr. Sartre, 1965): non vi è più dualismo, opposizione tra immanenza e trascendenza (tra tempo ed eternità), bensì vi è un essere sé solo nella tensione ad altro (la quale è appunto l'orizzonte prerequisito della stessa intenzionalità della coscienza).

2. METAMORFOSI DEL "COGITO"

Su questa base, Husserl non può accettare gli sviluppi dei suoi allievi all'insegna del metodo fenomenologico (soprattutto quelli di Scheler e di Heidegger); essi, a sua detta, vorrebbero liberarsi dalla soggettività trascendentale senza averla capita. Non entrerò qui nel merito di questa convinzione, ma si può osservare come il problema per Husserl non sia tanto quello di irrigidire in uno schema assoluto la soggettività, quanto quello di pensare l'universalità del significato. La soggettività è anzi flusso concreto, immanente e intenzionale e lo è solo in quanto intenziona sempre qualcosa che è fuori di sé (cfr. Bernet, 1979). A ben vedere, questo è il punto che rende la questione dell'io husserliana e la sua conseguente idea dell'uomo irriducibile non solo all'io penso (*Ich denke*) kantiano, ma anche all'*ego cogito* di Cartesio. L'intenzionalità pone l'io stesso nel flusso della vita, ritenendolo, come già si osservava, non una *res*, un pezzo di mondo, bensì una *stratificazione di possibilità* (cfr. Husserl, 1952, pp. 253–255, trad. it. pp. 645–646): l'io non è una cosa, una sostanza, ma un orizzonte di possibilità, un processo nel processo della vita e del mondo (cfr. Semerari, 1991). L'*io posso* si sostituisce all'io penso cartesiano e kantiano. Come si vede cade qui ogni enfasi spiritualistica e umanistico-retorica e l'uomo si presenta in tutta la sua oscurità e passività, dipendenza e negatività (cfr. *ivi*, pp. 129–30). L'io come possibilità (impulsività, razionalità, ecc.) non è mai attualità totalmente dispiegata del sapere. Esso immette di contro a orizzonti possibili di cambiamento, anche se quest'ultimo, per Husserl, procede secondo una *storicità trascendentale* e non empirica, nel senso che "riflette la genealogia e l'ordine logico-ideale secondo cui le facoltà si originano, si dispongono, si connettono a partire dall'io (*ibid.*; cfr. Marbach, 1974). Se l'io non è, come per Cartesio, un pezzo di mondo che detta leggi senza condizoni (come se fosse un monarca assoluto), tutta l'antropologia platonico-cristiana, che in Cartesio trova una versione moderna e secolarizzata, ma perfettamente in linea con le proprie istanze, viene destituita di fondamento. Non solo non si puó pensare l'io-sostanza come indipendente dal mondo; non si può neppure ritenere che l'uomo si identifichi semplicemente con la sua mente, quale luogo della sua estraneità metafisica al mondo stesso e indice della sua destinazione ultraterrena (cfr. Semerari, 1991). Pensare l'io come sostanza significa voler trasporre su di un altro piano la tradizionale tendenza obiettivista della metafisica. E in fondo è proprio questo ciò che fanno prima Cartesio e poi Kant allorché allestiscono un principio gnoseologico (l'io) atto a far confluire in sé quei caratteri di astrattezza, omogeneità, stabilità, formalità, eternità che hanno fin dall'inizio contraddistinto l'obiettivismo, astratto esso stesso, metafisico (cfr. *ivi*, p. 130).

Husserl, di contro, vorrebbe riaffermare il soggetto e la razionalità in ordine alla concreta esperienza fenomenologica, senza pregiudiziali nei confronti della variabilità, della mobilità, dell'eterogeneità che ineriscono all'esperienza medesima. Si tratterebbe, in ultima istanza, di portare alla luce un percorso metodologico che sveli la tipicità dell'esperienza contingente nell'orizzonte della coscienza e nelle operazioni intenzionali dirette non tanto verso il significato quanto verso il *senso*. Che cosa è infatti l'intuizione delle essenze se non l'esplicitazione del senso verso il quale si orienta la coscienza? Tale senso tuttavia non è reale come lo sarebbe un oggetto posto di fronte nella rappresentazione; a questo livello esso è *irreale* (cfr. Welton, 1983). "Della *specie* rosso – come è stato messo in evidenza – si può dire che non esiste realmente né in questa striscia, né in qualunque luogo del mondo, né soprattutto *nel nostro pensiero*, poiché esso stesso fa parte dell'essere reale, della temporalità" (Raggiunti, 1970, p. 26). Tali significati formano una classe di "oggetti-generali". Ma "ciò non vuol dire che esistano in qualche modo, almeno in un mondo iperuranico o in uno spirito divino. Il loro essere coincide con il loro essere irreali" (*ibid*). Tuttavia, poiché possiamo esprimere giudizi veri su numeri o entità logiche, siamo legittimati a parlare di numeri ed entità logiche come di oggetti (cfr. *ivi*, pp. 26–27). Per cui, a detta di Husserl, il principio del parallelogrammo delle forze è un oggetto quanto la città di Parigi (cfr. Husserl, 1984). Proprio per questo "la non esistenza di un oggetto, rappresentato o pensato come oggetto genuino, non può privare la relativa rappresentazione del suo rappresentato come tale" (Raggiunti, 1970, p. 47). Una cosa è l'albero percepito nella tesi naturale, altra cosa è l'albero *percepito come tale*. Scrive in merito Husserl:

> L'albero inteso *ut sic* può bruciare, dissolversi nei suoi elementi chimici, ecc. Ma il senso – il senso di questa percezione, cioé qualcosa che appartiene necessariamente alla sua essenza – non può bruciare, non ha elementi chimici, forze, proprietà reali (Husserl, 1952).

Ciò vuol dire che la filosofia, in quanto autoproblematizzazione delle scienze (naturali e dell'uomo), non puó limitarsi alla considerazione del campo dei significati e delle tesi naturali ad essi corrispondenti, ma deve interrogare il rapporto che essi intrattengono con il *senso d'essere* delle cose (cfr. Atwell, 1977), con il senso così come esso si offre per me:

> L'unico punto di partenza radicale . . . sono io, io con la mia presenza, io che non posso riconoscere alcuna significatività se non muovendo da me stesso, dai modi intenzionali secondo cui tento l'approccio a essa e la esperisco (Semerari, 1991, p. 126).

Tra senso e significato sembrerebbe intercorrere dunque una costitutiva differenza semantica (centrale nel pensiero di Heidegger). In realtà, per Husserl, come si vedrà, questa differenza è ancora interna al significato. Essa ha sì infatti luogo nei vissuti, nei fenomeni della soggettività (cfr. Masullo, 1992, p. 125); ciò accade però solo perché essa è pensata come condizione del significato (il senso è visto, cioè, come condizione della totalità del semantico). Si potrebbe notare come sia qui, in questo nodo teorico, che la fenomenologia husserliana perde la sua *chance* di pensare una forma di soggettività effettivamente alternativa rispetto a quella cartesiana.

Quanto detto trova un riscontro rilevante se si pensa alla riflessione husserliana sul linguaggio. All'inizio della sua ricerca, Husserl è convinto che una seria filosofia del linguaggio debba raggiungere una grammatica universale, sufficientemente attrezzata per enucleare e descrivere, anche enumerandole, le "forme di *significato*", intuitivamente date, in cui si originerebbe il fenomeno stesso della lingua (cfr. Edie, 1976). Ora, è vero che questa convinzione verrà via via perdendo consistenza, sovrastata dal riconoscimento dell'impossibilità *tout court* di ritenere qualsiasi lingua formale ed empirica traduzione imperfetta o variabile di una lingua pura e universale (cfr. Sokolowski, 1972). Ciò nonostante, e senza lasciarsi sfuggire lo sforzo husserliano di una radicale trasformazione della nozione moderna di soggetto, il quadro che viene delineandosi resta senza dubbio legato al primato dell'*ideale*, del resto necessaria conseguenza della riduzione del senso a modo del significato. Per lui, ad esempio, è l'idea, il soggetto trascendentale, che giudica la storia e non viceversa, per cui alla *Weltanschauung* di Dilthey oppone una eidetica della storia medesima (vi sarebbero essenze storiche dalle quali nessun lavoro storiografico potrebbe prescindere). Certo, lo si ricordava, anche le idee, e Husserl lo rileverà sempre in modo maggiore, mutano e la razionalità husserliana vuole anche essere dimostrazione dell'innegabilità del divenire. Senonché, divenendo, le idee si approssimano ad una *storia intenzionale* della coscienza – in questo senso ancora intrascendibile – lungo la quale si articolano un ordine e un senso, pur sempre riconducibili ad una struttura della coscienza stessa costitutivamente immodificabile.

Anche quando Husserl ammetterà che il relativismo storico ha il suo diritto incontestabile come fatto antropologico (cfr. Ales Bello, 1991), egli non eviterà di sottolineare con forza che esso resta mera esposizione di fatti empirici e non descrizione fenomenologica di essenze (di atti intenzionali, dai quali deriverebbe il loro senso). E da questa prospettiva Husserl può dire che la conoscenza storica convive con orizzonti di senso che resterebbero puramente incomprensibili se esplicitati soltanto in quanto storici.

Ora, come è riconosciuto anche da studiosi che intendono recuperare il pensiero husserliano nell'ottica di un umanesimo scientifico e razionale, ma non scientista, né razionalista (cfr. Semerari, 1991), l'invariabilità dell'*Io puro*, quale *indipendenza* da ogni possible trasformazione, indica con chiarezza la presenza di un atteggiamento dogmatico e sostanzialistico, sul quale peraltro Husserl fa convergere, peggiorando le cose, il bisogno di *radicalità* della filosofia:

Ciò che è vero è assolutamente vero, è vero "in sé"; la verità è unica e identica, sia che la colgano nel giudizio uomini o mostri, angeli o dei. Le leggi logiche parlano della verità in questa unità ideale, di fronte alla molteplicità di razze, individui, vissuti; e noi tutti parliamo della verità in questa *unità ideale* a meno che non siamo confusi dall'errore relativistico (*Logische Untersuchungen*, 1975, p. 125, trad. it. p. 132, corsivo mio).

La verità è così soltanto verità del *giudizio*. La svolta fenomenologica iniziata con l'interpretazione del *cogito* in chiave antiplatonca subisce in questa direzione un pesante contraccolpo, in ragione del quale si concederebbe forse troppo a Husserl qualora si vedesse nella riproposizione di un fondamento sostanzialistico un semplice residuo non espunto (cfr. Semerari, 1991, pp. 130–131). Così facendo, anzi, Husserl reintroduce la finzione dell'io epistemologico dalla quale intendeva liberarsi, ripristinando sia il separatismo cartesiano, sia il riduzionismo antropologico della metafisica classica, per il quale, appunto, la mente (lo spirito) costituirebbe l'essenza dell'uomo e il soggetto altro non sarebbe che sostanza intelligibile.

Anche la critica a Kant e al neokantismo rischia in questo modo di vanificarsi, giacché la molteplicità di io puri, monadicamente differenziati con i loro flussi di coscienza contrapposti alla vuota formalità dell'io penso kantiano, finirebbe col costituire non un *processo di unificazione* in divenire e mai circoscrivibile, bensì un'unità presupposta all'incontro fenomenologico con il mondo. Ne seguirebbe la riattivazione della versione secolarizzata del monoteismo ebraico-cristiano, tradotto in termini di un io trascendentale unico, sovratemporale e sovrapersonale (cfr. *ivi*, p. 131).

Reperire le ragioni di questa duplice configurazione del pensiero di Husserl, di questa oscillazione e ambiguità non risulta senz'altro agevole, anche se è lecito pensare che egli, identificando il destino della filosofia con il compito di una radicale fondazione di tutti i saperi (cfr. Costa, 1990), si rendesse conto di non poter cedere al soggetto uno spazio eccedente il puro "in sé" della coscienza. In caso contrario, nessuna scienza descrittiva e rigorosa del soggetto come io – flusso di esperienza – vita risulterebbe possibile e questo Husserl non voleva ammetterlo. Nella tematizzazione dell'io puro e invariabile (negazione della discontinuità del tempo relativamente alla coscienza) è dunque in gioco una posta che va al di là di qualsivoglia rassicurazione contro le incrinature o perdite dell'identità personale che l'io puro implicherebbe (cfr. Husserl, 1966). In gioco è piuttosto il futuro di un'umanità scientifica razionalmente orientata, nell'ambito di un processo infinito guidato dalla ragione stessa dell'uomo (cfr. Roth, 1960). E il problema da razionale diventa *etico*, sebbene per Husserl il pensiero non sia fin dall'inizio che la prassi medesima, ossia il tentativo di descrivere e illustrare l'autofondazione razionale della condotta umana e il significato globale dell'ente-uomo (cfr. Sancipriano, 1988; Sini, 1987). Ma l'etica conseguente alla desostanzializzazione del *cogito* e alla messa in crisi del riduzionismo platonico-cristiano mirava al conseguimento di una *Selbstverantwortung* (auto-responsabilità) mai conclusa in se stessa e aperta al principio del *Zusammengehören* (cfr. Semerari, 1991, p. 132 e ss.), cioè del riconoscimento della posizione del sé in quanto correlato al riconoscimento della posizione dell'altro (cfr. Theunissen, 1965; Waldenfels, 1971). Nella *Selbstverantwortung* l'uomo ritrova se stesso nell'altro e all'altro risponde delle proprie produzioni di senso. Di contro, l'assolutizzazione dell'io puro piega l'etica dell'autoresponsabilità verso il dover-essere fichtiano e kantiano, basato su regole presupposte al pensare e alla condotta pratica. Rispetto all'io puro, in cui kantianamente può tradursi la legge morale, gli io reali risulterebbero unicamente io empirici, il cui compito etico sarebbe quello di sollevarsi all'eterna purezza della coscienza. Il valore dell'altro cesserebbe di essere tale in forza della sua alterità costitutiva, divenendo semplice forma storica ed empirica da adeguare al modello dell'io puro. Prevedibili, a questo punto, diventano le conseguenze pratiche che tutto ciò porta o potrebbe portare con sé (ad esempio renderebbe legittimo il ricorso alla violenza in vista della presunta adeguazione di un io empirico al presunto modello puro di soggettività).

3. DA HUSSERL A HEIDEGGER

Le due anime che percorrono l'opera di Husserl sembrano qui condurre Husserl stesso di fronte a un bivio, sembrano chiamare ad una scelta, se è vero che nel 1935, ormai completamente isolato, annoterà per sé la seguente considerazione:

> Contestazione della filosofia scientifica – necessità della considerazione – la considerazione storica – qual è la necessità della storia? La filosofia come scienza, come una scienza seria, rigorosa, anzi apodittica – *il sogno è finito* (. . .). Io so naturalmente ciò che perseguo sotto il titolo di filosofia, so qual è lo scopo e il campo del mio lavoro. E tuttavia, insieme, non lo so. A quale autonomo pensatore è mai bastato questo "sapere", per quale pensatore, nella sua vita filosofica, la "filosofia" ha cessato di essere un enigma? (*Krisis*, 1954, p. 535).

Hans Georg Gadamer ritiene errato considerare tali parole come l'opinione propria di Husserl, il quale invece avrebbe, fino alla fine, combattuto il relativismo storicista come una "fatale rovina" (cfr. Gadamer, 1963, p. 53). Le considerazioni sulla storia, sul mondo-della-vita costituirebbero soltanto l'avvenuta presa di coscienza circa la necessità di una tematizzazione della storicità, senza che ciò comporti alcuna rinuncia al progetto di conseguire una visione scientifica del soggetto e dell'assoluto (cfr. *ivi*, p. 54). Su questo, del resto, la maggior parte degli interpreti sostanzialmente concorda. Resta altresì da pensare la portata, non riconducibile mi pare ad una mera opinione che Husserl non condivise e combatté come una fatale rovina, della domanda sul carattere strutturalmente enigmatico della filosofia, in nome del quale cessa ogni certezza fondativa ultima, ogni carattere che necessariamente converrebbe alla filosofia in quanto tale.

Come pensare allora i "supremi interessi" della civiltà umana, di una umanità scientifica e razionale se l'edificazione di una filosofia come scienza rigorosa è preclusa e lo è non perché altri possiedano convinzioni opposte, storiche, ma perché la filosofia è per se stessa enigmatica? Quando oggi Apel e Habermas si chiedono se sia possibile fondare un'etica razionale già limitano l'ampiezza della questione posta da Husserl nella *Krisis*. Si può parlare di una umanità scientifica filosoficamente fondata oppure la crisi della metafisica non può che coincidere con la cosiddetta "morte dell'uomo" e con l'abdicazione senza residui del soggetto? Anche rielaborando l'orizzonte ermeneutico e preriflessivo dell'intenzionalità (così come esso si prospetta in Heidegger), si tratterebbe qui di individuare, come è stato legittimamente affermato, il luogo di una identità sottratta alla "costruzione" del soggetto, e cioè oltre il mito di una soggettività fondante, e calata invece nell'orizzonte del mondo, dal quale soltanto può trarre la propria consistenza (cfr. Rovatti, 1986). Questa direzione mi sembra sia quella intrapresa da Heidegger – contro Sartre –, il quale nel seminario di Zähringen (cfr. Heidegger, 1977b) riprende la tesi esposta da Marx nei *Contributi alla critica della filosofia del diritto hegeliano*: "Essere radicale vuol dire cogliere le cose alla radice. Ma la radice, per l'uomo, è l'uomo stesso" (Marx, 1953, p. 216). L'uomo di Marx, osserva Heidegger, coincide con la produzione; e la società, che produce se stessa, non è altro che l'autoproduzione dell'uomo come essere sociale (cfr. Heidegger, 1977b; Lukács, 1951). Rappresentare il pensiero secondo il modello della produzione significa però ancorare l'essenza dell'uomo – l'espressione è di Heidegger – ad una metastorica e metafisica sicurezza, dimenticando che essa è di volta in volta in gioco nell'esistenza. E per Heidegger l'esistente concreto, l'esser-ci, è quell'ente sulla base di cui ne va dell'essere stesso (cfr. Heidegger, 1977b).

Sempre nel protocollo del seminario di Zähringen si legge poi:

l'uomo inteso come esser-ci, come e-statica insistenza nella radura dell'essere, si contrappone a ciò che asserisce la proposizione marxiana. Si può allora dire che per Heidegger l'esser-ci sarebbe la radice dell'uomo? No. Il concetto di "radice" rende impossibile portare al linguaggio il riferimento dell'uomo all'essere (cfr. *ivi*, trad. it. p. 166).

Togliendo all'esser-ci il suo carattere di apertura, si nega all'esistenza la possibilità di implicare un orizzonte di senso estraneo alla circostanzialità significante dell'esistere stesso. Questo è ciò che Heidegger intende escludere, visto che il senso verrebbe perciò a delinearsi, cosa che in Husserl non accade, come emotività originaria, come esperienza destabilizzante del tempo, non riducibile ai pubblici significati della ragione (cfr. Masullo, 1993). L'essenza dell'uomo dunque è in gioco nell'esistenza come apertura, ossia nell'incontro con il mondo che la cura dischiude, a partire dal quale in Husserl si delinea l'orizzonte già richiamato del *Zusammengehören*, dell'implicanza reciproca, del rinvio costitutivo entro il quale si legano io, alterità e mondo. Tale rinvio resta peraltro una pura possibilità che non riconduce l'uomo nell'ambito di una decisione libera da condizionamenti, bensì lo colloca di fronte alla non pienezza, alla negatività, alla dipendenza, che pure costituiscono percorsi irrinunciabili della stessa soggettività (cfr. Rovatti, 1988). E' forse questo il punto di massima convergenza ma anche di massima distanza tra la fenomenologia husserliana e Heidegger. La definizione del *Zusammengehören* diventa centrale per l'ermeneutica della fatticità e per un'etica dell'interpretazione, le quali così diventano indissolubilmente debitrici nei confronti della critica alla concezione metafisica del trascendentale (da Cartesio a Kant e ai postkantiani) come obiettivismo travestito. Per converso, l'idea di un destino teleologico della ragione in cui culminerebbe la rilevanza dello stesso *Zusammengehören*, chiude l'uomo in se stesso e lo circoscrive alle proprie gestualità intenzionali, protese al significato indipendentemente dal senso. Attraverso questa operazione, Husserl intende ancora dogmaticamente il principio di autoresponsabilità come legittimazione del senso prodotto dalla coscienza.

In Heidegger, invece, il problema dell'autoresponsabilità si configura come dimensione etica in cui l'uomo rinuncia a norme, valori, giudizi – alle codificazioni della morale – per ritrovare, al di sotto della quotidianità del *Man*, dell'oblio dell'essere e dell'imposizione della tecnica, l'originario interesse che lo costituisce (ossia la cura per l'esistenza – cfr. Masullo, 1993). Rispetto ad esso non c'è filosofia pratica, ragione teleologica, progettualità politica che non sia astratta costruzione di interessi fittizi. L'intersse per l'esistenza (in quanto dimensione dell'essere sempre coinvolti in un mondo) è qui il vissuto profondo e originario dell'esser-ci, se l'uomo è gettato e quindi sempre inserito in un contesto che lo precede (cfr. Heidegger, 1977a). In questo interesse ha luogo la domanda sul senso e quindi l'avvio di una *identità problematica* dell'ente-uomo, la quale non solo non viene meno con la crisi delle prospettive metafisiche tradizionali, cedendo al relativismo, ma abbisogna di essa. Vi è infatti un piano in cui – cosa che Husserl non vide intrappolato dalla dicotoma sapere assoluto-relativismo – l'universalità (non più intemporale, ma radicalmente storica) si caratterizza come capacità di porre domande, di corrispondere a ciò che si dà a pensare (cfr. Heidegger, 1977b). Di fronte alla domada circa il proprio destino, l'esser-ci non può che arrestarsi. E tuttavia l'uomo non può non porre tale domanda (il *chi* del soggetto post-metafisico è l'interrogante stesso). Se non lo facesse o se rispondesse – sapesse rispondere – non sarebbe più tale. L'esser-ci, infatti, non può rendere conto della sua esistenza, né a maggior ragione dell'essere intenzionato dal suo domandare. Ciò nonostante, l'uomo non può evitare di interrogarsi sul senso, perché non sarebbe "gettato" nel mondo se non si desse l'orizzonte problematico del "donde" e del "verso dove" (cfr. Heidegger, 1977a).

Mantenersi nel domandare vuol dire allora aprirsi alla dimensione etica del senso e dell'interpretazione, seguendo l'esser-ci nella sua singolare fatticità, mai, così, astrattamente ricoperta e disattesa. Il programma fenomenologico delle origini risuona precisamente in questo imperativo: svelare ciò che di volta in volta appare nel modo in cui appare. In quest'ottica non l'uomo muore, ma l'immagine che di esso ha dato la onto-teologia, con il suo prolungamento nella tecnica, ma rispetto alla quale ha luogo, in Heidegger, un'apertura etico-ontologica non del tutto estranea alla fenomenologia husserliana ed anzi fondamentalmente ispirata ad essa.

University of Perugia
Italy

RIFERIMENTI BIBLIOGRAFICI

AA. VV., *Effetto Foucault*, a cura di P. A. Rovatti, (Milano: Feltrinelli, 1986).

AA. VV., *Razionalità fenomenologica e destino della filosofia*, a cura di A. Masullo e C. Senofonte (Genova: Marietti, 1991).

Ales Bello, A., "Per una fenomenologia come archeologia della cultura," in AA. VV., 1991, *Razionalità fenomenologica e destino della filosofia*, *op. cit.*, pp. 207–218.

Atwell, J. E., "*Husserl on Signification and Object*," in *American Philsophical Quarterly* 6, 1969, ora in *Readings on Edmund Husserl's Logical Investigations*, ed. by J. N. Mohanty (The Hague: M. Nijhoff, 1977), pp. 83–39.

Bernet, R., "Bedeutung und intentionales Bewußtsein. Husserl's Begriff des Bedeutungsphänomens," in E. W. Orth (Hrsg.), *Studien zur Sprachphänomenologie, Phänomenologische Forschung* 8 (Freiburg-München: Alber, 1979), pp. 31–64.

Brand, G., *Welt, Ich und Zeit. Nach unveröffentlichten Manuskripten Edmund Husserls* (Den Haag: M. Nijhoff, trad. it. di E. Filippini, *Mondo Io e Tempo nei manoscritti inediti di Husserl* Milano: Bompiani, 1960).

Carr, D., *Interpreting Husserl. Critical and Comparative Studies* (Dordrecht, Boston, Lancaster: M. Nijhoff, 1987), pp. 227–246.

Costa F., *Introduzione* a E. Husserl, *La filosofia come scienza rigorosa* (Pisa: ETS, 1990).

Derrida, J., *Introduzione a Husserl. "L'origine della geometria,"* a cura di C. Di Martino (Milano: Jaca Book, 1987).

Edie, J. M., *Speaking and Meaning. The Phenomenology of Language* (*Studies in Phenomenology and Existential Philosophy)*, (Bloomington, London: Indiana University Press, 1976).

Gadamer, H. G., *Die Phänomenologische Bewegung*, in *Kleine Schriften*, III (Tübingen: Mohr, 1963); trad. it., *Il movimento fenomenologico* (Roma-Bari: Laterza, 1994).

Heidegger, M., (14). *Sein und Zeit* (1927) (Tübingen: Niemeyer, 1977). trad. it. a cura di P. Chiodi, *Essere e tempo* (Torino: Utet).

Heidegger, M., *Seminari*, trad. it, a cura di M. Bonola e F. Volpi, (Milano: Adelphi, 1977; 1992).

Held, K., "Das Problem der Intersubjektivität und die Idee einer phänomenologischen Transzendentall philosophie," in *Perspektiven transzendental-phänomenologischer Forschung*, pp. 3–60, hrsg. von M. Claesges und K. Held (Den Haag: M. Nijhoff, 1973).

Husserl, E., *Cartesianische Meditationen und Pariser Vorträge*, "Husserliana" I (Den Haag: M. Nijhoff, 1950); trad. it. a cura di F. Costa, *Meditazioni cartesiane. Con l'aggiunta dei discorsi parigini* (Milano: Bompiani, 1989).

Husserl, E., *Ideen zu einer reinen Phänomenologie und phänomenologische Philosohie. Zweites Buch: Phänomenologischen Untersuchungen zur Konstitution*, "Husserliana" IV, cit., trad. it. a cura di G. Alliney integrata da E. Filippini, *Idee per una fenomenologia pura e una filosofia fenomenologica* (Torino: Einaudi, 1952; 1976 (3)).

Husserl, E., *Die Krisis der europäischen Wissenschaften und die transzendentale Phänomenologie. Eine Einleitung in die phänomenologischen Philosophie*, "Husserliana" VI, cit. trad. it. a cura di E. Filippini, *La crisi delle scienze europee e la fenomenologia trascendentale* (Milano: Il Saggiatore, 1954; 1987 (7)).

Husserl, E., *Zur Phänomenologie des inneren Zeitbewußtsein* (1893–1917), "Husserliana," cit., trad. it. a cura di A. Marini, *Per la fenomenologia della coscienza interna del tempo* (Milano: Angeli, 1956; 1982).

Husserl, E., *Philosophie der Arithmetik. Mit ergänzenden Texten* (1890–1901). "Husserliana," cit., hrsg. von L. Eley, 1970.

Husserl, E., *Logische Untersuchungen. Erster Band: Prolegomena zur reinen Logik*, "Husserliana," cit., trad. it. a cura di G. Piana, *Ricerche logiche* (Milano: Il Saggiatore, 1975; 1988 (3), vol. 1).

Husserl, E., *Logische Untersuchungen. Zweiter Band: Untersuchungen zur Phnänomenologie und Theorie der Erkenntnis. Zweiter Teil*, "Husserliana," 2, cit., trad. it. a cura di G. Piana, *Ricerche logiche* (Milano: Il Saggiatore, 1984; 1988 (3), vol. 2).

Landgrebe, L., *Der Weg der Phänomenologie. Das Problem einer ursprunglischen Erfahrung, Gütersloher Verlagshaus Gerd Mohr, Gütersloh*, trad. it. a cura di G. Piacenti, *Itinerari della fenomenologia* (Torino: Marietti, 1963; 1974).

Lukács, G., *Existentialismus oder Marxismus* (Berlin: Aufbau Verlag, 1951; 1963).

Kern, I., *Husserl und Kant. Eine Untersuchung über Husserls Verhältnis zu Kant und zum Neukantianismus* (Den Haag: M. Nijhoff, 1964).

Marbach, E., *Das Problem des Ich in der Phänomenologie Husserls* (Den Haag: M. Nijhoff, 1974).

Marx, K., *Die Frühschriften* (Stuttgart: Kröner, 1953; 1968 (2)).

Masullo, A., "Il pensiero radicale e l'indicibile tempo," in AA. VV., 1991, *Razionalità fenomenologica e destino della filosofia*, cit., pp. 3–19.

Masullo, A., "Epistemologia dell'irreversibile ed etica del tempo," in *Paradigmi* XI, 31, 1993, pp. 125–152.

Melandri, E., *Contro il simbolico* (Firenze: Ponte alle Grazie, 1989).

Merleau-Ponty, M., "Il problema delle scienze umane secondo Husserl," in AA. VV., *Fenomenologia e tempo*, a cura di M. C. Franza (Roma: Edizione dell'Ateneo, 1982), pp. 21–32.

Paci, E., *Il senso delle parole (1963–1974)*, a cura di P. A. Rovatti (Milano: Bompiani, 1987).

Pöggeler, O., "La crisi del concetto fenomenologico di filosofia (1929)," in AA. VV., 1991, *Razionalità fenomenologica a destino della filosofia*, cit. pp. 41–60.

Putnam, H., *Reason, Truth and History*, Cambridge University Press, trad. it., *Ragione, verità e storia* (Milano: Il Saggiatore, 1989 (2)).

Raggiunti, R., *Introduzione a Husserl* (Roma-Bari: Laterza, 1970; 1991 (5)).

Ricoeur, P., "Phénoménologie et herméneutique," in Id., *Du Texte a l'action* (Paris: Seuil, 1975; 1986), pp. 39–73.

Riverso, E., "Le scienze a cinquant'anni dalla 'Krisis'," in AA. VV., 1991, *Razionalità fenomenologica e destino della filosofia*, cit., pp. 261–267.

Roth, A., *Edmund Husserls ethische Untersuchungen* (Den Haag: M. Nijhoff, 1960).

Rovatti, P. A., "I luoghi del soggetto," in AA. VV., 1986, *Effetto Foucault*, cit., pp. 71–65.

Rovatti, P. A., *La posta in gioco. Heidegger, Husserl, il soggetto* (Milano: Bompiani, 1988).

Sancripriano, M., *Edmund Husserl. L'etica sociale* (Genova: Tilgher, 1988).

Sartre, J. P., *La Trascendance de l'Ego*, Paris, Vrin. trad. it. di N. Pirillo, *La trascendenza dell'ego* (Napoli: Arturo Berisio Editore, 1965; 1971).

Schütz, A., "Das Problem der transzendentalen Intersubjektivität bei Husserl," in *Philosohische Rundschau*, V, 1957, pp. 81–107.

Semerari, G., "Auto-responsabilità. L'idea di uomo in Husserl," in AA. VV., 1991. *Razionalità fenomenologica e destino della filosofia*, cit., pp. 126–149.

Sini, C., *La fenomenologia e la filosofia dell'esperienza* (MIlano: Unicopli, 1971; 1987).

Sini, C., "Introduzione," a E. Husserl, *L'idea della fenomenologia* (Roma-Bari: Laterza, 1992).

Sokolowski, R., *Presence and Absence. A Philosophical Investigation of Language and Being* (Bloomington: Indiana University Press, 1978).

Theunissen, M. *Der Andere. Studien zur Sozialontologie der Gegenwart* (Berlin: De Gruyter, 1965).

Tilliette, X., *Breve introduzione alla fenomenologia husserliana*, a cura di E. Garulli, (Lanciano: Editrice Itinerari, 1983).

Vattimo, G., *La fine della modernità* (Milano: Garzanti 1985).

Vattimo, G., *Oltre l'interpretazione* (Roma-Bari: Laterza, 1994).

Vitiello, V., "Heidegger: l'Oggettività del mondo," in AA. VV., *Heidegger e la metafisica* (Genova: Marietti, 1992), pp. 277–294.

Waldenfels, B. *Das Zwischenreich des Dialogos. Sozialphilosophische Untersuchungen in Anschluß an Edmund Husserl* (Den Haag: M. Nijhoff, 1971).

Welton, D., *The Origins of Meaning. A Critical Study of the Thresholds of Husserlian Phenomenology* (The Hague, Boston, Lancaster: M. Nijhoff, 1983).

SILVANA PROCACCI

COMPLEXITY AS THE NODAL POINT FOR A PHENOMENOLOGY OF LIFE

1. INTRODUCTION

The study of the birth and evolution of the Universe and its connection with life is one of the most interesting inquiries of physics and philosophy.

Contemporary science is quite sure that, even if all the basic laws could be subsumed to a single set of equations, there would still be the question of why there is a Universe.[1] In particular the cosmic mystery that follows on our physical and philosophical research is that of whether the emergence of life was almost inevitable or a rare fluke.

The purpose of this article is to enrich the consideration of the problem of life with an exploration of the concept of complexity. The latest studies of complex systems suggest that order (and in particular intelligent life) could be an unavoidable effect of physical and chemical laws themselves. In fact it seems that there is a tendency to order, to self-organization, to complexity in the Universe. Through the development of three strictly interconnected themes, we will try to arrive to an interpolation between the human and the natural sciences with the aim of sketching some reflections for a phenomenology of life. We will examine some philosophical-scientific subjects to make a connection between complexity, consciousness and the evolutionary laws of the Universe. This project will develop the following themes: 1) the meaning of complexity and living organization in the Universe despite the apparent simplicity of the natural laws; 2) the place of man in nature; and 3) the moral dimension emergent with the particular condition of human being.

It is possible to connect these arguments considering (fundamentally) the nature of human creativity and our numerous capacities. In fact the human condition can be understood best if it is organized inside a philosophy which reflects the natural context in which it is immersed. For this reason we will explain the implications of a philosophical and scientific debate the main contribution of which is the foundation of a new humanism within contemporary culture.

The human condition, the natural order and organized complexity

A-T. Tymieniecka (ed.), Analecta Husserliana LIX, 233–247.

support a philosophical theory that expresses a possible cosmic meaning to be ascribed to living reality. Completing the vision of the horizon opened by modern scientific research with provocative speculations – that originate in their content in the philosophical tradition of our century – it is possible to interpret the phenomenon of life in all its forms as the result of a cosmic drive to complexity.

The points that will be developed in this paper are strictly connected with the arguments of A-T. Tymieniecka in her phenomenological analysis of life and the human condition. Nevertheless, the present perspective is also inspired by Bergson's and Teilhard de Chardin's philosophical reflections.

2. THE HERMENEUTICS OF COMPLEXITY: TWO HISTORICAL ANTECEDENTS

The success of the mechanistic interpretation of nature led to the hope that all natural phenomena were knowable reducing them to the mutual motion of elementary particles. In this methodological approach the natural reality is naturally deprived of any justification with respect to the capacity for coherence, co-ordination, unity, freedom and self-organization present in proportion to the complexity of natural phenomena. This deficiency is especially severe for such systems as living beings, in which it is possible to find many levels of organization: in these systems the perception of the highest level of organization is necessary to understand the entire state of the phenomenon, while the aim of the simplest component appears to interact harmoniously to realize the whole. The mechanistic model has shown undeniable predictive power as well as its limits, as reported in recent years not only in the human sciences, but even in the natural sciences. Due to a reconstruction of reality reduced to simple deterministic laws, man is fallen from the privileged and unconditional sphere he occupied until the Copernican and Darwinian revolutions. The modern epistemological debate over the ruling model in scientific culture before the beginning of twentieth century has been followed by a valorization of the contingent and historical aspects of the cosmos. This philosophical approach has raised many traditional problems such as the relation between knowledge and being, the secret of life's existence and lastly the relation between man and Universe. A phenomenological analysis of nature shows us that the tendency to complexity seems to be the principle or the essence of cosmic structure, verifiable in all its human and natural expressions. Already Bergson and Teilhard de Chardin attempted to investigate some natural laws that allow us to explain natural phenomena as the result of a cosmic push towards complexity. In these well-known theories, it is possible to maintain the existence of a wisely orientated evolutionary process that presides over the rise of order and of complex structures.

Bergson in *L'èvolution créatrice* supports an idea of cosmological evolution in which it is possible to grasp the presence of a natural tendency to complexity. The *èlan vital* represents the principle of this movement, which in its deep sense is a creative process, free because unpredictable, but still directed toward the creation of new forms. The thought pays tribute to this force that pushes matter toward complexity. For this reason awareness represents the highest result of the cosmos's tendency. Bergson's organicist philosophy tries to find the conceptual and methodological elements that can show the order and the complex structure of nature.

Pursuing this aim, Teilhard de Chardin tries to build an anthropogenesis that is an epilogue of the entire universal evolutionary effort. In the philosophy of Teilhard de Chardin there is a developing of the whole natural process in the Universe by means of laws that guarantee complexity and organization as in the case of human thought. In fact Teilhard investigated the way in which cosmic evolution is connected with the direction of organic evolution, which individualizes the prominent scientific theme of the connection between Universe and life. In Teilhard's vision the Universe exerts an influence over the living sphere, a sort of action of the whole on the parts that is visible in the increase in biological complexity. In the opinion of J. Piveteau, the philosophy of Teilhard de Chardin asserts that "La vie n'est point une combinaison fortuite d'éléments matériels, une accident de l'histoire du monde, mais le forme que prende la matiére à un certain niveau de complexité".[2]

Teilhard defends an evolutionary movement turned, in its inner changes, to a purpose, corresponding to more and more organized hierarchical levels, which points to a real cosmic orthogenesis. He develops a working hypothesis that tries to settle the relation between Universe and living sphere, attributing to the evolutionary process a precise direction or a project: the increasing of complexity. The ponderous philosophy of life that emerges in this vision, develops a real science of the human future observing a universal law of matter which points to the human phenomenon. Universal evolution increases in complexity going from simple particles to human consciousness, through a long but relentless course toward an Omega Point epilogue of the universe evolutionary effort. Human biological evolution has become cultural and, for Teilhard, it will find its accomplishment in a global spiritual process.

In this way, intelligence, as the architect of this marvelous edifice, becomes the very complex final result of natural interactions. Due to this capacity man is not only at the centre of the Universe, but he becomes the highest result of this universal inclination to complexity. For Teilhard, these philosophical reflections – that exalt anthropological discourse – have an objective foundation because they carry out a precise principle of universal development: the law of complexity-awareness. In this law Teilhard sees a progressive trend of matter toward organization. In this trend we can see a firm attempt to reconstruct the ancient unity between man and nature, broken by the birth and the development of modern science.

Teilhard's position is sustainable assuming a causal action of the whole on the parts, in other words supposing a "care", an "interest" on the part of the Universe for life. Due to this holism, these reflections advance themes and research paradigms that have been debated through the last decades by philosophers of science.

3. THE SELF-ORGANIZED UNIVERSE: THE ANTHROPIC PRINCIPLE

The new humanism that is opened through Bergson's and Teilhard de Chardin's philosophical positions is based on the consideration of mankind as the supreme product of natural evolution able to express the growing combination of universal circumstances. Teilhard's cosmogenesis adopts a holistic attitude justifying man's place in Nature. In this section we will try to describe how life and natural order may be generated out of chaos and complexity through the universal laws.

Over the centuries, science has developed an extensive system of answers to account for how the enormous diversity of life on Earth has arisen. It's very interesting to note that some themes of modern research in bio-cosmological science are characterized by an emphasis on the principle of complex organization in the Universe. The recent formulation of the Anthropic Principle takes up again themes that, on the contrary, a certain neo-positivistic mentality rejects as metaphysical knowledge. In contrast this principle posits a strong nomothetic relationship between macrocosm and microcosm, repositioning man (and life in general) at the center of the Universe. The Anthropic Principle says that we can deduce the values of various fundamental constants in nature from the fact that we are around to ask what it is. But we still do not understand the physical reasons for such a value to occur unless we suppose that the constants have the values that they have for the purpose of making intelligent beings (and life) possible. J. D. Barrow and F. Tipler assert that the "observed values of all physical and cosmological quantities are not equally probable but they take on values restricted by the requirement that there exist sites where carbon-based life can evolve and by the requirement that the Universe be old enough for it to have already done so".[3] In the *Review of Modern Physics*, physicist R. Dicke has said that the physical constants are "not random, but conditioned by biological factors"[4] because the organisms must exist in order that these constants can be measured.

Barrow and Tipler add that "the Anthropic Principle in each of its various forms attempts to restrict the structure of the Universe by asserting that intelligent life, or at least life in some form, in some way selects out the actual Universe from among the different imaginable Universes: the only 'real' Universes are those which can contain intelligent life, or at the very least contain a form of life". This definition should not appear unreasonable because "those properties of the Universe we are able to discern are self-selected by the fact that they must be consistent with our own evolution and present existence".[5]

Starting from the enigmatic presence of the life, the Anthropic Principle tries to explain the cosmic dynamic assuming the conditions required for the preservation of the biosphere to be fundamental and restrictive. For this reason we can describe the Anthropic Principle as a methodological standard that for its argumentations begins with the observation that natural laws permit with incredible precision the existence of cosmological conditions favourable to life. In fact cosmological science has confirmed that if the cosmos' most important physical and chemical properties were changed or were to change in the future, this movement would be disastrous for life as we know it due to its extreme sensitivity to any change in its fundamental parameters.

It has been noted that life is possible because the constants and parameters for the Universe and for the planet Earth – such as the strong and weak nuclear forces, the electromagnetic coupling constant, the fine structure constants, the distance of the Earth from the Sun, etc. –

lie within certain highly restricted ranges.[6] In fact complex evolution should be possible only in a Universe where the constants, the numbers of dimensions, etc. . . . have favourable values.

We can see that in an inexplicable manner these values are those actually observed, with the important consequence that life can establish itself. The incredibly exact balance of the complex organization of the biosphere challenges whoever thinks of neglecting life's position in framing the cosmological hypothesis given the narrowness of the apparent – and not assured – spatial and temporal window of opportunity. With the help of the Anthropic Principle it is reasonable to suppose (or at least it is a hypothesis that even the fiercest enemies of teleological discourse have to take seriously into account) that the Universe is structured to make possible the living order that from the simplest particle advances to the conscious dimension. For this reason the Anthropic Principle proposes itself as a scientific and philosophical argument able to speak about the human condition in the Universe. The Anthropic Principle can provide scientific support for a re-estimation of life (above all its intelligent forms) so that it does not appear to be an insignificant fluctuation in matter but the expression, although localized and fragile, of the sense of universal evolution. The Italian physicist A. Masani notes that with the Anthropic Principle,

> rientrava nell'ambito della scienza il problema dell'uomo e vi rientrava como processo strettamente legato a tutto il fenomeno evolutivo dell'universo in se stesso. Quell'uomo che la metodologia galileiana aveva relegato nella posizione dell'osservatore in un Universo che evolve e procede secondo leggi proprie, le quali con l'esistenza e la peculiarità dell'uomo non hanno nulla a che fare, torna adesso sulla scena cosmica come condizionante l'evoluzione stessa dell'universo tramite la propria facoltà conoscitiva di *physicist*.[7]

4. FROM DISORDER TO ORDER: AN INTERPRETATION OF ORGANIZATION

At one time man perceived the world as total chaos but, as generations passed, the regularity of the solar day, the repetition of seasons, and the regularity of the sidereal year introduced man to the notion of predictability and order. This was a notion which expanded and led him to believe in a completely ordered and predictable Universe. Specifically, as emphasized by Laplace, the Newtonian dynamic which governs all the macroscopic world is deterministic, meaning that the past and the future follow from its present state.

The mechanistic description of nature gives us the possibility of understanding the world in a scientific manner because it concentrates on simple, highly controlled systems. Physics deals with invented, simplified materials, reducing what is possible from the complex to the simplex. Mid-twentieth century man believed in the triumph of determinism at both the microscopic and macroscopic levels. Unfortunately, this methodological model is no longer convincing. In some fields of scientific research it has become clear that complexity prevails in many more natural and social systems than had been thought. In addition, Chaos Theory tells us that simple laws can have complicated – indeed, unpredictable – consequences. Simple causes can produce complex effects: the most startling revelation of contemporary dynamics is the discovery that Newtonian mechanics can break down into chaos in a system having only three degrees of freedom. For this reason, in order to study of these systems knowledge of the every single particle is no longer sufficient, it being necessary to consider the system as a whole. This methodological model issues a challenge to the classical reductionistic view – supported in Western scientific thought – and it outlines reality according to levels of organization, as is typical in organic (holistic) and ecological philosophy of nature.

The paradigm of complexity therefore is not only a *weltanschauung* by which to explain the ontology of the natural world, but proves to be a powerful instrument by which to analyze phenomena. The amazing success obtained in the molecular sphere through applying the analytic-reductionistic model have not decreed the end of certain claims of the organicist approach. In fact, the living organism displays a behaviour that does not reveal itself in material and inorganic reactions. The ancient vitalists were right to uphold the peculiar aspect of the living being, but they isolated this difference in an unknowable element, in a vital strength, that evaded (by definition) all attempts at scientific conceptualization. In contrast, the study of living systems in an analytic or mechanistic manner requires the isolation of parts (sub-systems) with the consequent destruction of the organic or holistic character of biological phenomena. The requirement of a conciliation between the evident difference between biological phenomena and the pretensions of their scientifically acceptable description, seems to be satisfied by the concept of organization.

G. Nicolis says that "Living beings are undoubtedly the most complex and organized objects found in nature, both in view of their morphology and their functioning. They are literally historical structures, since they have the ability to preserve the memory of forms and functions acquired in the past, during the long period of biological evolution".[8] To describe these systems adequately requires a holistic method, one that provides an interpretation unlike any in the mechanistic line. The concept of organization, one that posits a distinction between living organisms and inanimate objects on the basis of complexity in structure and processes, was already singled out by Teilhard de Chardin by means of a fortuitous intuition of an orthogenic development guided by the abovementioned law of complexity-awareness. More recently, this idea has been taken and formalized again in so-called Complex Systems Theory, acquiring greater scientific power. Although this line of research embraces very different fields – from cybernetics to ecology, from chemistry to psychology – on this occasion we suggest the contribution of this new model to the problem of the self-positioning of life. What emerges from this approach is the interpretation of complexity as an explicative paradigm for a natural philosophy able to revalue aspects of life – such as freedom, historicity, order and evolution – neglected or ignored in the reductionistic model.

In complexity theory evolution the process of evolution is not controlled by blind chance, but represents an unfolding of order and complexity. Though single processes still are not predictable, there is a verifiable natural movement toward organized complexity, as Teilhard de Chardin pointed out.[9] Thus we can speak of an organized disorder, which recalls ancient Heraclitean intuitions. Very interesting research has been carried out by Prigogine that tries to replace the idea of the development of living being through the interaction of chance with another evenmore, fascinating principle, a concept of life as a potentially preformed phenomenon, the expression of a virtuality enrolled in the natural code. With this scientific hypothesis, we can say that nature's dice are only slightly but nonetheless purposefully loaded. The scientific task will thus be to determine the loading and its purpose. In particular the theory of complex systems suggests a tendency for the same natural laws to produce organizations during special far-from-equilibrium thermodynamic conditions, pointing out that in many situations it is impossible to appeal to any deterministic explication because the ineluctable indeterminateness of the initial condition values increased during such times. The complexity paradigm began as an original methodological instrument which does not dismiss the qualitative characteristics of an object as fallacious, since it seeks understanding that object only through reference to the organization of a system. To understand complex systems, we do not refer to a composition of parts, but to a unitary organized interaction, endowed with properties unknown at the level of its isolated factors. This new model has very interesting consequences with respect to the ontological dimension: the concept of system replaces the notion of element and form-essence is not something static but an organized complex unity always in movement – a dynamic state of equilibrium – the product of a holistic interaction of elements inside and outside the system.

The theory of the self-organizing systems helps us to reflect in a new manner on the problem of time. In fact, where in classical mechanics the arrow of time is reversible and considered a parameter among others, the science of systems has discovered dissipative structures – or systems far-from-equilibrium – that produce order as long as there are relations between various circumstances. In studies of complexity, irreversibility is found to be involved in new phenomena of organization. This is exactly the contrary of what is assured in the classical deterministic paradigm, which holds the temporal arrow to be a fiction framed by our psychological approximations and in our ignorance, while nature is conceived to be fundamentally determined and reversible. In the exciting new perspective, on the contrary, life, order, and individual organisms are unpredictable results of particular breaks of symmetry occurring in complex systems: in these conditions we are always faced with an ample possible future developments, and the temporal dimension confirms the irreversible distinction between one physical state and another. With the arguments of Complex Systems Theory and the Anthropic Principle we can, for the first time since the Darwinian revolution, discover the possibility of a teleological vision in which nature is endowed with a free self-organizing capacity, is able to move into unpredictable states.

The presence of phenomena that, due to their complexity, elude the deterministic efforts of science is very important for the autonomy of nature and the human condition. In fact we can propose a rational solution to the ancient problem of the human will, one actually proceeding from the results obtained by the scientific research of complex systems. Natural systems show the existence of behaviours that are not knowable deterministically although universal laws are present: this indeterminateness makes it possible to advance the hypothesis that a considerable space exists in which to apply the principle of self-determination – which in the human being coincides with will.

5. THE PHENOMENOLOGY OF LIFE AND THE HUMAN CONDITION

In this last section, we finally attempt to define precisely the theoretical elements to consider in framing a paradigm of complexity which are able to complete the program of a phenomenology of life.

The secret of order and organization, the multiplicity of forms proceeding from the inorganic to the organic become the central argument of A-T. Tymieniecka's phenomenology of life. In particular, we have tried to find in the individualization of the tendency to complexity an essential parameter prompting a reconsideration of the place of man in nature. Tymieniecka centers her reflections fundamentally on the creative, reflective and inventive capacities of the human being. These functions are decisive for specifically human existence as it takes part in the universal process of nature. Every living being seems in fact related to all others in the struggle for life and for its own self-individualization through links, historical nets and laws that carry nature toward organized complexity. On the relevance of complexity within the cosmic process, Tymieniecka does not want to erect a phenomenology of life isolating the human being from the natural links which surround it, but considers the importance of the meaning of natural selection, one that can avoid the fallaciousness of an anthropocentric vision. On the other hand, the necessity of a conception of human nature that undergoes examination in its existential environment is considered to be of primary importance for culture, whether scientific or humanistic. We think that a philosophy of complexity can develop a connection between the two cultures, finding a common point in research seeking to define the place and role of man in the vast universal schema.

In her fundamental conception of the phenomenology of life, Tymieniecka notes "the human being immersed within his existential network together with all other living beings, partaking of the universal forces which carry life as such, suffering the restrictions which the struggle of life within a full network imposes and reaching his own stature and level of accomplishment".[10] The global evolutionary course is therefore understood as a process proceeding through disordered states – that are the greater part – to more ordered states, in which form and information is preponderant. Life is described as a self-individualization process of evolution, in which occurs the differentiation of species and the development of humanity. So, the universal progress of life proceeds through individual unfolding and through progress in complexity. With the human mind, we can say that life is advanced in flexibility, selectivity and innovation through the explosion of enormous creative capacities.

It is possible to find a prime fundamental concept and a common point of view shared by the paradigm of complexity and Tymieniecka's phenomenology of human condition. In the first three sections above an intelligible meaning for the development of the universe emerged from following the advance seen in an ascending line of phenomena. The question needing answer therefore is how and in virtue of what does an increase in complexity occur. The challenge for the scientific research of the next years will be, we think, to explain the nature of the complexity of which man is the most important product because he contains a particular set of virtualities. Conscious reflection is considered to be the chief point of the self-individualization of life: in fact man asks himself the meaning of this capacity and in this way "the human being, the knower, the one who asks questions and seeks to investigate things is and always remains at the center of the inquiry".[11] In the perspective introduced above, the problem of the observer's functioning is not limited to humanistic science, but concerns the world view given by natural science too: after all, according to what we have said, man is in many aspects in a privileged position. With the help of the Anthropic Principle, we have seen that there is a hypothesized close relationship – a epistemological and ontological one – between the observer and the observed Universe. In the global evolutionary development that advances from the simplest to the more complex, the human being "is the culmination of this advance of life's self-individualization".[12]

Secondly, the attempt to explain order with the help of the concept of the process of organization through disorder can yield a fundamental principle describing the role of the human condition. The human being in his functioning is a very complex expression of the entire self-organization process of nature (that Tymieniecka defines as the *Logos of Life*). This notion – that we can take into account with the Teilhard de Chardin's hypothesis of an orthogenetic development of nature – is so characterized because it tries to justify the richness of forms in all that exists, the infinite reservoir of fulgurating proficiencies present in the living organism, the many possibilities for differentiation, the tendency to order and organization.

Through its disclosing of potentiality, the natural reality arrives at a particular complexity with the human being, introducing into the biosphere what Tymieniecka delineates as "the inventive/creative proficiency, the *imaginatio creatrix* and the will".[13]

Through these capacities, man establishes a particular situation within nature. Among all these abilities, self-awareness especially expresses in "a creative orchestration of the entire human functioning" the tendency to complex differentiation of the *Logos of Life*. As a self-aware agent, the human being is transformed as "one who projects himself the significance of his existence".[14] He builds the world through his projects and employs all his faculties which therefore make him responsible for eeverything.

Tymieniecka gathers the meaning of human life from three factors:

1) By means of man's physical ability to know the natural world, he has related himself to its objects through an aesthetic/poetic sense. As that sense functions, he assimilates and elaborates the information coming to him from the external world. In this way man inserts himself into the inner natural structure, in experience with an aesthetic/sensible dimension.

2) The main instrument at man's disposal is his intellectual ability. In fact the human being uses his intellect to discover an objective world, and so he becomes creator, user and administer of natural things.

3) Building his world through the unfolding of these properties, man realizes the conditions of all other living beings on the Earth. For this reason, he becomes responsible for the positive or negative consequences of his deeds. The moral aspect of his actions leads him to occupy a key position in the world process, possessing a unique role in the administration of life.

Will – which in the theory of complexity can find own rational justification – opens man to responsibility towards himself and towards others. It is precisely the moral sense that moves us to behave out of a sentiment of benevolence, derived from a sense of responsibility for the vital interests of other living creatures. In fact man can balance his own desires and the interest, the inclinations and the feelings about things of other human beings. For this reason man can form a societal sphere, measuring each individual interest against the benefit of others. Though his moral and intellectual capacities, the human being recognizes the interdependencies, the ties and the strengths that bind him to the rest of nature. In this way, as Tymieniecka says, "the moral sense makes the human being CUSTODIAN OF EVERYTHING THERE IS ALIVE".[15]

CONCLUSION

This article has been an opportunity to compare reflections that, although having a different origin and history, show a common intention: to understand the role of life and human being in the nature. We tried to develop an analysis in which it should emerge how a part of modern scientific research is re-evaluating the function of complexity (such as that involved in human thought) in the Universe. The aim of this work has been to reveal the interconnection between the laws of nature and the complex systems that are living phenomena. Teilhard's idea of a cosmic evolution, the selective and hermeneutical function awarded to life by the Anthropic Principle, the holistic vision implied in the science of complexity and Tymieniecka's phenomenology of life are attempts to explain the meaning of the production of order in the Universe. At the end of these observations we can say that there are a large number of elements supporting the hypothesis that the course of cosmic evolution is strictly interwoven with the course of its organization, of life and of awareness. It is hard to believe that such a link can be the result of a random process. Rather, it may be due to the presence of a fundamental natural law (to this date unknown). If this law really exists, then we could have a coherent picture of the meaning of being. In fact, only by assuming the improvement of complexity to be an ascendent arrow in the movement of the Whole, can we avoid the loss of meaning connected to any vision entirely deprived of natural finality.

Istituto di Filosofia
Università di Perugia

NOTES

1 "And these equations, even if we know them and could write them down, would help us not at all in tackling most of the challenging questions we can ask about the natural world. These mainly involve old-fashioned atomic and nuclear physics. The subnuclear world and its uncertainties are generally irrelevant to large-scale phenomena, just as the atomic structure of liquids provides no practical clues to the unexplained complexities of turbulent flows in the air and ocean. We are not dealing with a tower or ladder where weakness at the base imperils all deductions about higher scale structures. We can in principle write down the equations (essentially Schrodinger's equations) governing any everyday physical process but we cannot solve these equations even for single molecules, let alone for anything larger. Nor, even if we could, would the initial data the accurate enough to permit useful predictions. So even reductionists, who believe all

phenomena can be reduced to physical fundamentals cannot hope to be constructionist, in the sense of deriving a full understanding of complex systems from their atomic constituents. Sciences will always be in a hierarchy where each level of structure entails new irreducible concepts," in Martin Rees (1991), p. 672.

[2] P. Teilhard de Charin (1956), pp. 11–12. "La Vie," Teilhard says, "non point anomalie bizare, sporadiquement florissant sur la Matière – mais la Vie exagèration privilégiée d'une propriété cosmique universelle – la Vie, non un epiphénomène, mais l'essence même du Phénomène," p. 27.

[3] J. D. Barrow and F. J. Tipler (1986), p. 16. The Anthropic Principle tries to describe the remarkable relationship between Universe and Life through holistic schema of causation, asserting that order must be taken into account to explain the fundamental parameters of the Universe. For a synthesis of the Anthropic Principle, see S. Procacci (1995).

[4] R. H. Dicke (1957), pp. 355–362.

[5] J. D. Barrow and F. J. Tipler (1986), p. 510.

[6] For an exact description of the anthropic restrictions on cosmological values see N. Dalla Porta and L. Secco (1991); J. D. Barrow and F. J. Tipler (1986), pp. 219ff.; G. Coyne and A. Masani (1989).

[7] A. Masani (1987), pp. 4–5.

[8] G. Nicolis (1989), p. 325. See also G. Nicolis and I. Prigogine (1987), I. Prigogine (1980), I. Prigogine and I. Stengers (1979).

[9] For a description of these arguments see F. Capra (1982). For a synthesis of complexity theory see G. Bocchi and M. Ceruti (1985). For chaotic systems, determinism and existence-uniqueness remain present, as they must, however predictability is lost: chaos eliminates the Laplacian fantasy of deterministic predictability.

[10] A-T. Tymieniecka (1990), p. 9.

[11] Idem, p. 8.

[12] Idem, p. 9. Tymieniecka notes that "when we envisage the inward functioning of the self-individualization of life along the entire scale of its progressively developed stages, we discover that there is a progress within the steps of this functioning which accounts for the progress in complexity or which is identical with it," p. 12.

[13] Idem, p. 12.

[14] Idem, p. 13.

[15] Idem, p. 16.

REFERENCES

Barrow, J. D. and F. J. Tipler, *The Anthropic Cosmological Principle* (Oxford: Oxford University Press, 1986).

Bergson, H., *L'èvolution crèatrice* (Paris: Alcan, 1907).

Bocchi, G. and M. Ceruti (ed.), *La sfida della complessità* (Milan: Feltrinelli, 1985).

Capra, F., *The Turning Point* (New York: Simon and Schuster, 1982).

Coyne, G. and A. Masani, "Il principio antropico nella scienza cosmologica," *La Civiltà Cattolica*, No. 140 (1989), 16–27.

Dalla Porta, N. and L. Secco, "Il principio antropico in fisica e in cosmologia," *Il futuro dell'uomo*, No. 2 (1991), 61–110.

Dicke, R. H., "Principle of Equivalence and Weak Interaction," *Rev. Mod. Phys* 29, No. 3 (1957), 355–362.

Masani, A., "Il principio antropico," in G. V. Coyne, M. Salvatore and C. Casacci (eds.), *L'uomo e l'universo. Omaggio a P. Teilhard de Chardin* (Città del Vaticano: Vatican Observatory, 1987) pp. 1–26.

Muratore, S., "Oltre l'antropocentrismo cosmologico," in *La creazione e l'uomo, Studi Religiosi* (Associazione Teologica Italiana, Padua: Edizioni Messaggero, 1992), pp. 189–258.

Muratore, S., *L'evoluzione cosmologica e il problema di Dio* (Rome: AVE, 1993).

Nicolis, G., "Physics of Far-from-equilibrium Systems and Self-organization," in P. Davies (ed.), *The New Physics* (Cambridge: Cambridge University Press, 1989), pp. 316–347.

Nicolis, G. and I. Prigogine, *Exploring Complexity* (New York: Freeman, 1987).

Prigogine, I., *From Being to Becoming* (New York: Freeman, 1980).

Prigogine, I., *Le leggi del caos* (Bari: Laterza, 1993).

Prigogine, I. and I. Stengers, *La Nouvelle Alliance* (Paris: Gallimard, 1979).

Procacci, S., "Il principio antropico: nuovo teleologismo o principio di selezione?," *Annali della Facoltà di Lettere e Filosofia*, Università degli Studí di Perugia, Vol. XXX, nuova serie XVI (1995), pp. 199–216.

Rees, M., "Will the Universe Expand for Ever?," *Memorie della Società Astronomica Italiana, Third Venice Conference on Cosmology and Philosophy, Venice, December 15–16, 1989* 62, No. 3 (1991), 653–673.

Teilhard De Chardin, P., *La place de l'homme dans la nature. Le groupe zoologique humain* (Paris: Editions Albin Miscel, 1956).

Tymieniecka, A-T., "Phenomenology of Life and the New Critique of Reason: From Husserl's Philosophy to the Phenomenology of Life and of the Human Condition," in A-T. Tymieniecka (ed.) *Reason, Life, and Culture: Phenomenology in the Baltics. Analecta Husserliana*, Vol. XXIX (Dordrecht: Kluwer Academic Publishers, 1990), pp. 3–16.

ZOFIA MAJEWSKA

PHENOMENOLOGY AND CUBIST SPACE

"We know now that art is not the truth. Art is a lie which allows us to come closer to the truth, or at least one that is recognizable to us."[1] Is it possible to find a more suggestive contradiction than that between Pablo Picasso's words above and Husserl's aspiration for the absolute truth whose criterion was supposed to be apodictic obviousness? Still, did not Husserl himself write:

> if one likes paradoxes it is possible really to say, especially if one comprehends the multiple meaning well, it is then possible to say in accordance with the truth that "fiction" is a source of phenomenology as well as of any eidetic discipline, that "fiction" is a source from which cognition of "eternal truths" stems.[2]

Different styles notwithstanding, one may notice that both paradoxical formulations express a belief in the cognitive value of art and imagination which is, however, differently understood. On the other hand, can one draw far-fetched conclusions from out-of-context sentences?

The link between phenomenology and cubism is not going to be the focus here. It has been previously formulated by historians of art and philosophy. What one can do, though, is relate the postulated analogies, provide new ones, possibly point out differences. The following considerations will concentrate on the phenomenological theory of space and the issues of artistic space in cubism.

THE ARTISTIC PROGRAMME OF CUBISM AND HUSSERL'S PHENOMENOLOGY

Hans Rainer Sepp attempted to compare phenomenology and artistic trends in a very interesting way as early as at the beginning of the twentieth century.[3] In his work he analyzed statements made by Edmund Husserl as well as by painters (the Cubists, Wassily Kandinsky, Piet Mondrian); he used a wealth of quotations trying to prove the affinity of their ideas. Sepp avoided the question of influences, but stressed parallel intellectual attitudes. He also concentrated on what draws Husserl and painters close to each other (abandonment of naturalism and

A-T. Tymieniecka (ed.), Analecta Husserliana LIX, 249–262.

historicism, interest in a world given in many ways to man as opposed to a world in "oneself"). Needless to say, his striving to show similarities in the presentation of reality results in his passing over the idiosyncracy of philosophical and artistic expression as well as in the obscuring of the differences in the analogies.

The affinity between Husserl's phenomenology and the artistic programme of cubism was also discussed by Ksawery Piwocki[4] who, however, limited his comparison to a very short time span; comparing the Husserl of 1901–1913 and the Cubism of 1908–1914. Such an approach makes it difficult to prove a flow of inspiration from one to the other. Piwocki treated Husserl's philosophy and Cubism as autonomous phenomena. He did not consider Cubism an illustration of phenomenological philosophy, but believed it to be an independent creation that had arisen more as a result of the situation in fine arts than in any willingness to live up to philosophers' intellectual tenets. Being methodologically very cautious as he was, Piwocki still managed to observe certain parallel phenomena in Husserl's phenomenology and Cubism. As the theoretical explication of such an affinity he cited Alois Riegl's theory of creative will, Erwin Panofsky's notion of semantic interpretation, and finally, Theodor Lipps' statement about a work of art's being an equivalent of feelings and thoughts.[5] He wrote; "Works [. . .] do not denote particular facts, they do not formulate themselves into unambiguous semantic expressions, but they have a bearing on the artist's whole orientation towards the world and man, and these are issues that expressly refer to philosophy's concerns."[6] He also went on to say:

> the civilizational unity of the contemporary world, speed and availability of information which we take in from various fields both consciously and unconsciously, the identity of the perspectives that astound us and the horror of dangers suspended over our heads, the savouring of these very same fruits of culture and the abandoning of similar armours or shells of tradition [. . .], these all create the above-mentioned atmosphere and consequently related attitudes.[7]

Piwocki stressed that the parallelism of motifs one could observe in phenomenology and Cubism was to a large extent a result of their polemical attitude towards philosophical and artistic traditions. Husserl fought psychologism; Cubism struggled with the subjectivism of vision represented by Impressionism. Both positions are related by means of a critique of sensorial experience.

The motifs which allow, according to Piwocki, for the comparison of Cubism and phenomenology are as follows: Cubism's scientific ambitions; a desire to move from sensorial phenomena to the intellectual matters of consciousness; simultaneity of presentation; and the ambitions of capturing the essence of objects.[8]

Another argument for the juxtaposition of phenomenology and Cubism is constituted by the role they played in twentieth-century culture. Their inspirational power in philosophy and art is unquestionable.

One striking fact in Piwocki's rather cautious argumentation is that he did not refer to the Cubists' works, but to the formulations of the artistic programme's creators (Guillaume Apollinaire, Albert Gleizes, Daniel-Henry Kahnweiler, Jean Metzinger). He also cited opinions formulated *post facto* by art historians and juxtaposed them with phenomenology. One should bear in mind, though, that the essence of an artwork does not mirror programmatic articulation. A comparison of artistic and philosophical achievements would respect the autonomy of art and philosophy to a far greater extent. One at least should expect a remark on to what degree the Cubists' works realized their explicated programmatic manifesto. This appears to be even more important in that Picasso on principle refused to formulate any programme and distanced himself from Gleizes and Metzinger, whom Piwocki quoted so often. Art historians unanimously stress that the works of these two were secondary to those of the likes of Pablo Picasso and Georges Braque.[9]

One can suggest a different perspective, though. The question could be whether Husserl's philosophical tenets were in any way analogous to the Cubists' artistic practice. Only then would the comparison involve autonomous disciplines, parallel in a sense that each articulates respective issues by means of its own inherent methods. Let us look at what Cubism experts have to say.

CUBIST SPACE

In his all-embracing characterisation of Cubism Mieczysław Porębski stresses that the question of artistic space was the prime *innovation* that the style introduced.[10] In the history of Cubism, Porębski distinguishes three periods: the pre-Cubist phase (1906–1909), analytical Cubism (1909–1911) and synthetic Cubism (1911–1912). He also mentions "Demoiselles d'Avignon" (1907) as a work which anticipated Cubism.

A new formula of an object was the essence – the substitution of a geometrical complex of triangular facets, broken edges, distinct points for a fortuitous optical look. Hard ribbed armature, spiky corners, stretched-to-the-limit surfaces signified the resistance which a mass generated against a two-dimensional surface. In order to overcome this resistance completely one needed something more: the ultimate disintegration of the cohesion of an objective shape, its disaggregation into loosely-composed fragments, the abandonment of limitations, the obliteration of univocality and continuity of contours. What follows – the definitive severance from object-enveloping, both homogenous and continuous, space which also separates the object from the observer; the construction of a different space that would allow for an immediate and multiple approach to the object, from different sides and directions simultaneously.[11]

Porębski also mentions that the *a priori* treatment of space was questioned in the sciences as early as the nineteenth century and that the Renaissance perspectival vision had been challenged in paintings prior to Cubism, even though it was theoretically still binding upon painters. Nevertheless, it was Cubism that made the construction of idiosyncratic artistic space an issue.

Apart from his disagreement with the rigidity of Kantian perspective forms the author does not refer in his considerations to philosophical concepts of space, but instead to the development of sciences in the nineteenth and twentieth century. What he seems to aim at is a reconstruction of the intellectual atmosphere from which the new artistic phenomena arose. Porębski does not treat Cubism as a style reflecting the theoretical achievements of the sciences, but presents it as rooted in the situation of the fine arts and in Picasso's and Braque's explorations.

To quote Braque:

Traditional perspective did not satisfy me. It seemed to be something mechanical and never allows for the full possession of an object. It cherishes just one point of view and sticks to it. But this point is meaningless. It is as if somebody was drawing profiles all their life making other people believe a man has only one eye. [. . .] When we started to think like that, everything changed – you can't imagine how much.[12]

The abandonment of perspective, of foreground and background, of an object as seen from one point, and, finally, of its surroundings led to the filling of the entire surface of a picture with this object, now seen from different points and constructed in a multiplicity of projections onto the surface. The simultaneity of presentation influenced not only vision, it also had an ambition to display properties linked to the other senses, and even the entirety of knowledge about the object.

Braque wrote: "First of all, I started to paint still lifes, because in a still life space is touchable, almost tangible. [. . .] This coincided with my desire, which I have always felt, a desire to touch things, not only to see them."[13] That is why he painted instruments so often; the space in which they were located was a tactile space: "a musical instrument as an object has the quality of coming to life when touched."[14]

This is how Porębski comments on Picasso's still lifes:

> He applies paint thickly, he expresses boldly – what he sees? No, what he knows about things which he sees before him. He knows that a table is square and pots – round, that in these two qualities hides the essence of their objective form, of their existence in space, of their similarity independent of accidental placement.[15]

Our knowledge of an object is greater than what one can directly perceive with one's senses; it is co-created by thought and imagination.

Analytical Cubism portrayed an object through the multiplicity of its projections upon a surface. Synthetic Cubism only suggested an object, its fragmentary approximation and splinters, postulating that "one cannot really see a model, or exhaust it."[16] Fragments of real surface were introduced into pictures, which resulted in significant complication in art theory, one leading to a disappearance of the division between painting and sculpture.

Cubism's analytical phase (an attempt at presenting a three-dimensional object on a surface), as art historians claim, did not have its equivalent in sculpture since its creator had always used three-dimensional shapes freely. It is no coincidence, then, that Cubist sculpture developed only after 1912.[17]

Compared to painting, sculpture is relatively neglected in both the theory and history of art, and aesthetic perception. Henry S. Moore writes:

> Contact with a sculpture depends on the ability to react to a spatial shape. That is probably why sculpture has been considered the most difficult of all arts. Doubtlessly it causes more difficulties than works in reaction to which one needs the ability to comprehend flat forms constructed in only two-dimensions. There are far more people "blind to form" than to colour. When they learn to look, children first distinguish only two-dimensional shapes and they are not able to perceive distance and depth. It is only later that wishing to secure their personal safety and satisfy their practical needs they must develop (through touch to a certain extent) the ability to realize three-dimensional distances. However, having fulfilled the demands dictated by practical necessity, most people do not move on. Even though they can attain significant mastery in perception of flat forms, they do not undertake the further intellectual and emotional effort which is instrumental in receiving forms in the entirety of their spatial existence.[18]

Psychology points out the absence of a receptor vital to perception of the third dimension of space. Experiments conducted on infants and animals suggest the innate ability for spatial vision. Some psychologists regard seeing angles and lines as one such ability. It is also stressed that perception of regular shapes is easier than that of complex baroque contours. Psychology distinguishes two levels of the organization of perception: the sensori-motor level (involving the perception of shapes) and the operative-semantic level (involving the perception of objects). Passivity (the use of the immobile sensory organ) leads generally to inadequate perception while activity (the involvement of sensory organs put in motion and of body movement) ensures a more appropriate image of an object. Objective perception involves not only the senses, but also thought and imagination as well as the organism's practical behaviour. The differentiation of colour within the surrounding reality plays an important role in the perception of three-dimensional objects; the same can be said about motor activity of one's body (the space involved is not only visual, but also aural, touch-related, or finally, "the space of activity").[19]

This only proves phenomenologists' and the Cubists' conviction that our knowledge of objects is not limited to narrowly-defined sensation. Moreover, such facts also postulate greater activity in the perception of sculpture, as proposed by Moore.

Cubism constituted a "purification" of sorts of artistic space through the geometrification of shapes. It brought artistic space down to the forms of our most basic spatial experience and eliminated baroque overgrowth in favour of regular shapes.

Some believe that the Cubists took their inspiration to simplify forms and reduce a mass to its geometrical coordinates from African sculpture. Others claim that the latter only served to confirm Cubist experiments that the artists had developed on their own.[20]

In their descriptions of and comments on Cubist sculpture art historians stress the following points:

- the expansion of sculpture's range;
- the introduction of new materials;
- the revolution in sculptural techniques;
- the new concept of an object as an artistic sign instead of a recreation of a model;
- the presentation of an object in its objective truth (not only in its shape, but also its colour);
- the disintegration of a homogenous sculptural mass;
- the inversion of spatial relationships ("negative form");
- the equal treatment of an object and its environment.[21]

Hence, the most essential elements of Cubism's artistic revolt can be summarised by seeing them as elements of a new concept of sculptural space.

The Cubists split the homogenous mass, created a "transparent sculpture" (Kahnweiler's description) which "does not respect the continuous skin of the solid bodies it wants to reproduce, but instead invents artistic signs exposing these bodies in their openness."[22] That is also where "negative form" comes in. It was already anticipated by Rodin's chiselled surface of the mass. In Cubism the mass's content, its continuity, the harmonious arrangement of elements was broken into pieces in favour of contrasting juxtapositions of these same elements.

> The arrangement is built here on contrasts, on the operation of directional tensions, on the large-scale use of convex and concave, positive and negative forms. In Cubist sculptures these strongly-rhythmed forms supplement each other in the opposing rhythm of mass and air, light and shadow. The extreme consequence of concave, hollowed forms was found in the drilling through of a sculptural mass, the operation of the "negative profile," the creation of active aerial space, an aerial form, which organised artistically not the mass only, but also the negative space.[23]

Henri Laurens said: "Hollowed form in a sculpture must mean just as much as full form. A sculpture is first of all a definition of space, space limited by forms."[24]

It was the Cubists who first noticed the interchangeability of protrusions and hollowness, their equal artistic quality and consequently, the ability to act aesthetically through the reversal of spatial relationships.

"The reversed form" was first introduced by Picasso in his 1912 painting "Guitar and Bottle." The extreme consequence of such a reversal was the appearance of a hole, a hollow form, inside the outer contour of a person's head in work by Alexander Archipenko. The artist commented on his Cubist period of creation in the following way: "Starting in 1912 I merged in one form what was elusive: space, transparency, light and reflection. Together with the concave (negative) form it gave rise to modern sculpture. All that is negative may become positive in a creative sense."[25]

Cubism eliminated the subject from sculpture in favour of an artistic object. Traditional, conventional motifs, so far reversed for sculpture, ceased to matter. Commonplace, everyday object were elevated to the position of artistic motifs. The human figure, the main subject of older sculpture, remained, but the Cubists abandoned its anatomical and psychological depiction. They treated it as an autonomous, simplified, geometrised artistic object. The rules of composition were more important than mimetic accuracy. In some cases it was a mere vertical arrangements that suggested the shape of a human figure, as in the work of Jacques Lipchitz. Even the sculptures of heads underwent far-fetched simplifications (see "Professor Gosset's Portrait" by Raymond Duchamp-Villon, "Head of a Man" by Joseph Csaky, Laurens's portrait busts).

Some Cubist sculptures are characterised by a developed kinetic quality. The artists were not interested in the relation of movement and the object, and unlike the Futurists they did not strive to recreate biological motion. They noticed, though, that the very artistic shapes, geometry of forms might suggest motion by use of rhythmic forms or contrasting juxtapositions (see "The Horse" by Duchamp-Villion, and the work of Archipenko, and Rudolf Belling). In his "Circus Medrano" relief Archipenko combined suggested and actual movement. The dynamism of the mass and facets was also amplified by the rhythm of colours – polychrome (see "Carrousel-Pierrot" by Archipenko, Juan Gris's sculptures, some of Lipchitz's reliefs).

Cubist sculpture combined cuboids, cones, cylinders, as well as flat surfaces with curved ones, simple angles with acute ones or arches, vertical elements with diagonal, curved, or elliptical ones. The use of predominantly simplified shapes and their combinations did not result in a monotony of sculptures; quite the opposite, it represented the multiplicity of choices inherent in these combinations. Hence the variety of works: from those preserving natural shapes to almost abstract ones, from frontal sculptures and reliefs to those which require the observer to move, from the rough architectonics of forms to the lightness of colour compositions, from the static seriousness to more dynamic play of elements. Applying the geometrification of mass and the use of convex and concave shapes, some artists strived to avoid straight lines as too stiff and devoid of lyricism, instead introducing subtle curves and arches (see Ossip Zadkine).

The selection of shapes was as important as their arrangement. A geometrical form does not dissolve in the multiplicity of ornaments, a shape devoid of decoration amplifies artistic expression. It was Cubism that discovered the artistic merit of the pure form of masses and facets. A non-representational sculpture affects the observer's imagination more strongly than one formed after a model. The former simply requires more concentration and activity on the observer's part.

All generalizations potentially entail simplification, and this is particularly so when the subject being discussed is Cubist sculpture. Its variety is best proved by the range of artistic phenomena it inspired as seen by comparing Kazimierz Malewicz's architectonics and constructivism and Moore's sculptures. Each of these used and explored one of the tendencies which originated in Cubism. Constructivists followed geometrical abstraction; Moore retained links with nature in his works with wire. In both cases the question of sculptural shape is most vital.[26] These are not the only conclusions twentieth-century artists have drawn from the Cubists' experience, either. One should mention new materials (iron, glass, plexiglass), the widespread use of actual motion, or the application of light and spatial forms. Let us remember once again that it was Cubism that broke the sculptural mass. However, those who came after them abandoned even the surface as a centre of sculptural space formation leaving only line and line-determined spatial forms (Richard Lippold for example).

In Poland only one Cubist sculpture was created – Xawery Dunikowski's "Bolesław Śmiały" (Bolesław the Bold). Cubist resonances, however, can be detected in Formism (as seen in works by Zbigniew Pronaszko and August Zamoyski), and in Katarzyna Kobro's works.

The constant evolution of the very notion of sculpture remains Cubism's most essential contribution. Having once broken the boundaries, sculpture does not allow itself to be closed within its old conventional forms. The definition itself has been enriched by Marcel Duchamp's "ready mades," the shift of attention from an object to an event (see Władysław Hasior), and finally, environmental art. Aware of the limits of a sculptural object's miniaturization and monumentalization, artists experiment with staging events, utilising not only visual, but also aural means, incorporating the elements (earth, fire) in their works, demanding from the audience not only contemplation, but particular behaviours.

Let me quote Andrzej Osęka: "Conflicts of shapes are most often convert or overt conflicts of disparate ideas or constructs of reality. The question seems to be: how to make sculpture? The real question is eternal: where have we come from, who are we, where are we going?"[27] and "Sculpture does not speak, it states, it is a form, an object enhancing a myth, a worldview."[38]

THE PHENOMENOLOGY OF SPACE

It might seem that the interest in philosophical issues declared by artists constitutes a reliable starting-point for a comparison of phenomenology and Cubism. However, classical phenomenology's reserve towards so-called "the philosophy of the worldview" and the concept of philosophy as a science seem to belie the apparent facility of such a comparison. Art's worldview-related ambitions do not wear well in the search for easy and effective analogies. In order not to falsify art's functions one must not forget the idiosyncrasy of its methods and its autonomy. Were only the transfer of intellectual content involved, one would write an essay and not struggle with recalcitrant materials. But art appeals not only to the mind, but also to the senses, emotions, and imagination.

Bearing in mind the differences between art and philosophy, let us try to find parallel motifs.

The role which the issue of space occupies in transcendental philosophy has been discussed many times.[29] In Husserl's phenomenology, located in the realm of that philosophy, this role assumes the shape of a relationship between consciousness and spatiality. Husserl faced a key problem: how can the phenomenon of space develop in a consciousness devoid of spatial qualities? And then, even where space itself is not a subject of phenomenological analysis, one can come across notions implying the issue of spatiality.

The very concepts of intentionality and immediacy require the existence of space. Since consciousness is intentional, since it focuses on an object in an relevant act, one must assume the existence of space as a background. Consequently one faces a paradox: a non-spatial being requires in its essence the existence of space. Any phenomenological description intended to have immediacy for a reader utilizes terms relating to spatiality (horizon, field, stratum, outlook, centre, background, "close-up," "distant shot," filling, hiatus). In the language of phenomenology these terms are often interpreted in such a way that they not only refer to spatial objects, but acquire more general connotations. Such notional modifications additionally complicate the discussion of space.

It turns out that spatial metaphors play a significant role in phenomenological cognition and that the description of any cognitive process must refer to space. *A priori* experience of mathematical (and geometrical) objects is also a kind of cognition. The knowledge of such objects is entirely positive as it is based on apodictic certainty and an analysis of their essence. Any eidetic discipline is interested in essential, not random forms.

Philosophy's and art's idiosyncrasies notwithstanding, phenomenology and Cubism are related by means of their orientation towards space. Phenomenology uses geometrical shapes and masses as objects for epistemological and ontological consideration, and in Cubism they are a means of constructing artistic space. The former is interested in the issues of space constitution, the latter in the aesthetic effect of a structured space.

Husserl turns to pure consciousness and its objects. A variety of impressions on the noetic side are juxtaposed with the object's noematic unity. What is present in an experience does not reduce itself to impressionistic content. An object, even though observed from different perspectives and sides, from far and near, is still perceived as one and the same. Experiences of consciousness are synthetic as they superimpose objective meaning over hyletic data allowing for the cognition of an object's essence. Cognition is observation of a real or imagined object, but it is not limited to the experiences of the observational or imaginative sensorium.

Husserl differentiates between the existence of consciousness and the existence of objects on the grounds of the different modes of their immediate perception. An object is transcendental in relation to consciousness and present in external perception – in the multiplicity of manifestations based on sensory data. An infinite number of presentations from different perspectives is inherent to the essence of an object. Such, admittedly limited, multiplicity of manifestations was a goal of Cubism. An experience of consciousness, on the other hand, is not revealed in a manifestation – "where there is no spatial being, there is no point in discussing perception from different perspectives."[30] Every experience contains what is given and mediated (the reverse side, the object's interior), but "what is currently perceived is partly comprehended and partly enveloped in a vaguely-realised horizon of indeterminate reality."[31] This horizon constitutes a potential field of perception. Wasn't Cubism, by showing objects from different sides and splitting their facets, a presentation of potential experiences of consciousness when confronted with these objects?

Phenomenologists did not confine consciousness to sensory data, but also noticed that visual impressions contain more than visual qualities (one can see the softness or hardness of an object without touching it – Wilhelm Schapp). The Cubists, in turn, tried to construct tactile space using artistic means. The theoretical notions of phenomenologists were, then, parallel to the artistic practice of the Cubists here.

Consciousness superimposes objective meanings – cognitive, practical, axiological – over objects' manifestations. That also applies to spatial forms; experience integrates particular emotions or moods into them.[32] Space around us appears to be flexible, susceptible to the meaning of consciousness.

It needs to be stressed that Husserl was interested in the ways one experienced space, not in abstract objective space. As a matter of fact, the latter constitutes itself through the exclusion of subjective ways of receiving objects as well as their definite identification due to mathematisation. Husserl created the fundamentals of a non-mathematical science of space, particular space being constituted in experience. His analyses were to show how a spatial object manifested itself, to point out the neotical-noematical multiplicities conditioning its interpretation as a synthetic unity. The immediate explication of emerging continuity was at stake. The space around us is orientated and our body is its zero point which cannot be constituted perspectivally. Cognition of others through empathy (experiencing their living body) is essential for the constitution of an intersubjectively oriented space. A solipsistic subject in motion would not be able to constitute a given objective space in an immediate way. The constitution of a spiritual world encompasses the constitution of a cultural space, with what is close and distant in it (personal and alien).[33]

Cubism and Husserl's phenomenology were parallel autonomous developments, and apparent analogies between them emerged independently. The intellectual atmosphere of the early twentieth century as well as the polemics of each against artistic and philosophical traditions resulted in similarities in their treatment of space. These affinities, however, were not spontaneous reactions to the cultural situation. To the contrary, they were achieved through artistic experiments and individual intellectual reflection. These analogies are discernible:

- between Husserl's theoretical postulates and Cubism's artistic programme;
- between phenomenological analyses of experience and Cubist methods of spatial construction;
- between the distance taken in each instance from the traditional radical contradictions. Phenomenology abandoned the notional opposition

of subject and object, of what is internal and external; Cubism abandoned the opposition of mass and surrounding space, and rigid criteria in the classification of art);

– between the roles they played in twentieth-century culture.

However, one needs too to consider the differences between phenomenology and Cubism:

– Husserl referred to unmediated experience while the Cubists also respected theoretical knowledge of objects;
– Husserl's phenomenology was explicitly Eurocentric while Cubism remained open to other cultures.

Max Scheler's philosophy and contemporary phenomenology continuations overcome Eurocentricism. Suppressing rigid spatial notions phenomenology and Cubism contributed to the belief that the world around us is flexible and susceptible to our creative activities.

Marie Curie University
Lublin

NOTES

[1] Mieczysław Porębski, *Kubizm. Wprowadzenie do sztuki XX wieku* [Cubism. Introduction to the 20th century art] (Warsaw: 1966), p. 6.

[2] Edmund Husserl, *Idee czystej fenomenologii i fenomenologicznej filozofii. Ks. 1* [The ideas of pure phenomenology and phenomenological philosophy] (Warsaw: 1967), p. 222.

[3] Hans Reiner Sepp, "Annaeherungen an die Wirklichkeit. Phaenomenologie und die Malarei nach 1990," in *Edmund Husserl und die phaenomenologische Bewegung. Zeugnisse in Text und Bild*, ed. H. R. Sepp (Münich: 1988), pp. 77–93.

[4] Ksawery Piwocki, "Husserl i Picasso" [Husserl and Picasso], *Estetyka* 3 (1963), pp. 3–22.

[5] *Ibid.*, pp. 16–17.

[6] *Ibid.*, p. 17.

[7] *Ibid.*, pp 17–18.

[8] *Ibid.*, p. 9ff.

[9] See Porębski, *Kubizm, op. cit.*, pp. 51–53; Elźbieta Grabska, *Apollinaire i teoretycy kubizmu w latach 1908–1918* [Apollinaire and the theoreticians of Cubism in 1908–1918] (Warsaw: 1966), pp. 74–79.

[10] Porębski, *Kubizm, op. cit.*, pp. 45–62; Mieczysław Porębski, *Kubizm, Futuryzm* [Cubism, Futurism] (sine loco Centralne Biuro Wysław Artystycznych, sine anno), pp. 3–27; idem., "U początków sztuki współczesnej" [The roots of the contemporary art], *Materiały do Studiów i Dyskusji z Zakresu Teorii i Historii Sztuki, Krytyki Artystycznej oraz Badań nad Sztuką* 6, Nos. 1–2 (1955), pp. 165–196.

[11] Idem., *Kubizm, Futuryzm*, p. 14.

[12] As quoted in *ibid.*, p. 15.

[13] As quoted in *ibid.*, pp. 15–16.
[14] *Ibid.*, p. 16.
[15] Porębski, "U początków sztuki współczesnej," *op. cit.*, p. 180.
[16] Porębski, *Kubizm, op. cit.*, p. 62.
[17] See Adam Kotula, Piotr Krakowski, *Rzeźba współczesna* [Modern sculpture] (Warsaw: 1985), p. 83.
[18] As quoted in Adam Kotula , Piotr Krakowski, *O nowej rzeźbie* [Of new sculpture] (Krakow, 1961), pp. 5–6.
[19] See Tadeusz Tomaszewski, ed., *Psychologia* [Psychology] (Warsaw: 1978), pp. 226–246; Włodzimierz Szewczuk, *Atlas psychologiczny* [Dictionary of psychology] (Warsaw: 1979), pp. 110–117.
[20] See Kotula and Krakowski, *O nowej rzeźbie, op. cit.*, pp. 24–25; *idem., Rzeźba współczesna*, p. 86.
[21] See *ibid.*, pp. 82–92.
[22] *Ibid.*, p. 87.
[23] *Ibid.*, pp. 88–89.
[24] As quoted in *ibid.*, p. 89.
[25] As quoted in *ibid.*, p. 94.
[26] See Antoine Pevsner, Naum Gabo, *Manifest realistyczny* [Realist manifesto]; Henry Moore, "Uwagi o rzeźbie" [Remarks on sculpture]; Elźbieta Grabowska, Hanna Morawska, *Artyści o sztuce. Od van Gogha do Picassa* [Artists on art. From van Gogh to Picasso] (Warsaw: 1969), pp. 334–340, 448–458.
[27] Andrzej Osęka, "Przemiany formy" [The changes of form], Andrzej Osęka, Wojciech Skrodzki, *Współczesna rzeźba polska* [Contemporary Polish sculpture] (Warsaw: 1977), p. 7.
[28] *Ibid.*, p. 8.
[29] See Stanislaw Judycki, "Niektóre problemy husserlowskiej teorii konstytucji przestrzeni" [Some issues in Husserl's theory of space constitution], *Roczniki Filozoficzne* XXIX, 1 (1981), pp. 125–163; Andrzej Półtawski, *Świat, spostrzeżenie, świadomość´ Fenomenologiczna koncepcja świadomości a realizm* [The world, cognition, consciousness. Phenomenological concept of consciousness and realism] (Warsaw: 1973), pp. 152, 272–274.
[30] Husserl, *Idee . . . , ks. 1, op. cit.*, p. 134.
[31] *Ibid.*, p. 88.
[32] See Roman Ingarden, *O poznawaniu dzieła literackiego* [On the knowledge of a literary work] (Warsaw: 1976), pp. 173–182; *Wykłady i dyskusje z estetyki* [Lectures and discussions in aesthetics] (Warsaw: 1981), p. 67.
[33] See Edmund Husserl, *Idee . . . , ks. 2* (Warsaw: 1974), p. 93; *Medytacje kartezjańskie* [Carthesian meditations] (Warsaw: 1982), pp. 172–176, 195–186, 202.

PART III

NATURE AS ALIVE

Marcello Sanchez Sorondo and John J. McGraw.

KAZIMIERZ KLOSKOWSKI

IS THE ESSENCE OF LIFE A NATURAL OR PHILOSOPHICAL PROBLEM?

1. INTRODUCTION

It is obvious that there is enormous abundance of living things. This quantity of living forms is regarded as a fundamental feature of the living world and it is owing to this that the living world persists.

Furthermore, we should consider the fact that the biosphere constitutes only one ten-milliardth part of the mass of the Earth. It comprises the lower part of the atmosphere – hydrosphere, and surface part of the Earth's crust – the lithosphere which forms a sphere one kilometer thick on the surface of half a square milliard kilometers. If the world was the size of a disc globe and we observed its surface at an angle, at the distance of an outstretched arm, we would not notice even a trace of the biosphere. However, life spread, owing to the many millions of living creatures which make up the so-called bio-diversity.[1] Despite this bio-diversity, life does not "reveal" its "true nature".[2] It only "reveals" its exterior, some phenomenal conditions. The essence of life is pertinaciously "hidden" whether before direct or indirect observation. As a consequence the question of the essence of life becomes an interdisciplinary problem. Thus, despite biological reflections, philosophical analyses, cognitive theory, the methodological and ontological become complementary.

In this frame, science looks for answers to the questions: what is life, what is its nature. Here we should remember that the term "life" is ambiguous, hence, the large number of interpretations and attitudes, when scientists discuss the topic of the essence of life.

I intend to describe natural attitudes first, and then proceed to philosophical approaches.

2. THE ESSENCE OF LIFE IN THE BIOLOGICAL DEFINITION

Research in the natural sciences on the question of the essence of life is conducted within either an analytical, abridged reductionist or organist and holistic framework. We conduct the first kind of research by sorting

A-T. Tymieniecka (ed.), Analecta Husserliana LIX, 265–275.

out the basic elements of a life form and then explaining their interaction by means of chemical and physical laws. The latter approach envisages an organism on the basis of its integrity, complexity and always as a system, integrated internally.

2.1. *Reductionist Propositions*

In this research perspective, one emphasizes the fact that living organisms consist of autonomous elements, which preserve their properties. These elements form one complexity, in which they interact. This interdependence intrigued I. B. Lamarck, who noticed that a living organism is an orderly static construction of organs, which resembles a clock work mechanism. On the other hand I. Loeb considers an organism to be a chemical machine the structure and function of which could be recognized by its being broken down into simple elements.[3]

P. Handler shares a similar opinion. For him "life can be understood in the terms of the laws which control and the phenomena which characterize the inorganic world. In fact, life can only be explained in the language of Chemistry."[4]

Furthermore, life in the reductionist approach, is treated as pure material. This is the result of the continuous development and evolution of matter. A confirmation of these conclusions is to be found in I. D. Bernal's opinion: "life is the manifestation of continuous, progressive, multiform and moderate in co-operation as the self-realisation of possible electronic interatomic conditions."[5]

Life should also be considered as a process. I. D. Bernal specifies that "life constitutes a self-reproducing process restricted to a specific space" and that a living thing, "is characterized by an order and a set of processes occurring at a higher level than a structural feature of its inert matter."[6] On the other hand, J. Loeb argues: "To the question of what animated matter is, we could give the same answer as we give to the question of what controls the phenomena of automatic development and reproduction. If all of life's phenomena are reducted to clearly chemical phenomena, then we shall find the answer within the framework of one or numerous sequences of definite chemical reactions."[7]

Transformations in the modern biology of the fifties, related mainly to the development of molecular biology, changed the previous attitude toward the nature of life. We could notice that in the statements made by F. Crick and James D. Watson, Nobel Prize winners.

Let us present for example J. D. Watson's statement that until not long ago, heredity seemed to be one of the most mysterious phenomena of life. The fact that knowledge of the DNA structure now enables us to understand, practically all the phenomena which accompany heredity at the molecular stage is of substantial importance. We can see that the laws of chemistry are sufficient not only for arriving at understanding of the structure of proteins, but also that all the known mechanisms of heredity remain consistent with these laws. At present, biochemists admit almost unanimously that the other functions of living organisms could also be understood fully as the manifestations of co-ordinated interactions of small and large particles. We know a great deal of less complex processes and we can get the impression that further research made with such verve as has been observed recently in genetics will afford all around knowledge of all the fundamental mechanisms of life. These achievements have proved that research into biological phenomena at the molecular stage (molecular biology) will lead to the early understanding of all the basic features of life.[8]

This research strategy undoubtedly contributed to the discovery and justification of the common features of living organisms. It refers, first of all, to the physical, and the cell structure, as well as to the chemical composition – the presence of a large amount of water, the appearance of macromolecules, and the universality of the genetic code. We have recently observed that there is a tendency to preserve certain structures and functions in the world of living creatures. Hence, since they came into existence histons and cytochrom C have not changed essentially in either their structure or function. Here, the mechanism of the heat shock proteins (Hsp) is worth mentioning.

The recent biochemical research, particularly with reference to the homologous sequences of DNA, shows that the aforementioned mechanism of stress response has been observed in all organisms analysed from archbacterium to eubacterium as well as from saccharomycetes, plants, invertebrates, vertebrates, to people.[9]

It is easy to observe that more features common to all living organisms are continuously being discovered, enabling us to approach the reductionistic strategy in terms of defining life, in a more realistic way than that of J. D. Watson. We should therefore look for solutions through the application of a different research perspective, for any explanation of the essence of life on the basis of the laws of physics and chemistry is insufficient.

2.2. *Holistic Solutions*

The principal thesis of the holistic definition of life is as follows: "Phenomena, processes and living objects make spacious and transient systems of mutually linked and interrelated elements. The most important are those holistic interrelations which take the shape of complex structures of regular relations and interdependencies."[10] Ludwig von Bertalanffy is considered to be the founder of this approach. In the opinion of such a noted scientist a living organism "is a hierarchic organisation of open systems, preserved in a state of conversion of constituents, owing to the holistic conditions present."[11] The holistic theory of life is notable for its very strong emphasis on the fact that life is an open material system at a certain level of complexity which is characterized by feedback. In such a context it is not surprising that for some scientists[12] the characteristic features of living organisms are: homeostasis (self-preservation and adaptability), a characteristic structure (a complex structural organisation, in which there are sub-systems ordered hierarchically demonstrating stability of form and structure despite the continuous conversion of constituents), integrity (the wholeness of a living system is something more than just the simple sum of its elements), conversion of matter and energy, metabolism, execution of movement and spontaneous mobility, ability to respond to impulses, reproduction of similar forms, gradual change in the complexity of a living system in its ontogenesis, a tendency of descendant systems to increase in complexity and homeostatic ability. Generally speaking, "a living system (i.e., a living organism) is the relatively individual homeostatic system [. . .] of a very complicated structure of hierarchically subordinated subsystems, usually rebuilding in time, endowed with the potential ability to reproduce similar forms of higher homeostatic ability."[13]

Another scientist, T. Ganti,[14] claims that life is simply a complex process taking place in a material system, simultaneously emphasizing that we have to deal with a living system when certain specific complex processes take place. Hence, it is not surprising that one can distinguish the so-called absolute and potential criteria of life.

The absolute criteria of life include:

1. A living system must be an inseparable whole;
2. It must manifest metabolic functions;
3. It must be stable – the stability of a system despite continuous functioning and changes in its environment;

4. It must include a sub-system which preserves and transmits information of use to the whole organism;
5. It must possess mechanisms self-regulating and self-steering the processes it contains.

The POTENTIAL CRITERIA of life include:

6. The living system must be able to grow and reproduce;
7. It must display hereditary changes;
8. It must be capable of developing;
9. It must be subject to the laws of nature, an important feature when considering the living world as a whole. But the life of an individual may cease without death at the commencement of the "fission" process.

Despite such clear definitions of life, there are still doubts as to what its essence is.

I thus feel the necessity to study the essence of life from a yet different perspective, i.e. research within the framework of biophilosophy.

3. BIOPHILOSOPHICAL AND METHODOLOGICAL-EPISTEMOLOGICAL REFLECTIONS[15]

A philosophical attempt to analyse the problems connected with the essence of life is based primarily on what refers to philosophical cognition. This type of cognition enables research into life on the basis of the living structure and its final conditions. Complementary to this are methodological and epistemological reflections concerning the heterogeneity of the designations of the definitions of life and in general a sense of the question concerning the essence of life.

3.1. *The Classical Concept of Saint Thomas Aquinas*

Saint Thomas Aquinas in his proposal takes into consideration the contemporary knowledge of Natural Science. He also refers to his own observations of the phenomena which occur in the world of plants and animals. When he points out at the same time, the multiplicity of the living world, he stresses the manifestations of the vegetative, sensate, and intellectual life, i.e.: feeding, growth, mobility and will.

Furthermore, he stresses the fact that characteristic of living organisms is their plasticity, spontaneity and also that it is the internality and spontaneity, as a matter of fact, which expresses in the most precise way the relations occurring between life and mobility. A manifestation of these relations are inbred functions. A living organism is capable of the mobility generated and persisting in the subject that is active.[16] When Saint Thomas makes an attempt to designate the finality of the inbred functions mentioned, he refers to substance as the basis for the functions of life. Consequently, when we indicate life, we mean the living substance (the ontic element framing the nature of its material existence), and not just the living functions of an organism. In other words, life does not refer to a function but to the substance constituting the source of vital manifestations.[17]

"Neither thinking nor sensation live, but existence which thinks and feels by the senses lives. Nature lives, understood as the source of life functions and owing to its nature an existing object feels, grows, thinks and has impressions. Life functions are the manifestation of life, in other words – life in action".[18] It is not difficult to notice that there are two terms for life in Saint Thomas' texts: One refers to inbred functions, the other to the substance in which they have their source.[19]

It is necessary that we should comment here on the claim very often used by Saint Thomas: "Life in living objects is existence itself" ("vivere autem viventium est ipsum esse eorum").[20] Life is therefore identified with existence. "The principle of life in all living objects is a substantial form which determines which living object is which and to what species it belongs. It is the principle of a living existence."[21] In other words, the living existence is the way the substance exists. Hence life in the aspect of existing (existential), is a specific, concrete act of existence dependent on the presence of a concrete substantial form. I think that it is also important to stress that life is defined by Saint Thomas analogously. Our author refers the concept of life to all hierarchically ordered organisms, from plants, animals to man and even to the life of God. If we considered the solutions mentioned above, we could define life after Saint Thomas as "a particular, more perfect than in non-living objects, kind of existence which belongs to a substance able to accomplish the inbred life functions, owing to the fact of being a composite of basic and suitably perfect matter as well as the substantial form adjusted to it".[22] In the opinion of K. Kłósak[23] who analysed the definition, Saint Thomas did not prove that in living objects there is a different way or kind of existence than in non-living objects. The quoted definition indicates the essence of living objects, but not their existence.

Therefore the question we should consider is that of whether this definition should be implemented metaphysically.

3.2. *Methodological and Epistemological Comments*

The difficulties connected with the solution to the question of what life is seem to relate to the problem of the heterogeneity of the designations of the concept of life. And this, in turn, requires an attempt to answer the most detailed questions asked in both the natural and philosophical perspectives.[24]

Thus, we ask if life is: an object or part of an object, a feature of an object, the feature of variability of an object, a feature of a feature or a feature of a relation, a feature of a living object, an element of a living structure.

If we assume that life inheres in the feature demonstrating a structure of existence, the questions again arise: whether life is a non-defined, primary feature or a complex one, and finally whether life is the order of that structure.

Our considerations have led us to the conclusion that an attempt to answer these questions, including that of the essence of life, should be limited to searching for the answer to the question of what life is, of what the state of the matter called life is. The concept of "life" is used because "this is convenient, but the adjectival form 'alive' should be used [. . .]. The essential answer to the questions as to what life is needs to define the concept 'alive'."[25] As a matter of fact what is "alive" is an object endowed with features characteristic of non-living objects.

4. CONCLUSIONS

These last considerations prove clearly that so far we have not elaborated an acceptable definition of life. This is because there are several different options in attitudes to take toward the question of life. Those who favour the first attitude suggest that it is senseless to formulate any definitions of life, since we can not treat life as a generality, that when we speak about life, we should do so only in relation to a given living organism.

The second attitude favours the thesis that life can not be defined of its nature, because it is a primary term, non-defined which functions in biology by rights as the term "set" does in mathematics. Another attitude stresses the fact that the state of scientific research does not permit the definition of the essence of life precisely enough. In my opinion, we may and should make attempts to define life. However, we should take the results of biological research into account and in the systems approach mentioned earlier, we may expect some positive solutions. I assume, consequently, that life is:

1. The way of existence ond functioning of living organisms;
2. The most specific property of organisms comprising an enormous variety and complexity of biological phenomena;
3. A continuous and complex process organizing a system, which in turn is characterized by a metabolism, the ability to preserve and convey genetic information, and the ability to adjust to surroundings and undergo evolution.

The precise definitions that have been based on the analyses elaborated in Section 2 of this article, arouse some doubts in me. There are a few reasons for this. First of all, the term "life" seems to still be susceptible to various interpretations. Secondly, the development of science and not only the biological sciences, makes it possible that we will learn new features common to all living organisms – the HSP mechanism is worth mentioning here. And thirdly, we cannot exclude the possibility that the new properties of life continually discovered may not embrace all organisms. Furthermore, the features may not occur in combination, as is the case in system formulations, which allow some exceptions. Hence, I decide that it is important to attempt to analyse life in accordance with the principle of conversion from the property of living organisms to the essence of life at the biophilosophical and methodological-epistemological levels.

I would like to define here the concept not of what life is in its deepest essence, but of what constitutes the basis of existence for the phenomena and properties of life.

I think it is necessary to emphasize here that the term "essence" in the expression "essence of life" is often interpreted differently. Firstly, it has an empirical sense which indicates a set of the most important features occurring commonly in and defining a given object. Secondly, it has an ontic sense which stresses the substance of an existing object, its deepest essence, or in other words – the form of existence. Therefore, it is unacceptable to claim that biological science researches only life's phenomena, while philosophy analyses the essence of life. Conversely, a living system at both the philosophical and biological levels is treated as consisting of ordered elements which constitute a basis for the processes of life. Speaking more precisely, at each level there are terms. Those which are of key importance for the defining of life at the biological level are: stable structure, open system, and the fulfillment of specified functions. And the terms having key importance at the philosophical level are: substance, existence and activity.

Academy of Catholic Theology
Warsaw

NOTES

[1] E. O. Wilson, *The Diversity of Life* (New York-London, 1993), pp. 15, 31, 35.
[2] J. Moraue, *Problèmes et pseudo-problèmes de déterminisme physique, biologique, psychologique* (Paris, 1964), p. 169.
[3] Sz. W. Ślaga, "Zycie – ewolucja," in M. Heller, M. Lubański and Sz. W. Ślaga, eds., *Zagadnienia filozoficzne współczesnej nauki. Wstęp do filozofii przyrody* (Warsaw, 1980), p. 346; K. S. Thomson, "Reductionism and Other Isms in Biology," *American Scientist* 72, No. 4 (1984), 388–390.
[4] Cited after L. Kuźnicki, *Życle*" in *Słownik filozofii marksistowskiej* (Warsaw, 1982), p. 389.
[5] I. D. Bernal, *The Physical Basis of Life* (London, 1951).
[6] Cited after T. Ganti, *Podstawy życia*, tłum. z węgierskiego T. Kulisiewicz (Warsaw, 1986), p. 36.
[7] J. Loeb, *The Dynamics of Living Matter* (New York, 1906).
[8] J. D. Watson, *Biologia molekularna genu*, transl. (Warsaw, 1975), pp. 72, 74; K. F. Schaffner, "The Watson – Crick Model and Reductionism," *The British Journal for the Philosophy of Science* 20 (1969), pp. 325–348.
[9] K. Kloskowski, "*Diversity and Unity of Life," Studia Philosophiae Christianae* 32 (1996), p. 1.
[10] J. A. Stuchliński, *Problems of Choosing Methodological Strategy in Contemporary Biology*, (Warsaw:1979), p. 13; V. Pratt, "Explaining the Properties of Organisms," *Studies in History and Philosophy of Science* 5, No. 1 (1974), pp. 1–15
[11] L. Von Bertalanffy, *Problems of Life* (London, 1952), p. 129.
[12] J. A. Chmurzyński, *W poszukiwaniu istoty życia* (Warsaw, 1973), pp. 46–64; F. Crick, *Life Itself. Its Origin and Nature* (London-Sydney, 1981), pp. 49–62; T. Owen, "Life as a Planetary Phenomenon," *Earth – Orient Space Technology* 4, No. 1 (1984), pp. 31–38; H. D. Försterling, H. Kuhn and K. H. Tews, "Computermodell zur Bildung selbstorganisierender Systeme," *Angewadte Chemie* 84, No. 18 (1972), pp. 862–865.
[13] Chmurzyński, *op. cit.*, p. 64.
[14] Ganti, *op. cit.*, pp. 82–85; Pratt, *op. cit.*
[15] In my analyses I quote Prof. Sz. W. Ślaga's considerations and reflections. For detailed questions look up the footnotes.

[16] Sz. W. Ślaga, "Próba uściślenia Tomaszowego określenia istoty życia," *Studia Philosophiae Christianae* 10, No. 2 (1974), pp. 81–82.
[17] Nam vitae nomen sumitur ex quodam exterius apparenti circa rem, quod est movere seipsum; non tamen est impositum hoc nomen ad hoc significandum, sed ad significandum substantiam cui convenit secundum naturam suam movere seipsam, vel agere se quocumque modo ad operationem. Et secundum hoc vivere nihil aliud est quam esse in tali natura; et vita significat hoc ipsum, sed in abstracto; sicut hoc nomen cursus significat ipsum currere in abstracto. Unde vivem non est predicatum accidentale, sed substantiale – St. Thomas Aquinas, *Summa Theologica*, (Turin, 1922), I, q. 18, a. 2.
[18] L. Kaczmarek, *Zarys psychologii neotomistycznej* (Poznań, 1958), p. 38.
[19] Sz. W. Ślaga, "Odrębność żywej materii na przykładzie wirusów," *Roczniki Filozoficzne* 11, No. 3 (1963), p. 104.
[20] St. Thomas Aquinas, Summa Theologica, I, q. 18, a. 2; I, q. 54, a. 2.
[21] Ślaga, "Życie – ewolucja," *op. cit.*, p. 372.
[22] *Ibid.*, p. 373.
[23] The discussions on this topic took place at Prof. K. Kłósak's seminars, which I attended in the years 1979–80.
[24] Ślaga, "Życie – ewolucja," *op. cit.*, p. 339.
[25] Chmurzyński, *op. cit.*, p. 65.

REFERENCES

Bernal, I. D., *The Physical Basis of Life* (London, 1951).

Bertalanffy, L. Von, *Problems of Life* (London, 1952).

Chmurzyński, J. A., *W poszukiwaniu istoty życia* (Warsaw, 1973).

Crick, F., *Life Itself. Its Origin and Nature* (London-Sydney, 1981).

Försterling, H. D., H. Kuhn and K. H. Tews, "Computermodell zur Bildung selbstorganisierender Systeme," *Angewadte Chemie* 84, No. 18 (1972), pp. 862–865.

Gaint, T., *Podstawy życia*, tłum. z węgierskiego T. Kulisiewicz (Warsaw, 1986).

Kaczmarek, L., *Zarys psychologii neotomistycznej* (Poznań, 1958).

Kloskowski, K., "Diversity and Unity of Life," *Studia Philosophiae Christianae* 32, No. 1 (1996).

Kużnicki, L., "Życie," in *Słownik filozofii marksistowskiej* (Warsaw, 1982). pp. 387–390.

Loeb, J., *The Dynamics of Living Matter* (New York, 1906).

Moraue, J., *Problèmes et pseudo-problèmes de déterminisme physique, biologique, psychologique* (Paris, 1964).

Owen, T., "Life as Planetary Phenomenon," *Earth – Orient Space Technology* 4, No. 1 (1984), pp. 31–38.

Pratt, V., "Explanining the Properties of Organisms," *Studies in History and Philosophy of Science* 5, No. 1 (1974), pp. 1–15.

Schaffner, K. F., "The Watson – Crick Model and Reductionism," *The British Journal for the Philosophy of Science* 20 (1969), pp. 325–348.

Stuchliński, J. A., *Problems of Choosing Methodological Strategy in Contemporary Biology* (Warsaw, 1979).

Ślaga, Sz. W., "Odrębność żywej materii na przykładzie wirusów," *Roczniki Filozoficzne* 11, No. 3 (1963), pp. 87–108.

Ślaga, Sz. W. "Próba uściślenia Tomaszowego określenia istoty życia," *Studia Philosophiae Christianae* 10, No. 2 (1974), pp. 67–99.
Ślaga, Sz. W. "Życie – ewolucja," in M. Heller, M. Lubański and Sz. W. Ślaga, eds., *Zagad nienia filozoficzne wapółczesnej nauki. Wstęp do filozofii przyrody* (Warsaw, 1980), pp. 297–423.
St. Thomas Aquinas, *Summa Theologica* (Turin, 1922).
Thomson, K. S., "Reductionism and Other Isms in Biology," *American Scientist* 72, No. 4 (1984), pp. 388–390.
Watson, J. D., *Biologia molekularna genu* (Warsaw, 1975).
Wilson, E. O., *The Diversity of Live* (New York-London, 1993).

JOHN G. McGRAW

FRIEDRICH NIETZSCHE: EARTH-ENTHUSIAST EXTRAORDINAIRE

This essay examines some cardinal reasons why Nietzsche's philosophy can be regarded as the quintessential advocacy of and apology for earthism (mundialism) insofar as it pre-eminently champions the sovereignty and sanctity of the earth against all its myriad adversaries. Since Nietzsche envisages humans as entirely earth-bred and bound, their self-transcendence, to wit, their self-overcoming, is accomplished only by an immanence that is solely geo-centric. Consequently, he construes the human condition as being one of homelessness until and unless the earth is honored as the exclusive realm for the realization of all physical, mental and spiritual capabilities and aspirations in a manner that enables all earthlings, to invoke his idiom, to become who and what they are. Accordingly, this presentation sketches six interlocking ingredients of Nietzsche's earthism, namely, his life-philosophy ("Lebensphilosophie"), ecologism, animism, panpsychism, proto-existentialism/anti-essentialism and naturalism. It concludes with his recommendations for the reclamation of the earth as mankind's one and only homeland.[1]

The first component of Nietzsche's earthism, his life-philosophy, signals the philosophy by which he extols life and all things earthen and denounces their detractors. However, his philosophy is not only geo-centric but bio-centric and not only bio-centric but "bio-fierist" in that life is understood as a continuous becoming, an "ocean of becoming," as it were.[2] This life-philosophy affirms, indeed exalts, the earth as a Heraclitean stream of life, one to be savored more as turbulent than tranquil. Nietzsche's earthism qua life-philosophy is dramatically incarnated in his alter-ego, Zarathustra, the unprecedented and unparalleled "earth-blesser" and nemesis of all "earth-blasphemers," *sans pareil.* Zarathustra demonstrates in word and deed that the first and foremost function of the philosopher is to embrace and enhance life in its totality and to assail its denigrators.

However, Nietzsche excoriates most philosophers for being life-negators and criticizes the entire history of philosophy as "a revenge on reality" insofar as it is "a secret raging against the preconditions of life, against the value-feelings of life, against partisanship in favor of

A-T. Tymieniecka (ed.), Analecta Husserliana LIX, 277–306.

life."[3] Furthermore, Nietzsche generally blames traditional philosophy, by which he means epistemological rationalism and metaphysical idealism, for having undermined the basis of what is known today as environmentalism and, thereby, having paved the way for sundry sorts of pillaging and poisoning the planet, considerations which lead to the second aspect of Nietzsche's earthism, his ecologism/ecology.

Although he is rarely associated with matters ecologic, a persuasive case can be made for Nietzsche's being a prominent proto-ecologist, both as a pioneer in the philosophy of ecologism and in its application, ecology. While he did not use ecologic terminology and obviously could not avail himself of the yet non-existent ecologic disciplines, nonetheless, their tenets and themes find themselves woven into the foundation and fabric of his thought in a manner which anticipates some crucial components of contemporary environmentalism.[4] Indeed, in his early writings Nietzsche often approves an earth-paradigm in which humans are portrayed as preservers and perfectors of nature, a position today promulgated by revisionist or reformist ecologists. These so-called "shallow" ecologists consider humans to be, at the very least, mentally superior to other earthlings such that they are obliged to be the enduring caretakers, stewards, managers, custodians, trustees and the like of nonhumans.

However, Nietzsche's later thought is more evocative of radical or revolutionary ecologists, who stress that humans are but equal and quite likely ephemeral members of the life-community with no internal or external mandate to guide, guard or otherwise unduly involve themselves with the earth's evolution. In fact, it is precisely human interference with the earth's processes which these so-called "deep" ecologists specify as the culprit in the alleged current eco-crisis. Of course, in concert with both types of ecologists, Nietzsche would find reprehensible any exploitation, domination and desacralization of the earth as well as any kind of "imperialist" ecologism that would decree safeguarding the earth solely or predominantly for human designs.

The third ingredient of Nietzsche's earthism is his animism, which consists in the ensoulment of all earthlings such that they are alive in some actual or virtual manner, with the result that there is no qualitative difference between the organic and inorganic realms. Since there is a living continuity between the lowest and highest earth-dweller, there can be no strict differentiation between the organic and the environmental nor between the bio-centric and the geo-centric.[5] Immersed in his animism is Nietzsche's fourth feature of earthism, namely, his panpsychism, which is illustrated by the fact that he envisages every entity as not only living but as thinking.[6] As well, all beings, even those customarily considered inert, are vested with at least some sensory capacities such as the ocular or oral. In this manner, Nietzsche predicates sight of lakes and speech of stones.[7] What is more, he sustains a panpsychism of self-determinism by his admission, for instance, that whenever nature is mute and "tongue-tied," it is so voluntarily.[8] For his part, Nietzsche insists that if nature is to speak to humans, then the latter must learn to hear and, above all, listen to it. This attentiveness to nature is something which demands the development of diverse arts of auscultation and attunement that sensitize one to both the sounds and silences of all of nature, including one's own deepest drives and instincts.[9]

However, Nietzsche's panpsychism is particularly patent in his *Thus Spoke Zarathustra* wherein, for example, several dozen different animal species speak to the protagonist, each with its distinctive voice, be it barking, roaring, cackling, croaking, braying, whining, shrieking or the like. What is more, the animals converse with Zarathustra on an equal if not a superior basis, the latter case being typified by the serpent's terrestrial and the eagle's celestial wisdom. Both animals maneuver, as it were, in a circuitous manner, thereby signifying the globular and global circular motion of that most inclusive of Nietzsche's wisdoms,[10] the eternal recurrence of all things. This Nietzschean doctrine is paradigmatically emblematic of the ecological concern for the preservation/conservation of the earth's diverse resources. In addition, Zarathustra is careful in the pursuit of his wild, Dionysian wisdom to protect the habitats of both fauna and flora. As well, he enters the animals' wilderness not to domesticate them, according to anthropomorphic and anthropocentric objectives, but simply to appreciate them as living beings in and for themselves. All the above underscores Nietzsche's recognition of the necessity and desirability of what today is designated as bio-diversity, preservationism, interspecies equality and environmental or eco-ethics.

Nietzsche's aforesaid reflections are consonant with the findings of current scholarship, which documents that cultures which are animist and panpsychist tend to have little or no environmental degradation. In these societies, non-human beings speak, speak for themselves and speak out. On the other hand, if nature is regarded as deaf and dumb, then it need not be heard or spoken to except in an imperative and imperialist manner, which is generally the policy and practice of non- and anti-ecological societies. Such cultures render the nonhuman a silenced and subjugated entity to be addressed only for human entertainment and exploitation. In his opposition to such anti-natural practices, Nietzsche is mirrored by Michel Foucault, who contends that "social power" operates through "a regime of privileged speakers" whose voices and modes of discourse are taken "seriously" in contrast to the unimportant and "meaningless" discourse or muteness of the less powerful.[11]

Similarly, Christopher Manes argues that Western culture forces nature to be "voiceless and subjectless," in order to protect humankind's exclusive "prerogative" as "a speaking subject." Contrariwise, in non-Western cultures, Manes continues, it is "not just people, but also animals, plants, and even 'inert' entities such as stones and rivers" that are presumed to be "articulate and at times intelligible subjects, able to communicate and interact with humans," all of which has just been witnessed in Nietzsche's scenario.[12] Moreover, as the studies of Hans Peter Duerr indicate, humans are generally far less inclined to ecologically neglect and/or abuse a nature whose members can speak to and with them.[13]

This said, Nietzsche, ever the educator, emphasizes that if one wishes to guide a young person on "the path of true culture," the latter's "immediate and personal relationship with nature" must not be disturbed. In order that the youth "unconsciously and gradually feel the metaphysical unity of all things in the great image of nature," words that typify his ontological commitment to what is today titled "eco-organicism," Nietzsche proposes that the "woods, the rocks, the winds, the vulture, the flowers, the butterfly, the meads, the mountain slopes, must all speak" to the youth "in their own language."[14] Consequently, Nietzsche bewails the fact that the teaching of the living union of all things has been forsaken with the result that the life-sciences, including "the physiology of plants and animals," have been severed from their "poetical phantasmagoria" and vitalized metaphysical moorings. He further protests that a life-draining mathematicization has been imposed upon the life-sciences in order that they can be employed by technology and its "lethal instruments" to mechanistically "subdue nature." On the other hand, Nietzsche lauds the contemplation rather than the control of nature as being indicative of a "truly cultured" person.[15]

A fifth sense by which Nietzsche professes his earthism are his proto-existentialism and anti-essentialism. Beyond his aforesaid insistence that philosophy itself must rank life as the supreme value, is his existential motif that philosophy is not a career, occupation or a vehicle to success but a calling to be lived ("une philosophie vécue"). Appropriately, Nietzsche admits to "profiting from a philosopher only insofar as" the latter "can be an example," one that "is supplied by his outward life."[16] Consequently, Nietzsche is convinced that the only genuine "critique of a philosophy [is] whether one can live by it."[17] In order to become a philosopher, then, one must not only think but live, Nietzsche avows, as "a real human being" in contrast, for example, to being a mere scholar.[18]

If, therefore, a philosophy is to be authentic, it must, Nietzsche asserts, be nothing less than "the personal confession of its author" and a testimony to life and to the philosopher's life in particular.[19] In Nietzsche's reckoning, however, if one is "to live at all," one must live dangerously since it is relatively easy to reflect on things but terrifying to be them, that is, to live them.[20] For a philosopher, living perilously entails "dancing" on abysses and, correspondingly, foregoing faith, fatalism, fanaticism and all other comforting and consoling credos, underlying all of which is what Nietzsche describes as a "*disease of the will.*"[21] He, therefore, appreciates a philosopher insofar as the latter demonstrates by his actions that "love of truth is something fearsome and mighty" in contradistinction to a merely academic philosopher whose inglorious epitaph ought to be that his thought "*disturbed* nobody."[22] Accordingly, Nietzsche conceives his own mission to consist in the welding of philosophy to existence "in such a way that philosophy is not superficial and the life of the philosopher does not become mendacious."[23]

Concerning his conflict with essentialism, Nietzsche is particularly critical of that form which will be designated in this presentation as "logocentricity," the philosophy in which reason becomes self-mesmerized by its hyperlogicality as well as by its hyper-reflexivity and search for self-transparency. Logocentricity is reason's self-apotheosis such that it, its progeny and minions – predominantly rationalism, intellectualism, logicism, scientism and technologism – variously appoint themselves as the ground, gauge and goal of reality. Now, it must be underscored that Nietzsche is not an irrationalist who is dedicated to the abolition of reason along with its logical and technological derivatives. Rather, he is determined to insert reason and its offspring into their proper ecological (extramental) and eco-logical (mental) environments, the latter being the province of the unconscious, the instinctive, the sensory and sensual as well as of the affective aspects of life. In so doing, Nietzsche provides the grounds for removing the kind of logocentric models and myths of rational consciousness that "eco-existentialists," as one might call them, regard as injurious if not ruinous to the life-world and life-sciences.

Of course, for Nietzsche, all the sciences are life-sciences in some manner or other and all have their raison d'être in the enhancement of the life-world. As for the restoration of life and liveliness to the philosophical disciplines, Nietzsche would revivify the ethics devitalized by logocentric, absolutistic thought by means of its naturalization and secularization, the latter being the same as sacralization for him. With respect to epistemology, it would be reanimated via "a perspective theory of affects," one that would involve, first, their transfiguration, that is, their disciplining, refinement and sublimation; secondly, their hierarchization; and, thirdly, a revelation of their "superior order" and "spirituality." With respect to metaphysics, it would be enlivened by "the theory of eternal recurrence."[24]

If reason is to be rehabilitated and re-installed in its proper eco-logical and ecological niches, Nietzsche contends that it must administer to the body, the instincts, the drives, the unconscious and the emotions, provided one can control them so that one can ultimately accede to them.[25] In other words, the pre-reflective Dionysian must obey the commands of the reflective Apollonian in order that the post-reflective Dionysian can be set free in earthly self-mastery.[26] At this last stage, the Dionysian has incorporated but transcended the Apollonian "freedom under the law" to become freedom as a law unto itself (autonomy).[27] This process entails the conquering "of the chaos one is" and compelling it "to become form"[28] such that one has "disciplined himself to wholeness" and has thereby "*created* himself" into a "personality."[29]

If, moreover, rational consciousness is to become eco-rational, it must acknowledge that "every passion [has] its quantum of reason," and, indeed, its measure of wisdom.[30] Whereas passion may be deemed life at its most intense, dynamic and concentrated, reason's thoughts are "but the shadows" of one's feelings – they are "always darker, emptier and simpler" than one's emotions.[31] By means of the emotions and their own kind of "reasonality," one lives more *down to earth* such that undergoing them assists in one's self-overcoming. In this vein, Nietzsche stipulates that "Genius resides in instinct; goodness likewise."[32] Here as elsewhere, the deeper one goes into the cave of the passional and the instinctual, the more fully can one soar to the sun (pace, Plato). Of course, to fuse depths and heights is to build one's home, as Nietzsche characterizes it, on the "slopes of Vesuvius" or to sail "into uncharted seas"; in any case, it is, once again, to live precariously.[33]

Consequently, Nietzsche alleges that the conceptual, discursive, technical and linguistic modes of rationality, largely under the aegis of sundry sorts of aseptic and ascetic idealism, Platonic or otherwise, have damaged if not destroyed their natural internal and external habitats by degrading or "castrating" the passions. However, to belittle or betray the emotions amounts to an assault on life.[34] Therefore, Nietzsche aspires to dethrone intellectualistic and rationalistic reason, being convinced that divorced from the instinctual and passional, it endeavors to be an end-in-itself, a thing-in-itself, in short, the in-itself. Arrogating all power to itself, rationalistic reason "tyrannizes" all else, including the passions, until it "becomes 'passion-in-itself'."[35] And whenever there is the postulation of an in-itself, it portends, Nietzsche avers, an attendant attack on life.

Additionally, Nietzsche censures logocentricity's modern offspring, technologism, with its unparalleled "hubris," as evidenced, he says, by the way nature is "violated" through "machines and the heedless inventiveness of our technicians and engineers." This invasion of nature reaches its climax in "homo technologicus," to use present parlance, and the latter's totalizing and technophiliac type of culture.[36] Nietzsche warns that technologism alienates life by replacing the latter with its own, appropriately named, "deus ex machina," that is, "the god of machines." He cautions that the ramifications of this mechanization of nature include stripping life of its mysteries and conceiving it as nothing more than a series of solvable technical problems.[37] In brief, Nietzsche assails logocentric essentialism for transmuting reason, logic, science and technology from means to ends and, worse, to absolute ends, whereas they ought to be instruments of life, that is, servants of the unconscious, instincts, emotions, senses and the body with its superior wisdom, its "great reason."[38]

This said, Nietzsche avows that what ordinarily counts for consciousness, especially among subjectivists and idealists, is itself not only "unreasonable" in its assumptions and inferences but is perhaps best interpreted as "a pathological state."[39] Consistent with this criticism, Nietzsche considers Plato nothing less than "a coward before reality"[40] and unquestionably "Europe's greatest misfortune."[41] Plato's faintheartedness, Nietzsche adds, generated his "ideomania," his "almost religious madness," as displayed in his predilection for the de-earthed esoteric etherealities of a putative world of unchanging Forms.[42] For Nietzsche, the Forms are dead and a dead-end, the antitheses and antagonists of life itself. Platonism is simply reason gone berserk in a metaphysical mirage of cadaverous concepts. To the extent that such otherworldly insanity succeeds in losing its mind to such disembodied ideas, all embodied beings, to wit, all earthlings, needlessly suffer. Plato's Forms, therefore, constitute logocentricity's ultimate treason to the earth and reason's final testament to its fatal attraction for otherworldly *being*. Contrariwise, Nietzsche requires reason to devote itself to understanding and supporting the life of this world, that is, the earthly province of *becoming*.

Nietzsche traces the genesis of conscious acts, including the cerebrealities of philosophical musings, to the instinctive. In fact, he proposes that "our most sacred convictions, the unchanging elements in our supreme values," are but "judgments of our muscles," that is, products of our physiological constitution.[43] Logic itself "and its seeming sovereignty of movement" are based, Nietzsche tell us, on "the physiological demands" of one's drives to preserve or promote a particular perspective on life.[44] He assigns "the aberration" of logocentric philosophy to the fact that, "instead of seeing logic and the categories of reason" for what they are – pragmatic devices for adjusting to the world ("an adaptation" that is, at bottom, "an expedient *falsification*") – logocentricity conceives the categories as self-validating and possessing the sole and sufficient criteria of "truth and *reality*."[45] Correlatively, Nietzsche locates the "naiveté" involved in this process in the assumption that an "anthropocentric idiosyncracy" is "the measure of things [and] the rule for determining 'real' and 'unreal'." Whence, he claims that this supposition begot the "greatest error that has ever been committed, the essential fatality of error on earth," namely, the postulation of a hidden but "true" otherworld versus an "apparent" but false world, the earth-world.[46]

Nietzsche professes that the belief in otherworldliness is now collapsing under the weight of its own delusions and lies,[47] principally, the beliefs in God, man's oldest mendacity, and the metaphysics of being as well as the ground of these two dogmas, an absolutistic, good-versus-evil morality.[48] In turn, the disintegration of otherworldliness causes a passive or negative nihilism, which occurs when "*the highest values devaluate themselves*,"[49] when the "will to nothingness has become master over the will to life"[50] and when one "judges of the world as it is that it ought *not* to be, and of the world as it ought to be that it does not exist."[51] But the culmination of nihilism is reached when all is said to be "necessarily false" in and of the earth-world "because there simply is no *true world.*"[52]

Nietzsche concludes, therefore, that it is "faith in the categories of reason" that is responsible for subverting the concern for and confidence in the earth-world. By means of this inversion, logocentric reason manufactures a system of concepts by which it concocts a perfect but non-existent ideal world, and then appraises the imperfect but existent earthly world in light of such perfectionized and contrived abstractions, an evaluation which ends with the earth's degradation. Hence, for Nietzsche, these conceptual categories or forms of reason "*refer to a purely fictitious*," transcendent, ideated or ideized world, which is subsequently transmuted into the idealized and ultimately an idolized world.[53] Thus, by means of these rational categories the earthly world is voided of all ultimate value. Thereupon, Nietzsche queries: "The 'beyond' – why a beyond, if not as a means for besmirching *this* world?"[54] In effect, the fabrication of the otherworld is spawned by a life-reviling no-saying to the earth. The dishonoring of and revulsion for the earth-world is a synonym, Nietzsche summarizes, for "nonbeing, non-living, not wanting to live."[55]

Nietzsche names philosophy as one of the three main conspirators in the defamation of the earth in conjunction with religion[56] and non-naturalistic morality.[57] Philosophers, he proclaims, are simply obsessed with the loathing of animate becoming and the love of corpse-like being.[58] They "pass sentence on this world of becoming as a deception" and invent "a world beyond it, a *true* world."[59] Philosophers, he charges, "have never hesitated to affirm" another ideal world upon the condition that it contradicts the earth-world, a condition that provides them "a pretext for speaking ill of this world." Indeed, philosophy itself "has hitherto been the grand school of slander" of the known world.[60] As a result, Nietzsche incriminates philosophers for being "artists in abstraction,"[61] "concept-albinos"[62] and intellectualistic ghouls who are bent on valuing something only when they can "de-historicize" it "sub specie aeterni." Satisfied only with the mummification of concepts, these devotees of "theoretical man" let "nothing real escape their grasp alive." In a word, they are vampires who, having drained life out of existence, reduce it to "bones," that is to say, "categories, formulas, *words*."[63]

Not surprisingly, Nietzsche also rejects the Cartesian mind/body dualism in which life is predicated of the former and diminished if not denied vis-a-vis the latter. In contrast to Descartes' "cogito, ergo sum," Nietzsche declaims, in proto-typical existentialist fashion, "Sum, ergo cogito," which is to say that existence precedes and circumscribes thought, that the self *is* its experiences and that experience is earth-bound. Contrastingly, the Cartesian ego is but a dogmatic and devitalized fiction that only hyper-reflexively cogitates *about* its experiences, the latter being detached from the cogitator even as the self is segregated from its body and life.[64]

Speaking further to these issues, Nietzsche denies the validity and emphasizes the dangers of such dichotomies that allegedly obtain not only between self and the world, the self and its acts, mind and body, subject and object but also between theory and practice, thought and action, essence and existence, internal and external relationships as well as the thing-in-itself and its appearances, not to mention the bifurcation between value/interpretation/perspective and fact. Also, Cartesian logocentricity canonizes clarity, distinctiveness and certainty as the touchstones and terminations of truth. Contrarily, Nietzsche maintains, first, "to suppose that clarity proves anything about truth is perfect childishness"[65]; secondly, what constitute distinctiveness and indistinctiveness are merely the preferred foci of an individual perspective;[66] and, thirdly, certitude is simply an illusory but comforting preference for what is a "hundred times" easier, namely, a life fixated on fictional, stable and secure units of being as compared with the challenge of facing the fluidity that is reality as becoming.[67]

Logocentricity, according to Nietzsche, be it Platonic, Cartesian or Kantian, eu-*logizes* consciousness and then dooms, denies, ignores or trivializes all that which is not rationally conscious and, a fortiori, all that which is non-conscious.[68] In so doing, it strives to ignore if not eliminate the singular, the unique, the living, in a word, all becoming, while it struggles to enshrine being in funereal, uniform, universal essences. Logocentricity, therefore, is the opponent of all those who would become who or what they are in their singularity and particularity.[69] It blindly believes that the " 'world of truth' can be mastered completely and forever with the aid" of what Nietzsche scathingly calls "our square little reason." He asks, "Do we really want" the world to be so "degraded" that it is "reduced to a mere calculator and an indoor diversion for mathematicians" such that existence is shorn of its rich diversity and "ambiguity?" In the final analysis, Nietzsche denounces the logocentric interpretation of existence, including its mechanistic legacy, as being "a crudity," "an idiocy" and a veritable "mental illness."[70] In his judgment, an egregious example of such imperious, logocentric conceit is crystallized in the Kantian claim that reason "does not create its laws from nature, but dictates them to her. . . ."[71]

In disclaiming the primacy of idealism, rationalism and intellectualism in their various forms, Nietzsche wishes to destroy the ethical, metaphysical and epistemological foundations which buttress logocentricity's despotic rule over nature, that is, its logocracy or the hegemony by which reason tyrannizes and terrorizes all other earthlings. In this connection, logocracy, particularly in its scientistic guise, being "spurred by its powerful illusion" of having a divine-right omniscience and omnipotence, "speeds irresistibly towards its limits where its optimism, concealed in the essence of logic, suffers shipwreck," having destroyed the basis of all existence, life itself.[72] It is no wonder, then, that Nietzsche dubs the invention of reason "the most arrogant and mendacious minute of 'world history'."[73] In short, he imputes the responsibility for the modern estrangement from nature to the control that logocentrism and technocentrism wield as a consequence of their dictatorial re-structuring or exclusion of the primordial modes of consciousness and to their imposing on nature their own moribund mentality.

The above reflections lead to the sixth ingredient of Nietzsche's earthism, his naturalism, which entails two pivotal principles. The first is that the world functions not by any supra- or supernatural power(s) but entirely by means of its own immanent power(s), and, specifically, "the will to power." The second principle is that the world's comprehensibility, insofar as it is understandable at all, is so by purely natural means. So construed, there is no functionality or intelligibility that surpasses or supersedes natural capabilities. Moreover, the two aforesaid principles imply a psycho-ethical position, to wit, that any self-transcendence is obtained only by naturalizing oneself in the process of self-overcoming.[74] Therefore, Nietzsche bemoans the "all-too-human" tendency to be oblivious to the fact that when speaking of "Nature" one forgets that "we ourselves are Nature, *quand même*."[75]

Since man is wholly nature, both his most "noble" and "terrifying" aptitudes are engendered by the same "fertile soil" of the earth, whereas otherworldly idealism falsely prides itself on not being soiled by anything earthen.[76] However, the belief that humans are not wholly nature, or that they can somehow go beyond their earthly bonds and boundaries, inevitably induces the misunderstanding and mistreating of both their internal and external environments. This being so, Nietzsche finds postures such as "man *against* the world," man being a "world-negating principle," and man being "judge of the world who in the end places existence itself upon his scales and finds it wanting" as revealing mankind's "monstrous insipidity" and its catastrophic stupidity. Likewise, Nietzsche considers the juxtaposition of "'man *and* the world'" to be preposterous, as if the two could be "separated by the sublime presumption of the little word 'and'." But what is even more grievous in his eyes is that all these poses indicate mankind's own contempt for itself as well as for earthly existence *in toto*.[77]

Although Nietzsche accentuates autonomous individualism as a principal objective of his naturalism, he nonetheless emphasizes the ecological union (or "homonomy," to use Andras Angyal's term) of all earthlings, since they are intrinsically and irreversibly interrelated and interdependent. This all-encompassing interconnection, or what can be denominated the ultimate, post-sectoral eco-system, is discernable, Nietzsche avouches, from the fact that man or any other natural being "is not only a single individual but one particular line of the total living organic world."[78]

In a statement that also startlingly prefigures contemporary ecological and naturalistic thought, Nietzsche proposes that life is nothing but a "multiplicity of forces, connected by a common mode of nutrition," that is, by one and the same food chain.[79] Even more radically, he proffers the thesis that the differences among animals, humans included, are mostly if not exclusively due to the nature and quality of their food.[80] So viewed, Nietzsche's "cosmology" is his ecology insofar as he postulates that the world never begins or ends because it "lives on itself: its excrements are its food."[81] Therefore, the catena and cycle of ever-changing life is forever fastened to the ubiquitous and unending necessity for food.

In fact, Nietzsche contends that life itself consists in but one factor, "the will to power,"[82] with philosophy being its "most spiritual" expression.[83] Indeed, the continuous cycle of life is constituted by this will to power, since the latter is "the ultimate ground and character of all change. . . ."[84] So conceived, the will to power is both the content and form of existence and as such it is the constitutive and regulative principle of the universe. As the animating force of all reality, the will to power is "the will of life."[85] Consequently, Nietzsche argues that the universe is but one enormous energy-system in which "every displacement of power" at a given point affects everything else such that "together with sequential causality," there "would be a contiguous and concurrent dependence."[86] In sum, the universe is, Nietzsche essays, but a single, animistic environment which constantly transforms itself in growth and decay, comprised, as it is, of a flow of energy that is everlastingly reused and recycled in an eternal recurrence of the same.

Whatever the scientific status of the doctrine of eternal recurrence may ultimately have been for Nietzsche, it assuredly represents his love of his peculiar destiny ('amor fati") and the centerpiece of his Dionysian "lived-philosophy":

> Philosophy, as I have hitherto understood and lived it, is a . . . Dionysian affirmation of the world as it is, without subtraction, exception, or selection – it wants the eternal circulation: – the same things. . . . The highest state a philosopher can attain: to stand in Dionysian relationship to existence – my formula for this is "amor fati."[87]

More specifically, the eternal recurrence implies, for Nietzsche, that one must love the earth and life as they are, despite the incalculable and often insuperable pain and suffering which imbue existence and which are the essential motives for seeking solace in otherworldliness.[88]

The theory of the eternal recurrence further involves living the moment in contrast to a hedonistic and non-joyous living for the moment. It necessitates that one love the infinitesimal present as infinitely repeated such that the moment is held to be sacred and self-validating. To live thusly is to divinize, that is, to fully humanize existence in such a manner that the moment of becoming and the eternity of being coalesce, so to speak, in their endless recurrence. To cherish the world (and oneself) as it is and yet to labor to improve it is emblematic of a higher-level or synergic merging of radical, non-interventionist ecology and revisionist, interventionist ecology, respectively. It is an undertaking that parallels Nietzsche's desire to transform static being into ecstatic becoming, a conversion that signifies the centrality of his transvaluation or re-valuation of all non-earthly values into those that are earthly.

It is crucial to realize that Nietzsche's fieristic naturalism excludes any kind of "meaning-in-itself" or any "facts-in-themselves," since meaning or value "must always be projected into them before there can be 'facts'."[89] Nor is there a purely objective world or world of objects; there is only the life-world suffused with interacting living beings with innumerable perspectives. Correlatively, there are, for Nietzsche, no value-free disciplines either in their origin, their application or even, it would appear, in their content. Derivatively, every discipline contains implications and ramifications for the universe, which, to re-emphasize, is one seamless ecological totality or one all-inclusive eco-system.

As exemplifying the aforesaid metaphysical "ecological holism," Nietzsche regards the "properties of a thing" as "affects on other 'things': if one removes other 'things,' then a thing has no properties." In other words, he claims that "there is no thing without other things, that is, there is no 'thing-in-itself'."[90] Otherwise stated, reality "consists precisely in [the] action and reaction of every individual part towards the whole."[91] Consequently, it is mere dogma, Nietzsche asseverates, to subscribe to the view that things 'possess a constitution in themselves quite apart from interpretation and subjectivity," that is, apart from their perspectives.[92]

Nietzsche holds, therefore, that existents are structured by their relatedness to everything else such that their internality is comprised of their externality, that is, their environment. In this way, the fundamental forces or units of existence, Nietzsche's "dynamic quanta," manifest "a relation of tension" to all other energy-quanta.[93] Similar views have surfaced in contemporary ecologists such as J. Baird Caldicott, who depicts organisms as being the sum of their relationships and as conduits and configurations of universal energy. He states that from

> the perspective of modern biology, species adapt to a niche in an ecosystem. Their actual relationships to other organisms . . . and to physical and chemical conditions . . . literally sculpt their outward forms, their metabolic, physiological, and reproductive processes, and even their psychological and mental capacities. A specimen is, in effect, a summation of its species' historical, adaptive relationship to the environment . . . one might say that from an ecological perspective, relations are "prior" to the things related, and the systemic wholes woven from these relations are "prior" to their component species because the nature of the part is determined by its relationship to the whole.[94]

Of course, for Nietzsche, the organism is not only passively or reactively related to its environment, since it can actively alter the latter either directly or by changing itself and its perspective on both its inner and outer milieus.[95]

Because dynamic inter-relatedness is the defining hallmark of an organism, Nietzsche repudiates the existence of an "atomistic anarchy" of isolated beings and adopts the position that all entities share a life-community, one, however, that betokens a functional union rather than a structural unity.[96] Hence, while the ever-fluctuating centers of force which make up the universe are not circumscribed, indivisible, unchanging atoms, neither do they form one-and-the-same numerical organism such that the universe is a superordinate whole which obliterates all lesser wholes. Though all organisms enjoy a relatively sovereign existence, they exert different kinds of power upon one another, due to which a hierarchization prevails among them.[97]

Therefore, while these dynamic forces or energy-centers are interconnected and interdependent, they remain individual entities, striving to augment their power and to achieve their vital uniqueness and autonomy. For these reasons, Nietzsche rejects not only a mechanistic but any animistic monism in which the universe would be but a single unitary entity endowed with life, such as propounded by the proponents of the "Gaia Principle."[98] In rejecting both an atomistic/monadic as well as a monistic model of the universe, Nietzsche again foreshadows eco-philosophers like Caldicott who maintains that for

> both contemporary ecology and quantum theory at their respective levels of phenomena, the oneness of nature is systemic and (internally) relational. No undifferentiated Being mysteriously "manifests" itself. Rather, nature is a structured, differentiated whole. The multiplicity of particles and living organisms, at either level of organization, retain, ultimately, their peculiar, if ephemeral, characters and identities. But they are systemically integrated and mutually defining. The wholes revealed by ecology and quantum theory are unified, not blankly unitary; they are one more as organisms are one, than one as an indivisible, homogeneous, quality-less substance is one.[99]

Nietzsche observes that nature produces such an abundance and diversity of life that it evinces an exuberant extravagance. Prolific to the point of prodigality, it absurdly "squanders" itself.[100] Alas, nature is but a naive and inexperienced "economist," whose "expenditure is much larger than the income it procures. . . ."[101] Therefore, and in agreement with revisionist ecologists, Nietzsche observes that nature is needful of managers and directors for its living household, a position, it might be said, that entails the enlightened alliance of *eco-logy* and *ecology* with *eco-nomy*. To this end, nature enlists the caring comportment of the philosopher, artist and saint. Since they are cognizant of its "trade secrets,"[102] nature engages the philosopher for planting its seeds and the artist for culturing them.[103] Finally, the saint, nature's moral exemplar, serves to demonstrate and inculcate a "profound feeling of oneness and identity with all living things."[104] In these three roles, nature's precious and precocious few act as visionary and revisionist ecologists insofar as they are its defenders and perfectors.

Nature, therefore, basks in its bountifulness, in its variegated biodiversity, one might say, just as Nietzsche's magnanimous man radiates largesse of spirit and largeness of heart. Thus, nature's divestment of its riches attests to its inherent generosity and innocence, all of which may seem foolhardy when perceived from a pessimistic and pusillanimous perspective. But Nietzsche is convinced, despite his non-teleological and non-providential view of reality, that nature is wise in the ways of its world and so he increasingly sides with the more optimistic view of nature to which radical ecologists adhere. The latter countenance a "nature-knows-best" policy and, therefore, oppose any tampering with natural processes, since it is precisely human intervention that they cite as the cause of the current putative eco-crisis.

Additionally, Nietzsche advocates the humanization of nature so that humans feel themselves "more and more masters within it" but, of course, not of it, since humanization or self-mastery occurs to the extent that one becomes naturalized, a necessary condition of which is that one enslave no other earthling.[105] Whereupon, Nietzsche adjudges the humanization of nature to follow from its naturalization, a feat which would signal humanity's completion rather than its inception.[106] Therefore, while mankind's naturalization necessitates a return to nature, it does not mean "a going back," for example, to any kind of Rousseauian noble savagery, where, Nietzsche thinks, mass mediocrity and superficiality would assuredly abound. Instead, the revisitation of nature that he urges is an ascent to a "naturalness where great tasks are something with which one plays."[107] The conquering of such heights requires, then, harmonizing the awesome terror of nature (without, on the one hand, being weighed down by pessimism and its intractable "spirit of gravity" and "ressentiment") with the playfulness and innocence of nature (without, on the other hand, being duped by the naïve optimism of Rousseau and the romanticism of "nature enthusiasts.")[108] Thus does Nietzsche oppose both the divinization and demonization of nature. In summary, one must neither glorify nor revile the world, which, considered in itself, is neither good nor evil, "let alone the best or the worst."[109]

In like fashion, Nietzsche instructs one to avoid the "much-too-human" penchant for sundry sorts of anthropocentrism and anthropomorphism, as if nature existed, respectively, for human projects and projections. In this vein, he derides what, for him, are the brazen beliefs that man is created in the image of God and that nature is then created in the image of man, dogmas at once both preposterous and perilous. Thereupon, Nietzsche ridicules those who pretend to want to live "according to nature" but, in truth, only wish to manipulate it according to their own agenda of exploitive greed and narcissistic self-glory. Similarly, and in a protestation that would endear him to eco-ethicists, Nietzsche chastises humans for their "abysmal arrogance" in imposing their version of morality and idealism upon nature. This transgression is inspired, he states, by "the insane hope" that, because they are adept at tyrannizing themselves, humans foolishly believe that "nature, too, lets herself be tyrannized."[110] In other words, Nietzsche alerts man to the fact that nature always has the last word or, in the vernacular, always "bats last."

Nietzsche thinks that it is because of its blunders about its "origin, its uniqueness, its destiny, and through the *demands* that have been advanced on the basis of these errors," that man has brought upon the world "an unspeakable amount of suffering, mutual persecution, suspicion [and] misunderstanding."[111] Concomitantly, he deplores mankind's proneness to evaluate the truth and existence of earthlings solely in terms of its own utilitarian needs, as if nature were but some infinite stockpile or storehouse for human consumption. Hence, Nietzsche is persuaded that to the extent that humans fancy themselves as the culmination of nature, to that extent will they devalue other earthlings. In these regards, he professes that each individual and species brings its own perspective to the "world," the latter being but a word for the totality of "every center of force" which is in continuous action and reaction to all other centers.[112]

Now, while every being apprehends and appraises the world from its own protective and pragmatic viewpoint, the human being tends to arrogantly absolutize its own perspective and relativize that of all others such that its preferences are promulgated as "universal" and "constitutive." Nietzsche considers this axiological/epistemological anthropocentrism and anthropomorphism to be indicative of one of "the hereditary madnesses of human pride,"[113] in that it posits itself as "the meaning and measure of the value of things."[114] In fact, so much so does mankind consider itself to be "the goal of nature" that at one time it was unquestionably "assumed that nothing could be disclosed by knowledge that was not beneficial" to humanity. If this were not enough, Nietzsche opines that mankind even has the effrontery to believe that things other than those which were "salutary and useful" to it "*could* not, *ought* not to *exist*."[115]

However, Nietzsche declares that the truth of the above matters is that the non-human world has its own intrinsic truth and value totally independent of human outlooks and valorizations. Consequently, he abhors man's setting himself up as the pinnacle of evolution and pontificating about the superior status putatively provided him by his vaunted reason, that faculty which supposedly distinguishes him from all other species and which he deems has deservedly catapulted him to such stupendous heights. In haughtily fancying his knowledge as accessing and then adorning the peak of the cognitive "pyramid," man proclaims to the rest of nature that he is "nature perfected." Nietzsche, however, discredits this boast as species-centric pomposity and, thereupon, avers that mankind's "knowledge does not perfect nature"; on the contrary, it only destroys his own nature.[116]

In words that might have been penned by the most fervid radical ecologist, Nietzsche states that "the vanity of the vainest individual" is minuscule in comparison to the conceit which the most modest person feels when he thinks of his position in nature and in the world as "Man!"[117] Nietzsche hypothesizes in all seriousness that perhaps the ant thinks itself just as unique as man does when the insect fancies itself as the "aim and purpose of the existence of the forest."[118] Therefore, and in addition to espousing the desirability and necessity of bio-diversity, Nietzsche anticipates, as previously implied, the principle of "ecological egalitarianism" by affirming that all living beings, including humans, have the same level of interspecies perfection. If anything, he considers man to be the most inferior animal, owing to his having "strayed most dangerously from his instincts."[119]

It merits mention that Nietzsche has been acknowledged as "the first philosopher to accept and use fully the implications of 19th century evolutionism."[120] However, it must be immediately appended that he considers the human species to be neither the goal much less the glory of evolution. Although he rebuffs all efforts to reduce humans to any kind of egalitarianism or equalitarianism and, instead, articulates an "aristocratic individualism," according to which "the goal of humanity [is] its highest exemplars," this intra-species elitism does not at all mean that Nietzsche has jettisoned his stance with reference to "bio-spheric egalitarianism" or his "eco-natural organicism and holism."[121] In fact, his species-egalitarianism is part of his naturalistic, eco-centric ethics, one that includes the outlines of, for example, an animal and even a plant ethic. For instance, while Nietzsche does not say that animals are free and responsible moral beings in the sense of being moral agents, he believes that they are moral subjects in that they are affected by the actions of moral agents. Then, too, Nietzsche argues that all the "Socratic" moral virtues are inchoatively animal, emerging as they do from the need "to seek food and elude enemies," drives which form the fulcrum of all ethics such that it is quite proper to construe "the entire phenomenon of morality as animal."[122]

As well, Nietzsche criticizes humans for taming and domesticating animals, a practice widely regarded as beneficial to them but which he derides as largely damaging to all parties.[123] He is especially incensed by mankind's torment and mutilation of its earthly siblings when this maltreatment is hypocritically mixed with maudlin sentiments.[124] On the other hand, he insists that persons with profound feelings have always had sympathy for animals and are "indignant at the sight of [all] senseless suffering."[125] His animism and panpsychism reinforce his position that all earthlings be treated with respect and a reverence-for-life, a position, of course, that predates Albert Schweitzer's injunction of the same.[126] It is clear, in any case, that a truly humane, versus a human, "all-too-human" treatment of animals, is a substantial element of Nietzsche's eco-ethics. In fact, if the "humanization" of mankind means anything, it requires, Nietzsche stipulates, that man first become fully the "splendid" animal he is. However, for this goal to be achieved man's customary cavalier and cruel behavior toward other animals must be reversed.[127] It is in this context that he counsels one to "love nature the less humanly it behaves."[128]

Nietzsche lists man's most outrageous boast as consisting in the belief that if he were to cease existing, so would the world. It is a wonder, Nietzsche muses, that man has not arranged for "a universal twilight of the world and the gods as the funeral ceremony of the last man."[129] The fact is, however, that should man vanish, "nothing would remain lacking in the world."[130] In observations that would win rousing acclaim from radical ecologists, Nietzsche continues his unflattering assessment of man by remarking that he is but "a little, eccentric species of animal" whose earthly appearance may well be but a "mere moment, an incident, an exception without consequences, something of no importance to the general character of the earth."[131] Man, then, is simply an animal whose duration will last but a minute, once the star on which he lives is "cooled and congealed."[132] In truth, mankind's cessation might well be a source of immeasurable relief and rejoicing for the rest of living beings. Nietzsche even doubts whether humans are considered admirable beings by other animals, who might well be saying to one another that mankind "'is a prejudice of which we animals at least are free'."[133] For their part, humans have much more to fear from their conspecifics than from any other animal,[134] since man, in Nietzsche's estimate, is the "'cruellest animal against himself'."[135]

Nietzsche's naturalistic ethic involves other facets of an applied ecologism, as the following examples exhibit. As indicated above, he manifests a profound concern for the preservation of natural habitats. Furthermore, he considers the shielding of nature's solitude as indispensable for human creativity, re-creation and for the maintenance and improvement of all forms of health.[136] Moreover, it is in solitude that Nietzsche himself finds many of his chief mentors, including "forest and fountain."[137] In fact, it is in the solitary mountains where his "wild wisdom" is bred, borne and nurtured.[138]

It is noteworthy that the practice of physical and spiritual solitude, along with courage, insight and sympathy, is one of Nietzsche's four cardinal virtues.[139] Additionally, he wishes to build the reflective and contemplative solitude provided by nature into the architecture of large cities.[140] Hence, he applauds sedate and spacious cities, like Turin, with its "sublimely clear" and dry air as well as its open squares where one has "a proud sense of freedom."[141] From another perspective, Nietzsche praises the Genoese for their love of life as shown by the fact that their homes are "built and adorned to last for centuries and not for a fleeting hour," as happens in contemporary "throw-away societies" with their practice of inbuilt obsolescence.[142] As well, he counsels avoidance at all costs of those places where pollution and general pandemonium prevail.[143]

In allied fashion, Nietzsche claims that the health of an environment depends on an array of moral, psychological and physical factors which function in continuous interaction. Thus, he remarks that the internal "mood" in which humans "usually exist depends upon the mood" in which they maintain their external "environment."[144] Consequently, the external and internal milieu are isomorphic; as a result, the sight of environmental ugliness "makes one bad and gloomy."[145] In fact, Nietzsche is convinced that most environments are so unhealthy that they "estrange" one from life's calling or even prevent him from hearing it.[146] In fact, Nietzsche criticizes man for being "a polluted stream," having been poisoned both by his penchant for external noise and internal "ressentiment" as well as by the insidious preachers of otherworldliness.[147]

As one who is both personally and philosophically keenly concerned about health, Nietzsche hopes that the future educators of mankind will urge better self-care by means of a "stricter diet" as well as by "air, sun [and] habitation." He is even sure that the vegetarians, with their regimen of eating "less and more simply, are of more use than all the new moral systems taken together."[148] For purposes of all-around health, he also prescribes regular and demanding exercise. Fittingly, he pronounces "the sedentary life" to be "the very sin against the Holy Spirit" and concludes that only "thoughts reached by walking have value."[149] (One might deduce from these reflections that if the peripatetic is the proper mode of philosophizing for this world, then the Cartesian supine manner is the appropriate one for otherworldliness.)

In brief, Nietzsche exacts of all human undertakings that they be fortified with sound diet, constant exercise and a clean environment.[150] As a case in point, he records that Julius Caesar defended himself against sickness and headaches by his "tremendous marches" as well as by his "most frugal way of life, uninterrupted sojourns in the open air [and] continuous exertion." Such regimens are, Nietzsche reports, the "universal rules of preservation and protection" against the "highest pressure" known as "genius."[151] Then, too, his entire philosophy is distinguished by his attention to various kinds of sickness and sickliness.[152] If space permitted, it could be detailed how Nietzsche was a pioneer in ecological, holistic and philosophical medicine, all of which contributed to his being a self-described "cultural physician."[153] Now, the above are but a few of the concerns of his ecology, ones regarding which he not only wrote but lived.

Nietzsche's practical ecologic interests contain his caveat to treat "with special respect those who are lowliest," a monitum that would apply not only to other humans but, it may be surmised, to all earthlings as well.[154] Correlatively, he indicts idealists (ideists) for their bad faith in believing that their loftiest causes do not demand "exactly the same foul-smelling manure that all other human undertakings require."[155] Indeed, as the consummate geo-phile, Nietzsche is proudest of his trans or re-valuation of the grandiose otherworldly values into those which are *mundane*. This reversal of values, on the one hand, requires restoring reverence to reality, namely, to the bodily and to all that which is *mundial*, and, on the other hand, directing a much deserved disdain to the soul and all else that is otherworldly:

If anything does us honor, it is this: we have transferred seriousness: we regard as important the lowly things that have at all times been despised and left aside. . . . We have discovered the "smallest world" as that which is decisive everywhere. The way our streets are paved, good air in our room, food – we grasp their value; we have taken all the necessities of existence seriously and despise all "beautiful soulism" as a kind of "levity and frivolity." – What was formerly most despised has been brought to the front.[156]

Nietzsche petitions humans to become adults, that is, citizens of this world rather than remaining children in order to enter the otherworldly "kingdom of heaven," as Christianity commands.[157] It is incumbent, he prescribes, that one become what he calls an "ecumenist," that is, one with goals "embracing the whole earth." This planetary or world-citizen must possess an earth or geo-centric conscience and a geo-gnostic consciousness. Indeed, if mankind "is to keep from destroying itself," it will need "a *knowledge of the conditions of culture*" in order to realize its "ecumenical goals," which entail fulfilling the instinctual yearnings of all the earth's inhabitants. Nietzsche further predicts that all these responsibilities will be "the enormous task of the great minds" of the next [twentieth] century."[158]

Accordingly, Nietzsche rebukes those who prevent mundialization by their provincialism and nationalism or, as he describes it, the "nationalistic neurosis" with its "particularism of *petty* politics." Perhaps prophetically, he declares that the primary world priority is to engineer "a unity out of Europe, a political and *economic* unity," in order to establish what he terms "a world government."[159] Once again, Nietzsche foresees what numerous ecologists and politicians have only recently begun to see as indispensable to the being and well-being of all earthlings, to wit, a new ecumenical or world-wide eco-economic/political order (including, surely, what today is termed "sustainable development"). This new order necessitates the founding of global institutions, preeminently a world political federation, to deal with a host of global problems.

Besides worldly and world-wide institutions, Nietzsche announces that what the earth desperately needs now is a new kind of moral hero or saint, one whose happiness "smells of the earth" and who praises "everything earthly."[160] These intrepid, all-too-few exceptions "*remain faithful to the earth*" in their rejection of the admonitions and advances of those who speak of "otherworldly hopes."[161] Like the radical ecologist who has an implicit trust in nature's wisdom, Nietzsche's saint places his faith in the salvation of and by the whole earth. The saint's holiness is from the outset holistic in that it resonates with a spiritual purity which is entirely fashioned from his earthiness. The moral hero has, furthermore, forsaken his nay-saying, his destructive and "de-constructionist" philosophy, in order to celebrate

> the cosmos with a joyous and trusting fatalism, in the *faith* . . . that all is redeemed and affirmed in the whole – *he does not negate any more*. Such a faith, however, is the highest of all possible faiths: I have baptized it with the name *Dionysus*.[162]

Nietzsche's being an exponent of conservationism is further witnessed in his identifying man as nature's "prodigal son" whom the earth will welcome back home by freely lavishing its treasures once man refrains from mercilessly ransacking it to the point of depletion and exhaustion. Only when humans value the earth for itself, therefore, will it bestow its most prized bounties,[163] including its "*heart*" which is made of "*gold*!"[164] When this *unearthing* does eventuate, Nietzsche is confident that humans can "create a meaning for the earth" now that the old ideals or idols are perishing from their sheer incredibility.[165]

While Nietzsche pronounces the overman ("Ubermensch") to be "the meaning of the earth," the latter is such only insofar as he ministers to the world.[166] In this role, the overman represents the staunchest exponent and embodiment of the Dionysian development of the earth's potentialities. However, it is Zarathustra's sagacity that presages the overman since the former, like an assistance (revisionist) ecologist, leads earthlings to themselves but, like a non-interventionist (radical) ecologist, does so in order that they can become themselves on their own terms. In this manner, the overman represents the devoted friend of the earth who encourages its indwellers to follow him if they "want to follow themselves," in accordance with the unique destiny fate has assigned each of them.[167] In the end, the overman is arguably but a model for each individual "to go it alone" in becoming himself.[168]

As for the earthly homecoming of that renegade rover of hinter- and netherlands himself, Nietzsche finds joy in the little things which bespeak his homeland, the earthland, itself having proved to be the most faithful of friends, the kind, alas, that he could not find in his life-long search for "star" friendships. In the final analysis, it is the humus, that is, the earthly which furnishes Nietzsche with the humility to realize that it is not the mendacities of otherworldliness – however majestic those transcendent travesties might seem – which bring bliss but rather the mundane for it is "*little* that makes the *best* happiness." With his solitary odysseys now at last completed, Nietzsche, the pioneer of philosophical earthism and its dauntless practitioner, rejoices that he is like a ship that has sailed into its stillest cove – now it leans against the earth, tired of the long voyages and the uncertain seas. Is not the earth more faithful? . . . Like such a tired ship in the stillest cove, I too rest now near the earth, faithful, trusting. . . .[169]

Concordia University
Montreal

NOTES

[1] Please note that the masculine, rather than the feminine or the generic, is the operative gender when considering Nietzsche for reasons which, if they were to be examined here, would exceed the space-constraints of this presentation.

[2] F. Nietzsche, *Daybreak*, trans. R. Hollingdale (Cambridge: Cambridge University Press, 1990), p. 157 #314, hereafter D.

[3] F. Nietzsche, *The Will to Power*, ed. W. Kaufmann, trans. R. Hollingdale and W. Kaufmann (New York: Random House, 1968), p. 253 #461, hereafter WP.

[4] Ecology did not become a distinct science until around Nietzsche's death in 1900 (see M. Hallman, "Nietzsche's Environmental Ethics," *Environmental Ethics* 13 [1991], p. 100).

[5] Ecology, of course, concerns the relationship between organisms and their environments. But if all beings are organisms in some sense, as Nietzsche holds, despite some isolated statements to the contrary, then the environments surrounding them are also living. From this perspective, the separation between organisms and environment vanishes; see note 6.

[6] F. Nietzsche, *The Gay Science*, trans. W. Kaufmann (New York: Random House, 1974), pp. 298–299 #354, hereafter GS; see WP, p. 329 #611. Although Nietzsche says that such thinking may take place "in primitive conditions (pre-organic)," the latter should not be construed as entirely passive, inert and lifeless (WP, p. 273 #499).

[7] F. Nietzsche, "Miscellaneous Maxims and Opinions," Division I, *Human, All Too Human*, Part II, trans. P. Cohn, Vol. VII, *The Complete Works of Friedrich Nietzsche*, ed. O. Levy (New York: Russell and Russell, 1964), p. 35 #49, hereafter HH II MO).

[8] D, p. 181 #423.

[9] For an analysis of these auditory arts, see J. McGraw, "Nietzsche: The Silence of a Solitary," *Lonergan Review* 2 (1993), pp. 1–28.

[10] F. Nietzsche, *Thus Spoke Zarathustra*, in *The Portable Nietzsche*, ed. and trans. W. Kaufmann (New York: Viking Press, 1965), pp. 136–137 #10; see p. 408 #1, hereafter Z.

[11] In C. Manes, "Nature and Silence," *Environmental Ethics* 14 (1992), p. 340.

[12] *Ibid.*

[13] In Manes, *op. cit.*, p. 340.

[14] F. Nietzsche, "On the Future of our Educational Institutions," trans. J. Kenndy, Vol. III, *The Complete Works of Friedrich Nietzsche*, ed. O. Levy (New York: Russell and Russell, 1964), fourth lecture, pp. 95–96, henceforth FEI.

[15] FEI, p. 96.

[16] F. Nietzsche, "Schopenhauer as Educator," *Untimely Meditations*, trans. R. Hollingdale

(Cambridge: Cambridge University Press, 1991), pp. 136–137, hereafter UM-S; see F. Nietzsche, *Philosophy and Truth: Selections from Nietzsche's Notebooks of the Early 1870's*, ed. and trans. D. Breazeale (London: Humanities Press International, 1992), p. 106 #44, hereafter PT.

17 UM-S, p. 187.

18 UM-S, p. 181.

19 F. Nietzsche, *Beyond Good and Evil*, trans. W. Kaufmann (New York: Random House, 1989), p. 13 #6, hereafter BGE.

20 UM-S, p. 144 #3 ff.

21 GS, pp. 288–289 #347. All italicization in quotes refers to Nietzsche's own emphases.

22 UM-S, p. 194.

23 PT, p. 134 #193.

24 WP, p. 255 #462.

25 See Z, p. 311 #4.

26 See WP, p. 207 #384.

27 WP, p. 539 #1050.

28 WP, p. 444 #842.

29 F. Nietzsche, *Twilight of the Idols,* in *The Portable Nietzsche*, p. 554, #49, hereafter TI.

30 WP, p. 208 #387.

31 GS, p. 203 #179.

32 WP, p. 243 #440.

33 GS, p. 228 #283.

34 TI, p. 487 #1.

35 WP, p. 329 #611.

36 F. Nietzsche, *On the Genealogy of Morals*, trans. W. Kaufmann and R. Hollingdale, in *On the Genealogy of Morals* and *Ecce Homo*, ed. W. Kaufmann (New York: Random House, 1989), p. 113 #9, hereafter GM.

37 F. Nietzsche, *The Birth of Tragedy*, in *The Birth of Tragedy* and *The Case of Wagner*, trans. W. Kaufmann (New York: Random House, 1967), p. 109 #17, hereafter BT.

38 See Z, pp. 146–147; see WP, pp. 347–348 #659.

39 GS, p. 305 #357.

40 TI, pp. 558–559 #2.

41 *Nietzsche: A Self-Portrait from His Letters*, ed. and trans. P. Fuss and H. Shapiro (Cambridge, Massachusetts: Harvard University Press, 1971), p. 94 #114, hereafter SP.

42 GS, p. 304, #357.

43 WP, p. 173 #314.

44 BGE, p. 11 #3.

45 WP, pp. 314–315 #584.

46 WP, p. 315 #584.

47 See WP, p. 9 #3; p. 10 #5.

48 GS, p. 283 #344.

49 WP, p. 9 #2–3.

50 WP, p. 217 #401.

51 WP, p. 318 #585.

[52] WP, p. 14 #15.
[53] WP, p. 13 #12 B.
[54] TI, p. 535 #34.
[55] WP, p. 322 #586 C.
[56] The second primary source of world-defamation is religion and particularly Christianity, the latter being defined as "the denial of the will to life become religion!" (F. Nietzsche, *Ecce Homo*, trans. W. Kaufmann, in *On the Genealogy of Morals* and *Ecce Homo*, ed. W. Kaufmann [New York: Random House, 1989], p. 320 #2, hereafter EH). Christianity's God, as the supreme concept of being, was "invented as a counter-concept of life." God is the notional thesis, the "gruesome unity" of "everything harmful, poisonous, slanderous, the whole hostility . . . against life" (ibid., p. 334 #8). God is simply "the declaration of war against life, against nature, against the will to live! God – the formula for every slander against 'this world'" and "for every lie about the 'beyond'!" (F. Nietzsche, *The AntiChrist*, in *The Portable Nietzsche*, pp. 585–586 #18, hereafter AC). In other words, "Life has come to an end where the 'kingdom of God' begins" (TI, p. 490 #4). As well, God symbolizes the nemesis of the "natural"; thus, whatever is natural has become synonymous with the "reprehensible!" (AC, p. 582 #15). Furthermore, Christianity identifies sin and evil with the worldly and natural but innocence and goodness with the otherworldly and supernatural (WP, p. 193 #351). Correspondingly, it schemes to "infect the innocence of becoming by means of 'punishment' and 'guilt'" and, therefore, it can be classified as "the metaphysics of the hangman" (TI, p. 500 #7). Likewise, the Christian God is the deification of "the spirit of gravity," the enemy of the joy and naturalness of Dionysian becoming; it is the implacable foe of "the innocence of becoming" and the playfulness of nature (WP, p. 299 #552). In addition, Nietzsche supposes that sin was invented "to make every elevation and nobility of man impossible" (AC, pp. 630–631 #49). However, now that "God is dead," the only sins are transgressions and treasons "against the earth" (Z, p. 125). Nietzsche also indicts Christianity for its disparagement of the body, its instincts and inclinations (GS, p. 236 #294). Correlatively, he chastises Christianity for teaching that only humans have souls or, at least, "divine" and "immortal" souls ("soul monads"), positions which he considers inimical to life as well as lunatic pretence and unabated hubris (WP, p. 401 #765). Thus, Christianity raises man above the earth, thereby judging it unbefitting his dignity with the result that he pays it "only a passing visit," like the proverbial Augustinian pilgrim whose heart is restless until it rests in God (D, p. 182 #425). Nietzsche, therefore, faults Christianity for declaring that its home is not the terrestrial; the latter is, as it was for Plato, simply a vehicle to the otherworldly celestial. He likewise avers that Christianity, while it fervently defends rationalism within the philosophical order, crazily canonizes anti-rationality by means of its "intoxicated" *faith* within the religious and theological realms (D, p. 52 #89). In the end, Nietzsche confesses that his wars with the "anemic Christian ideal" are designed not to eliminate it but to stop its "tyranny" and to make way for "new ideals, for *more robust* ideals," that is, for earthly, natural ideals (WP, p. 197 #361).
[57] Nietzsche's third major source of earth-denigration and its accompanying postulation and exaltation of otherworldliness are non-naturalistic types of morality. In essence, every "naturalism in morality – that is, every healthy morality – is dominated by an instinct of life." Whereas, he charges, every anti-natural morality vilifies the instinct of life (TI,

pp. 489–490 #4). Nietzsche singles out Christian morality as being especially anti-natural since it opposes, first, "the enjoyment of life [and] gratitude towards life"; secondly, "the beautifying, ennobling of life"; thirdly, "the knowledge of life"; and, fourthly, "the development of life, insofar as it sought to set the highest phenomena of life at variance with itself" (WP, p. 152 #266). Alas, Nietzsche declares that the battle of Christian morality "with the basic instincts of life is itself the greatest piece of immorality that has yet existed on earth" (WP, p. 156 #274). In its transvaluation of the legitimate values of nature and natural morality, Christianity's supernaturalism miraculously elevates the "misfits," the "misbegotten" and other "low-lifes" to the levels of health and holiness while it gleefully censures the natural aristocrat against whom rage, revenge and "ressentiment" are delivered by the aforesaid collection of "zeroes" (see GM, pp. 120–129 #13–16). It is doctrines such as these that Nietzsche deems so disastrous to earthism and which induce him to lambaste Christian morality as the corrupter of humanity and "the most malignant form of the will to lie." But what he finds even more "utterly gruesome" about Christianity is that, as the archetype of "*anti-nature*," it has garnered for itself "the highest honors as morality" so much so that it "was fixed over humanity as law and categorical imperative" (EH, p. 332 #7).

58 See WP, p. 377 #708.

59 WP, p. 13 #12 A; see TI, pp. 482–484 #4–6.

60 WP, p. 253 #460.

61 WP, p. 277 #513.

62 AC, p. 585 #17.

63 GS, p. 333 #372; regarding the instincts and philosophizing, see BGE, p. 11 #3.

64 WP, pp. 267–269 #483–485; see p. 338 #635.

65 WP, p. 291 #538.

66 WP, p. 285 #528.

67 WP, p. 291 #538.

68 In anticipating Freud, Nietzsche depicts consciousness as a "surface," one both quantitatively and qualitatively inferior to the unconscious (EH, p. 254 #9). In fact, Nietzsche proclaims that one must "seek perfect life where it has become least conscious (that is, least aware of its logic, reasons, its means and intention, its utility). . . . The demand for a virtue that reasons is not reasonable" (WP, p. 242 #439).

69 "What does your conscience say? – 'You shall become the person you are'"(GS, p. 219 #270).

70 GS, pp. 335–336 #373.

71 F. Nietzsche, *Human, All Too Human*, Part I, trans. M. Faber (Lincoln: The University of Nebraska Press, 1984), p. 27 #19, hereafter HH I.

72 BT, pp. 97–98 #15.

73 PT, p. 79 #1.

74 See Z, pp. 225–228.

75 "The Wanderer and his Shadow," Division II, HH II, pp. 356–357 #327, hereafter HH II WS.

76 F. Nietzsche, "Homer's Contest," an 1872 fragment in *The Portable Nietzsche*, p. 32.

77 GS, pp. 286–287 #346.

78 WP, pp. 359–360 #678. However, it must be underlined that for Nietzsche, the

individual "is a hundred times more than merely a unit in the chain of its members; it is this chain itself, entirely; and the species is a mere abstraction from the multiplicity of these chains and their partial similarity" (WP, p. 361 #682).

79 WP, p. 341 #641.

80 D, pp. 20–21 #26.

81 WP, p. 548 #1066. It must be understood that Nietzsche regards "the total character of the world" as a complete "chaos" rather than a cosmos since it lacks "order, arrangement, form, beauty, wisdom" and all other names we assign it by "our aesthetic anthropomorphisms" (GS, p. 168 #109).

82 BGE, p. 21 #13; see WP, p. 148 #254.

83 BGE, p. 16 #9.

84 WP, p. 364 #685.

85 GS, p. 292 #349; BGE, p. 48 #36.

86 WP, p. 340 #638.

87 WP, p. 536 #1041.

88 See WP, pp. 310–311 #579; Z, p. 143

89 WP, p. 301 #556.

90 WP, p. 302 #557.

91 WP, p. 305 #567.

92 WP, pp. 302–303 #560.

93 WP, p. 339 #635.

94 J. B. Caldicott, "The Metaphysical Implications of Ecology," *Environmental Ethics* 8 (1986), pp. 311–312.

95 See WP, pp. 332–341 #618–639.

96 WP, p. 303 #561.

97 Ecologists Karen Warren and Jim Cheney present a position akin to Nietzsche's when they state that each being struggles to "organize itself into discrete and relatively disconnected or autonomous holons and hierarchical levels of organization" ("Ecosystem, Ecology and Metaphysical Ecology: A Case Study," *Environmental Ethics* 15 [1993], p. 112).

98 GS, p. 167 #109.

99 Caldicott, *op. cit.*, p. 313. As Alistair Moles reports, Nietzsche held both "a primitive quantum theory of force" as well as a kind of relativity of space and time as early as the 1880s (*Nietzsche's Philosophy of Nature and Cosmology* [New York: Peter Lang, 1990], p.14).

100 GS, p. 292 #349; see WP, p. 344 #650.

101 UM-S, p. 178.

102 PT, p. 6 #24.

103 UM-S, pp. 177–178.

104 UM-S, pp. 160–161. Nietzsche reasons that not only are the philosopher, artist and saint indispensable for the community of nature as a whole but their production and cultivation ought to be the chief individual and communal human objective (*ibid.*).

105 WP, p. 329 #614. This position, however, does not contravene Nietzsche's division of morality into master and slave; rather, it assumes this classification.

106 WP, p. 73 #120; see GS, p. 169 #109.

107 TI, p. 552 #48. One's true nature lies "immeasurably high" above oneself (UM-S, p. 129).

[108] WP, p. 186 #340.
[109] HH I, p. 33 #28.
[110] BGE, pp. 15–16 #9.
[111] D, p. 182 #425; see GS, p. 174 #115.
[112] WP, p. 305 #567.
[113] WP, p. 305 #565.
[114] WP, pp. 13–14 #12 B; see GS, pp. 285–287 #346.
[115] D, pp. 181–182 #424.
[116] F. Nietzsche, "On the Uses and Disadvantages of History for Life," UM, p. 108, hereafter UM-H.
[117] HH II WS, p. 349 #304.
[118] HH II WS, p. 194 #14; see D, p. 182 #425.
[119] AC, p. 580 #14.
[120] C. U. M. Smith, "'Clever Beasts Who Invented Knowing': Nietzsche's Evolutionary Biology of Knowing," *Biology and Philosophy* 2 (1987), p. 65.
[121] UM-H, p. 111.
[122] D, p. 21 #26.
[123] See WP, p. 214 #397 and TI, p. 502 #2.
[124] D, pp. 151–152 #286.
[125] UM-S, p. 157.
[126] See HH II WS, pp. 225–227 #57.
[127] WP, p. 538 #1045.
[128] GS, p. 342 #379.
[129] HH II WS, p. 194 #14; see D. p. 182 #425.
[130] WP, p. 169 #302.
[131] WP, p. 169 #303.
[132] PT, p. 79 #1.
[133] D, p. 162 #333.
[134] Z, p. 137.
[135] Z, p. 330.
[136] See "The Solitary," GS, p. 54 #33; EH, p. 234 #9; SP, p. 74 #91.
[137] HH II MO, p. 166 #356.
[138] Z, p. 309.
[139] BGE, p. 226 #284.
[140] See GS, pp. 226–227 #280.
[141] SP, p. 116 #134.
[142] GS, p. 233 #291.
[143] See Z, pp. 295–298.
[144] D, p. 151 #283.
[145] GS, p. 233 #290.
[146] EH, p. 240 #2.
[147] Z, p. 125.
[148] PT, p. 106 #44.
[149] TI, p. 471 #34.
[150] See WP, p. 78 #126.
[151] TI, pp. 532–533 #31; see D, p. pp. 121–123 #202–203.
[152] See D, p. 33 #52.

[153] See GS, Preface for Second Edition, pp. 33–38 #2–4.
[154] EH, p. 258 #10.
[155] HH I, p. 235 #490.
[156] WP, pp. 524–525 #1016.
[157] Z, p. 428.
[158] HH I, p. 31 #25.
[159] EH, p. 321 #2.
[160] Z, p. 229.
[161] Z, p. 125.
[162] TI, p. 554 #49; see WP, p. 536 #1041.
[163] BT, p. 37 #1.
[164] Z, p. 244.
[165] Z, p. 144.
[166] Z, p. 125.
[167] Z, p. 135.
[168] Z, pp. 190 and 307.
[169] Z, pp. 388–389. Nietzsche writes: "We must be as near to flowers, grasses and butterflies as a child. . . . He who would have a share in all good things must understand at times how to be small" (HH II WS, p. 224 #51).

W. KIM ROGERS

IMAGINING; THE INVENTION OF NEW ENVIRONMENTS

A New Ecological Interpretation of the Imagination and Its Role in the Economy of Human Life

From 1913 until his death in 1955 Ortega y Gasset was engaged in developing what he called a philosophy of vital reason. In this philosophy he presented "my life" as the fundamental reality. Life is, in Ortega's view, from its beginning and fundamentally a "unitary duality,"[1] it is "the mutual and reciprocal existing of man and world."[2] However, "my life" cannot be described just as the coexistence of human being and world. That is too static. Coexistence seems to imply that I and the world only stand beside one another. Rather, "my life" is "a dynamic dialogue between I and my circumstance."[3] "Things happen to me just as I happen to them, and neither has a primary reality other than that determined by this reciprocal event."[4]

In Ortega's philosophy of vital reason there appears for the first time what I have described elsewhere as an ecological approach in philosophy.[5] The term "ecological" as used to describe Ortega's approach draws its meaning from the description which biologists give of the subject matter of ecological science or ecology. For instance, consider the way Begon, Harper and Townsend define ecology: "the scientific study of the interactions between organisms and their environment."[6] Also, we have this statement by the biologist and anthropologist J. Bennett: "Ecology . . . expresses in a single word, the idea of all components of a milieu in reciprocal interaction with each other."[7] Ortega described "my life" in the same way, in very nearly the same terms. However, hitherto the ecological approach begun by Ortega has not been followed, nor well understood I believe, even by his closest students, such as Marias.

It should be noted that in a completely independent development in the 1970s, some American cognative psychologists started exploring what they explicitly called an ecological way of thinking about the mode of existing of living beings and their environment. To begin with, this new development in psychology was led by J. J. Gibson and was focused upon problems related to perception, especially visual perception. Gibson took as the basis for his studies of perception the mutuality of living beings and environment which "make an inseparable pair. Each term implies the

A-T. Tymieniecka (ed.), Analecta Husserliana LIX, 307–322.

other."[8] And in 1984, a British social psychologist, John Shotter, stated that he shall follow Gibson and "adopt essentially an *ecological* approach"[9] in which "everyday social life" will be seen to be "a whole mosaic of interlocking activities with the function of mutually supporting and mutually defining on another. . . ."[10] However, neither Gibson nor Shotter have considered what the role of the imagination could be if human beings and their environment are to be regarded as existing only in reciprocal interaction with one another.

To summarize the ideas thus far presented, an ecological approach means that one takes the connection and interaction between living beings and their environment as the basic framework for all of one's studies. Each sort of life is viewed as a particular, permutable system, involving all of its components in reciprocal interactions with each other. The relations of living beings and their environments are seen as the mutual complementing and completing of each in and by their interactions.

Let now turn specifically to the interpretation of the imagination when one adopts an ecological approach.

Beyond a few passing remarks on the role of imagination in his earlier works,[11] Ortega provides only two suggestive paragraphs in his last – unfinished – major work, *The Idea of Principle in Leibniz and the Evolution of Deductive Theory.* He wrote there of a human being's "*freedom to imagine* in the face of sense-perceptible things. . . ."[12] One can "fabricate for himself . . . a structure of fancies organized into an imaginary world . . . but that is not exact: he can fabricate innumerable worlds."[13] And mathematical and scientific reason "*is only one way among many in which imagination may function.*"[14] Although Ortega by no means developed a complete interpretation of the role of imagination in the economy of human life, what he said is enough to indicate his view of the character and importance of it as the world-fabricating capability of human beings.

Alfred Schutz deserves especial mention here for his development of William James' idea of multiple realities.[15] In his *Principles of Psychology* James wrote about there being several, perhaps an infinite number of various orders of realities, each with its own unique style of existence. Schutz preferred, however, not to speak of orders or sub-universes of reality but instead of finite provinces of meaning upon each of which may be bestowed the accent of reality, "because it is the meaning of our experiences and not the ontological structure of the objects which constitutes reality."[16]

According to Schutz, by the varying of tension of our consciousness founded in a turning away of our attention from everyday life, and a specific *epoche*, one can break with the "vivid present" and leap into other provinces of meaning.[17] These latter include the worlds of fantasy, dreams, art, religious experience, and scientific meaning.[18]

The above could as well be a statement of one part of my own position on our experience of many regions of being, except for the primary role which he gave to varying the tension of consciousness. But that is a very significant exception, since Schutz did not see that these many worlds themselves as well as the objects found in each are all imaginative inventions, and cannot then be found just by modification of our consciousness and the working of appropriate types of *epoche*. Further, Schutz does not recognize that objects in these other worlds can, once these worlds exist for us, address us, call us to respond with some fitting activity. And Schutz plainly rules out the coordinating of some affairs or activities in one of these other worlds with those in an another or within the world of everyday life.[19] Yet this, as we shall see, is one of the more important imaginative activities of which we are capable. To deny that such coordination can occur would, for example, make the invention and use of calendars and maps impossible.

Contemporary phenomenological treatments of the imagination, such as those by Casey[20] and Murray[21] consider it as the producer of images, metaphors, symbols, or, as with Sartre,[22] of unrealities. However, these treatments of the imagination, by focusing upon the nature of images or other objects which owe their existence to the imagination, do not come close to adequately describing its role in the economy of human life. Murray does, taking up at one point an idea of Heidegger's, declare in so many words that the imagination projects horizons within which entities can appear and in which we function,[23] but he goes not one word further than that. Braun's *World of the Imagination*[24] seems to promise more, but "world" only appears in the title, the text treats rather of images and not of what Ortega sees as the world-fabricating capabilities of the imagination.

Along the latter lines, Ricoeur offers at least the promise of a general theory of the imagination (see *From Text to Action*, 1991). Ricoeur reinterprets image as not something enclosed within the mind which has been derived by imagination from sensory experience, but as having "a distinctive intentionality, namely to offer a model for *perceiving things differently*, the paradigm of a new vision."[25]

Imagination is to be recognized as the capacity or power for letting new worlds shape our understanding of ourselves. This power is conveyed "by the emergent meanings in our languages. Imagination would thus be treated as a dimension of language."[26] However, the truth in this case would be, I think, the reverse, for language itself can be considered a work of the imagination, as I shall show hereafter.

Ricoeur says we should see in imagination "an aspect of *semantic innovation*," of a form of discourse characteristic of the metaphorical use of language "whose use engenders the imaginary."[27] It is "when a new meaning emerges out of the ruins of literal predication that imagination offers its specific mediation."[28] "Imagining is above all restructuring semantic fields. It is . . . seeing as . . . and it gives an image to an emerging meaning."[29]

The image-producing imagination performs a kind of *epoche*, it introduces, Ricoeur says, a note of suspension with regard to the "thesis of the world," "an effect of neutralization, in short, a negative moment, thanks to which the entire process is placed within the dimension of the irreal . . . in the element of fiction."[30]

Fiction provides one with the power to redescribe reality. "Because it designates the non place in relation to all reality," it has what Ricoeur calls a new reference-effect, and so the "capacity to open and unfold new dimensions of reality by means of our suspension of belief in an earlier description"[31] (1991, p. 175).

By the "subversive power of the imaginary" the mode of being of the world is opened up to the mode of the possible, or of the "power-to-be."[32] The imagination has the power to put us in a state of "non involvement" with respect to the world of perception and action, to allow for the "free play of possibilities. . . . It is in this state of non involvement that we try out new ideas, new values, new ways of being in the world."[33] I strongly agree with the latter apprehension and would want only to now add to the list new modes of (trans-sensorily-met-and-immediate) experiences and (trans-bodily) actions. I shall comment on his interpretation of possibilities below.

But Ricoeur here goes on to see an essential connection between imagination and human action. "Imagination . . . applied to action, has a projective function that is part of the very dynamism of acting. . . . Without imagination, there is no action. . . . This is so in several different ways: on the level of projects, on the level of motivations, and on the level of the very power to act."[34]

Zaner affirms that "being able to *free myself from* what is actually before me" is correlatively "to *free myself for* what is possibly otherwise than the actual,"[35] that is, for what is the non actual. Like Ricoeur, Zaner sees a strong link between imagination and the mode of the possible, in fact, according to Zaner, "this leap into the possible . . . *is* the field of imagination. The imagination pertains strictly to the possible" (*Idem*). Not only does imaging, as Ricoeur sees it, make action possible, but for Zaner, "imagining is what makes thinking itself possible" (*Idem.*).

It should be noted that in a later article by Zaner this leap into the possible is called phantasying and is there distinguished from imagination ("Examples and Possibles," 1973). The core sense of the imagination, in Zaner's later view, is rather "to release . . . to distance myself from the actual, from the immediately surrounding circumstances . . . *to free myself from the press of the actual*, in order to search for 'ways out'." "It is what may be called the act of *possibilizing*,"[36] in terms of which the actual can then be seen as one instance of the possible, that is, as capable of being different from what it is or of being changed. Although the break with the world of immediately and sensorially given affairs and correlative actions is indeed, I think, the beginning of imagining, it does not seem to me that stepping from the real to the irreal or fictional (Ricoeur) or from the actual to the possible (Zaner) is the only or necessary account of what happens in the act of imagining.

Because of the importance given by Ricoeur and Zaner to connecting imagining and possibilizing, notice should be given here to the several meanings which they give to the latter. Ricoeur speaks of the mode of the possible as the power-to-be, and similarly Zaner of imagining as that which makes (thinking) possible. Our language expresses this meaning of possibility in the usages: "can be, can do," or less strongly, "could be, could do." Ricoeur speaks also of the free play of possibilities when we enter a state of non involvement with the world of perception and action and this seems to be identical to what Zaner later calls phantasying.

Zaner additionally speaks of the act of possibilizing in which by releasing or distancing oneself from the actual world it comes to be treated as (one) possibility. That is, I am "able to *free myself from* what is actually before me and correlatively to *free myself for* what is possibly otherwise than the actual." This is the central feature of what Husserl called fictive variation, and it presupposes that one can meaningfully ask "What else can be?".

But how can human beings ask that question if they have had no experience of something other than the actual? No philosopher that I know of has pursued this inquiry and consequently a fundamental aspect of the act of possibilizing is unaccounted for. However, the experience of possibilities can be found in every choice by human beings of their actions. In determining what one will do one anticipates the conjoining of one's actions and one's experiences in one's circumstance in a certain way, that is, the relating of one's possible actions to possible affairs.

Possibilities in this instance are acts and affairs conditionally do-able or find-able. They are first given to one in and through certain patterns of readiness effectually present in the human organism's bodily predispositions and concomitant actions.[37] These are subsequently augmented by one's experience of life and one's imagination as one experiences in one's negotiations with one's circumstance some of one's actions and some affairs as the border of a range or ranges of other activities and affairs not done or not yet here. In this connection, Popper in 1988 at the World Congress of Philosophy in Brighton spoke of the world as "an unfolding process of possibilities already folded in and of new possibilities thus made." Such non present acts and affairs can be the "same" but can also be "otherwise" than what is done or experienced at present and are able to be changed or replaced by something different, so that something new can begin to be done or encountered by one.

To return to our main theme, Johnson gives a much wider role to the imagination than Ricoeur and Zaner, closer to what Ortega suggested and what I will hereafter discuss. Johnson writes: "Visual imagery, icons, visual memory, dance, language, and knowledge are all fundamentally matters of the imagination."[38] This view of imagination is not focused upon "static images, but rather on dynamic patterns of processes or activities that are basic to our inhabiting a shared, meaningful world."[39] "Imagining so understood, thus includes the full range of organizing activities, from the forming of images (in different sensory modalities), to the execution of programs, to the manipulation of abstract representations, and even to the creation of novel orderings."[40] However, given Johnson's focus upon organizing processes, he concentrates upon image schemata or metaphors rather than upon the invention of different finite provinces of meaning (Schutz) or regions of being, as I term them.

The imaginative process should be seen as a creative process which adds something new to reality, which expands the contexture of the actual in my view. This approach to an understanding of imagination was introduced first after Ortega by Tymieniecka, who offers us a philosophy of the human being as creator, inventor, 'imaginator'. The creative act of the human being, she says, "makes him human,"[41] it "underlies all his specifically human functioning . . . first within nature, and then within our societal framework."[42] I couldn't agree more. Further, as she says, invention and creativity expand life's possibilities "*into possible worlds of life*." [43]

However, up to now she has focused her studies of the creative activity of the imagination not upon the invention of new realms of being ("worlds of life") and so the expansion of the Real but on the shaping by human beings of themselves through processes of self-individualization and the introducing of objects of their creation into the world, thus bringing about new forms of experience. This latter occurs "in man rebelling against the interpretation of the Real that the present phase of the constituted world gives us . . . in giving a new shape to his personal and intimate vision of the Real."[44] This visioning of a new universe "aims at finding or inventing a new form for grasping the Real more adequately and embodying it in the form of an object that can enter into the present phase of the constituted world . . . in order to transform it"[45] and also thereby transforming himself.

Creation in all instances, even the most material transformations such as chipping a flint or making pots out of clay – in which we put off activities which would bring present satisfaction for the sake of a future goal-fulfilling activity – involves our working in other regions of being which we invent along with the objects that occupy them. Tymieniecka has not yet explored that area of the human imagination's activities, but it is exactly this area that I am going to explore now. By focusing, as Ortega put it, on the world-fabricating activities of human beings, I shall in this way carry further the study of their imaginative and inventive capabilities, a project so well begun by Tymieniecka.

In the following pages I will develop an ecological interpretation of the imagination using as my starting point the brief but suggestive ideas of Ortega mentioned in the beginning of this essay.

Through a human being's imaginative capabilities his circumstance comes to be comprised of many different environments or regions of being, and thus many different kinds of interactions with his circumstance come to be at his disposal. Of course, we do not invent these as empty regions of being, but as I said above, they are invented along with an object or objects that occupy them. Each of these new regions of being is an invention of – and for – new modes of acting and experiencing. In a manner like that of adding new rooms to a house, their imaginative invention adds to the circumstance of human beings diverse new regions of being, new environments containing fields of action for certain episodes which are character-cultural specific, and in these environments also are landscapes which have a mnemonic role suggesting these activities.

Using his imaginative capability involves a human being's breaking his attention away from immediate and sensorially met reality. For example, through his imaginative capabilities a human being can order his actions and his experiences by "storying" them. By storying them, I mean a time-factoring and totalizing way of patterning his actions and experiences, independent of the specific details, in which a move is clearly made away from the region of sensorially, immediately met affairs. This time-factoring, story-ordering of activities can be found in every field of human activity including the most abstract, such as numbering and logical argumentation.

As Marshack said concerning human activities even as early as the Neolithic era, "there is a time-factored and time-factoring use of *story* that is already distinctively human. It has all been done by . . . performing, playing out, or communicating meanings related to the past, present, and future."[46] These temporally distinct regions of being are nonetheless experienced as connected and belonging to a totality of actions and experiences through their relation to the actor whose character or life program is the orienting preferential core of "his story." His life is the enactment of "his story," in the course of which diverse individually unique and socially common stories compose its successive episodes. A human life, that is, takes on the form of a specific sequence of actions, by both the stories he invents and the stories he learns from others, his community's repertory of stories. (This, of course, is not the same as "telling" or "listening to" a story, though all these depend upon his imaginative capability to break away from immediate reality and to invent regions of being distinguished from one another in temporal terms as "past," "present," or "future.")

Each of his "plans," that is, patterns of acting which anticipate the conjoining of his action and circumstance in a certain way, is storied and is "nested" in "his story." Every field of action, being composed of interlocking "plans," is capable of being story-ordered, that is, separate "plans" may be sequentially connected together giving a narrative patterning to actions in the landscape provided by that field. In a story a certain configuration of his circumstance is bonded to a certain time-factored series of acts. As one enacts "his story" in a certain landscape first one and then another "turning" shows up, obstacles and aids are revealed, perilous passages are recognized, and so on.

A human being is able to give an account of acts previously performed to himself or others (whether done in fact or fantasy), and to re-enact a series of such actions (as falling in the self-initiated range of actions this is not to be confused with imitating) by making these previous acts into a prototype to which other acts are comparable as more or less mutually conforming instances. Stories thus can provide "models" for his action. Likewise he can categorize affairs when he can tell similar or related kinds of stories about them or about himself and them. Thus he will be able to unify, compare, and to foretell these actions and affairs.

In his storying of his actions and his experiences, change, sequence, and connection are given together to a human being through the periodization of activities and correlative experiences in terms of temporally differentiated regions of being: the past, the present, the future.

Take as another example of the inventing of such new regions of being, the making of calendar. A calendar is not produced by adding day to day, week to week, or month to month, but by distinguishing one from another within a pre-established period. Consider also the example of maps. To just say that a map is a combination of symbols is like saying a skeleton is made up of bones. These statements are true, but it is only the particular articulation of their elements as belonging to a certain region of being that confers upon them their distinctive significance as being components of. . . . First comes the invention of the map, and then follow decisions as to certain relations of certain objects to be included in it or excluded from it. The same holds in general for all other such inventions of the imagination: various images, metaphors, symbols and their relations may be placed or found in them but these regions of being themselves are not produced by the connecting of such elements.

Further, a human being can invent regions of being where – in contrast to his experience of and action with respect to those sensorially and immediately met affairs concerning which he can, certain conditions being met, do something about – he by himself alone, independently of the power or operation of some other, apparently does and experiences something. These are regions of quasi-immediate affairs to which the name fantasy is commonly given. His negotiations with the quasi-immediate affairs in these regions of being take the following form: if he only could do this, his circumstance would be thus and so, or if his circumstance only could be this way, he would do thus and so.

A human being may give attention to an affair as this unique, particular affair with which he finds himself in a given circumstance, but, through an imaginative expansion of his circumstance, he also may give attention to the affair as being experienceable in this circumstance whether he is present or not. He may abstract himself "out of life." He thus comes to have an experience of the affair as a reality that does not need the particular someone whom he is to be experienced, an affair that with respect to himself is independent and self-sufficient and that remains throughout its existence identically itself for all possible observers. He comes to attend to it as "something in itself" and so as an "abstract" object, that is to say, as an object for no one and everyone. The recognition of particular affairs as representatives of types, classes, species, the comparison of affairs as being more or less, better or worse examples of something, and the development, too, of a technical language all involve an attending to such "abstract" affairs. This same abstract attitude can be adopted also in reflection turned towards his own actions, feelings, and experiences. He can temporarily suspend his involvement with other affairs and reflectively turn to his own proximate or remembered actions, feelings and experiences as themselves comprising a distinct part of his circumstance about which he then can do something.

By adopting this abstract attitude, he can, as Goldstein said, suspend some of the immediate claims made upon him by particular parts of his circumstance, and so he becomes able to shift from the patent aspects of his circumstance to the latent, to choose which will be figure and which ground, to attend simultaneously to various aspects of affairs, to deal with two aspects which are not immediately or directly connected, to grasp various aspects as a whole, break a whole into parts, isolate parts or combine them into wholes.[47] This abstract attitude is the basis of any theoretical approach to an "objective" reality and is the professional attitude adopted by the natural scientist.

Before he can invent different regions of being to act in and experience, there must be a starting point – a human being must be already installed in another region of being. A new region of being is to be found in and through a human being presenting it to himself as one who is already somewhere and who, to begin with, is installed in the region of sensorially and immediately met reality where he is related to other affairs in a bodily way. From that region of being in which a human being is already installed, however, he can then move to be installed in another region.

Through his capability for imagination a human being not only invents new regions of being but also coordinates these with other regions, for example, such regions of being as a map of Tennessee, or pure mathematics, or *Hamlet*, or *Don Quixote* are in some ways, or at least can be, coordinated with that of sensorially met affairs.

To elaborate on just two of these examples, everyone recognizes that in the region of sensorially met affairs a map is a separate affair from the physical area it depicts, but as a map it also comprises a separate region of being whose meaning is to be found precisely in the coordination which a human being makes between affairs appearing in it and those appearing in the region of sensorially met affairs. Another example of such coordination of regions of being is to be found in the theatre where particular human beings and physical objects immediately and sensorially met by an audience are coordinated with other persons and circumstances existing in the distinctly different region of the play being performed. (In what way can Richard Burton be like or different from "Hamlet" or even be a possible "Hamlet"?)

There need not be any similarity or correspondence between the affairs so coordinated. For example, human beings learn to combine certain auditorily discriminated markers or phonemes into commonly used instruments or morphemes by which they give patterns of distinction and connection to and apprehend certain patternings of experienced affairs, and they are also taught to use some of these morphemes in taxonomical ways to express distinction or connection between other morphemes. Now between phonetic combinations or morphemes and the affairs with which they are coordinated no isomorphism is usually to be found. The morpheme "kat" in no way resembles the animal which we experience.

The invention of symbols or the recognition of an affair as a symbol are particular examples of the coordinating of different regions of being. In using symbols, a human being is turned from one of the affairs found in a particular region of being to an affair or affairs in another region. (New symbols can, in a given historical situation, be evocative of new ways of seeing and relating affairs.) Take as an example the use of sounds as symbols in language. For phonemes to be used to form morphemes they must function as symbols: they point away from themselves and beyond the region of sensory affairs towards the region of morphemes as those affairs that "mean" other affairs.

In similar fashion morphemes may on occasion be experienced as pointing to affairs belonging to yet other regions of being. However, morphemes are commonly signs through which some affair begins to present itself and becomes immediately operative in the area of a person's own activity. Another example would be the mark "+," which can be experienced as a symbol of a mathematical activity, namely, adding, or can become a signal for the practicing mathematician who upon the appearance of this sign will at once begin the performance of that operation by which one comes to a sum.

Furthermore, some verbal "images" give us new expectations, metaphorically structuring our experiences in new ways, as when one speaks of "grasping an idea," "raising a question," "standing for office," "the growth of the economy," "the winds of war," "the key to success." Consider here also such actions as the erection of a boundary stone in ancient Greece and Rome, and the initiation of young men in many societies. These ceremonies should be viewed as actions that take the form of metaphors and metaphors taking the form of actions, that is to say, as act-metaphors.[48] "Images" from one region of being are projected into different regions of being so that, to take the last instance mentioned above, rites of initiation can be seen as a birth process, or better, as the death of the ways of childhood and a re-birth into the social world of adults.[49]

Human beings tend at first to try to give meaning to what occurs in new regions of being by interpreting these as being parallel with affairs in "the everyday world," and especially with one's body parts and movements. But one is able also to see the affairs of "the everyday world" as being parallel to, or even as being a projection or translation into this region of common action of some affairs belonging to another region (e.g., an affair which is a god in another region is a stone or a bear, etc. in "the everyday world"). Furthermore, affairs in one region of being can be transposed by a human being into other regions of being, as when the dabs of paint on a canvas are transposed into the region of a work of art, or when mathematical models are transposed into the region of sensory affairs. Human beings can therefore determine their action in the one region in terms of the way they understand affairs to be related in the other.

For example, my youngest child once told me of something she had seen in the woods which she wanted me to see, too, and I asked "Where is this interesting affair to be found?" She cannot tell me where it is but can retrace with me the path by which she discovered it. After a rambling twenty minute walk we arrive at the right spot and I recognize that the location is within hailing distance of our starting point. Now question and answer are not the same for me as they are for my child. Let us consider each of these in turn.

The affair which excited my child's interest was an unusual florescence of Lady Fingers unlike any she had seen before, and she had come upon it unexpectedly in her wanderings in the woods. The possibility of having such an experience is connected for her with the actions which had preceded its discovery and the places where they had occurred. From her point of view, the answer to my question ("Where?") was the consequence then of the possibility of a series of actions in particular circumstances being repeatable, i.e., of her finding it again.

For me the affair was locatable at one of an indefinite series of spatial points through the use of vectors established by reference to commonly accepted landmarks. The possibility of having such an experience was tied up for me with my ability to go to where it was after finding it in two different regions of being, one of sensorially met things and the other of geometrical abstractions, locating it first in terms of the latter and then in the former by the coordinating of these two regions. The answer to the question must be for me complex, for it is concerned with my eliminating all but one of the various places where the affair could be and of the diverse routes to it within a twofold landscape of sensuous and geometrical affairs.

When coordinating a new region of being with other regions of being which a human being has experienced already, what he finds in these may also assist him in discovering some of the possibilities within this newly postulated region of being. Note that in postulating these new regions of being, he does not invent possibilities as such, but ranges of possibilities can be found in them just as in the region of sensory affairs.

Think of Cervantes discovering other possible events in the lives of Don Quixote an Sancho as were finally told in the second volume of his chronicles of their adventures. Consider also the discovery by Brunelleschi of the possibility of mathematical perspective in a picture, or the discoveries in the region of numbers themselves of other possible kinds of numbers: those which are odd, or even, prime, irrational, none of these ("o", "∞"), partial (fractions), negative, any of these ("x" or "y"), all of these ("n . . ."), and so on.

Thus by the inventing and coordinating of new regions of being, that is, new environments, human beings can have new possibilities of experience and action. Moreover, discoveries in these new regions of being may lead in turn to awareness of new possibilities and to new forms of action and experience in those regions previously acted in and experienced.

To conclude by way of a summary: various interpretations of the role of the imagination have been briefly examined, including those of Ortega, Schutz, Ricoeur, Zaner, and Tymieniecka. As they have all affirmed, using one's imaginative capabilities involves first of all one's breaking of one's attention away from immediate and sensorially met reality. But then, by the imaginative inventing and coordinating of new regions of being, that is, new environments, human beings come to have new arenas of acting and experience.

As one's circumstance comes to include many new environments or regions of being, many new kinds of interaction within one's circumstance come to be at one's disposal. Moreover, discoveries in these new regions of being may, through coordination of these with regions previously acted in and experienced, in turn lead to awareness of new possibilities and to new forms of action and experience in those regions too.

Imaginative activities on the part of a human being appear in the use of new human capabilities, in what human beings are newly able to do and to experience. A human being can story (a time-factoring and totalizing way of patterning actions and experiences done or to be done) by which a certain configuration of one's circumstance is bonded to a certain series of acts. One can give an account of one's actions and re-enact a series of actions in fact or fiction when that is called for. One can recognize sequence, change, and continuity in terms of temporally differentiated regions, that is, past, present, and future, and consequently can put off a present satisfaction for a future goal. One can abstract oneself "out of life," recognize abstract objects, and see particular affairs as representatives of types, classes, species. One can categorize, unify, compare, and foretell actions and affairs. One can invent new means of communication, e.g., dance, music, visual images, language and with the latter, symbols, metaphors, signs or words. One can shift from one aspect of an action or affair or sets of actions and affairs to other aspects, choose what will be for one figure and what ground, deal with two or more aspects which are not immediately or directly connected, grasp various aspects as a whole, break a whole into parts. In other words, one can perform those activities which the West has identified with reason and science.

These capabilities, rooted in the activities of the imagination, are precisely those which in the past have many times been appealed to as constituting the humanness of the human organism. As Tymieniecka has already clearly seen, it is the inventative and imaginative activities of human beings then that make them human, which underlie all their clearly human functioning.

East Tennessee State University
U.S.A.

NOTES

[1] Ortega y Gasset, J., *Phenomenology and Art*, trans. Philip Silver (New York: Norton, 1975), p. 55.
[2] Ortega y Gasset, J., *Historical Reason*, trans. Philip Silver (New York: Norton, 1984), p. 82.
[3] Ortega y Gasset, J., *Phenomenology and Art*, p. 55.
[4] *Ibid.*, p. 67.
[5] See my article "Ortega's Development of an Ecological Philosophy," *The Journal of the History of Ideas* (1994).
[6] Begon, M., J. Harper and C. Townsend, *Ecology: Individuals, Populations, and Communities* (Sunderland, Mass.: Sinauer Associates, 1990), p. x.
[7] Bennett, J., *The Ecological Transition* (New York: Pergamon Press, 1976), p. 163.
[8] Gibson, J., *The Ecological Approach to Visual Perception* (Boston: Houghton-Mifflin, 1979), p. 8.
[9] Shotter, J., *Social Accountability and Selfhood* (Oxford: Basil Blackwell, 1984), p. 90.
[10] Shotter, J., "Social Accountability and Self Specification," in *Social Construction of the Person*, K. J. Gergen and K. E. Davis, eds. (New York: Springer, 1985), pp. 167–168.
[11] See, for example, *Ideas y Creencias* and *An Interpretation of Universal History*.
[12] Ortega y Gasset, J., *The Idea of Principle in Leibniz and the Evolution of Deductive Theory*, trans. Mildred Adams (New York: Norton, 1971), p. 134.
[13] *Ibid.*, p. 135.
[14] Idem.
[15] Schutz, A., *Collected Papers*, Vol. 2 (The Hague: Martinus Nijhoff, 1967), p. 230.
[16] Idem.
[17] *Ibid.*, p. 233.
[18] Schutz, A., *Collected Papers*, Vol. 2, pp. 232, 341.

[19] *Ibid.*, p. 232.
[20] Casey, E., *Imagining: A Phenomenological Study* (Bloomington: Indiana University Press, 1976).
[21] Murray, E., *Imaginative Thinking and Human Existence* (Pittsburgh: Duquesne University Press, 1986).
[22] Sartre, J. P., *Psychology of the Imagination* (New York: Citadel Press, 1963).
[23] Murray, E., *Imaginative Thinking and Human Existence*, p. 64.
[24] Braun, E. T., *World of the Imagination* (Savage, Md.: Rowman and Littlefield, 1991).
[25] Ricoeur, P., *Hermeneutics and the Human Sciences* (Cambridge: Cambridge University Press, 1981), p. 292.
[26] Ricoeur, P., *From Text to Action* (Evanston: Northwestern University Press, 1991), p. 172.
[27] *Ibid.*, pp. 171, 172.
[28] *Ibid.*, p. 172.
[29] *Ibid.*, p. 173.
[30] *Ibid.*, p. 174.
[31] *Ibid.*, p. 175.
[32] Ricoeur, P., *Hermeneutics and the Human Sciences*, p. 292.
[33] Ricoeur, P., *From Text to Action*, p. 174.
[34] *Ibid.*, p. 177.
[35] Zaner, R., "The Leap of Freedom: Education and the Possible," *Maincurrents* 28 (1972), p. 178.
[36] *Ibid.*, pp. 177–178.
[37] Bransford, J., J. Franks, N. McCarrel and K. Nitsch, "Towards Unexplaining Memory," in *Pereceiving, Acting, and Knowing* (Hillsdale: Lawrence Erlbaum Assoc., 1977), p. 435.
[38] Johnson, M., "The Imaginative Basis of Meaning and Cognition," in *Images of Memory: On Remembering and Representation*, Kuchler, S. and Melion, W., eds. (Washington: Smithsonian Institution Press, 1991), p. 74.
[39] *Ibid.*, p. 75.
[40] *Ibid.*, p. 79.
[41] Tymieniecka, A., *Logos and Life: Creative Experience and the Critique of Reason*, Vol. I (Dordrecht: Kluwer, 1988), p. 6.
[42] Tymieniecka, A., *Man's Self-Interpretation-in Existence: Phenomenology and the Philosophy of Life* (Dordrecht: Kluwer, 1990), pp. 4, 5, 6.
[43] Tymieniecka, A., *Logos and Life: Creative Experience and the Critique of Reason, op. cit.,* p. 6.
[44] *Ibid.*, p. 26.
[45] *Ibid.*, p. 27.
[46] Marshack, A., *Roots of Civilization* (New York: McGraw-Hill, 1972), p. 115.
[47] Goldstein, K., *Language and Language Disturbances* (New York: Grune and Stratton, 1948), p. 6.
[48] See M. Polanyi and H. Porsch, *Meaning* (Chicago: University of Chicago Press, 1975), pp. 117–120.
[49] See M. Eliade, *Rites and Symbols of Initiation* (New York: Harper and Row, 1965), p. xii.

RUURDTJE BOERSMA AND WILLEM BEEKMAN

UNDERSTANDING NATURE AS LIVING: A CHALLENGE FOR ACADEMIC EDUCATION

INTRODUCTION

Education and training in the life sciences in modern universities is mostly based upon a reductionistic, mechanistic, deterministic and causal-analytical approach to nature. From that viewpoint, life is an epiphenomenon of physio-chemical processes, not yet fully understood, just as consciousness is seen as being a metaphor for mechanistic brain process.

An advantage of this approach was that it gave a tremendous power to dominate nature and exploit it in modern technological agriculture. However, immediately disadvantages became apparent: large environmental problems, loss of biodiversity and a feeling of alienation between ourselves and nature occurred.

The scientific methods which were used have been directed towards destruction of life instead of understanding life as being an intrinsic property of organisms.

In recent years, both within agriculture and with society in general, development towards sustainability has been promoted. The Brundtland Report (World Commission on Environment and Development, 1987) certainly gave extra impulse for the implementation of this concept.

The concept of sustainability results in the need for conceptual and educational tools for understanding nature as living, supporting its sustainability and respecting its intrinsic values. Nature, in this respect, is regarded as taking in all from single organisms to ecosystems, even at the global level, as developed in the Gaia-hypothesis of Lovelock (1979) and others (Sheldrake, 1990).

We distinguish different aspects of approaching nature as living:

- ecological thinking (ecophilosophy);
- ecological attitude (ecopsychology);
- ecological acting (organic agriculture).

The Department of Ecological Agriculture of Wageningen Agricultural University is developing methods of education and training focused on these three aspects. We use a holistic view of nature and education, which

A-T. Tymieniecka (ed.), Analecta Husserliana LIX, 323–331.

implies not only recognition of high level systems betokening the integration of living entities (agro-ecosystems), but also recognition of the fundamental unity of man and nature in a global realm (Edelglass, 1992; Maturana and Varela, 1984).

This illustrated in Figure 1:

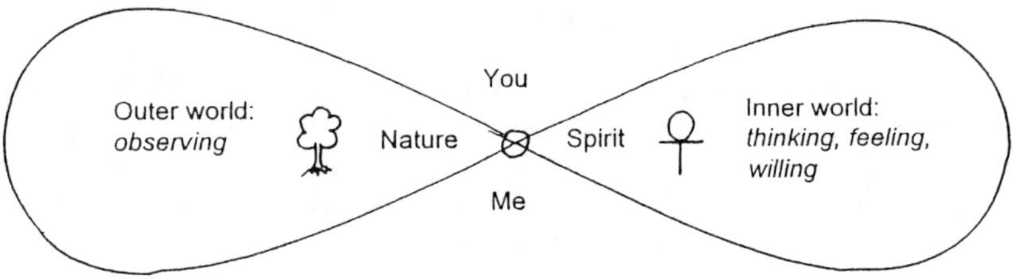

Figure 1. Inner and outer world.

The centre of the lemniscate represents human consciousness, which tries to understand nature on as many levels as possible (not only cognitive, but also psycho-motor and in its effects). In this act of knowing two direction are visualized: one is the out direction towards the phenomena in nature, the other is the inner direction towards the realm of personal development: intention, intuition, imagination and inspiration. The latter we call the spiritual realm. Our understanding of holism envisions the reciprocal enrichment of these two directions of epistemology when understood as being two sides of the same coin; this means understanding nature as living. In this way ecophilosophy and ecopsychology are linked together in order to develop students' skills for improving a future organic agriculture with a spiritual component.

EXAMPLES FROM EDUCATION

We will here give examples from two training courses to illustrate the above approach. This is not meant as an exhaustive review, but only a preliminary overview.

1. *Goethean Phenomenology*

Students are trained in this method of understanding nature is living in a four week practicum. One of the basic principles of understanding nature is to dwell on phenomena. This creeping into plants or animals is in a way comparable to a surrender of self in order to let nature "speak" for itself. The first scientist to practice this method was J. W. von Goethe, as can be seen, for instance, in his *Metamorphosis of Plants* (1974). According to this method the alienation between man and nature (between subject and object) is temporarily removed in favour of an intimate bond with nature. In the very process of making contact with nature, letting all prejudices and biases fade away, being as completely as possible directed towards the natural object, students discover new questions, new ways of looking at nature, which results in promising research questions that normally would not have evolved out of the use of conventional analytical methods. For instance, the notion has arisen that weeds in crops have the tendency to resemble the crop in which they grow. As a consequence, these questions can be put under the scrutiny of normal scientific methods, in order to communicate the results to the scientific community. Part of this method is reflection on pre-scientific notions of science, nature, agriculture, plants and other subjects. Only after thorough reflection on personal prejudices and biases, can students surrender themselves into the time-consuming and enveloping Goethean phenomenology.

In practice this means a diversity of observational training methods, such as,

- daily observations of garden plants, using accurate drawings;
- weekly observations of trees, with mental exercises in between in order to sharpen the power of imagination (memory). This harks back to the Goethean concept of exact sensory fantasy;
- exercises with colours, aided by watercolour paintings in order to develop a feeling for colour qualities in nature;
- daily landscape observations, assisted by colour paintings;
- mental exercises in concentration and focusing on a chosen and very simple object, such as a coin or a spoon. This daily exercise resembles an object-directed meditation, which aims to sharpen the powers of exact thinking.

All these exercises are embedded into a theoretical framework covering the development of Western scientific thinking, the role of phenomenology in science, and especially the meaning and application of Goethean phenomenology in the life sciences and agriculture.

As an example we will give our view on leaf metamorphosis. One of the fundamental discoveries of Goethe (1749–1832) was the metamorphosis of leaves. On the basis of this and other discoveries, Goethe founded a scientific school of phenomenologists, to which the authors belong. In particular, the Swiss botanist Bockemühl (1981) has done a lifetime's research on leaf forms and the ever changing shapes of leaves in the growing process of flowering plants. One of our students once had a kind of "eureka-insight" and formulated the metamorphosis as follows: plants are growing what we are thinking. See Figure 2.

The leaf sequence of an ordinary weed *Senecio vulgaris* (Groundsel) is presented. The leaves are arranged in a kind of spiral, which is purely artificial, but represents the array of leaves spiralling around the stem of the plant. From left to right the leaves grow higher up the stem: the first leaf is the lowest, the last leaf the highest. the remarkable thing is the smooth and regular change of form. Each next leaf can be developed from the previous one without any problem. But there are no intermediate forms, although theoretically, one could imagine thousands

Figure 2. Leaf metamorphosis of Senecio vulgaris (groundsel): "Plants are growing what we are thinking".

of them. The overall change of form is called by Goethe a "Fliessgestalt", a flowing form of which the individual leaves are the raw material. This metamorphosis is law-like: each groundsel plant is doing this in almost the same way. However, the crucial thing is that each plant is slightly different in detail, depending on growth conditions and other environmental influences. In fact you can see the metamorphosis as a "personal" diary of the plant reflecting its own history in time and space. In reverse you can look at the metamorphosis and obtain information about the growing conditions. It could even be used as a kind of indicator of environmental influences during the growing season.

How does each next leaf "know" what form is needed just to make a tiny step forward in the overall growth pattern. If it is a function of the genes, how do they know? Could it be some "memory field" in the sense hypothesized by Rupert Sheldrake (1988) in his theory on morphogenetic fields, or an "etheric body" as proposed by many spiritual scholars? We hold the view that Goethe's "Fliessgestalt" is an old word for some principle that modern science is trying to identify. We think that this form principle is strongly related to our own mental process, a notion which is in a similar way expressed by Bateson (1979). The only being that can recognize the metamorphosis as a flowing process is the human. Only in our imagination can we re-create this growing process and make the inner connections between the different forms. The plant is making this inner connection too, but the outer result is only stepwise, because of the limitation that matter puts to the plant. The only organism that indeed can grow in this totally flowing way, without the need for steps in between, are animals and humans in their embryological development and growing processes over a lifetime. There you never find older stages of development because the whole organism is changing in a time process. If the growth of plants is, in essence, related to our thinking process, than we have a very deep inner connection with the plant life around us; they grow what we think. The individual embryology of the groundsel leaves is an astonishing detail in metamorphosis. From leaf bud to mature leaf they too follow a metamorphosis (mostly invisible and hidden in the buds), but this process has a direction opposite from that of the major metamorphosis along the stem. So the very primitive leaf form is reflected in the last leaf on the stem, and the older the leaf becomes, the more its form descends along the stem to the lower final forms. It seems as if the plant has two time sequences at once: one sequence is the visible one as shown in the figure and reflecting the time arrow from the past to the present. The other is the embryological sequence reflecting the reverse process from future to the present. The presence of the past and the presence of the future in one organism at the same time is no less than an everyday miracle.

2. *Integral Agricultural Science*

In the second example from the training courses in which we approach nature as living, the central theme is that of our "basic attitude towards nature". The concept of "basic attitude" gives an indication of the character of the fundamental relation between human beings and nature. The characters of this relation can be distinguished, for example, as follows (Zweers, 1995):

- the *despot*, who subjects nature and for whom there are no limits to growth;
- the *enlightened ruler*, who still dominates nature, but realizes also our own dependence on nature;
- the *steward*, who manages the earth as a property of God or the next generations, but may use the rent for his own;
- the *partner*, who considers nature as a worthy partner;
- the *participant*, who views nature as a totality of which he is part and experiences a strong connection with nature;
- *Unio Mystica*, mystical unity with nature.

In current agricultural education students learn about the relations between plant and soil, between plant and plant or between plant and animal, but not about the relations between humans and soil, plant or animal. The distance between humans and nature has become so wide, that it is difficult to understand nature as living. In order to reduce this gap it is important to understand something of the philosophical and psychological dimensions of the concept of "basic attitude", theoretically as well as experientially.

The series of twelve weekly workshops is called "Integral Agricultural Science", because of the direct connection we make between the inner world of humans (students or farmers), with the outer world (especially the agricultural). The "basic attitude", from the inner world, influences the outer world, agricultural and scientific practices. To get an understanding of these influences it is essential to become acquainted with the history of the relationship between humans and nature, the theoretical approach, but it is also important to become conscious of one's own attitude, one's own way of standing in life, i.e., to be more experience orientated. To achieve this, the programme of the course has three parts. In the first part the students learn about the history of the Western way of thinking and about new approaches in science and farming practice, in which the basic attitude towards nature is less dominating. This is done mainly by reading and discussing texts.

In the second part the students have to form an idea of their own attitude towards and relation with their environment. This part is, in fact, not strictly separated from the other two parts. In the whole programme attention is given to one's own way of thinking, feeling and doing. We use various methods for this part of the programme, some of them applied in the workshops themselves, others in homework assignments between the weekly workshops. Some of them are:

- considering one's own view of humankind and world view, writing about this and exchanging ideas;
- reading articles about farming or research practices, or philosophies advocating that the human-nature relation be tighter, for example, articles about communication with plants and about the work of the scientist Barbara McClintock (Keller, 1983), who is known for her discovery of the spontaneous transposition of genes;
- making a piece of work, by drawing, writing, modelling (or whatever creative activity the student chooses), about the Greek Elements of nature. The aim of this assignment is to realise what earth, air, fire and water mean in essence for one personally;
- dramatic expression to help feel the relation between mind and body;
- exercises in visualisation for becoming aware of one's own capacities for intuitive observation, aimed at developing the capacity for communication with plants;
- sowing beans, giving loving attention to one half of them, then observing for some weeks what happens.

When the students have some idea of their own attitude, we look at the relation between other persons and their environment in the third part of the course. On a day excursion we visit two farmers with different ideas and ways of farming, for example, an organic farm and a modern technological farm. In these visits students have to use and to sharpen their insight into the influence the basic attitude of the farmer has on the farming practice. This is an exercise in observing the phenomenon of the farmer and of the farm, while being aware of one's own thinking and feeling.

We do the same exercise while visiting a scientific researcher. Observing the research itself and the research environment is difficult in a short visit, but in a discussion with the researcher the students can get a picture of the relation between the way of thinking of a researcher and his or her work.

An important aspect of the course is the continuous interaction between learning about the outer world, in theory as well as practice, and learning about ones own inner world, i.e., becoming conscious of ones own thinking, feeling and willing.

HOW DO STUDENTS RESPOND?

The two training courses described are exceptional at Wageningen Agricultural University, both for the subject matter and the way of teaching. They are optional, but in some study programmes they are commended. So in general students choose the courses very conscientiously. As teachers we notice that students are looking for an education which makes a connection between their own knowledge and experience and new theoretical or scientific knowledge. Students are frustrated by the current scientific education. They feel that their intuition and creativity is shrivelling up in this alienating form of education, in which almost no life can be recognized.

Students who are following our training courses are very much involved in the learning process. Many of them indicate that they take these courses with pleasure and satisfaction, in contrast to other courses. They are also thinking and talking about subjects from the courses outside the workshops and the time spent doing their homework. So the social environment is also in some way involved in the learning process.

The courses are eye-openers to students and confirm their own ideas that there is more on earth than the scientifically known material world. Students also get confirmation that they themselves matter personally, and that understanding nature as living matters.

Some students, when they come to follow our course, become hesitant about their whole study programme, whether they will continue in it or not, and which direction to choose. Many of them are clearer about what they want afterwards. The courses make an important contribution to the personal development of the students, which should be, in fact, a task of every form of education.

CONCLUSION

Developing conceptual and educational tools by which to understand nature appears not only to be meeting a need to develop sustainability and encouragement for students who are looking for real learning, it also is a challenge for academic education. Participative learning is also participative teaching, which means that the teacher as well as the student is involved in the learning process. Thus the teaching method is as important as the relation between the teacher and the student.

Developing this kind of educational concepts and methods is part of our activities exploring paradigm shifting concepts and is incorporated into the methods used by the Department of Ecological Agriculture at the Wageningen Agricultural University.

Wageningen Agricultural University
The Netherlands

REFERENCES

Bateson, G., *Mind and Nature* (Wild Woodhouse, London, 1979).

Bockemühl, J., *In Partnership with Nature* (Wyoming, R.I.: Bio-Dynamic Literature, 1981).

Edelglass, S., G. Maier and H. Gebert, *Matter and Mind* (Edinburgh: Floris Book, 1992).

Goethe, J. W., *The Metamorphosis of Plants* (Springfield: Bio-dynamic Farming and Gardening Association, 1974).

Keller, E. F., *A Feeling for the Organism. The Life and Work of Barbara McClintock* (New York: Freeman, 1983).

Lovelock, J., *Gaia: A New Look at Life on Earth* (Oxford: Oxford University Press, 1979).

Maturana, H. R. and F. J. Varela, *The Tree of Knowledge: The Biological Roots of Human Understanding* (Boston: Shambala, 1984).

Sheldrake, R., *The Presence of the Past* (London: Fontana, 1988).

Sheldrake, R., *The Rebirth of Nature* (London: Century, 1990).

World Commission on Environment of Development, *Our Common Future* (New York: Oxford University Press, 1987).

Zweers, W., "Ecological Spirituality as Point of Departure for an Intercultural Dialogue," in *Agricultural Spirituality, Essays from the Crossroads Conference at Wageningen Agricultural University* (Utrecht: International Books, 1995).

GABRIELE BOSELLI

A PHENOMENOLOGICAL PERSPECTIVE ON EDUCATIONAL PLANNING

For some decades phenomenology and its derivations have been favourably represented by the varied scientific research work carried out under the aegis of Anna-Teresa Tymieniecka and her World Institute for Advanced Phenomenological Research and Learning and its review Analecta Husserliana.

Every view of scientific progress in the world of physics and especially in the human sciences evidences how "the Phenomenological Movement" has made available to every form of knowledge of a language, a type of approach and a method useful for any science when carrying out the task of individualizing meaning, correctly interpreting experiences.

In this report I will deal with a field of research which phenomenology has generated in the pedagogic sector by some scholars who, during the 1980s elaborated a theory of pedagogic planning, namely "Post Programming" in opposition to the leading theories of that period, which emphasised pre-programming.

The practice and theory work were performed in the militant school field, with, however, the active support and encouragement of Enciclopaideia, a group of scholars that since 1987 are active in Bologna under the leadership of Piero Bertolini, who since the 1960s is the authentic Master of Italian studies in educational phenomenology.

1. THE PEDAGOGIC EFFECTS OF ECLIPTIC MODERN PROCESSES

The fact that man and project (hopefully also expectancy, dreams and fantasy) are practically contiguous terms is well known. Man has always planned, searched for ways of reducing his own and others' tension in order to achieve fulfilment. Teachers have always planned and been involved in determined goals, guided by pre-categorical experience and culture. This rapport between subject and other (other being both the subject and the exterior) in a cultural context is a topic that educational theology has dwelt on in the history of Western pedagogy. There has always been this concern wherever an educational objective has been

A-T. Tymieniecka (ed.), Analecta Husserliana LIX, 333–342.

proposed: this could be declared openly or not, known or not, considered or completely ignored. In the present, past or future of subjects, it involves them in trans-formative processes, in metamorphosis, in passsing through new lands.

With Modernism, recognizably starting with comenius and terminating with the expansion of programmed studies and experiences, we have a decline in the role of theology and educational practice is progressively restricted only to education and confined to detailed syntagmatic structures definitely closed. The emergence of the New and Other is forbidden; The educator is unable to seize opportunity.

Nowadays we are at the height of a bureaucracy that has invaded not only offices but classrooms so much so that many believe it is impossible (some say non-scientific) to imagine educational planning in terms other than programming. The name (and substance) of this restriction in planning goals, of this limitation of subsystemic aims, of this failure to recognize tradition is therefore Programming. The word dominates schools in developed countries hiding behind the well known terms: Reason, Technology, Economy, Power. It well suits the needs of having certainties, information and efficiency, which are increasingly required in our end of the millennium society.

I trust however that in the middle distance (historical times) other configurations of Postmodernism will be operative: the system of late modernism will become a constellation, and controllable complexity, hypercomplexity, that history now partially suspended from technology will once again gain headway in general awareness. In theory this is already widely recognized: in the general context of the phenomenological approach, the theory of complexity/hypercomplexity and hermeneutics is becoming the most important code for interpreting planning phenomena.

Years ago I wrote these things in a book *Postprogrammazione* which attempts to propose through the phenomenological perspective educational planning on the basis of the new ideal openings of the Postmodern, which echo too the ancient paths of idealism.

2. THE SOURCES

Husserl

Together with Husserl (and Rosenzweig, author of *The Star of Redemption*), we have listened to the words of Israel, a people deprived of its native land for 2000 years, to the second roots of the Western world. With the author of the *Crisis*, we have read and held a lesson on education as a rejection of any fact-study technique, namely, the erroneous mining of events without a subject, the mere accumulation of systemic elements.

Educating is, on the contrary the sustaining of life, a transcendental act whereby man is shaped when encountering the Other, recognizing him, since recognition and awareness is also participating in the forms of historically constituted intersubjectivity (the disciplines). Above all we have learned the art of radical questions, learned how to avoid being afraid when coping with enormous prejudice and bias.

Programming is virtually a mountain, but it will be sufficient to eliminate some of the ideological theories of modernism, and it will disappear.

Heidegger

With Heidegger (and Gadamer) we have meditated on the technical ideology as a system capable of managing any practical activity, from the producing of washing machines to the formation of Man. We have seen man as the subject (not object) of knowledge information, that knowledge (such as didactics) is essentialy narrative and not descriptive, intersubjective an not objective, epistemological and not epistemic. Being a subject, becoming a subject, helping others to become a subject entails waiting, listening, interpreting, planning oneself and planning in general. In other words, by going beyond dependency, one becomes what one authentically is and what one can become.

3. THE PROS AND CONS OF A CONCRETE PHENOMENOLOGICAL PROJECT

A rejection of the hegemony of the technical aspects of economism has led us to erase from pedagogic practice "the corollaries": thus no hierarchic pre-ordered objectives, no executive protocols, no verifications. Yes, to the aims to the creativity of the relation, to the interpretation of the effects. No to bureaucratic practices, to technicism, to the triviality deriving from cultural segmentation. Above all no to the non-recognition of the subject. *Recovery of the transcendental point of view* is prime. Re-elaboration of the discipline.

In "Post programming" an evaluation without a verification claim is involved to valorize the subjectivity evaluating and open the way not to objectivity (impossible) but a possible intersubjectivity.

A. *The Cons*

No to Technology. Programming is a mythical expression of late modernism: Technology. It is a way of putting one's competence in educational processes into practice, knowing how to employ means and events for an economic carrying out of qualitative and quantitative objectives; in short it is *a series of norms for the managing of practical activities*. A restricted form of rationality like technology is nowadays (mainly owing to the seduction of the products deriving from it) identified as an aim in itself. The excessive power of late modernism's technology has led to its becoming an ideology, a syntagma operative in every field of human activity. Moreover, it is an ethic; what can be called technical is good and rational. Technique is almost a religion that from early morning to late night guides the actions and thoughts of the subject at the end of millennium. It was impossible to imagine that technology would not colonize, among other fields, the school too, ignoring its peculiarity, its high values with respect to other enterprises which would not serve its belonging to the rest of the system. And so it came to be regard as a site for production like any other, with teachers being called "operators" or "professionals" and with the modest vanity of headmasters being flattered by calling them "managers".

In my opinion technology should not ruin the charm of the game played by adults and children, should not interfere with determination and objectives, as often happens in every system which has acquired too much power. The power of technology makes deserts grow and reduces woodland and forests, it creates enormous areas of wasteland on the face of the earth. Scholastic programming, a modest form of technological planning can contribute to the growth of internal deserts in the human soul.

No to Objectives. Programming entails acting according to objectives relevant to the action itself, classified by hierarchy and succession, checked through behaviour congruent with these standards. (Tyler, De Lanscheere, Filograsso)

Objectivity can be a valid criterion when it is concerned with producing objectives, when what is important is not the producer, but the object. If this is positive when it comes to material things, it is not so with a spiritual thing like education. Objectivity may represent a criterion for action on the part of soldiers or managers or agents, but educating is a completely different realm of fact. To set objectives and work hard to have them carried out *crushes the intentionality of the Other* and leads to neglect of the aim itself. As with many aspects of the contemporary way of thinking, only the immediate counts, that which is made explicit and imposed as a goal to be achieved within a productive or distributive structure. Only the target counts, but we do not know where it will lead us. Having objectives means neglecting opportunities, ignoring unforseen events, occasions which have not been programmed, and those fragments of the universe that can not be pieced together. And so most of the world itself and its organic chaos are considered interference.

It is reassuring to note that the masses of objectives written (or more often transcribed) in school registers frequently represent a tiresome, but harmless pretense. The common sense of teachers and professors does not allow us to overlook the necessity of Sense, namely, the authentic intentionality which has not been subordinated to structures.

No to (the demand for) Verification. In every system for the production of objectives there is a structure for the qualitative and quantitative control of the product. Since the production of objects is concerned, this control can only be "objective".

The recent modernized world, in its technocratic and ideological economizing development ordered according to efficiency, efficacy productivity has extended its criterion to all fields of human activity, including the school and all its scientific activity.

Of little importance is the fact that epistemology has actually now abandoned any demand for verification; little attention is paid to contexts, to history. We go ahead just the same, like Japanese bulldozers in the sacred forests of Indonesia, only concerned with the tons of wood produced. Countries rich in pedagogic traditions now have evaluation systems based on the international ideology of verificationism.

I think it is necessary to note that:

(1) In schools no objects are produced. The formation of subjects is involved.

(2) The object as such has no recognizable truth. The truth can presumably derive from its recognition by a subject (the subjectivity of verification) or by a plurality of subjects (the intersubjectivity of the at of "verifying"); but it would be the subject's truth (or better still, that of the anonymous Heideggerian "one") that make itself known, not that of the object. However, there is no "objective" verification just a possible declaration of recognition as true by one or more subjects. There may be operations of recognition "prone to regard as true" a phenomenological hypothesis.

(3) The evaluation (regulations connected with the subject's own values and those of the assessing system) in any case precedes its findings since it pre-comprehends verification, transforming the results according to conscious or unconscious intentional deformations.

B. *The Pros*

Yes to the Subject and to a School Based on Experience. Despite useless clatter (TV, the press, sometimes the school itself) and the consequently lessened frequency of authentic words, the subject is not yet extinct. He has given up changing the world, but he is still be to prevent the world from changing him according to planned systems.

Those who are responsible for education should reinforce what there is left of the authentically subjective, to help those pupils in the initial stages of their formation. With this aim in mind, they should find direct access to the fonts of culture and propose them to students outside of mediations, e.g., programming, or even worse, didactic guides.

The subject, namely the one without predetermined limits to his existence and the master (teacher or manager) should cooperate in the formation of subjects who in time will not finish finding their limits. This is why the teacher feels he is a contemporary of all educators of all times who have coped with their pupils. He is an autonomous subject helped by the quality and variety of his own intellectual network such as the non-defining of his own models of interpretation and of the design of which he is a part and of which he makes others a part.

He is free . . . , free from . . . and free to . . . (and he himself finds the words with which to fill in the gaps). He has no objectives. He has a sense of the aim, the horizon and the "beyond". He knows that to educate is to transcend and help others to transcend, not through common ways, but through paths laid down in the passing through of other generations.

Yes to Relationships. In the world of production, interaction among operator comes about according to schemes where in what counts is not their identity, but their status and role. There is no reciprocity, role exchanges or emotional participation in the experiences of other actors. Only the achievement of the organization's goals counts and the lives and relationships or interaction of subjects are of secondary importance.

In education what counts is the subjects' well-being, their happiness, their encountering difficulties connected to knowledge and existence. In post-programming there are no pre-established schemes; participants play the game by exploiting opportunities. The relationship, a bond that liberates, helps to draw oneself and others out of inauthenticity, establishes connections with the other that is found in others and in ourselves. In this perspective the classroom is a tent temporarily set up in a sort of no man's land by those sharing a path and giving importance to being together, heeding the voices of texts from other times and other places. This tent will be taken down the following morning, since it is important to move on.

Yes to Disciplines (Under Certain Conditions). If by disciplines we mean the mere sedimentation of an age, the simple coding of knowledge, incontrovertible description, the direction of a strong subject, an episteme which dominates, I believe any discipline is miseducational. Instead of leading out (*e-ducate*) "drawing out" we have, the subject presented with a pre-established view of the world.

In my opinion Discipline is not a sort of revelation, but an unprogrammable voyage inside the enigma of the world. Discipline, then, is linked to seeing, being and "indicating". Thus in post-programming discipline is marked by:

(a) the representation of phenomena considered in the fortuitousness of their occurrence;
(b) a compromise between experiences and subjective/intersubjective intentions;
(c) a weak decree to gather opportunities to find evidence of physical of cultural phenomena;

(d) the historicization of phenomena;
(e) the differentiation of events;
(f) questioning, reciprocal questioning between teacher and pupil;
(g) common interpretation of a subject who has experienced an organic confrontation with an area of knowledge.

Yes to Interpretation. The evaluation, which is never objective, is always subjective. A subject can be single person or a ministry but always a subject who assesses from his own point of view. I think it is important – and honest – that the subject, attempting to respect the object as an object declares his subjectivity and realizes and admits that his assessment will not be a mirroring but a portrayal of the encounter between the evolving world and the assessor. Then we will have an evaluation made from subject to subject, a project that both help to outline. In such a hermeneutical model of evaluation:
(1) The assessor and the assessed should both realize that they are in question;
(2) There should be exchanges of meaning regardless of different power status;
(3) Faces should not be disguised behind norms since evaluating does not mean keeping at a distance, but approaching the Other, his world, his history;
(4) Unintentional results, those not considered in the project should be recognized and assessed;
(5) Hopefully there should prove to be developments.

I think that from a phenomenological point of view theory and practice in pedagogy (and not only the pedagogy) may in the future evolve towards the following schematic aims:
(a) The renunciation of interventionism;
(b) The perception of limits beyond which action for becomes action/agitation on;
(c) Knowledge of foundations, not fundamentals: the status of fundamentals has irreparably deteriorated; dynamic foundations are more resistant to unstable conditions;
(d) Developing by eliminating the frenzied urge to act and to subordinate thinking to action – the ability to meditate and ponder;
(e) Conceiving formation not along the lines of a model useful to the

system (the industrial and company approach), but as a "faint" helping towards an autonomous shaping of self and others;

(f) Avoidance of homogenization, the safeguarding and exploitation of differences, projecting suitable routes;

(g) The building up of awareness of the historicity of all planning: knowing that students are not insects, that they have a long memory, that of humanity;

(h) Recognition of the importance of the teacher: taking entropathy (imagining, while remaining oneself and not presuming too much, how the other may live the proposal we are about to make) as the starting point of planning;

(i) Recovery from the transcendental point of view: we cannot plan from the point of view of the subject-other (impossible), nor from a personal point of view or society's (alienating), but according to the type of relationship that the subject can develop with elements in the field;

(j) The renunciation of models (syntagmatic figured offered offered for others to replicate) and the invention instead of situations (paradigmatic contexts built up as unexemplary, but e-ducative stories);

(k) The exploitation of expectation: the future in pedagogy cannot be imagined only as a project but also as expectation that something will develop even extraintentionally, without this meaning worry that such will have a negative effect;

(l) Invitation: openness to acting in "our" existential and professional space, letting happen and allowing the removal of obstacles to the self-achievement of the subjectivity to be found in ourselves and in others.

From a didactic point of view we can think of a development of theory characterized by:

(a) Thinking without being afraid (one should never fear and anyhow there is no reason to fear) of pedagogic knowledge tied to philosophy and science. The teacher from the nursery to university, has the duty and right to take an interest in the whole theoretic field and not just what concerns daily practice;

(b) Viewing didactics as narrative knowledge, a history of ordinary stories and extraordinary education;

(c) Plans in didactics that can be clearly interpreted;

(d) The identification of problems as problems for . . . someone: the realization that problems change according to the identity and the context of the person involved;

(e) The raising of opportunities to provoke and not impose;

(f) The renunciation of taking decisions for others (not even for minors) and of being objective (judging others according to our standards or worse the standards of others);

(g) The renunciation of objectives, namely, working for immediate results;

(h) Inspiring grand ideas for daily planning and action;

(i) Bringing up questions and declaring one's point of view;

(j) Renouncing demands for an objective verification (for ethical, epistemological and practical reasons);

(k) Interpreting the consequences of our actions according to an inter-subjective point of view.

CONCLUSION

The school should never become an alienating "total quality control" factory. It is a place for the spirit, of peace, of transcendence. It is a sincere invitation to further investigation. Postprogramming has been given the years of phenomenological reflection on the part of a small group of scholars, with the school in mind. It entails a cultural opening of the school to the past and future, a responsible answer to the queries of many teachers, the constructive development of plans looking at the future.

The appreciation of few eminent university professors (in particular, besides the aforementioned Piero Bertolini, Fabrizio Ravaglioli and Cesare Scurati) and of Ministry of Public Instruction administrators has enabled us to outline a theoretical and practical construct as yet impossible to carry out. It will perhaps be in place in the near future, since there are clear signs of deterioration, not only in the theoretical universe of education, but also in the schools' capacity to maintain ideologies and systems of application which are by now completely alien.

Ministry of Public Instruction
Italy

REFERENCE

Boselli, Gabriele. *Postprogrammazione* (Florence: ed. La Nuova Italia, 1991).

ORAZIO CIANCIO AND SUSANNA NOCENTINI

THE NEW SILVICULTURE
Epistemological Considerations

> In the first place, there is a unity in things so that each thing is one in itself, it is made up of itself, and is consistent with itself. In the second place, there is a unity through which each creature is united with the other and all parts of the world comprise a single world.
> – Pico Della Mirandola

1. INTRODUCTION

Silviculture, *ad literam* cultivation of the woods, of the forest, is, according to Alfred Moeller,[1] agriculture's younger sister. It has developed through the application of farming techniques, productive arrangements and economic evaluations that are typical of the older sister. Against the background of a conceptual apparatus this can be translated as 1) the identification of farming procedures in order to obtain a constant, maximum annual product or yield; 2) the definition of standards to optimize land yield or output; and 3) the preparation of a cost/benefit analysis similar to the one applied to intensive, automated industrial agriculture.

According to the dominant concept of modern humanism, the forest is a machine for producing wood and services, or, if we prefer, direct and indirect utility. The race to produce more in the shortest possible time has and still constitutes a categorical imperative for industrial and post-industrial society. Technicism, that takes advantage of an aura of authority derived from the protective cloak of official science was and is the tool of an often futureless development. The forest has changed its nature: from a complex biological system it has been modified into an over-simplified agronomic-type system – an agrisystem – comprising a group of identical trees, ordered in time, arranged in space, and ready to produce wood. In this race we have not taken into account the effects related to the destruction of the original, primitive forest. This transformation into an artificial forest has marked the shift from extensive silviculture in harmony with nature to intensive silviculture in conflict with nature. *Hybris*, violence as an end unto itself using the ancient Greek term,

A-T. Tymieniecka (ed.), Analecta Husserliana LIX, 343–357.

has conquered: utilization for necessity has been transformed into exploitation for the sake of exploitation.

Pope John Paul II[2] says that:

> sacrifices [. . .] presuppose a high level of consideration of the material world and can be viewed as an antidote to excess and greed, contrasting that *sense of having* and of *using at any cost* that prompts man to become the absolute master of his surroundings. And how can it be denied that, among the negative results of this "culture of domination", there is also a distorted use of nature, that destroys its face, compromises its equilibrium, and does not even stop in the face of the threat of *ecological disaster?* Ascetic virtues [. . .] help [mankind] see material things in the proper perspective. They teach how to *use without abusing*, to utilize in a united and not a selfish manner, focusing not only on immediate use, but also on the future. [Man] has the right to use other parts of creation. But this does not authorize him to rule over nature, much less to destroy it.

All this is significant. In forestry, according to Orazio Ciancio,[3] the conceptual apparatus that provides the background for the change from forest ecosystem to agrisystem, is the theory of *economic realism*, based on the predominance of economic over biological and cultivation aspects. There is more: economic realism is flanked by the corollary of using nature for pure profit. This conceptual apparatus has many limits and they can be summarized as follows: 1) configuring forest yield almost exclusively in terms of the quality and quantity of wood; 2) under-estimating he magnitude and form of the services demanded of the forest by a society in continuous economic expansion; 3) an inability to predict the real scope of the role played by environmentalism as that phenomenon that has marked the culture of our era more than any other.

If it is true that we cannot totally reject economic realism – especially since it allows us to see the forest microcosm for what it really is and not as what we would want to it be – it is also true that, in order to interpret the forest and behave responsibly, we must overcome the above limitations. We must realize that biology is not physics, that the forest is not merely an outdoor factory. Instead, it is a biological system subject to a complex of internal and external influences. Like all living systems, the forest is always far ahead of what the establishment thinks it knows.

Various forms of silviculture have developed through the years. Here, we are attempting to a) verify the link between epistemes and practice; b) examine the evolutive process with the aim of prefiguring the outlook for the future.

2. EPISTEMES AND PRACTICE

For Aristotle,[4] philosophy, mathematics and physics lead to a conceptual understanding of cosmic phenomena – θεωρία τοῦ κόσμου – and comprise theoretical science, the ἐπιστήμη θεωρητική. Θεωρία is based on the observation of phenomena and represents the essence of science. Therefore Aristotle makes an important distinction: he places the ἐπιστήμη θεωρητική on one side, and science with the aim of 1) achieving practical goals and 2) poietic goals on the other. This subdivision leads to a different attitude towards nature. In the case of the theoretical sciences nature is viewed as a totality. In the case of the practical and poietic sciences, instead, θεωρία is at the service of the needs of life. Silviculture, undoubtedly, falls into this latter category. The development of this discipline, which refers to bio-ecology and economics, was pursued in accordance with the guidelines of physicalist reductionism rather than those of biological complexity. It is a shadow area, difficult to understand and even more difficult to remove.

Forestry sciences have always, or rather since the scholastic period, been considered experimental. Knowledge derives from and increases through applied research. Time in the forest is long, very long, and there is no possibility of shortcuts: following field research over a period of time is an inescapable condition. Technical-scientific innovation is the result of the development of suitable research protocols and the results of experiments. Without this *modus operandi* forestry sciences would have never made the quality leap essential for entry into the broader realm of the natural sciences.

Experimenting implies the use of a scientific model of reference. Forestry has adopted the mechanistic one based on the experimental method. All this is common knowledge. It is part of our culture, a culture that has distant roots and that has marked the rhythms and pace of technological development. This paradigm, however, has recently become an object of debate in the scientific community itself. Some have pointed out how it presupposes a clear distinction between man and the forest. This is a separation that does not allow the former to be integrated with the medium in which he works. They maintain that knowledge acquired under these conditions is a conventional datum and not real. Others, more pragmatically, raise the issue of the intrinsic difficulty of applying this paradigm to biology, and in this specific case, its adaptability to bodies such as forests which have their own peculiarities.

The new silviculture implies a new way of cultivating, but also, and above all, a new way of interpreting the forest. The subject goes into territories that are not actually part of the discipline. It touches various scientific fields and attracts scholars of different cultural extraction. For example: it concerns the historian who explores the past to give memory a meaning by which to understand the present; it concerns the futurologist to lay the foundations to guide the advancement of knowledge: and it concerns the epistemologist in developing the exegesis of forestry. But, is a type of knowledge, rooted in categorized research sufficient for giving an effective response? We believe not. However, the importance of the forestry issue and its related problems make it necessary to highlight the relationship that can be created between man and the forest without hiding the shadow areas that do exist and that we normally tend to leave hidden. This is a phenomenon that deserves more attention from those active in the field, be they scholars, technicians or administrators, and that has made us enter a territory where even angels fear to tread.

3. THE FOREST, A CULTURAL AND ENVIRONMENT ASSET

It is a difficult territory. The obstacles are visible as soon as we ask what, in general opinion, is meant or should be meant by the term "culture of the forest", and they manifest themselves to their full extent as the answers begin to flow, as each one differs so markedly from the next. And it could not be otherwise. Fernando Pessoa[5] said that everything lives because it is opposed to something. Each response is the result of a personal re-elaboration of what has been acquired through study and experience. Notions from a pure form of erudition become the essential component of an individual's way of being and thinking. We launch a process that is based on values, on religious and spiritual convictions, on artistic activities, on esthetic tastes. . . .

The forest has been a symbol of mystery and of a harsh, wild landscape; a place of refuge and of worship; a protective entity and a biological filter; source of knowledge and irreplaceable resource. In brief, for a long time it was both a *reserve* and a *resource*. Starting from the Enlightenment, however, it was considered more and more a *resource* and less a *reserve*. Over the past few decades, in line with the theory of historical recurrences, we have witnessed an inversion of this trend.

The aggregate of historical, archeological, artistic, environmental and scenic, archival and book assets first received international legal sanction from UNESCO in 1949. The forest fully meets the criteria of the category that was the defined as *biens culturels*: it belongs to culture, that is, to the history of civilization, of traditions. . . . The forest is a strong and vital system, part of our lives; it is a cultural and environmental asset. And yet, according to Generoso Patrone,[6] it is still a "little known remnant of the protocosmos". It is an unknown world: it conceals enigmas that man tries to reveal and categorize. However, this laudable and difficult intent is offset by a perverse desire to destroy. This is not alarmism, it is fact.

4. THE *GENIUS LOCI*, OFFICIAL AND REAL SILVICULTURE

Knowledge in forestry has been defined and accepted by the scientific community on the basis of concepts, principles, theories, propositions and techniques that refer back to and link up with the Newtonian or Cartesian paradigm. This accumulation of knowledge took place within the alveolus of a physicalistic and mechanistic vision. Divisions and subdivisions were the working method. The adoption of this paradigm led to the classification of the forest in categories and to considering it an *instrumental entity*, that is, an entity capable of producing services and satisfying society's needs. The current trend is to consider the forest as an *entity of value*. This term defines an extremely complex *subject* which, as such, lives an autonomous life, influenced by countless factors. Research on the *uniformities* that govern the behavior of complex systems is a field that is yet to be explored. *Uniformity* must be defined with *phenomenological* characteristics.

As we have already said, the old technical, that is mechanistic, apparatus did indeed give a decisive thrust to the development of forestry sciences. This was possible because the goal was univocal: to maximize output. This technical apparatus, however, has outlived its usefulness and the time has come to turn over a new leaf as it were. The results of research and experiments are often not applied or are even rejected. There are many reasons for this and they vary in complexity. We can put forward only a few hypotheses: 1) the results are not very evident; 2) the message is not convincing; 3) the in-formation is not sufficient.

In addition, the data given to the researchers often conflict with local traditions that are the fruit of knowledge and long experience. The *genius loci* plays a decisive role in this context. This also helps us understand why the results may seem incomprehensible and at times not very credible. The fact is that the existing situation can be defined as paradoxical, and that is an understatement: *real silviculture does not correspond to the official silviculture.* Traditional silviculture no longer impacts reality, or if it does, the impact is only partial and/or very insignificant. Perhaps it is appropriate to ask why this situation has developed. The response is multiple, but we can pinpoint some aspects.

The first aspect is technical. Rationalization of forest management is configured in a network of rigidity and schematics that is difficult to apply and to maintain in a time span as broad as the one we have when dealing with forests. It is precisely for this reason that it is often rejected. It is considered an unreasonable demand that has fallen from the sky, as it were. Albert Einstein[7] said that intuition is truly the precious factor. The refusal to accept and apply management forms that tend to oversimplify the forest's complexity is more intuitive than rational. Anyone who lives is close contact with nature develops his intuitive abilities to the utmost, and exploits them when needed. The results of this divergence between the official and real positions are here for everyone to see. Deforestation is increasing geometrically, and reforestation is proceeding at an infinitely slow rate.

The second aspect is scientific. Science, in general opinion, only has value if it can explain and predict. This creates an eternal problem for the researcher, to deduce the future from the past and the present. The dominant scientific culture aims at programming the future on the basis of data acquired through experiment and observation. Current knowledge in biology and silviculture does not provide any certainty that changes of particular conditions do not affect the results. The uncertainty derives from the fact that we work in a changing environment. This involves methodological problems which are easy to overlook rather than carefully ponder.

The third aspect is epistemological. There are two main issues in the field of the biological sciences when it comes to defining a context for research work. The first is related to the nature and scope of the teleology. It comes up when the researcher, in addition to asking "how" – and this is the norm – starts asking "why". This complicates matters enormously. The latter question implies a finalistic concept that biologists have always rejected. Someone very shrewdly remarked that teleology is a lady whom the biologist cannot live without, and yet he is embarrassed to be seen in public with her.

How can we resolve this problem? The fundamental principle of science is to know that we are ignorant. Jacques Monod[8] says that natural systems – a cell, an organism, an ecosystem – are teleonomic, that is, systems seeking an end. The forest is a system that has never been completed. Within the context of its organization, structural changes are an invariable. Time brings with itself something that is unpredictable, undefined, uncertain. This awareness leads to the disappearance of the finalism implicit in the question "why", and with it the presumption of directing the present by starting from the future.

The second problem concerns the scientific model. In the *The Structure of Scientific Revolutions* Thomas S. Kuhn[9] says that in order to be accepted as a paradigm a theory must seem better than those competing with it; it does not, however, have to explain all the facts with which it may be compared, in fact, it never explains them. It is obvious that we must change our attitude towards the forest. We must make radical changes in the scientific approach. Even if we cannot eliminate it, we must reduce the gap between the viewpoints of the scientific community and those of postindustrial society.

The fourth aspect is esthetic. Paul Cézanne[10] wrote that at times he wondered whether the country people know what a landscape or a tree actually is. He believed that a farmer knows what the weather will be tomorrow but that trees are green and this green is a tree, that this soil is red and this red mound is actually a hill; he truly believed that most people don't feel, since they know nothing beyond their own unconscious inclination towards what is useful.

Joachim Ritter[11] expresses a similar thought: for the country dweller nature is always that of his birthplace, inherent to his working life: the forest is wood, the earth land to be farmed, water a source of fish. Anything that lives beyond this limited area remains foreign; there is no reason to "go out" to seek "free" nature as such and to contemplate it. The landscape only becomes nature for those who leave it (*transcensus*) to participate in nature as a living and present whole, from the outside, through the pleasure of contemplation, According to Massimo Venturi Ferriolo[12] the *hyle* – the woods, the forest, the material in nature, *the indeterminate that form determines* (Authors' emphasis) – can be both the land that the *techne* uses and transforms in view of an end, and Virgil's *genius loci*.

Ritter[13] also maintains that the landscape is nature that reveals itself esthetically to those who observe and contemplate it with feeling. Venturi

Ferriolo[14] says that this "esthetic revelation" manifests itself when man turns to nature without practical purpose, without utilitarian ends. This is the man-nature relationship without history's mediation. Nature becomes "free", released from exploitation. The world of technology, the territory of the poietic sciences affects nature and modifies it to the point of destruction. This process is *history*, the world of man, of the useful, a "second nature".

The fifth aspect is ethical. It is configured as a *trait d'ésprit*: respecting nature means respecting oneself. It is for this reason that in ethical and cultural terms the relationship between man and nature becomes dual. To paraphrase Aldo Leopold[15] we can say that we must think like the forest. It is a legal entity with rights and therefore man must act accordingly. Problems must be recognized and possibly resolved if we do not want to remain – or worse – be placed at the edges of knowledge. If this is what we want, all we can do is join hands, form a circle and encourage each other. We must not forget that the issue of the rights of living systems, and hence of the forest, is part of the current agenda all over the world.

Italo Calvino[16] also dealt with the issue. In "Il barone rampante" the baron acquired fame via the *philosophes* (Voltaire, Diderot, etc.) for certain "politically correct" treatises he wrote on subjects such as republican constitutions and social contracts. But one of these, entitled *Progetto di Costituzione per Città Repubblicana con Dichiarazione dei Diritti degli Uomini, delle Donne, dei Bambini, degli Animali Domestici e Selvatici, compresi Uccelli Pesci e Insetti, e delle Piante sia d'Alto Fusto sia Ortaggi ed Erbe* [Outline of a Constitution for Republican Cities with a Declaration of the Rights of Men, Women, Children, Domestic and Wild Animals Including Birds, Fish and Insects and Plants be they Trees, Vegetables and Grasses] is ignored. And yet, it was a beautiful work that could have helped all governments; instead no one took it into consideration and it remained dead.

According to Robert P. Harrison[17] Cosimo's treatise was ignored because his era was only interested in the declaration of the rights of man – the rights of human subjects and not of nature's species and objects. Today, we are witness to the consequences of these unilateral declarations of the rights of a single species, neglectful of the natural rights of all the other species. In this sense Cosimo's treatise was ahead of its times – and even ours with respect to this issue.

Can we but not agree? Is there anything to add? Any comment would be superfluous. However, there is one thing to emphasize. Poets, artists, men and women of letters, especially of Calvino's caliber, are always trailblazers: they generate culture. Technicians need only take heed; they must give practical content and form to those intuitions and syntheses, especially if they are elegant and harmonious.

5. REDUCTIONISM, HOLISM AND ECOLOGISM BETWEEN PHILOSOPHY AND SCIENCE

The old, Cartesian-Newtonian scientific paradigm we mentioned earlier is based on the concept of the neutrality, objectivity and predictive capacity of science. Descriptions of phenomena depend on the observer. The metaphor of knowledge is that of the building with foundations, bricks, etc. The attitude with respect to the object of study – and in this case the forest – is that of domination and control. The experimental approach is reductionist. Research and experiments are base on the conviction that in a complex system the behavior of the whole can be deduced from the behavior of its individual parts. The scientific method is deterministic. According to Ciancio and Nocentini,[18] the theoretical structure is based on a *rationalistic, analytical, reductionist and linear mode of thinking*.

The new scientific paradigm is based on the concept of the intersubjectivity of science. Descriptions of phenomena depend on the observer. The metaphor of knowledge is the network of relationships. There is no hierarchy or foundation. The properties of the parts can be understood in light of the dynamics of the whole. The scientific method is based on *approximations in sequence*. The process of knowledge is based on the systemic vision. The theoretical construction relates to an intuitive, synthetic, holistic and non-linear way of thinking.

Humberto R. Maturana and Francisco J. Varela[19] note that the essential purpose of biological research is to understand the organization of living systems in relation to their individual character. In recent years, the position that considers the forest as an *autopoietic system*, that is, a system with the property of *autonomy*, has gained strength. This system can subordinate structural changes to the conservation of its own *organization*. The system's components comprise an interconnected network of relationships. The properties of the whole cannot be deduced from the properties of the parts. With the new scientific model it is not so much the practical approach that is being debated as the theoretical approach.

It is a difficult transition that presupposes a different viewpoint and a new cultural dimension. The culture of domination and control over nature must be substituted not by the culture of submission to nature but by the culture of respect for nature.

In science confusion often arises when talking about reductionism and there is no reason for surprise. Scientists are not philosophers and when they talk about reductionism they almost always refer to the aspect related to the art of scientific research: that is, research strategies and experimental design. In order words, in scientific practice reductionism becomes a research method with which to analyze parts for a better knowledge of their single and specific aspects. In this sense reductionism has made an extraordinary contribution to technological development. The scope of these results, however, is a completely different issue. If, however, reductionism is understood as a philosophy for comprehending nature, the issue is totally different. Physicalism, the philosophy of physical-chemical laws, and even physicists are saying the same thing, is not capable of giving a sufficient impetus towards the comprehension of living systems. These cannot be observed, studies and understood by the reductionist method alone.

Much water has passed under the bridge since 1926 when Jan Christian Smuts[20] introduced the term "holism". However, the disputes over scientific philosophy, between mechanicism and vitalism, between casuality and finality, between reductionism and holism, have not ended even though there have been highs and lows. Every now and then they reappear with vigor. This is happening today with the reawakening of ecological sensibility: the problem is related to scientific research as a process for learning about the fundamental phenomena of life. Worster[21] maintains that Smuts stated that *all* reality is aggregative, contemporaneous and emerging and that what we call evolution is the progressive development of the wholes created in each phase – from the most rudimentary, imperfect and inorganic to the most highly developed and organized wholes; therefore, it is clear that Whitehead was not alone in his search for an alternative to mechanistic materialism. Worster maintains also that philosophic idealism seemed to be reborn, albeit at times in the shadows, as for example, the Hegelian concept of a nature which, in the struggle for final perfection, undergoes a series of metamorphoses.

The holistic view, along with the concept of *superorganism, or complex organism* as expressed by Frederick Clements,[22] made an important contribution to the development of ecology and these characterized an era of change in the scientific world. According to Worster,[23] in general holism, organicism and integration had their roots in the old idealistic tradition, in the idea of the one that incorporates various elements in a single common body; in fact, idealism never accepted the exclusively physical world, and ever since Plato, it nurtured the hope of attaining the world of eternal pure and immaterial Ideas. Worster[24] maintains that the search for transcendence is still alive in organicism: Alfred North Whitehead[25] defined religion as the vision of something beyond, underneath and within, the fleeting flux of immediate of things. The view of organic relationships satisfies a basic moral need and put us on guard against the conceit of man in relation to nature. The task which is considered most pressing today is that of breaking up the dualism that isolates mankind from the rest of nature. According to Worster,[26] Krutch[27] was quite astute in understanding that ecology without respect and love could not become anything other than the shrewder exploitation of what we should instead admire, enjoy and share.

From the philosophic standpoint even holism can be considered a form of reductionism. If, on the one hand, we cannot understand the whole from an examination of its individual parts, on the other, attempts to understand the whole without analyzing its parts is just as hazardous. The holistic principle has little meaning for anyone engaged in scientific pursuits. But this, as we know, is wrong. The confirmations come from molecular biology: the lair of the reductionist discipline par excellence. François Jacob[28] states that the "convergence of analysis" between the simple and the complex is the right system for achieving significant scientific progress. Arthur Koestler[29] considered the holon – the two-faced Janus – the *trait d'union* between reductionism and holism. The holon is both part and whole: it is part of a vaster system but at the same time it is an organized and autonomous system. Both reductionism and holism complement each other in our observing and understanding nature: they are the two sides of the same coin. It is sufficient to understand their shape and not be so presumptuous as to generalize about the results obtained through the analysis of parts.

"Ecologic thought" is certainly holistic. But, according to Orazio Ciancio,[30] it is also something more. On the one hand, it emphasizes the need to look at the whole instead of the parts – and up to here it concurs with holism – on the other hand, it considers the whole as part of increasingly larger entities. The difference, regarding this significant aspect, is substantial. In fact, we can say it is decisive. For example, there are the problems related to the study of ecosystems. There is even more. Ecologic thought, as the vehicle of the holistic outlook, considers the functional interrelationships within a given entity, that is, among the parts comprising the whole. However, it also takes into account the relationship with other wholes, as well as their relationships with the global environment. In other words, the network of internal relationships interconnects with an increasingly large network of external relationships. It is this process that involves the principle of uncertainty, indefiniteness and incompleteness which is typical of everything that is complex. The ecologic view is broader, more complete and mainly more indefinite than the holistic standpoint. It is precisely for this reason that it allows us to acquire an awareness of the interdependency among all natural phenomena, and of the way they interact at the global level. The totality rooted in increasingly larger totalities maintains intrinsic and extrinsic functional relationships. The practical and theoretical implications are enormous, and presuppose a new scientific and social paradigm. This leads to modifications in the structure of concepts in the dominant culture.

6. THE PREFIGURATION OF FUTURE HORIZONS

Man and the forest: a never ending story; the history of mankind with all its good and evil; the history of its relationship with the forest, and in more general terms, of its attitude towards nature. It is emblematic: continuous doing and undoing which comes through more in the undoing than in the doing. The aphorisms of François R. Chateaubriand, "Les forêts précédent les peuples, les déserts les suivent" (Forests come before people, and deserts follow them.) and Giambattista Vico, "L'ordine delle cose umane procedette: che prima furono le selve, dopo i tuguri, quindi i villaggi, appresso le cittá, finalmente le accademie" (The order of human things proceeded thus: first came the forests, then the hovels, the villages, the cities and finally the academies.) should make us stop and think.

The world of forestry must cross the border of attaining the maximum *direct and indirect utility*, which is often translated as *hybris*. The German word, *Raubwirtschaft*, economy of exploitation or rather plunder, perhaps best conveys the idea. In a society in which arrogance is considered a symbol of power, defacement a sign of strength, and contempt an aphrodisic emblem, we must change our attitudes towards the forest. This change in the forest-man relationship means adopting the new scientific paradigm we discussed earlier. If it is thus, then the new silviculture is not only biology as some would have it, nor merely economy as others would prefer. It is, instead, bioeconomy. And since the forest is a biological system, the new silviculture is definable as the *bioeconomy of the forest system.*

What future is there for the forest-man relationship? Over the years silviculture has modified attitudes vis-à-vis the forest, shifting the horizon ever forward. From financial silviculture we have moved on to naturalistic and then to the phytogeograhpic silviculture on ecological bases. Currently, the trend is to broaden the horizon to speak of systemic silviculture which is nothing other than the bioeconomy of forest systems we mentioned above. This is a research project whose development cannot be easily predicted. It is precisely for this reason that Imre Lakatos[31] defines it as *progressive.* It is a project in which man works in the interest of and on behalf of the forest system. And it has a dual goal: to respect the organization and not predetermine its structure. Man has the right to work in this context since he is part of the system with which he interacts. All action is knowledge and all knowledge is action, maintain Humberto Maturana and Francisco Varela. This can be paraphrased as acting with "wisdom" in the sense of working a unity between science and ethics.

In industrialized countries the forest is no longer threatened by abuse in order to satisfy primary needs, it is threatened instead by a process that has neither face nor soul, a pseudoculture that knows the price of everything but the value of nothing. It is a pseudoculture that makes the useless necessary, and the superfluous indispensable. *Sic stantibus rebus*, how can we convince people to make choices the results of which can only be vaguely perceived in the distance? The only way is by persuasion and education. If the *informative* process is important, the *formative* is decisive. As Renato Dulbecco[32] writes, scientific education in schools does not seem to achieve its goal because it does not convey the excitement of the discovery of unknown things which is the most important part of science. What we need is a "new deal", a new way of thinking, a new philosophical outlook with regard to nature. Or, in other words, what we need is new way of seeing the forest.

We must shift over from a lowbrow culture to a highbrow culture. We must consider forest management not only from the practical standpoint, but also in esthetic, metaphysical and ethical terms. When technical disputes would arise among foresters and they tried to settle them, they used to say: let's go into the forest and discuss it in the field. This was a method for putting the reasons of the heart ahead of the passions of the mind. The wisdom of the forester is manifested through contact with the forest; in its presence, the reasons of the heart prevail over the passions of the mind; the common feeling, respect for nature and forestry ethics come into the picture, this is the *esprit forestier*. If this is truly the case, then along with Aldous Huxley[33] we can say that it is vitally important to participate in the solution of these problems, not only as technicians and experts, or as persons who want to use products made of wood, but as human beings with a moral nature, with an esthetic nature and with a philosophic inclination. The culture of the forest is, therefore, the prefiguration of the future horizon.

Accademia Italiana di Scienze Forestali
Piazza Edison 11 – Firenze
Istituto Sperimentale per la Selvicoltura
Arezzo
Italy

NOTE

* Orazio Ciancio is chair of Forest Management at the University of Florence and General Secretary of the Accademia Italiana di Scienze Forestali in Florence. Susanna Nocentini is with the Istituto Sperimentale per la Selvicoltura in Arezzo. The Authors have contributed equally to the drawing up of this paper.

REFERENCES

1 Moeller, A., *Der Dauwerldgedanke. Sein Sinn und seine Bedeutung* (Berlin, 1922).
2 Giovanni Paolo, II., *Omelia del 24 marzo* (L'Osservatore Romano, 25–26 marzo, 1996).
3 Ciancio, O., "La Selvicoltura oggi," *L'Italia Forestale e Montana* 46(1): 7–20 (1991).
4 Aristotele, *Metafisica* (Milano: Rusconi Libri, 1994).
5 Pessoa, F., *L'ora del diavolo* (Roma: Biblioteca del vascello, 1992).
6 Patrone, G., "Stravaganza quinta e somma – Un avanzo poco conosciuto del protocosmo: la foresta," *Annali dell'Accademia Italiana di Scienze Forestali* XXX: XIX–XL (Firenze, 1981).
7 Einstein, A., Preface in *Where is Science Going?* by Max Plank (London: Allen & Unwin Ltd., 1933).
8 Monod, J., *Il caso e la necessità* (Milano: Mondadori, 1970).
9 Kuhn, T. S., *The Structure of Scientific Revolutions* (University of Chicago Press, 1970). Trad. it. *La struttura delle rivoluzioni scientifiche* (Torino: Einaudi, 1969).
10 Cézanne, P., in Gasquet J. – *Ce qu'il m'a dit* (Paris, 1926).

[11] Ritter, J., . *Paesaggio. Uomo e natura nell'età moderna* (Milano: Guerini e associati, 1994).
[12] Venturi Ferriolo, M., "Joachim Ritter e la teoria del cosmo come 'fondamento del paesaggio'," in *Paesaggio. Uomo e natura nell'età moderna* (Milano: Guerini e associati, 1994).
[13] Ritter, J., *Ibidem*, 1994
[14] Venturi Ferriolo, M., *Ibidem*, 1994
[15] Leopold, A., "The Conservation Ethic," *Journal of Forestry* (1933).
[16] Calvino, I., *Il barone rampante* (Torino: Einaudi, 1957).
[17] Harrison, P. R., . *Foreste* (Milano: Garzanti, 1992).
[18] Ciancio, O. and Nocentini, S., "Il paradigma scientifico, la «buona selvicoltura» e la saggezza del forestale," in *Il bosco e l'uomo* (a cura di Orazio Ciancio). (Firenze: Accademia Italiana di Scienze Forestali, 1996).
[19] Maturana, H. R. and Varela, F. J., *L'albero della conoscenza* (Milano: Garzanti, 1992).
[20] Smuts, J. C., *Holism and Evolution* (New York, 1926).
[21] Worster, D., *Nature's Economy. A History of Ecological Ideas* (Cambridge: Cambridge University Press, 1985). Trad. it. *Storia delle idee ecologiche* (Bologna: Il Mulino, 1994).
[22] Clements, F., *The Development and Structure of Vegetation* (Lincoln, 1904).
[23] Worster, D., *Ibidem*, 1985.
[24] Worster, D., *Ibidem*, 1985.
[25] Whitehead, A. N., *Science and the Modern World* (New York, 1925). Trad. it. *La scienza e il mondo moderno* (Torino: Bollati Boringhieri, 1979).
[26] Worster, D., *Ibidem*, 1985.
[27] Krutch, J. W., "The Paradox of Humanism," in *The Modern Temper* (New York, 1949).
[28] Jacob, F., *Evolution and Tinkering* (New York: Science, 1977). Trad. it. *Evoluzione e bricolage* (Torino: Einaudi, 1978).
[29] Koestler, A., *Janus. A Summing Up* (London: Hutchinson, 1978).
[30] Ciancio, O., "Pensiero olistico e pensiero ecologico", *L'Italia Forestale e Montana* 50(2): 117 (1995).
[31] Lakatos, I., *Lezioni sul metodo*. In *Sull'orlo della scienza. Pro e contro il metodo* (Milano: Cortina, 1995).
[32] Dulbecco, R., *Ricerca, educazione e società*. Scienza e società (Venezia: Marsilio, 1995).
[33] Huxley, A., *La condizione umana* (Pavia: Liber, 1995).

PART IV

TRANSITIONS: LOGOS AND EXPERIENCE

Daniela Verducci, Louis Houthakker and Daria Carloni the panorama of Macerata.

LILIANA ALBERTAZZI

FORMS OF EMOTIONS

> The place of psychology is among the biological sciences, since it studies those "phenomena of life" which, in fact, are phenomena of the "psychic life".
> – Vittorio Benussi, Proceedings of the IVth National Meeting of Psychology, Florence 1925

INTRODUCTION

The theory of the emotions has played a key role in phenomenological theory since its beginnings, given that it refers, at least in the first instance, to a specific and fundamental *class of psychic phenomena* together with presentations and judgments. In Brentano's classical formulation, emotional phenomena are ontologically founded on (i.e., *dependent* on) presentations, i.e., there are no emotions in absence of presentations. Like presentations, they too are *intentional*.[1]

The term *intentional presentation* which characterizes Brentano's thesis of inner perception indicates the *mind's reference to something*, to an object reality of some kind.[2] According to Brentano's theory of intentionality, we have evident perception of things only in the *psychic phenomena* that recognize them as *existent*, in the *now*, and in a certain *mode*.

As to emotions, in Brentano's earlier texts every inner perception is also emotionally connoted, since every psychic act brings along a sense of positive or negative feeling.[3] However, Brentano soon came to realize that the dependent role assigned to emotions in his classification of psychic activities raised problems, and, accordingly, he later ascribed a particular status of independence to emotions.[4]

In the subsequent phenomenological and related literature, the problem was addressed by the Graz school of Meinong, by the Munich school of Pfänder, by the Berlin school of the Gestaltists and by the Leipzig school of Klages, to stick to the *scientific* side of phenomenology.[5]

Moreover, Höfler, Klages and Köhler verified the theory of emotions on the basis of laboratory experiments, and this aspect became domi-

A-T. Tymieniecka (ed.), Analecta Husserliana LIX, 361–378.

nant in Benussi, assuming the features of an *experimental phenomenology*.[6]

1. EXPERIMENTAL PHENOMENOLOGY

What is an experimental phenomenology? The term has been coined by Bozzi, and it belongs to a tradition which derives in large part from developments (almost all of them *non*-experimental) of Husserlian phenomenology.[7]

Experimental phenomenology, then, is not only a pure *theory of consciousness*, but is also a *science* which could be defined as a *descriptive psychology* of perception.[8]

Experimental phenomenology sustains the (ontological) independence of the phenomenological level of perception (i.e., it claims that the phenomenological level is *not reducible* to the physiological or to the physical one) and uses an empirical and experimental method.[9]

Finally, it claims that the analysis of the phenomena of experimental phenomenology goes in two directions, *descriptive* and *morphogenetical*, i.e., it analyses phenomena (i) as they are given to consciousness in the *now* and (ii) in their *actual genesis*.

Actually, experimental phenomenology has had mainly a descriptive development, what we could define as the *strictly phenomenological*, excluding every relation with physiology. From this point of view, experimental phenomenology seems to have abandoned the postulate of *isomorphism*, which guided the experiments and the researches of the Berliners, Wertheimer, Köhler e Koffka in particular, which work to a certain extent was accepted also by Benussi, a member of the Graz school.[10]

Actually only cognitivists seem to show interest in physiological researchers. More than by conscious choise, however, this seem to be a product of the scientific paradigm adopted in certain specific disciplines.

Anyway, even within the same range of experimental phenomenology some dissonant voices are heard: Bozzi, for example, was projecting experimental research on the electrophysiology of the perception of apparent movements. Briefly, as in the case of the reversibility of shape in the Jastrow duck-rabbit ambiguous figure, Bozzi's proposal was to analyse the phenomenon of reversibility between stroboscopic and gamma movements.[11] In the same direction are to be recalled Wallach's experiments on after-effects and the hypothesis of a higher order isomorphism by Shepard.[12]

Of the various *conceptual categories* of experimental phenomenology, three are important:

(1) the notion of the *observable*, elaborated by Bozzi; that is, whatever lies within range of the eye and the hand and is amenable to manipulation in the actual duration of time.

Observables, in the first instance, carry with them at least one determination of colour and of place, as well as a determination of direction and velocity. Observables can be, from time to time static or dynamic objects, parts of the perceptive field or hand-made by paper and pencil, as in the example given by Brunswik or by Kanizsa;

(2) Vicario's notion of *event*, which, in general, indicates that the objects of perceptive fields have duration, and that the perception of events is therefore tied to the perception of *temporal structures*.[13]

(3) the notion of the *invariant*, developed mainly by the Gestalt psychologists, which concerns the essential characters of the observables.[14]

Given these premises, a basic point for a theory of emotions is therefore to understand *what kind of observables and/or events they are*.

1.1

One of the fundamental problems of experimental phenomenology is the following: How can one pass from a set of *observables* given in actual perception to identification of true perceptive *objects*? In philosophical terms, do there exist material and formal *invariants* of events, which justify their recognition as objects, on the basis of their salience against a background, their profiles, etc.?

If we strip the invariants of perceptive events to their bare essentials, according to a correct phenomenology, we may summarize them in: 1) temporal determinations, 2) spatial determinations; 3) qualitative determinations (of filling); 4) change; 5) movement; 6) shape; and 7) form.

I propose to distinguish shape (silhouette, outline, profile) from form, the first being static and spatial, the second being dynamic and temporal, including perception of change and/or movement.

In particular, forms are temporal events, integrated totalities whose parts contribute to the formation of wholes. A classical example of form is a melody, a classical example of shape is a triangle. Forms, in their turn, can be *shaped*, if considered in a static instant of their process: think, for example, of a photograph of a jumping rabbit.

An analysis of emotions should show how they are *shaped* by the *invariants* of perception.

1.2

Analysis of the objects of perception has: (i) a *descriptive* aspect which consists in their perceptive recognition as *objects*; (ii) a *genetic* aspect; which consists in analysis of the *stages* and *phases* through which the constitution of objects in the *now* (or in the time of presentness) is realized.

Identification of the perceptive objects requires: (a) a theory of *perceptive continua*.[15]

Genetic analysis requires: (b) a theory of the *consciousness* or of *intentionality*.[16]

These two aspects are strictly correlated, since intentionality arises on the basis of perception. What, with reference to the latter point, does *a theory of intentionality* mean? It means relating the objects of intentional presentation to their *genesis in the actual duration* of time, to that which exists *here and now* in the present moment.[17]

We owe an *experimental* theory of the actual genesis of the objects in the duration of time to the Leipzig school. One of the main assumptions of the Gestalt view is that the perception of form comprises the functional prevalence of the Gestalt quality of the whole over that of its individual parts. However, the generality of this assumption has almost invariably obscured the fact that there exists a *time of the development of forms*, and specifically that the *appearance of a form occurs at a certain temporal point* of consciousness. From this point of view, the individual Gestalten are *sub-wholes* of a larger whole – that is, sub-wholes of the entire content of consciousness.[18]

Husserl's theory of dual intentionality was one of the first attempts to construct a theory of intentionality which takes account of both the *structural framework* of the time of the development of forms and of specific *individual forms* in the actual now.[19] However pioneering, Husserl's theory is unclear. Benussi's experimental analysis, by contrast, is a scientific development of theses of phenomenology, which from this point of view are more like the theses of modern constructivists in cognitive science.[20]

But, how is it possible to observe the *genesis* of *Gestalten*? It is, indeed, made possible by reducing the stimulation, for example, by presenting optical figures by means of a tachistoscope, or by increasing brilliance against a dark background, or by shrinking or enlarging the normal dimensions of visual objects. All these experiments, obviously, provide windows on the life of consciousness of *few seconds* and make *univocal* the conditions of experience.

In this case analysis concerns *pre-Gestalten*, which are characterized by marked instability of the moments connecting the parts and by an equally marked tendency for them to assume stable form.

Emotions play an important functional role in *pre-Gestalt* experience, as showed, for example, by the role played by the tertiary or ecological qualities.[21]

2. CONTINUA OF FORMS: EMOTIONS

We have already said that *identification* of the perceptive objects requires a theory of *perceptive continua*.[22]

The perceptive continuum arises from the experience of a *qualitative change* in a psychic state; in other words, it springs from our certain and diverse mode of presentation of objects in the actual duration of time.[23]

The perceptive continuum has qualities, possessed by objects to a greater or lesser extent, and which admit to a further gradation or increment, like from red to blue, from Do to Re, from short to long, from dark to bright, from void to full, etc. Moreover, perceptive continua have, to various degrees, *orientation, velocity* and *position*.

A characteristic feature of perceptive objects is their emotional connotation. As with sounds and colours, emotions are phenomenological *primitive categories* of the perceptive continuum.

Now, the question is the following: from the point of view of an experimental phenomenology, is it possible to develop a *geometry of the emotions* like a geometry of colours or an analysis of sound in space?[24] In other words, is it possible to detect a pleasure/pain continuum like the white/black continuum in the spatial order of colours – as the axis around which the spatiality or all emotional situations rotates?[25]

If this is indeed possible, we should also be able to detect a passage, a boundary in the emotional continuum, as well as a *polarity* between a *plus* and a *minus* of emotions aligned on a span whose extremities are pleasure and pain. In the constant variation of emotions, moreover, it should be possible to discern a further type of variation: that between evidence and absence (one should not in fact confuse emotional *content* with emotional *function*) arranged along a rectilinear continuum *superimposed* upon the emotional one. And one should also be able to discern within this continuum the same features of *velocity, direction* and *orientation* displayed along other perceptive continua.[26]

Inquiry of this type (which, for what concerns emotions, approaches the limit of an experimental mystical, as Benussi says) has actually been conducted within experimental phenomenology, and specifically by Benussi himself.[27]

2.1

Benussi's experiments, in particular, investigated the *autonomy of emotional functions*. It is necessary, however, to specify Benussi's premises. His reconstruction of the *genesis of the forms of the emotions* proceeded from the assumptions of a descriptive psychology, namely:

(i) *constituting an object* reality is the characteristic note of all psychic phenomena;

(ii) psychic processes *transform* non-reality into reality, given that every non-real object, of the past or of the future, may become *mentally present*;

(iii) *perceptive presence* may be external (environment) or internal (subjective attitude);

(iv) *perceptive* presence is characterized by *assimilative* functions and *synthetic* schemata; *mental* presence is characterized by *reproductive* functions and *additive* schemata; and

(v) words are *conveyors* of reality: to understand a word is to experience the *mental* presence of an object.[28]

Benussi's experiments focused on two aspects of the problem in particular: (i) do there exist *emotional* functions which are *distinct* and *autonomous* from intellective functions? and (ii) do *object-less* emotional functions exist?

Consequently, Benussi's experiments set themselves these tasks: (i) to prove that psychic functions can be *isolated* and *decomposed*; (ii) to induce an *object-less emotional state* by means of suggestion; and (iii) to establish whether a corresponding *somatic* expression can be detected in the form of breathing or by means of graphic or metrical analysis of chest movements registered and shaped by a pneumograph.

In Benussi's experiments there are to be distinguished: (i) a *hypnic attentive state* inducing a base sleep (WL); (ii) a *base hypnic sleep* (BS);[29] (iii) a *quotient of velocity* of inhalation and exhalation spiration (Qv); (iv) a *quotient of capacity* of air consumption in a unity of time (Qc); and (v) a *surface area* of the average respiratory curve (S).

From a general point of view, Benussi's experimental method was to produce a hypnosis-induced transitional stretch and then give a task to complete to the subject of the experiment by means of *words*, for example, "now you will experience a state of anxiety, or of well being, or of fear", etc., without specifying the object or the reason for such state of well being, anxiety, fear, etc.

The base sleep (BS) condition was induced via an attentive induced stretch (WL).[30] The induced *base sleep* (BS) lasted 3–5 seconds which ensured the uniformity of experimental conditions and greater control over effects. In the base sleep a *pure* emotional situation was triggered.

The breathing movements of the thorax were recorded on carbon paper by means of a lever drum. The instruments is provided with a strap so that it can be hung from the subject's neck, thereby preventing any movement at the point of application. It is also fitted with an adjustable belt so that it can be attached firmly to the thorax. Expansions and contractions in the rib cage are thus recorded.

Recordings of the type of breathing that marked the onset of the two different hypnic states immediately revealed the following:

(i) a correlated *stability* between the subjective state of the *situation* of consciousness as present in base sleep and the *shape* of breathing as registered by the pneumograph;[31]

(ii) the *shapes* plotted for the induced attentive state and base sleep are different: in fact, corresponding to the induced attentive state was a *convex* shape, and to base sleep a *concave* shape;

(iii) the induced attentive state and base sleep have different breathing *velocity* quotients.[32]

As well as the velocity quotient, Benussi also calculated the *quotient of capacity* (Qc) proportional to air consumption in a unit of time, and the surface area of the average respiratory curve (S). The shapes are plotted by joining the points marking: 1) the onset of inhalation; 2) the depth reached at mid-inhalation; 3) the maximum point of inhalation; 4) the reflux; and 5) the point of maximum exhalation.

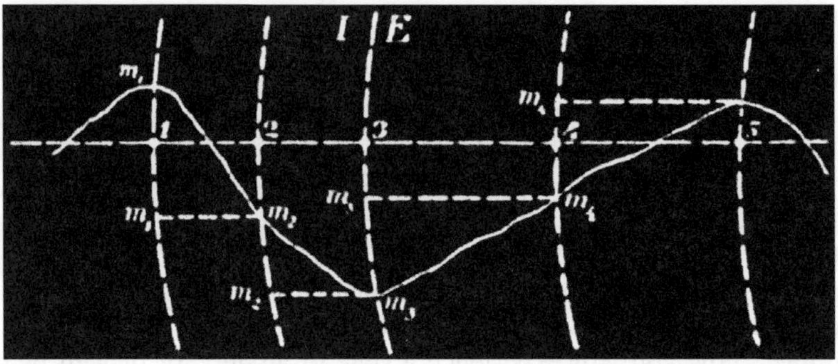

Figure 1. Recording by Marrey's pneumograph.

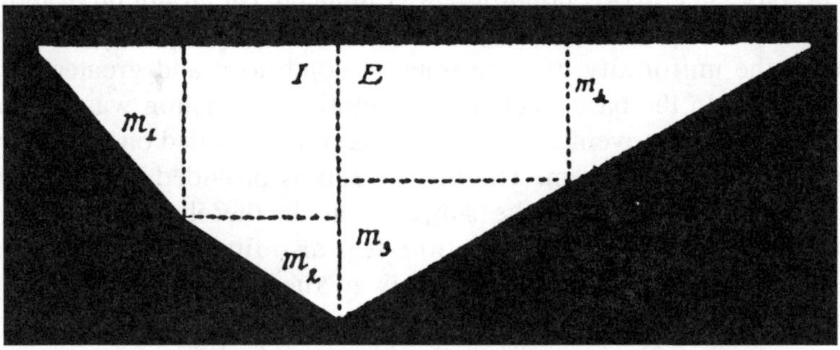

Figure 2. Respiratory shape.

(iv) It is to be noted that: in BS all elements of *spatial* orientation were annulled, while the *temporal* orientation remained intact.

A first finding, therefore, is that *erasing the contents* of thought does not annull the activity of consciousness or of inner perception.

2.2

In detail, Benussi's experiments proceeded as follows: after an initial state of wakefulness (W), following a first stage of attentive hypnic state (WL), different but equally *object-less* emotional states were triggered in a BS state of intellective (i.e., object-less) sleep. The procedure was then reversed until the state of wakefulness was reached.

The different phases of the process, therefore, followed each other in this sequence:

$$W - WL - SB\ (BS + E)\ (BS - E) - WL - W$$

The pneumograph registered two interesting facts during the process:

(i) transition from one phase to the next place of the emotional continuum took place *abruptly* (with a sudden rather than gradual *change of direction*): i.e., the boundaries between the phases were very clear;

(ii) each transition was characterized by its *specific shape, surface areas* and *velocity quotients*.

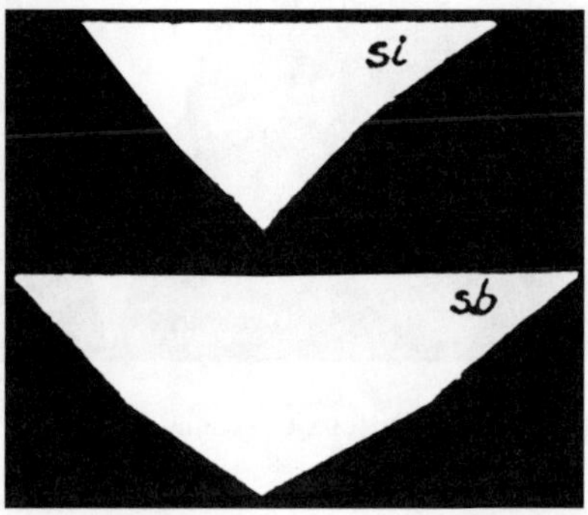

Figure 3. Different shapes of WL and BS.

In short, the experiments showed that each transition is a *boundary* in an emotional continua depicted by the respiration graphs, the so-called *respiratory shapes*, as *pentagons* whose superior side showed the time coordinate.

Specifically, BS has a *bi-concave* symmetrically-shaped shape, with a low capacity quotient (Qc) and average surface area (S).

Transitions from one state to another are characterized by a breathing profile of biconcave-rectilinear symmetrical shape, with high quotient of capacity (Qc).

Another distinctive feature of emotional forms is that, although they are externally imperceptible from a somatic point of view, they internally differ from the intellective functions expressed by the respiratory mimicry. In short, *emotional forms* are *internal invariants* of the perceptual continua.[33]

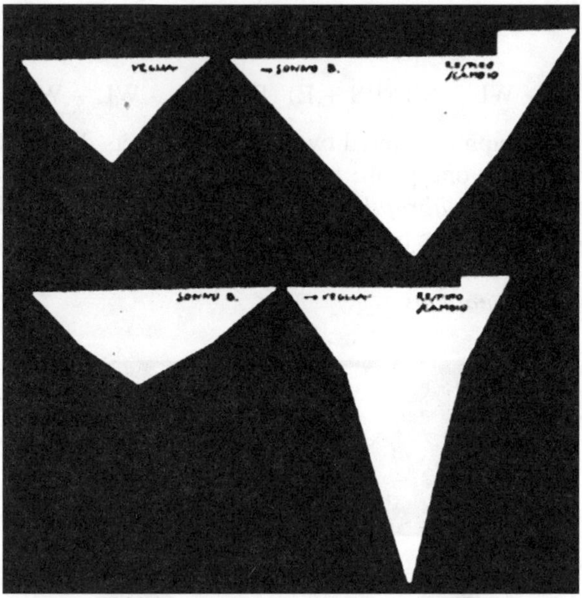

Figure 4. Different exchange breaths.

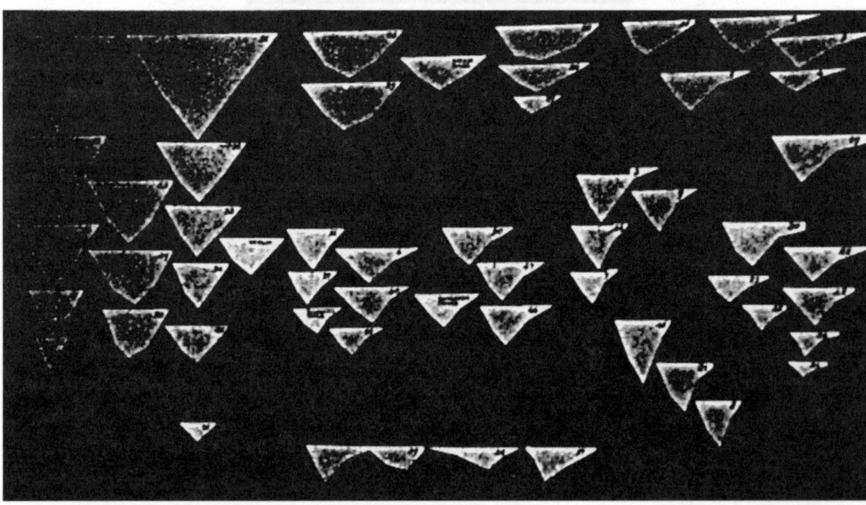

Figure 5.

The fifty emotional situations analysed by Benussi confirmed the following hypotheses:

(i) The *emotional functions are autonomous*: object-less thoughts exist, and emotional states exist in a consciousness devoid of thoughts and/or images, according to the theses of the Würzburg school;[34]

(ii) Different emotional states can be *graphically expressed* by two types of shape, biconcave and biconvex;

(iii) Specifically, the process from W to BS can be depicted by *biconcave shapes* (rectilinear to concave), usually of large surface area; the process from BS to W can be depicted by *biconvex shapes* (rectilinear to concave) of smaller surface area;

(iv) Since BS is a state of maximum well-being (describable as an ecstatic state), its relative graphic forms are progressively concave or convex.[35] The graphic profile of physical pain is the reverse shape to that of physical pleasure; that is, their velocity quotients (inhalation and exhalation) are inverse. In other words, pleasant states all have similar shapes, and the same holds for unpleasant ones.

(v) The continuum of bi-rectilinear profiles expressing the pleasure/pain continuum therefore seems to correspond to the white/black continuum in the perception of colour.

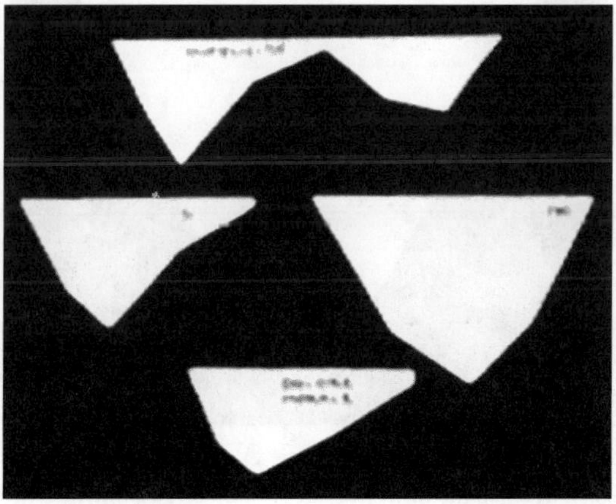

Figure 6. Shapes of unhappiness, yes, no, moral sufferance.

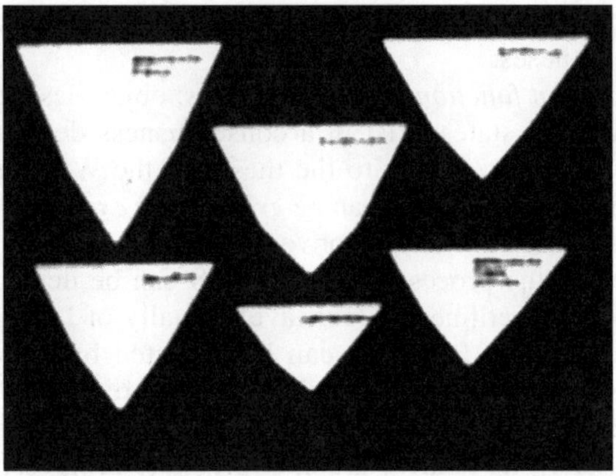

Figure 7. Shapes of physical pleasure, sympathy, evidence, relief, contentment, well being.

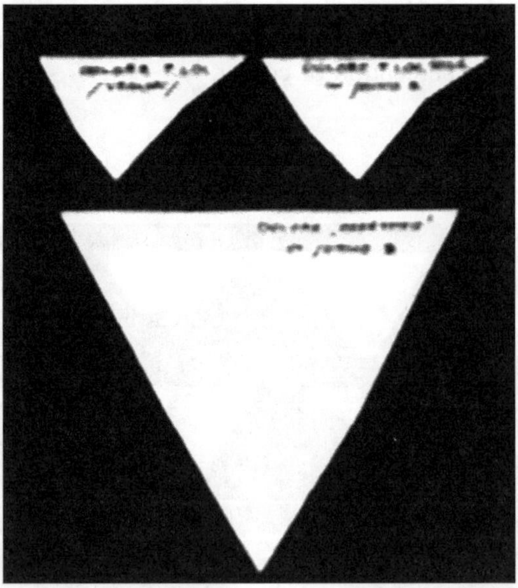

Figure 8. Shapes of local pain in W, of local subjective pain in BS, of objective pain in SB.

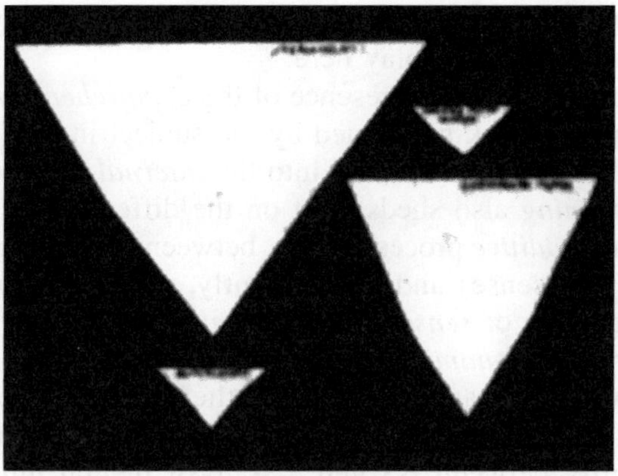

Figure 9. Shapes of hope, diffuse physical pain, desperation, physical pleasure.

3. TOWARDS AN EXPERIMENTAL METAPHYSICS

What conclusions can be drawn from this *geometry of emotional states*?

Benussi's real analysis – *real* because it allows effective variation in the conditions internal to the onset of psychic phenomena and consequently considers *souls* as *res*, as *observables to manipulate* – demonstrates:

(i) the existence of similar *perceptive invariants* in diverse kinds of perceptive continua;
(ii) the correlation between psychic *situation* and respiratory *shape*;
(iii) the *uncopling* of the emotional functions from intellective ones, and the independence of functions (or of aggregates of functions) from each other;
(iv) the existence of *phases* of development in the continua of consciousness (and in the strata of consciousness) of which we are usually unaware;
(v) the possible *elision* of conscious reflexes to certain external stimuli while all the others (combinations of orthogonal movements) are preserved;
(vi) the difference between (grasping) the *sense* and (understanding) the *meaning* of a sign or a word. Sense, indeed, is grasped by experiencing a *concrete situation* devoid of discourse elements and

constituted by *fragments of behaviour*: that is to say, emotional-affective states are in play here.[36]

(vi) the fact that the mental presence of the *comprehension of the sense* of a word (the task performed by the subject in BS) is realized in a transformation of the sense into the *internal perceptive presence of the meaning* also sheds light on the difference between *assimilative* and *additive* processes (i.e., between perceiving and thinking in the strict sense) and consequently, according to (vi), on the morphogenesis of *sense* and *meaning*;

(vii) the *expressive dynamic nature* of the emotional states depicted by the respiratory graphs shows that they are behavioural in character; that is, the *schemata* or diagrams are of facts and/or actions;

(viii) the identity between BS and a state of mental union or mystic ecstasy which allows experimental analysis of these phenomena;

(ix) since Benussi's findings have been corroborated in intellective situations involving states of evidence, negation, doubt, etc., the transformation of *mental presence* into *perceptive presence* gives rise to states of extreme lucidity which may even assume the character of reality, which is an important step in the comprehension of altered psychic states too.[37]

To conclude: one of Benussi's findings of keenest interest to an *experimental phenomenology* is the *correlation among the fifty respiratory profiles* plotted with the emotional situations analysed. Independently of the individual emotional states that generated them, the straightforward *formal grouping of the profiles* evidences that situations of a certain type belong to the pleasure/pain continuum or vice versa. Specifically, biconcave curves belong without exception to states of well-being, and biconvex ones to states of malaise – to varying degrees.

Theoretically, this finding points to important conclusions: firstly, that erasing the thought of objects does not entail either erasing consciousness or its intentional character or the so-called *sensus sui* which accompanies it. This is an important point for a *phenomenology of life*, since it throws new light on the problem of *intersubjectivity*. In fact, if bodies, other people, etc. are surely important for the constitution of meaning of life, the experimentally tested *autonomy of intellectual and emotional functions* shows that bodies are *bearers* of minds, but that minds are autonomous entities, even if ontologically founded on them.[38]

To use a strong expression, *the mind is an autonomous mistique corps.*

Secondly, from a psychological point of view, this finding points to the fact that emotions are not so much a specific class of psychic phenomena as a *form of colouring, a tertiary quality of the actual genesis of the Gestalten* corresponding, as Metzger said, to the *momentary structure of the impulse* of the phenomenal self towards their transformation.[39] It is in fact, a sort of *functional* autonomy more than a *psychic* autonomy of emotional states.[40] Giving even greater significance to the thesis is the fact that the *priority of the whole* over its parts corresponds morphogenetically to the *priority of the emotional qualities* over the whole.[41]

Università degli Studi di Trento
Italy

NOTES

[1] Cf. Brentano 1874; Albertazzi 1996.
[2] The correct term should be *soul*, indeed, since Brentano's descriptive psychology is non intellectualistic.
[3] Cf. Brentano, 1973, p. 218.
[4] Cfr. Brentano, 1971, p. 139.
[5] Cf. Höfler, 1897; Pfänder, 1924; Köhler, 1929; Klages, 1927.
[6] Cfr. Benussi, 1925a, 1925b and 1927a.
[7] Cfr. Bozzi, 1989.
[8] Cf. Brentano, 1973.
[9] The 'empirical' is understood in Brentano's sense, as the description of psychic facts present to consciousness in an evident way: Cf. Brentano, 1973. The date of the birth of experimental phenomenology can be traced to the discovery of the *phi* movement by Werthcimcr.
[10] Cfr. Benussi, 1925a; Koffka, 1935.
[11] From a private conversation. Cfr. also Mecacci, 1988. The point under discussion, for what concerns isomorphism of structures, is not he distinction between *molecolar* behaviour of the so called *quantitative* sciences vs. the *molar* behaviour of the so called *qualitative* sciences, but (i) from one side the *reduction* of qualitative to quantitative sciences; (ii) from the other side the absolute deny that also physics and physiology may have intrinsic molar aspects. However, some recents developments of quantum physics (for ex. in Bohm), would contrast this second opinion and molar phenomena are at home in biology. Last, it is not that accepting the postulate of the isomorfism molecolar phenomena have to disappear, but, as already Koffka observed, they cease to be independent events, and become *local* events within a more wide field which determines them.
[12] Cfr. also Shepard, 1981
[13] Cf. Benussi, 1913; Vicario, 1972.
[14] Cf. Koffka, 1912; Zimmer, 1997.
[15] Cf. Brentano, 1976; Musatti, 1926; Metzger, 1941.

[16] Cf. Husserl, 1966; Klages, 1927; Krueger, 1926; Sander, 1930.
[17] Cf. Meinong, 1904.
[18] Cf. again Klages, 1927; Krueger, 1926; Sander, 1930. Strict descriptive experimental phenomenologists, however, would not agree with this point. Cf. Vicario, 1993.
[19] Cf. Husserl, 1966.
[20] Cf. Benussi, 1913; Albertazzi, 1995 and 1998a; Michon and Jackson, 1985.
[21] Cf. Gibson, 1979.
[22] Cf. Brentano, 1976; Musatti, 1926; Metzger, 1941.
[23] Cf. Bonaventura, 1929; Albertazzi, 1993.
[24] Cf. Benussi, 1922–25; Piana, 1989; Albertazzi, 1993 and 1997.
[25] Cf. Benussi, 1922–25.
[26] Cf. Brentano, 1979; Albertazzi, 1998a; Benussi, 1922–25.
[27] Cf. Klages, 1927; Benussi, 1925b.
[28] Regarding to this latter point, it is worth noticing hat Benussi's experimental analysis is one of the few that experimentally deals with the relation between *signifying* and *expressive* functions of signs from a genetic point of view, being a sort of experimental analysis of the Brentanian theory of language. Cf. Benussi, 1925b; Kanizsa, 1991. On this, Albertazzi, 1995.
[29] Attentive hypnic state and base sleep are both phenomena of suggestion.
[30] More precisely, there were *two* passages before reaching a state of BS. On this cfr. Wallon's criticism in Wallon, 1926.
[31] Benussi learned this method already in Graz.
[32] In fact, whereas the ratio between the velocities of the first and second stages of inhalation was practically the same as the ratio between the velocities of the first and second stages of exhalation (Qvi = 1.40; Qve = 1.56), the *velocity quotients* for base sleep were Qvi = 1.24; Qve = 0.80. In this latter case, therefore, *changes in velocity* were symmetrical.
[33] Cf. Albertazzi, 1995.
[34] Incidentally, the hypothesis of autonomous emotional functions is akin to the hypothesis of the existence of different cognitive modules in contemporary cognitive science.
[35] Benussi was considering the affinity of his findings, on the report of Dr. Signorelli, her subject of experiments, with the descriptions of ecstatic states made, for example, by St. Teresa of Avila (the so-called *prayer of quiet*) and St. John of the Cross.
[36] The meaning of a sign is understood by *actuating particular objectual functions* directed towards to "higher-order objects": meaning depends on (written or spoken) *discourse situations*. Cfr. Hornbostel, 1927; Musatti, 1931.
[37] A critical stand towards some aspects of Benussi's experiments is in Musatti 1987, but specially in Wallon's review of Benussi's works on the topic. Cf. Wallon 1926 and Benussi's reply in Benussi, 1927b.
[38] Cf. Hartmann, 1948.
[39] Cf. Metzger, 1941.
[40] Cf. Benussi, 1927b, 341.
[41] Cf. Metzger, 1941; Koffka, 1912; Krueger, 1926; Werner, 1919; Volkelt, 1930; Klages, 1927.

REFERENCES

Albertazzi, L., *Strati* (Trent: Reverdito, 1989).

Albertazzi, L., "Psicologia descrittiva e psicologia sperimentale: Brentano e Bonaventura sul tempo psichico", *Axiomathes* 3 (1993), pp. 389–413.

Albertazzi, L., "Il campo figurale della lingua: esprimere, indicare, rappresentare", in Negrini (ed.) 1995, pp. 9–43.

Albertazzi, L., "Strutture temporali e ontologia", in Albertazzi, Cimino, Gori-Savellini, 1998a.

Albertazzi, L., "Time in the Brentanist tradition", in *Brentano Studien* 7 (1997), forthcoming.

Albertazzi, L., "Form Metaphysics," in L. Albertazzi (ed.), 1998b.

Albertazzi, L., (ed.), *Shapes of Form*, Kluwer, 1998b.

Albertazzi, L., G. Cimino and S. Gori-Savellini (eds.), *Francesco De Sarlo e il laboratorio fiorentino di psicologia* (Bari: G. Laterna, ed., 1998), forthcoming.

Benussi, V., *Die Psychologie der Zeitauffassung*. Heidelberg: Winter, 1913.

Benussi, V., *Introduzione alla psicologia sperimentale*. Lezioni tenute nell'anno 1922–25 dal prof. V. Benussi e raxccolte dal Dott. C. L. Musatti assistente (Padua, 1922–25).

Benussi, V., *La suggestione e l'ipnosi come mezzi di analisi psichica reale*, Bologna: Zanichelli, 1925a.

Benussi, V., "La suggestione e l'ipnosi come mezzi di analisi psichica reale," *Atti del IV Congresso nazionale di psicologia* (Florence: Bandettini, 1925b, pp. 35–63).

Benussi, V., "Zur experimentelle Grundlegung hypno-suggestiver Methoden psychischer Analyse," *Psychologische Forschung* 9 (1927a), pp. 197–274.

Benussi, V., "Sur l'autonomie fonctionelle émotive (A propos d'un compte rendu de M. H. Wallon)," *Journal de Psychologie normale et pathologique* 24 (1927b), pp. 341–343.

Bonaventura, E., *Il problema psicologico del tempo* (Milan: Soc. Analitica Editoriale Scientifico, 1929).

Bozzi, P., *Fenomenologia sperimentale* (Bologna: Il Mulino, 1989).

Brentano, F., *Psychologie vom empirischen Standpunkte* (1874); Reprint Hamburg: Meiner, 1973.

Brentano, F., *Die Klassifikation der psychischen Phänomene* (1911): Reprint Hamburg: Meiner 1971.

Brentano, F., *Philosophische Untersuchungen zu Raum, Zeit und Kontinuum* (Hamburg: Meiner, 1976).

Bühler, K., *Sprachtheorie* (1930); Reprint Stuttgart: Fischer Verlag, 1965.

Gibson, J. J., *The Ecological Approach to Visual Perception* (Boston: Houghton-Mifflin, 1979).

Hartmann, N., *Zur Grundlegung der Ontologie*, 3rd ed. (Mesenheim am Glan: Hain, 1948),

Höfler, A., *Psychologie* (Vienna und Prague: Tempsky, 1897).

Hornbostel, E. M. von, "Laut und Sinn," *Festschrift Meinhof, Sprachwissenschaftliche und andere Studien* (Hamburg, 1927), pp. 239–248.

Husserl, E., "Vorlesungen zur Phänomenologie des inneren Zeitbewusstsein (1905)," *Husserliana*, Vol. **X** (1996).

Kanizsa, G. and N. Caramelli, *L'eredità della psicologia della gestalt* (Bologna: Il Mulino, 1988).

Kanizsa, G., *Vedere e pensare* (Bologna: Il Mulino, 1991).
Klages, L., *Zur Ausdruckslehre und Charakterkunde* (Heidelberg, 1927).
Koffka, K., "Experimentelle Untersuchungen zur Lehre vom Rythmus," *Zeitschrift für Psychologie* 52 (1912), pp. 1–109.
Koffka, K., *Principles of Gestalt Psychology* (New York: Harcourt, Brace & Co., 1935; Reprint London: Kegan Paul, 1963).
Köhler, W., *Gestalt Psychology* (New York: Liveright, 1929; Reprint London: Bell, 1930).
Krueger, F., "Zur Einfürung – Über psychische Ganzheit," *Neue psychologische Studien* 1 (1926), pp. 1–121.
Kubowi, M. and J. Pomerantz (eds.), *Perceptual Organization* (Hillsdale, NY: Erlbaum, 1981).
Masin, S.C. (ed.), *Foundations of Perceptual Theory* (Amsterdam: North Holland, 1993).
Mecacci, L., "Perché non possiamo non dirci gestaltisti in neurofisiologia," *Kanizsa, Caramelli* (1988), pp. 203–225.
Meinong, A. (ed.), *Untersuchungen zur Gegenstandstheorie und Psychologie* (Leipzig: Barth, 1904).
Michon, J. A. and J. Jackson, *Time, Mind and Behaviour* (Berlin: Springer, 1985).
Musatti, C., *Condizioni dell'esperienza* (Florence: Editrice Universitaria, 1926).
Musatti, C., "Forma e assimilazione," *Archivio italiano di psicologia* 9 (1931), pp. 61–156.
Musatti, C., "Vittorio Benussi e i fenomeni ipnosuggestivi," *Mucciarelli* (1978), pp. 107–119.
Metzger, W., *Psychologie: die Entwicklung ihrer Grundannahmen seit der Einführung des Experiments* (Dresden und Leipzig: Steinkopff, 1941).
Mucciarelli, G. (ed.), *Vittorio Benussi nella storia della psicologia italiana* (Bologna: Pitagora, 1987).
Murchison, C. (ed.), *Psychologies of 1930* (Worchester, Mass: Clark University Press, 1930).
Negrini, G., *Categorie, oggetti e strutture della conoscenza*, Istituto di Studi sulla Ricerca e Documentazione Scientifica (Rome: CNR, 1995).
Pfänder, A., "Grundprobleme der Charakteriologie,"*Jahrbuch für Charakterologie* (Berlin 1924), pp. 289–335.
Sander, F., "Structure, Totality of Experience, and Gestalt," in Murchison (ed.) (1930), pp. 188–204.
Shepard, R. N., "Psychophysical Complementarity," in Kubowi-Pomerantz (1981), pp. 279–341.
Utitz, E. (ed.), *Jahrbuch für Charakteriologie* (Berlin, 1924).
Vicario, G. B., *Tempo psicologico ed eventi* (Florence: Giunti e Barbèra, 1972).
Vicario, G. B., "On experimental phenomenology," in Masin (1993), pp. 197–219.
Volkelt, J., *Versuch über Fühlen und Wollen* (Munich: Beck, 1930).
Wallon, M. H., "Compte rendu: V. Benussi, La suggestione . . . ," *Journal de psychologie normale et pathologique* 23 (1926), pp. 1040–1048.
Werner, H., "Rhytmik, eine mehrwertige Gestaltenverkettung," *Zeitschrift für Psychologie* (1919).
Zimmer, A., "What is a Form?," in Albertazzi (ed.), forthcoming.

MARGARETE DURST

PHILOSOPHY AND EMOTION

THE PHILOSOPHICAL APPROACH TO EMOTION[1]

The world "emotion" is associated with so vast a variety of phenomena that it is difficult to provide the main definitions, even when exclusively confined to a philosophical approach. This word is usually given a series of complementary terms which are no less vague and ambiguous and which contribute to heightening the elusive character of the concept they refer to. In spite of the lack of a precise semantic reference, emotion has given birth to many interpretations, and the different paradigms and methods of various disciplines have been applied to it with unvaried interest long since.[2] First of all medicine, before it developed its plentiful branches, psychology and anthropology, in one word, all the human sciences, at last epistemology and history of science, are deeply involved in finding out the traces of the aleatory and elusive psychological *quid* that is emotion, following the path of all apparently reliable knowledge. Within mythology, narrative and art, where emotion has always been an important element, a privileged role has been assigned to it. Thus, emotion has turned into a pivot of the uncountable relationships between earthly and heavenly things.

As far as the philosophical approach is concerned, after the Pre-Socratic period, during which the stongest and most radical emotions were called into play, that is, the love and hate which rule the world, interest in the emotional aspects of existence has gone through intermittent phases. This implies resistance to approaching a phenomenon which immediately turns out to be hard to rationalize. As matter of fact, on the one hand, classical philosophers speak quite a lot about emotion, minutely analyzing their different expressions; on the other, they tend to place it at the margins of their systematic elaborations. Therefore emotionality is at once the *primum movens* of knowledge, as suggested by Aristotle's appeal to wonder, and the most fallacious way to come closer to truth. Apparently, thanks to the Sophists, sense of the ambiguity of the philosophical approach to emotions evaporated as appeal to emotion became instrumental and knowledge of how to manipulate

A-T. Tymieniecka (ed.), Analecta Husserliana LIX, 379–400.

Angela Ales Bello and Margarete Durst.

human feelings through rhetorical and persuasive art was harnessed to the grab for power. In reality, the Sophists' interest in emotion cannot be reduced to a pure form of manipulative communication. The most well-known sophists pay particular attention to the bewitching power of emotions, that is, to their influence on people, because all emotions interact, giving birth to a stream of relations between different and distant people. Consequently, emotion, because of its inner superiority, which makes a ruler of it, may appear elusive. The only way to reach and dominate it, then, is to accept its role and to be subject to it. In any case, the main problematic question of the philosophical approach to emotion concerns the relationship between the emotional and the rational, independent of the means used to rationalize a phenomenon, which is so difficult to conceptualize. The need to apply human reason to this field also characterizes the whole history of philosophy.

The *iter* of the philosophical approach to emotions obviously involves many branches. Therefore almost every tradition gives a peculiar interpretation to emotion, differently pointing out the links with other fields of experience. Since the Platonic myth of the charioteen who controls the impetus of the two opposed winged steeds, philosophers have shown care in their approach to emotions, thus recognizing their power and influence on human life and their lack of adaptability to a conceptual approach. Therefore, the philosophical approach to emotion turns out to be characterized by a radical ambiguity inherent to the very nature of a knowledge which has as its pivot *philia*. Thus a prime emotion inevitably exceeds the limits of the *limen* which circumscribes the field and dominions of research.

Within the phenomenological and existential approach to philosophical research, an extremely important leading function has been assigned to emotion, turning it into a presupposition of the intuitive prefiguration of the world; which gives emotion an essential role in the existential orientation. The value recognized by Heidegger in *Befindlichkeit* is, from this point of view, one of the most meaningful demonstrations of philosophical appreciation of emotivity. It is also Heidegger's view, as some of his followers observe,[3] emotion or emotional nuance, remains entangled in a transcendental approach from which scholars have tried to set it free.

At this point we are facing the question of the complex relationship that phenomenology and existentialism entertain with criticism and with the specific matter of the transcendental deduction of the prior categories and forms of sensibility, that is, that space-time dimension which is the primordial condition of all existence. In the rationalistic perspective which goes back to and beyond Kant, until the analytic approach of Descartes' philosophy, the link between sensibility and intellect, here coming into play, receives the form of a subordination, one which, in a more or less direct way, refers emotivity back to rationality. Even though one may, critically, recuse from the functionalist and structural approach of Kant's schematic, it remains, given these premises, difficult to go beyond a reductionistic approach. Reductionism, applied to the psychic field, has now succeeded in configuring even consciousness in terms of a pseudo-problem, and thus it proposes its quick elimination. Furthermore, reductionism considers emotions to be at the same level as the hundredth version of the so-called "ghost in the machine" on physicalist terms. It, thus, rejects any qualitative approach to intellect, understood as a realm of the psychic sphere.[4] Some supporters of the modern "philosophy of the mind", in promoting this methodological approach, (moreover secondary), make use of different metaphors to show how the habit of referring to consciousness and to all the concepts linked to it, as though it and they represented meaningful realities, is obsolete. After all the field of psychological research on emotions has received from the philosophy of the mind great support, mainly because critics with theoretically and experimentally integrated clarifications, have added an innovative process started within the different disciplines methodologies.

Cognitivism itself, as it will be possible to observe in the following, tends to consider emotional factors to be determinant in the motivational dynamic and in our settling on goals; as such it aims to go beyond emotions exclusively rationalistic meaning.

But emotion, though it is part of the analytical and computational procedures of research, still has not thereby gained its own peculiarity, and the vast body of empirical knowledge acquired mainly by the empirical psychology in this field is still trapped in a theoretical perspective which is not much inclined to grant this recognition.

As for the philosophy which emphasizes the value of emotivity, this esteem has still to inspire a pure research investment, that is, a systematic study. When we want to give a determinant function to emotion, we usually tend to refer emotional expressions back to our vital field. We are usually inclined to avoid any kind of description. Therefore, once we assign to emotion a determinant and disclosing role in the construction of experience, we prefer to narrate rather than to describe it, in the way that Rousseau tells us about his inner and personal life, exhibiting and concealing it at the same time. All the philosophers who make a protagonist of their "I" simply narrate the flow of their emotions. This kind of work shows the ambiguity which is peculiar to these figures, who do not grant to the object of emotion full philosophical dignity. Probably only with Sartre is the narrative approach to existential problems considered to be a conscious philosophical choice, and according to this perspective, existentialism develops into an interpretation of emotion, one defined by Sartre himself as a project of "phenomenological psychology".[5] Particularly in Sartre's and Merleau-Ponty's studies on imagination (with the latter exclusively basing his research on perception), emotion becomes the object of an original and theoretical elaboration that is limited neither to narrative nor to Kant's and Husserl's transcendentalism. Once emotion is seen through the prism of its interaction with imagination, perceived in a dynamic way and not according to the schematizing typology going back to Kant, it has acquired a phenomenological priority. But the course of these studies (interrupted and later reconsidered without reaching a conclusion) confirms an insurmountable difficulty which is mainly ascribable to an unwillingness to transcend the threshold of first certainty, by separating and binding the conscious and the unconscious. Sartre's suspicion of Freud's unconscious, mirrors Husserl's attitude toward the whole of psychology. Furthermore, this attitude is typical of all twentieth-century philosophy,[6] wherein it is possible to perceive the influence of a rational interpretation of emotivity ascribable to a certain Kantianism which owes Descartes' analytical methodology much.

As a matter of fact, the necessity of rational control of emotion is explicitly stated by Kant, and knowledge of the emotions has been promoted with just that aim. That does not exclude that Kant's thought, under some other of its aspects, is open to recognizing the imponderability of emotion, and that it also considers this feature positively, finding in it an element which is typical of human nature, not only on a sensible-animal level but also in relation to the end human nature wants to achieve. Kant's works obviously offer different perspectives, either in relation to the diversities of the faced thematics or to the changing backgrounds they refer to. It is not surprising that Kant's works deal with emotivity both by taking the typical approach of formalism, aiming at a definition, and by describing its symptoms. The former adopts the classical differentiation between emotions, feelings and passions, based on a major and a minor conceptual nucleation of affection, whereas the latter follows the path of English and Scottish philosophers' research on feelings and passions. It is not surprising, then, that Kant talks about love, an undoubted strong emotion, as a feeling that can be put aside in the context of the authentic moral determination and also assigns love an essential function of support comparable to that of the air beneath the soaring dove. Anyway the need for a rational control, mirrors the worry that emotivity might break up the existential balance that the same need contributes to creating, at least because of its natural status. The positive contribution made by emotion, understood as natural inclination, interacting with the sphere of needs and desires, is in fact always influenced by its taking part in a well-ordered system which is ruled by reason. All that comes under the rubric emotional inclination bears a coefficient of risk higher than that to be forseen from its natural status; this is a risk that education, constantly playing a restraining role, explicitly shows.

The exhortation expressed by Kant in his *Anthropology*,[7] inviting people to develop their personalities through a deep interiorization of a rational ethic, in order rise to spirituality, recalls the image of a supported emotivity, one restrained and guided by the very bridles from which the philosopher himself would like to set everyone free. Kant's appeal is obviously subject to different interpretations, but it certainly encourages being very careful in approaching emotion. Thus the acceptance of emotivity apparently involves only its "rational" aspects, that is, those more easily brought back to the rational elaboration and which contribute to strengthening the fixed methodology of life, or at least to not wandering from it. By this approach it is possible to widen the gap between emotion and reason: the former interpreted in an intellectual and formal way, the latter understood as a revolutionary inclination in relation to rationality.

EMOTIONS, AFFECTION, FEELINGS

It is evident that in the history of philosophy the tendency to consider the milder aspect of affectivity – sympathy, benevolence, interest (feelings easier to rationalize than love, hunger and hate . . .) – proceeds, at the same rate as the definition of the cognitive parameters determining the scientific aspects of research; these parameters are more and more involved in rigorously limiting the areas of observation. The emotional universe is thus divided into different branches respectively connected to the field of feelings, passions, and affections, and so loses what gives it its coherence, that is, that element which is the basis of every discriminating approach characterized by emotion in that it cannot be assimilated into rational interpretations. We will now refer to an existential situation determined by a paradoxical, unitary and twofold rationality, which reveals itself mainly in an emotional way, that is, through a complete psychophysical involvement as well as in a conceptual way. We are talking about the primary narcissism based in the symbiosis and fusibility of the mother/child relationship during the first phases of life. Narcissism, being the core of pure emotivity, is the inner emotional constant factor of the whole existence and the source of every other emotional expression.[8]

While poetry, music and plastic arts addressing emotions, have always represented their genesis, evolution and effects, evoking their origins, philosophy, which is always critical towards subjects hard to conceptualize, has not taken into consideration this emotional matrix, which is despite that part of its thought. It is important to admit that the classical philosophical approach, aiming at separating the stronger and sharper tonalities within the emotional sphere, from the weaker and muted ones, made it possible to point out some nuances which would otherwise be difficult to perceive, although it might have damaged the chromatic vividness of the whole.

Among the many metaphors of emotion, that of a cloud, determining the coefficient of fluidity, condensation, brightness and temperature, apparently better corresponds to the framework of a changing and solid phenomenology of a reality such as the emotional. The image of an energetic and mainly unitary flow, though subjected to a vast range of variations which are difficult to be numbered, immediately evokes the idea that our sensitive and perceptive levels may fuse and interact with our fantasizing and creative ones. Furthermore, the emotional dynamic seems to subdue a net of obscure, uncertain and particular movements which are difficult to be rationalized and which surround the universe with an aura of imponderability and mystery. Poetry and the plastic arts are, surely more fit than philosophy to emphasize this almost magic element, which is clearly inscribed in their inapproachable nature. The philosophical approach, greatest merit of which is its outlining the filagree of the emotional phenomenon, can, by turning its attention more explicitly to this area, better emphasize its links with all the fields of existence; thus philosophy does not confine the emotional phenomenon to the sphere of the primal experiences. Furthermore philosophy, excluding the emotional origin which is at the base of prescriptive approaches to feelings and affections, comes, on the one hand, to limit emotivity only to a rational area and on the other, to exacerbate the gap between emotion and reason.

The term emotion underlines the dynamic aspect of the phenomenon, more or less clearly yet all the words applied to the emotional sphere, such as feeling, passion, affection, which are differently used in various linguistic contexts, always show the merely passive aspects of emotivity, whether passive on active. It is the active aspect through which it is possible to bring the whole range of emotional expression back to the particular psychophysical situation peculiar to narcissism, which predisposes one to feel and thus experience feelings, passions and affections. Philosophers have never paid any attention to this area, leaving it to poetry, literature, art – addressing exclusively instead the study of feelings and affections, that is, to say the area where it is easier to isolate a conceptual core.

Today there is a clear reverse trend which comes out of ever increasing study of strong feelings, that is, of passions or the so-called authentic emotions in order to point out their intensity and endurance. This new approach in research may be also due to the fact that today the sphere of feelings and affections is being intensively studied by the experimental psychologists as well as by anthropologists and sociologists. All this may be prompting philosophers to involve themselves in a remote area of the human psyche.

The fact that feelings and affections have in the past been more carefully and minutely examined than passion, certainly reflects our tendency to bring phenomena, by affinity or contrast, back to well-known contexts, thus selecting some of its aspects. All this refers to the interaction between the motivation and cognitive setting which guides research, the pure paradigm (with all its instruments, criteria, methods) and the specific operative situation. These are problems which have long since been the focal point of the epistemological debate where the social context with all its different dimensions, mainly political and economic, is involved with the scientific apparatus. All this can help us better understand how these problems might have influenced the studies on emotion, favouring, depending on the era and social contexts, some aspects of emotion at the expense of others. In the philosophical field there is a widespread tendency to focus on emotional expressions which seem to be linked to rational meanings so that they can be considered reasonable, confining the others to the field of an irrationality unknown to reflection. Nietzsche's intolerance of this approach is today being reconsidered by scholars of morality,[9] who seek a new exploration into the background of so-called moral feelings.

The acceptance given the moderate area of emotivity, reflects a need to counterbalance an excess of conflict as much as it stands for a desire for balance which searches in the middle situation for a contribution to peace. Remo Bodei, in his reconstructive description of the passage to modernity, shows how much passions have actually ruled history,[10] passions much deeper than the wished for feelings of prudence, quiet, benevolence, emphasized by some philosophers of that age. We can not forget how much the development of the new sciences influenced the tendency of philosophers to involve themselves in the study of the middle and temperate area of emotivity. We advert to the fact that the scientific revolution has strictly bound knowledge to rational control and, through that, to power. An economy open to trade and financial brokerage has always promoted a moderate interpretation of emotivity, wherein the interest of others and of oneself intentionally become the core of the system of affections. When Adam Smith wrote his essay on moral feelings,[11] the ethical and moral conjunction with economic interest already had its own history, wherein philosophers are to be assigned a place beside all the bureaucratic and military apparatus of nations. The process developed mainly in Holland, France, Scotland and England after having taken root in the Italian seignories, where it did not flourish in the same way. This history threw up obstacles to approaching emotion by favoring an analysis of the more constructive feelings, particularly those which, exhort to the virtue of tolerance, and lay foundations for pacific and propitious living.

Thus, the tie between feeling and the sphere of morality becomes stronger; this tie is influenced by the political necessity to grant a fixed stability to the community by appealing to common criteria of justice. As a matter of fact, the path of justice and morality is characterized by feelings which are frequently encouraged in the philosophical works preceding the revolutionary age – which, on the contrary, exalts its own passions. In that context the feelings of higher merit are those which can be directly associated with strong concepts, characterized by an objective and even institutional importance. As has been already said, above all ethical-moral, socio-political and legal-economic concepts, and the common emotive background which grants them "naturalness" provided them with a legitimacy beyond the theoretical level. Foucault has widely documented the interrelation between morality, rationalization processes and organizational structures; this is a dynamic which finds expression in the institutional systems wherein order and law, and not in a painless way, cooperate to give birth to feelings useful in the practise of power. But his detailed analysis, the last step of which concerns egoistic love,[12] produces an excessively tendential and negative vision which denigrates the authenticity of those feelings which do not coincide with the egoistic need for self-realization giving the label of loser to whomever has them.

Actually, rationalization has value other than its repressive force; even in the field of emotion the development of analytical and calculating procedures predisposes people to possibilist positions, more than qualitative approaches on a dialectical basis might allow. As a matter of fact, the latter approaches radicalize the opposition forces, where the rational elaboration carried out according to quantitative parameters, enables the modulation of a vast range of moderate positions between the extremes. This follows Locke's view, according to which the individualization process consolidates on structures of mastery, in which process feelings of shame and esteem both in the self and the other come into play. The opposition between these feelings, loses its dialectical form, typical of the classic context of the Greek tragedy, and takes the shape of a dynamic rich in combinations which leaves space for intermediary solutions.[13] Analogous analysis can be made where, socially speaking, honour is concerned. Honour is a feeling in which qualitative value has become a quantitative one over time, so that dishonour is less keenly felt, no longer being associated with an inner and essential aspect of personality.

But this consideration of emotions only in the form of feelings that are manageable on the basis of strategies of rational control, diminishes the dramatic force of a clear contraposition and also induces forgetfulness that the field of permitted variables, though vast and articulated, corresponds exclusively to a circumscribed section of a wider unity. Furthermore it contributes to considering the emotions' being seen as objects of reasoning, as something different from experienced emotions. This last difference come out of the seventeenth and eighteenth-century philosophers' works wherein, if it is not possible to find the conflicts perceived by some scholars, at least ambiguity and tensions, symbols of an unresolved uncertainty, are evident. Significantly different modern philosophers have considered these works in the light of a restored interest in moral feelings, which are, not by chance, in vogue again, within a society which wants to legitimate an idea of justice commonly shareable by everyone on a level of originating theory.[14] It is easy to think that feelings might represent the *humus* at the origin of the affinities and similarities among individuals, ethic groups and peoples who are not well-disposed to renounce their peculiarities as soon as they meet or interact.

This premise reflects the idea that feelings rely on a universal sphere of existence which goes beyond all the situations of diversity and which is thus flexible and free from any kind of influence duc to language, time and place. But, after a further analysis, this flexibility might turn out to be inconsistent, in that feelings, just because of their universality, may remain unchangeable at their root, beyond their superficial expressions. All that brings us back to the emotional origins common to all the different expression of emotion: and this approach goes beyond analytical studies of emotion.

COGNITIVISM AND PHILOSOPHICAL PSYCHOLOGY

The emotional world, divided into its sections and described on the model of dissection, is still largely unknown, but, thanks to many experimental psychological, psychiatric as well as neuro-philosophical studies, emotion has acquired a more and more important role at all the levels of existence: theoretical, pragmatic-behavioural, communicative, etc. Today, it is not only Spinoza's followers who assign this role to thought even in its most conceptual and abstract form.

In analysing cognition, as has been seen, it is possible to find all the problems of the philosophical study of the emotions, above all if they aim at proceeding the hard way, proposed in Freud's metaphor of draining the Zuiderzee, but also if the gentler procedure associated with Jung's metaphor of the island that runs the risk of having its territory flooded by the surrounding see is adopted. The impact of psychoanalysis deeply influences the philosophical interpretation of the emotions, even beyond the clear use of the psychoanalytical conceptions of emotivity. This approach has aroused great interest among philosophers, more than among psychiatrists and psychologists bound to psychiatric traditions. It favours an interest in the most radical and deep emotional manifestations, to which feelings and affections would then, because of their tenacity and duration, to refer, since feelings and affections extend beyond their sudden and evident manifestations and the explicit consent given to them. This way of referring to emotions, aiming not at erecting a division of the emotional world again, but at underlining its continuity provides an orientative criterion for a complex and scarcely objectionable phenomenology which can not be valued on the basis of exclusively quantitative coefficients. The experimental research in this field carried out by psychiatrists and psychologists further points out the intrinsic limits of a purely cognitive approach to the emotions, and it brings about a widening of the notion of acceptable emotion, including in it the vast area of unconscious motivations, in which the dynamic of desires and needs comes into play.[15]

At the origin of this modern trend of philosophical study of emotion there is a progressive disintegration of a methodological framework which had aggravated the fragmentation of knowledge and, on the basis of very selective criteria, had caused the exclusion of disorder and discontinuity from the research field. Thus, the antireductionist schema now finds a hearing, and the many disciplines are caught up in a new way of thinking, and not only those humanistic disciplines that have been deeply influenced by the hermeneutical trend in epistemology. In predisposing thinkers to a wider conception of rationality and consciousness, and in giving importance to the rational dynamic, hermeneutics has thus introduced the factors of mutability and incoherence into the field of research: factors which also influence emotion and which contributed to moving philosophical studies away from this field. As the need for rational control weakens, these features might induce an irrationalistic interpretation of emotions, one emphasizing the extraneousness to the scientific approach. It is thanks to the scientific culture itself if this has occurred only marginally. By accepting the inner emergency provoked by the factors of imponderability and irrationality which can not be brought back into the nets of the strategies of calculation, and by getting ready to face disorder and contradiction in the procedures of inference, analytical studies of formal logic have also contributed to create a more tolerant scientific context, one with a wider field of variables to examine and include, the emotional ones also.

The singular phenomenological, aleatory and inconstant aspects characterizing emotion seem to be currently accepted, at least as far as the dynamic of identity and individualization are concerned, even by cognitivism which has adopted motivation as its interest.[16] Once we introduce cognition into a dynamic of purposes, we have succeeded in encapsulating the definition of a core of emotional constants also: this seems to be typical of the "affective categories" used in the field of the dynamic psychology. The emphasis on the rhetorical and narrative levels of all speech, the effort to assign a new scientific dignity to metaphorical concepts, has contributed to setting further parameters of interpretation of emotion. For example, the analytical tradition of the philosophy of the language, through its mapping approach, implemented thanks to the use of information technologies, has, in order to go deeply into the semantics of natural language, succeeded in pointing out the variations of meaning implied by words with emotional timbre in different sentences. Thus it contributes to identifying the bounds between communicative context and variations of affect.[17] Modern psychologists, psychiatrists and philosophers of the mind and of language testify to an approaching process combining the cognitivist-computational and phenomenologic existential traditions and the new approaches of dynamic psychology. Up to a few years ago all this was inconceavable.[18] The gap between the qualitative and quantitative approaches to emotional phenomena has been almost completely filled.

The most innovative researches in all the fields of the human sciences are converging on their emotional territory. These researches are more and more absorbing the concept of an individual examined in his sphere of human relationships and as an actor gifted with a limited and multiple rationality. There is a tendency to take into consideration the affective sphere of the cognitive and motivational dynamic in relation to the genesis of values and the rules; this is the direction in which the latest Anglo-American cognitivism is developing.[19] On this path a systematic and microsystematic approach develops, the latter offering an articulated and modular model able to examine the irrational and disturbing factors and the interaction between affect and cognition. Philosophical and Anglo-American psychology are partly linked with cognitivism and have introduced emotions into a complex framework of different cognitive and motivational systems. Each of these has got its own inner criteria of coherence and apparatus of rules, whose eventual opposition or contradictoriness occurs only at an intersystematic level, that is, within a perspective involving different systemic unities.[20]

On the basis of this interpretation, emotiveness is an important factor for systemic structuring, in that some mental features, though evidently inconsistent with the apparatus of criteria and rules, are, if emotionally relevant, not eliminated, but lateralized, giving birth to new systems and sub-systems which make them acceptable, hiding their contradictory aspect. A contradiction eluded within a certain system, can emerge at an intersystemic level only if the subject is willing to endure the impact of the strongly opposed emotional urgencies. A similar model of the mind which involves procedures for drawing distinctions and correlations among a vast range of cognitive-motivational unities, has contributed to a calm incompatibility within the interpretative theory of the mind. Studies on irrationality, on the phenomena of akrasia, on multiple choice, etc., which have been carried out in this field, have emphasized a series of imponderable factors in the motivational dynamic. It is possible to trace all these factors back to the sphere of emotions.[21] The mere fact that the problems of choice include interest, puts forth emotions and cognitions, in that, independent of other variables, interest involves *ipseitas* in a too direct way to be examined exclusively at a rational level. But if the field of variables is widened in order to take into consideration the multidimensionality of the "I" and, on this basis, makes a choice excluding interest, the following increasing difficulties in the process of rational elaboration, may both resolve and increase indecisiveness and emotional tension. Independently of the chosen perspective, the administration of emotion is not easy, and the invitation to a judicious balance should be able to rely on deep emotional experiences in order to be authentically accepted.

Therefore, the modern development of studies on emotion, a symbol of an increasing interest in its theoretical and empirical aspects, is taking place through a great number of channels. But the different interpretative processes may always fall back into the forms of rationalization, transforming emotion into something different from itself, that is, into a concept or term of reasoning. Or, on the contrary, emotions may be excluded from research into their deepest and more lasting emotional features. If we rely on psychoanalytic studies on emotion, these risks may decrease, in that, psychoanalysts, in their effort to back the scientific nature of their approach, focus on deep emotion, to which they refer back all emotional experiences. One of their favoured instruments is emotion itself, that is the empathy experienced at a level of sensibility and intuition, which lays the ground for any more explicit communication.

When it comes to the knowledge that the psychoanalytic conception of emotivity has brought to the context of the philosophical studies, it is clear that, together with the interest and the careful regard (it is sufficient to think of Ricoeur), there has been evident resistance to acceptance of the concept of the unconscious that is the support of Freud's interpretation of the emotional phenomenon. This has restrained heuristic effort to effect an interaction between psychoanalysis and philosophy. The effort could also have influenced the more static psychology. This kind of resistance is evident in semeiology too, which is particularly interesting because it involves a recall to the way classical and modern philosophers consider emotion, while relying on the knowledge and the use of modern analytic methodologies.

THE SEMEIOLOGY OF PASSIONS

The semeiotic approach to emotions, which bears with it the risk of rationalizing the emotional phenomenon, is very interesting. The scholars of this field try to avoid any restrictive approach which might isolate the intellectual features of emotivity from those which apparently are more irrational, although they do not renounce conceptualizing emotions. In order to pull themselves back from this mistake, they keep to a descriptive reconstruction which shows only those elements of emotion deducible from given procedural principles. Herman Parret who focuses his semeiotic studies on the passions[22] in his effort to emphasize the gap between the modern renewed interest in emotion and the previous eclipse, above all, deplores the confinement of emotivity to irrationality; he rightly lays claim to study emotion in a scientific way without distorting its peculiar features. In the semeiotic perspective, emotions, studied in their different expressions over the whole history of culture, have lost their indeterminacy and they have assumed the most plastic shapes of passion. With this word we can recall the tradition of the ancients and seventeenth-century philosophy and separate our research from the changing field of emotivity, that is, from a word which emphasizes the elusive aspects of affect. Semeiologists, though keeping themselves apart from this matrix, do not separate the passions from it. They concentrate their analysis on the passions; in fact the world passion, vivid with intensity and depth of affect, projects us into the area of the most intense and deepest emotions. Semeiologists, in their effort not to weaken the peculiarity of their approach, emphasize their detachment from epistemic perspectives. In addition, these scholars, in their reconstructions of the universe of the passions also exclude the psycho-social dimension, relying only on some of its themes as correlate factors of the indicators which are at the basis of the description they attempt. The well-ordered sequences with which the passions are photographed or mapped according to typologies of frequency/absence, lack of any reference to the conscious/unconscious terminological double, indicating a feature which is indeed very important not only in psychoanalysis, but also in a psycho-social and epistemic approach to emotional phenomena.

It is possible to consider Descartes' work on the passions (which, despite his well-known pretence to start *de novo* a study in this field, summarizes a complete history of interpretations)[23] as a prototype of all descriptive analyses of the emotional phenomenon and consequently also of the model traceable in the sophisticated frameworks used by semeiologists to describe the grammar of the passions. If grammar has proved to be suited to language, which is not free from peculiar and uncertain elements, the grammar/passions combination seems legitimate. In this combination the latter are examined as if they were one single language differently articulated on the basis of the same generative process. The series of connections pointed out by Descartes between the six passions that he considers basic and those that he calls secondary differs from the anatomy of passions of semeiology and not only because of a clear contextual difference. The gap between the two constructions, which is more or less traceable to the turning point represented by Romanticism, involves the epistemological paradigms, in each of which, however, the same meaning is given to emotion, even though it is generically considered as a psychophysical situation disturbing the usual tenor of experience, (although the lack of disturbance does not automatically imply an absence of emotivity).

As a matter of fact, during the era of Romanticism emotion acquired an extremely important salience, but then, as a reaction to the prevailing intellectualism, its irrational features were exaggerated, all that excludes it from the scientific research. Herman Parret emphasizes the negative influence that Romanticism has had on the study of the passions and the break made by irrationalist and vitalist culture within the philosophical tradition of the seventeenth and eighteenth centuries. He considers the emotional phenomenon as he would any other object of study. According to this interpretation, the Romantics, in their effort to restore emotion to its so-called "natural" place, that is, to the field of art, poetry and life, exaggerated the irrational side of emotion, thus increasing the indifference of philosophical researchers to this whole sphere of human experience. This negative evaluation may come out of a too simplicistic approach, which does not consider those factors by which the Romantics, in their effort to assign a primary role to emotion, promoted research using new procedures of qualitative analysis.

It is necessary to underscore the fact that, behind the exclusion of emotivity from philosophical research carried out scientifically, increasing value is being assigned to the strong cognitive paradigms and models of observation able to provide exhaustive representations and explanations. These approaches do not enter into phenomena like the emotions, which have an implicit dimension that cannot be translated into explicit language. Referring to this last point, semeiology, understood to include any possible signifying dimension increases the range of variables susceptible of descriptive reconstruction. After all, the latter ratifies an objectifying method, which in its effort to indicate the tip of the iceberg, that is, the expression that is a sign of an inseparable submerged reality, excludes the submerged sphere from even indirect observation, intentionally carried out on the basis of the evident data. Once it is accepted that subjectivity, the genesis of which is being pursued, finds its roots in a dimension of the subject preceding any of its objectivations, semeiology is deprived of this primary situation, the origin and matrix of explicit emotions, and it even refuses to denominate it. This attitude exposes the persistence of the cognitive paradigm which excludes the emotional from the methodology of the approach to emotion. That prevents us from catching the peculiarity of the events of life, that peculiarity which suggests the use of empathy as a key instrument in research.[24]

The fact that emotion strictly concerns life turns out to be a feature susceptible of double interpretation. On the one hand, it means that emotion deals with an unintelligible interaction, both internal and external to the individual; on the other hand, emotion is considered as the starting point of the intentional dynamic to which the primary conditions of existence are to be traced back, such as the "self", expectations, reasons, practical deliberations and even the epistemological sequence. An example of the doubleness of the vitalistic meaning of emotion is also seen in the double evaluation given a determined feature of emotions, namely, changeability, which on the one hand is criticized in that it prevents us from dealing with the emotional phenomenon in terms of universality and foreseeableness and also because of it functions to destabilize every kind of rigidity. On the other hand, how can changeability be reconciled with the tenacity and resistence peculiar to emotion? One merit of semeiotics is its ability to emphasize these paradoxical aspects which philosophy closer to cognitivism has tended to dissimulate.

As the epistemic functionalist approach has begun to consider those perspectives concentrating on research into meaning, such as semiotics, all the features of emotion are actually gaining more and more relevant positive value. Thus, the habit of considering emotion as a paradoxical phenomenon which can not be brought back to standards of formal coherence is changing. As the paradoxical phenomenological features characterizing all emotional expressions have come to be accepted, it is becoming possible to overcome the dichotomy between the more rational emotions (thus medial and moderate and approachable by philosophical research) and the extreme and irrational emotions (which are, consequently, impossible to examine). This dichotomy is mirrored in the distinction between emotions and passions made according to the fleeting or sustained endurance of affection. Inserting emotion as a complex and indefinable phenomenon into "serious" philosophical approaches means taking all responsibilities on, the foreseen responsibilities of facing the difficulties of contradictory and paradoxical factors. Semiotics, in its effort to bring emotions back to methodologically oriented research, through applying disassembling/reassembling models, constitutes a formal and remarkably interesting "tendency" (so because it lacks a specific orientation), since it usually occurs with formalization that contradictory elements are excluded – though expressed in the shape of antinomies and paradoxes – from the "argumentative strings". In other contexts those features considered irrational are usually seen as if they were actually expected event. But it is not easy to take such factors into consideration, above all, if one does not simply seek to understand what is meaningful from what is not. A study of irrationality, if by it we do not mean a logical approach to a formal coherence, implies a negotiation able to put instruments and methods up for grabs. At this point the limitations of the semiotic approach to emotion are evident; this approach apparently tends to withdraw, almost in a prejudicial way, from interaction with disciplines, such as psychoanalysis, that use emotivity as a cognitive instrument and empathy as a fundamental key for interpretation. The emotional dimension represents precisely what the semeioticians want to exclude from their study of passion, thereby insisting on their structuralistic matrix, the symbol of the detached position they still have not abandoned. As far as empathy itself is concerned, semeiology examines its meanings within studies which represent milestones of research on the passions. Furthermore semeiology, as happens also in Kant's works, binds the whole dimension of pathos to empathy, recognizing in it the origin of the universe of passions. But this recognition occurs in an exclusively descriptive context, based on premises which obstruct any access to the field where the different affective modalities originate. Thus the *Einfühlung* of the passions is excluded from the framework of the theory of the passions not only because the word *ab quo*, the existential feature, though supposed, cannot be reconstructed, and is therefore doomed to remain unutterable, but because this unutterableness is not recognized as a significant sign of a primary condition of existence. Thus, emotion, though not recognized as an expression of an existential modality exclusively linked to pathos, precedes any predication, and it represents the origin of that empathic *Einfühlung* from which any other affect arises.

As is well known, emotional involvement indeed assumes an essential function not only in the psychoanalysis, but in all therapeutic work at a heuristic and cognitive level; above all empathy represents the best key to getting closer to the deepest spheres of emotivity. Following this perspective, making use of the empathic approach together with its conceptual criteria and instruments, has enabled some psychoanalysts to focus on emotional phenomena – first of all by considering its paradoxical aspects without interpreting it at a narrative level, *tout court*. This perspective also enables us to elaborate some interpretative models of emotion, and to bring its multiple manifestations back to the sphere of formally structured theory. The semiotic and psychoanalytic perspectives find a common point in their making a theme of the cognitive and pragmatic value of the emotions, the constitutive paradoxicality of which they both recognize.

University of Rome
Tor Vergata

NOTES

[1] The problem in hand in this part of our work is the subject of a monographic study, therefore notes and bibliographic references are limited to the essential. From the time of the James's question "What Is an Emotion?" (W. James, *Mind*, no. 9 (1884), pp. 188–205) emotion has been studied from many points of view. For an interesting philosophical approach see: *Explaining Emotions*, ed. A. O. Rorty (Berkeley: University of California Press, 1980); D. Parfit, *Reasons and Persons* (Oxford-New York: Oxford University Press, 1984); A. O. Rorty, "From Passions to Emotions and Sentiments," *Philosophy*, no. 57 (1982), pp. 159–172; S. P. Greenspan, *Emotions and Reasons. An Inquiry into Emotional Justification* (New York-Long: Routledge, 1988); R. Wollheim, *The Mind and Its Depths* (Cambridge, Massachusetts/London: Harvard University Press, 1993).

[2] See: *Emotions: Theory, Research and Experience*, ed. R. Plutchik, H. Kellerman (New York: Academic Press, 1980); *Theories of Emotion*, ed. R. Plutchik, H. Kellerman (New York: Academic Press, 1980); W. Lyons, *Emotion* (Cambridge: Cambridge University Press, 1980); *Emotion: Philosophical Studies*, ed. K. S. Irani and G. Myers (New York: Haven, 1983); *Approaches to Emotion*, ed. K. R. Sherer and P. Ekman (Hillsdale: L. Erlbaum Associates, 1984); *Experiencing Emotions: A cross-cultural Study*, ed. K. R. Scherer, H. G. Walbott and A. B. Summerfield (Cambridge: Cambridge University Press, 1986); *The Social Construction of Emotions*, ed. R. Harré (Oxford: Basil Blackwell, 1986), Italian transl. *La costruzione sociale delle emozioni* (Milan: Giuffré, 1992); *Psicologia delle emozioni*, ed. V. D'Urso and R. Trentin (Bologna: Il Mulino, 1988); J. D. Vicent, *Biologia delle passioni*, Italian transl. (Turin: Einaudi, 1988); H. Flam, "Emotional Man (The emotional "man" and the problem of collective action and Corporate actors as emotion motivated emotion mangers)" *International Sociology*, Vol. 5, no. 1 (1990); no. 2 (1990); Italian transl. *L'"Uomo" emozionale* (Milan: Anabasi, 1995); *Emozione e conoscenza. Prospettive filosofiche, psicologiche e cliniche*, ed. T. Magri, F. Mancini (Rome: Editori Riuniti, 1991).

[3] A. Masullo, *Filosofie del soggetto e diritto del senso* (Genoa: Marietti, 1990), p. 190ff.

[4] See *Philosophy of Mind*, ed. S. N. Hampshire (New York: Harper and Row, 1966); M. Di Francesco, *Introduzione alla filosofia della mente* (Florence: La Nuova Italia, 1996).

[5] J.P. Sartre, *Esquisse d'une théorie de émotions* (1938) (Paris: Hermann, 1975).

[6] Wittgenstein, for example, took the same position.

[7] I. Kant, *Anthropologie in pragmatischer Hinsicht*; Italian transl. *Anropologia pragmatica* (Rome Bari: Bibioteca Universale Laterza, 1985), parte II.

[8] B. Grunberger, *Il narcisismo* (Italian transl. Rome-Bari: Laterza, 1977); A. Green, *Narcisismo di vita, narcisismo di morte* (Italian transl. Rome: Borla, 1992), reprinted; H. Kohut, *Narcisismo e analisi del sé* *(Turin: Boringhieri, 1977); Idem, *La guarigione del Sé* *(Turin: Boringhieri, 1980, 1987); *La ricerca del Sé*, *(Turin: Bollati-Boringhieri, 1982); A. Appeau, *Développement du concept d'organisation narcissique, poits de vue topique, économique et épigénétique* (Lyon: Césura, 1990); *Modelli della mente* ed. A. Rothstein, *(Turin: Bollati Boringhieri, 1990); S. Tisseron, *La Honte. Psychanalyse d'um lien social* (Paris: Dunod, 1992, especially the part "Honte et narcissisme", pp. 13–14; R. Guy-Gillet, *La blessure de Narcisse* (Paris: Albin Michel, 1994), Chap. 1–2, pp. 11–60.

See also, along different lines, books well known outside psychology and psychoanalysis: C. Lasch, *La cultura del narcisismo. L'individuo in fuga dal sociale in un'età di disillusioni collettive* *(Milan: Bompiani, 1981); A. Lowen, *Il narcisismo. L'identità rinnegata* *(Milan: Feltrinelli, 1985); P. Zweig, *L'eresia dell'amor di sé. Storia dell'individualismo sovversivo nella cultura occidentale* *(Milan: Feltrinelli, 1984).

[9] Cfr., M. Durst, *Il filo di Arianna. Bene e giusto nel contesto della filosofia pratica* in *Etica e educazione*, ed. M. Durst (Florence: Le Monnier, 1993), (for a summary of contemporary moral philosophy's lines).

[10] R. Bodei, *Geometria delle passioni. Paura, speranza, felicità: filosofia e uso politico* (Milan: Feltrinelli, 1992), reprint.

[11] A. Smith, *Theory of Moral Sentiments* (London: 1759), Italian transl., *Teoria dei sentimenti morali* (Rome: IEI, 1992).

[12] M. Foucault, *Tecnologie del Sé. Un seminario con Michel Foucault*, ed. H. L. Martin, H. Gutman and P. H. Hutton, Italian transl. (Turin: Bollati Boringhieri, 1992).

[13] See the studies on Aristotle's moral philosophy: W. Fortenbaugh, *Aristotle on emotion* (London: Duckworth, 1975); *Essays on Aristotle's Ethics*, ed. A. O. Rorty, Berkeley: University of California Press, 1980). See also P. Foot, *Virtue and Vices* (Oxford: Blackwell, 1978); A. MacIntyre, *After Virtue, *Dopo la virtù Saggio di teoria morale* (Milan: Massimo, 1993); H. Jones, *Das Prinzip Verantwortung. Versuch einer Ethik fur die technologische Zivilisation* (Frankfurt a.M.: Insel Verlag, 1980), **Il principio responsabilità* (Turin: Einaudi, 1990); R. M. Hare, *Moral, Thinking: Its Levels, Methods and Points* (Oxford: Oxford University Press, 1981); Idem, "Come decidere razionalmente le questioni morali," Italian transl., in *Etica e diritto*, ed. E. Lecaldano and L. Gianformaggio (Rome-Bari: Laterza, 1986), pp. 45–58; J. Habermas, *Teoria dell'agire comunicativo* *, 2 Vols. (Bologna: Il Mulino, 1986); K. O. Apel, "Il problema della fondazione di un'etica della responsabilità nell'epoca della scienza," * in *Tradizione e attualità della filosofia pratica*, ed. E. Berti (Genoa: 1988); E. Levinas and A. Peperzak, *Etica come filosofia prima*, * (Milan: Guerini e Associati, 1989); P. Ricoeur, *Soi-meme comme un autre* (Paris: Seuil, 1990), Italian transl., *Sé come un altro* (Milan: Jaca Book, 1993).

[14] This trend can be found in all the lines of contemporary moral philosophy but principally in the Anglo-American tradition: T. Nagel, *The Possibility of Altruism* (Oxford: Clarendon Press, 1970); Idem, *Mortal Questions, * Questioni mortali* (Milan: Il Saggiatore, 1986); R. C. Solomon, *The Passions: The Myth and Nature of Human Emotion* (New York: Doubleday, 1976); B. Winters, "Hume's Argument for the Superiority of Natural Instinct," *Dialogue* n. 20 (1981), pp. 635–643; W. Neblett, *The Role of Feeling in Morals* (Washington, D.C.: University Press of America, 1981); J. Rawls, "Justice as Fairness: Political Non Metaphysical," *Philosophy and Public Affairs* 14 no.3 (1985); B. Williams, *Ethics and the Limits of Philosophy* (London: Collins, 1985); Idem., *Shame and Necessity* (Berkeley-Los Angeles-Oxford: University of California Press, 1993); D. Gauthier, *Morals by Agreement* (Oxford: Clarendon Press, 1986); C.E. Larmore, *Le strutture della complessità morale* *(Milan: Feltrinelli, 1990).

[15] S. P. Greenspan, *Emotions and Reasons. An Inquiry into Emotional Justification* (New York-London: Routledge, 1988); M. Miceli and C. Castelfranchi, *La cognizione del valore* (Milan: Franco Angeli, 1992).

[16] *Psicologia delle emozioni*, ed. V. D'Urso and R. Trentin (Bologna: Il Mulino, 1988); *Consciousness in Contemporary Science*, ed. A. J. Marcel and E. Bisiach (Oxford:

Clarendon Press, 1988); R. De Sousa, *The Rationality of Emotion* (Cambridge, Massachusetts/London: The MIT Press, 1990); *Emozione e conoscenza. Prospettive filosofiche, psicologiche e cliniche*, ed. T. Magri and F. Mancini (Rome: Editori Riuniti, 1991).

17 P. F. Strawson, "Intention and Convention in Speech Acts," in *The Philosophy of Language*, ed. J. R. Searle (Oxford: Oxford University Press, 1971), pp. 23–38. See also M. Mizzau, *Eco e Narciso. Parole e silenzi nel conflitto uomo-donna* (Turin: Boringhieri, 1988).

18 *The Unconscious Reconsidered*, ed. K. S. Bower and D. Meichenbaum (New York: Wiley, 1984); M. Conte and A. Gennaro, *Inconscio e processi cognitivi* (Bologna: Il Mulino, 1989).

19 See: K. M. Sayre, "Intentionality and Information Processing: An Alternative Model for cognitive Science", *The Behavioral and Brain Sciences* 1986, no. 9, pp. 121–138; P. N. Johnson-Laird, *Mental Models. Towards a Cognitive Science of Language, Inference and Consciousness* (Cambridge: Cambridge University Press, 1983), **Modelli Mentali* (Bologna: Il Mulino, 1988); N. H. Frijda, *The Emotions (*Cambridge: Cambridge University Press, 1986), **Emozioni* (Bologna: Il Mulino, 1990); K. Oatley and P. N. Johnson-Laird, "Towards of a Cognitive Theory of Emotion," *Cognition and Emotion*, no. 1, 1987, pp. 29–50.

20 We are referring to the positions of R. Wollheim, A. O. Rorty, and S. P. Greenspan.

21 D. Davidson, "Paradoxes of Irrationality", in *Philosophical Essays on Freud*, ed. R. Wollheim and J. Hopkins (Cambridge: Cambridge University Press, 1982); A. O. Rorty, "Akratic Believers," *American Philosophical Quarterly* 1983, no. 20, pp. 175–184; D. Pears, *Motivated Irrationality* (Oxford: Oxford University Press, 1984); *The Multiple Self*, ed. J. Elster (Cambridge: Cambridge University Press, 1986); Italian transl.*L'Io multiplo* (Milan: Feltrinelli, 1991); M. Black, *Perplexities, Rational Choise, the Prisoner Dilemma, Metaphor, Poetic Ambiguity and Other Puzzle* (Ithaca and London: Cornell University Press, 1990).

22 H. Parret, *Les passions. Essai sur la mise en discours de la subjectivité* (Brusels: Mardaga, 1993).

23 R. Descartes, *Les passions de l'ame*, Italian transl., *Le passioni dell'anima* in *Opere filosofiche*, Biblioteca Universale (Rome-Bari: Laterza, 1986), Vol. IV, Part I, Article I.

24 For bibliographies, see: M. Durst, "Empatia, socializzazione e sentimenti morali", *Scuola Democratica*, October–December 1995, no. 4, pp. 109–116; N. Lalli and N. Giacchetti, "Collaborazione e resistenza nella relazione terapeutica", *Attualità in Psicologia* X: 4 (1995), pp. 35–42.

KONRAD ROKSTAD

ON THE HISTORICITY OF UNDERSTANDING
A Phenomenological Interpretation of a Text by Husserl

In "*The Origin of Geometry*", one of the appendices to the *Crisis*[1] – the work in which Husserl gave his final interpretation of *transcendental phenomenology*, he characterises *history* in the following manner: ". . . history is from the start nothing other than the vital movement of the coexistence and the interweaving of original formations and sedimentations of meaning". He also speaks of "the universal a priori of history with all its highly abundant component elements" – which, he says, would result from such investigations, as they are "carried out systematically", with "such self-evidences", that he is laying a grounding for in that appendix.[2]

1. INTRODUCTORY REMARK

In this paper it is my intention to give an interpretation of this text on the "origin" of geometry while at the same time trying to explicate a phenomenological meaning speaking to the "*historicity of understanding*". It might seem redundant to explain why I bring the concept of understanding into a philosophical discourse – it is, of course, essential both in philosophy, science and human life quite generally. Yet, in Husserl's phenomenology "understanding" doesn't seem to have any prominent philosophical position (and Husserl might have had his reasons[3] for that!). Nevertheless, the concept obviously has to be of quite essential importance to any interpreter of his phenomenology. The problem, however, will be giving this "understanding" a substantial *phenomenological* meaning. The reason, then, for bringing it into our discourse interpreting this text[4] is at least a double one: a) The text itself might yield very relevant *material* for giving sense to such a concept – as it quite concretely demonstrates how "historicity" enters into our cognitive activity, and b) in our interpretational work, then, substantially we have got text-material from Husserl himself, that might *activate our* concrete understanding, in that way *phenomenologically* "giving" what might be meant by the "historicity of understanding". As this also might broaden the understanding of our own activity as interpreting

A-T. Tymieniecka (ed.), Analecta Husserliana LIX, 401–422.

Konrad Rokstad, Louis Houthakker, Martin Nkafu Nkemia and Velga Vevere.

philosophers in that *life-world* as being essential in (at least) the later philosophy of Husserl, who himself really struggled to present his transcendental phenomenology in a "final" way, this conception of the "historicity of understanding" might emerge as a key-clue to his *whole* philosophy.

These lines of thought represent, however, my general *motivational* horizon, and I can, of course, not go into this whole field of very extensive, substantial problems. In this paper I will not be interpreting the total corpus of Husserls philosophy, nor even the *Crisis* as such – at least not as my primary textual material. I'll stick to that appendix, however, being well aware of the total horizon of questions and problems that might be implied therein. Then, not particularly presupposing any advanced methodological procedure (such as "reduction"), I will let that relatively small piece of text "speak for itself" – leading us onto those "self-evidences" it might contain. But, of course, some matters have to be straightened out already at the outset. For instance, this is a text on the "origin" of the science called geometry, it is written by Husserl, it is related to the Crisis-complex etc., and our subject of investigation is what is called the "historicity of understanding". We might also add that the statement on history we quoted at the outset appears more towards the end of the text, after the main analysis on the origin of geometry has been carried out. Let us now look then into the beginning of the appendix.

2. TRADITION AND THE PROBLEM OF GEOMETRY AS A "FIRST" REALISING THE HISTORICITY OF UNDERSTANDING

At the outset Husserl is obviously relating what he is about to go into, to the main text[5] of the *Crisis* – to his analysis of Galilei, as he, the "entdeckender und verdeckender Genius"[6] of the Renaissance so to speak, is transforming history, bringing forward elements from the ancient tradition, transforming them into a new praxis, in that way laying the foundation for the modern age. Husserl then underscores that his analysis would have seemed strange to people like Galilei, and we might first simply ask: why? In the analysis of the main text he, of course, stresses the importance of the *life-world* as a more or less "hidden" precondition for Galilei's establishing a new praxis, and what Husserl most strongly underscores in the appendix, is that geometry is a *tradition*: "The geometry which is readymade, so to speak, from which the regressive inquiry begins, is a tradition". But in declaring that he also right away says something about tradition and human existence quite generally: "Our human existence moves within innumerable traditions. The whole cultural world, in all its forms exists through tradition".[7] Thematising that "aspect" of tradition (not really an aspect here at all, but rather a totality), would probably not only have seemed astonishing to Galilei, he might even have rejected its whole relevance. But what is then Husserls point? He is talking about all these "forms" of the whole cultural world as they belong to tradition, and stresses that they have not arisen "merely causally; we also know already that tradition is precisely tradition, having arisen within our human space through human activity, i.e., spiritually, even though we generally know nothing, or as good as nothing, of the particular provenance and the spiritual source that brought it about". This "not-knowing" of the historical particularities is, however, not without significance to Husserl. Quite on the contrary he says: "And yet there lies in this lack of knowledge, everywhere and essentially, an implicit knowledge, which can thus be made explicit, a knowledge of unassailable self-evidence".[8] Thus bringing the problem of geometry into this conception of tradition, Husserls first point, then, might simply have been to make the reader aware of this "lack of knowledge" – not really getting into it, but in that way setting the *horizon* for explicating that implicit knowledge with its possible self-evidences now explicitly related to potentialities for "evidence" that the historically constituted science of geometry lays open.

Reminding ourselves of what was said about history in our opening, already at this point, as Husserl is mentioning that "unassailable self-evidence", we might be looking in direction of that universal a priori of history, now then more specifically related to that "implicit knowledge" to be found everywhere, essentially related to our "not-knowing". That self-evidence – this way related to tradition and the *whole* cultural world – might give the clue, but, of course, all this still remains in a way a matter of "pure" generalities. For Husserl also makes this further point: "It begins with superficial commonplaces, such as: that everything traditional has arisen out of human activity, that accordingly past men and human civilizations existed, and among them their first inventors, who shaped the new out of materials at hand, raw or already spiritually shaped". Then at least we have got some of the generalities specified, and we might, of course, wonder – from where, out of which sources, do such "specifications" occur. Quite obviously – and now we need neither the *specific* philosophical nor the scientific tradition for an answer – they might well emerge from our everyday, common experience and understanding in life, as that is situated in our historical life-world. This, then, also indicates a deep-seated, rather obvious connection between our actual ordinary life and tradition, and maybe history too, that in a way, however, only quite generally realised. But then Husserl might continue his exposition: "From the superficial, however, one is led into the depths. Tradition is open in this general way to continued inquiry; and, if one consistently maintains the direction of inquiry, an infinity of questions opens up, questions which lead to definite answers in accord with their sense". Husserl then stresses the generality of what appears this way, and points out the applicability to "individually determined particular cases", so that they can be grasped – though, as he also underlines, "only that in the individual that can be grasped through subsumptions".[9]

Then looking into what he has been putting forward, what have we got other than a "scheme", or perhaps we should rather say "horizon", for interpreting history? But we have got it precisely through that dimension thetradition lies open as it primarily functions, quite generally constituted in the horizon of common, everyday experience and understanding. The particular cases (in history) might then, of course, be grasped – although not without a certain restriction, set by the form of the grasping as "subsumption"; the reason why should be obvious, since you can't bring "back" to life the individual particular cases of history, else it would not have been history. Nevertheless we obviously are living with history "present", and then again that dimension of tradition, extending into the whole cultural world, constitutes the "link". In this respect, "ready-made" geometry constitutes no exception, and this then presents a "new" dimension[10] for understanding both geometry and the concept of tradition.

It might be overinterpreting Husserl a bit, but what he really seems to be saying is that understanding geometry in *this* context of tradition, gives us the concrete *possibility* of "realising" history – in its spiritually vivid "moments" correlating understanding with *evidences*, which makes certain elements *concretely alive*. As our actual understanding of geometry functions, it vitalises spiritually those acquisitions that are ready-made in the past and also constitutes preconditions for new ones. That whole situation – including, of course, that general, life-worldly horizon tradition emerges through – taken into consideration, might then be explicated in Husserl's fashion: "Clearly, then, geometry must have arisen out of a *first* acquisition, out of first creative activities. . . . it is not only a mobile forward process . . . but a continuous synthesis in which all acquisitions maintain their validity, all make up a totality such that, at every present stage, the total acquisition is, so to speak, the total premise for the acquisition of the new level. . . . The same thing is true for every science".[11]

3. REFLECTING ON THE PROBLEM OF THE BEGINNING – DEEPENING THE UNDERSTANDING IN HISTORICITY

All that has been put forward, is, however, very "loaded" with meaning-implications, questions and problems, and thematising that would really bring us into the "depths", and we shall follow up the main lines in Husserl's exposition of these. But first we must remind ourselves of that "link" now established between history and geometry, giving something "essential" both ways, putting geometry within a horizon of understanding constituted through historicity, and putting the essentials of geometrical cognition within understanding history; then also history itself might be understandable through this phenomenologically explicated conception of historicity. This will, however, have to be a story including both "the obvious commonplaces" of life and "the depths", and what Husserl now starts out with is precisely a deeper "digging" into the problem of the *beginning* – as it quite naturally occurs in the horizon constituted through this analysis of tradition.

How, then, does this situation of problems and potentialities give the clue and conduct our first realising of the core-problem to be a genuine problem[12] of (and in) the beginning? If geometry, in its ready-made beginning was successfully realised with its "first" self-evidence, and it did not emerge out of the heavenly spheres, put into human cognition and existence by the gods, it must have emerged out of more "primitive" formations of culture, as they were present in human experience and praxis at that time – laying the soil open for the growth of such self-evidences that constitute geometrical understanding. This seems in its own manner "self-evident" to the "rational" mind not relying on the mercy of the gods to found our basic human knowledge. Nevertheless, a very perplexing problem might appear at this stage, as Husserl points out: "Self-evidence means nothing more than grasping an entity with the consciousness of its original being-itself-there. Successful realization of a project is, for the acting subject, self-evidence; in this self-evidence, what has been realized is there, *originaliter*, as itself".[13] What is it then, the problem that might be indicated in saying this?

First we have to realise that any successfully realised geometrical formation, necessarily involves that self-evidence constituting the formation, but as this evidence will precisely be the grasping of an entity with the consciousness of its original being-itself-there . . . , it seems as though one (already) has to have got the entity as a precondition for exactly this successfully being able to fulfill that self-evidence (then) "producing" the entity. If one then thinks about this puzzle within a rationality constituted within the horizon of history as a "one-way-advancing" process, then one has got a really perplexing problem, because the obvious "precondition" of this evidential "productivity" seems to conflict with both the purpose and the result of that same productivity. In the text, however, Husserl really does not stress and go into this problem;[14] what then is it that he is doing? First he underscores that ". . . now questions arise", and then, that what takes place in the beginning, occurs "purely within the subjectivity of the inventor . . . within his mental space. But geometrical existence is not psychic existence. . . ."[15] Confronting the problem that way, he is really not confronting it , but laying it open so that it might be brought into the total class of the spiritual products of the cultural world as it relates to the subjective, so that it is the horizon of historicity, and not that of history in the "objective" sense that now opens itself as the concrete field[16] for inquiry into the question of the "first" objectivity of geometry. As it was the concept of tradition that led into the problem, now it is "commonplaces" related to *language* that will lay a grounding for getting a better hold of it. But what then does this mean?

4. LANGUAGE, WORLD AND CONSCIOUSNESS AS THE FIELD OF HISTORICITY FOR THE EVIDENTIAL CONSTITUTING OF THE COGNITION AND UNDERSTANDING

According to Husserl, what he is speaking of as "'Ideal' objectivity", makes up "a whole class of spiritual products of the cultural world"; not only science then with such entities as the Pythagorean theorem, but also fine literature, . . . exists only once. . . ."[17] This class has its various expressions, oral utterances or written and other documentation, and these sensible utterances have, of course, spatiotemporal individuation in the world; this, however, is not true of the spiritual form itself: ". . . the idealities of geometrical words, sentences, theories – considered purely as linguistic structures – are not the idealities that make up what is expressed and brought to validity as truth in geometry; the latter are ideal geometrical objects, states of affairs, etc.".[18] Why then thematise this linguistic aspect related to the spatiotemporal? It is simply because, as Husserl puts it: "Our problem now concerns precisely the ideal objects which are thematic in geometry: how does geometrical ideality (just like that of all sciences) proceed from its primary intrapersonal origin, where it is a structure within the conscious space of the first inventor's soul, to its ideal objectivity? In advance we see that it occurs by means of language, through which it receives, so to speak, its linguistic living body (Sprachleib)".[19]

What we then might realise is precisely *how* Husserl so to speak transforms that problem of the "first" conceiving of geometry's ideality into this field of historicity wherein language has its quite obvious functions – functioning in a manner pretty similar to that dimension of the tradition the analysis started out from. This whole "new" horizon however, does not stress the distinction between the "first" in the historical sense and that horizon language actually constitutes for expressing what we might be cognising even as the historical, and this is done without presupposing the objectivity of the ideality in question. On the contrary, what now has to be done is the explication of something that might seem rather obvious about "the relation between language, as a function of man within human civilization, and the world as the horizon of human existence".[20] This analysis of that relation goes into the "real" depths of historicity constituting the preconditions of intersubjectivity, and the "result" is, then, something we, probably every speaking person, have some sort of knowledge about – that "Everything has its name, or is nameable in the broadest sense, i.e., linguistically expressible". But Husserl also adds right away the following: "The objective world is form the start the world for all, the world which "everyone" has as world-horizon. Its objective being presupposes men, understood as men with a common language".[21] Then having that "presupposition" stated and explicated with this rather common "self-evidence", Husserl returns to the "first" primally establishing geometer and says he can obviously, also express his internal structure. However, the real difficult problem still remains, for "How does the latter, in its 'ideality', thereby become objective? . . . how does the intrapsychically constituted structure arrive at an intersubjective being of its own as an ideal object which, as 'geometrical', is anything but a real psychic object, even though it has arisen psychically? Let us reflect".[22]

At this point the reflected analysis still is using experiences, going toward the common horizon most people relate to in their daily life, but it is at the same time also taking yet another step into the depths of our cognition. Now it is not primarily language, but the functions of our mind, such as memory, that "oscillate" into the depths of the problem Husserl is facing. These functions are, of course, interwoven into the field of language, and Husserl is still relating the problem to that "first" geometer, but really describing quite generally how "something" being there – even "given" as the "original being-itself-there, in the immediacy of its first production, i.e., in original 'self-evidence', results in no persisting acquisition at all that could have objective existence".[23] It passes, disappears and can nevertheless be reawakened. As to memory quite generally, this is something we all might experience, but it also probably has decisive functions grounding the possibility for having history altogether. What Husserl first and foremost is focusing on is, however, the possibility this gives for "self-evidences" now correlated to new activity that might occur founded in the common phenomena of our consciousness called "recollection":[24] "Now if the originally self-evident production, as a pure fulfilment of its intention, is what is renewed (recollected), there necessarily occurs, accompanying the active recollection of what is past, an activity of concurrent actual production, and there arises thereby, the original 'coincidence', the self-evidence of identity: what has now been realized in original fashion is the same as what was previously self-evident. Also coestablished is the capacity for repetition at will with the self-evidence of the identity (coincidence of identity) of the structure throughout the chain of repetitions". With memory, thus, correlating that *possible* objectivity with the *both* constituted and possibly constitut*ing* "ideality", then, the crucial point will, of course, be that of the *identity*; but we have to keep in mind that description is now related to the "subjective' – and also includes that capacity for *repetition* "at will". This is essential for, as Husserl then says, ". . . we still have no 'objectivity' given", and in his special manner of regressive inquiry, he again thematises that aspect of historicity that he just has been investigating, saying: "It does arise, however – in a preliminary stage – in understandable fashion as soon as we take into consideration the function of empathy and of language. In the contact of reciprocal linguistic understanding, the original production and the product of one subject can be *actively* understood by the other".[25]

What now follows, brings, in a way, the analysis to an end – indicating how that "first" entity produced in self-evidence might transcend subjectivity and enter into an inter-subjectivity constituted in communication, in that way being intersubjectivly realised as the *one structure* common to all. As Husserl puts it: "In the unity of the community of communication among several persons the repeatedly produced structure becomes an object of consciousness, not as a likeness, but as the one structure common to all".[26] And this is, of course, the essential "result" of the constitutional analysis of the subject of objectivity – as it has been constituted through those "evidences" produced by our understanding in this total field of historicity.

5. THE "NATURAL" FIELD FOR THE CONCEPT OF "UNDERSTANDING"; THE "EXAMPLE" OF GEOMETRY AND CRITICISM THROUGH REFLECTED HISTORICITY

There are, however, severe problems still remaining, and this is by no means unexpected in the perspective of historicity now being exposed; what has been opened and put forward all make it quite natural that questions related to such "trivialities" of human life as sleep(!) emerge into the whole picture, because this fundamentally subjective grounding poses the obvious question of *persistence*: "What is lacking", says Husserl, "is the *persisting existence* of the 'ideal objects' even during periods in which the inventor and his fellows are no longer wakefully so related or even are no longer alive. (. . . their continuing-to-be even when no one has (consciously) realized them in self-evidence)".[27] Then the whole analysis, in a way, has to start from a new beginning, explicitly bringing into the picture what has only implicitly been functioning – and, of course, is a quite common phenomena in our historical culture and life-surroundings, constituting an essentially important "space" for the human activity of cognition and understandings, that is, *written*, documenting linguistic expression. As Husserl put it,

> The important function of written, documenting linguistic expression is that it makes communications possible without immediate or mediate personal address; . . . Written signs are, . . . straightforwardly, sensibly experienceable; and it is always possible that they be inter-subjectively experienceable in common. But as lingusitic signs they awaken, . . . their familiar significations. The awakening is something passive; . . . what is

> passively awakened can be transformed back, so to speak, into the corresponding activity: this is the capacity for reactivating that belongs originally to every human being as a speaking being. Accordingly, then, the writing-down effects a transformation of the original mode of being of the meaning-structure, . . . It becomes sedimented, so to speak. But the reader can make it self-evident again, can reactivate the self-evidence.[28]

This "new" situation that has been historically situated even before the "first" geometer, then lifts the communalization of man to "a new level", now making it understandable how those entities produced originally in the self-evidence of subjectivity, quite obviously might obtain their persisting existence. And solving that problem might also be regarded as the solving of the problem that might seem most perplexing in such a fundamentally "subjective", that is 'intentional-historical', strategy as Husserl has chosen. But then, of course, all these problems commonly related to the concept of the understanding come onto the field, as they obviously have been there all the time, ever since that "new level", in the very far past, was brought into historical existence. Making this level explicit, we are thus facing that "space" where the problem of understanding normally (as it, as tradition tells us, has been a long-reflected problem, especially in the Humanities) will be situated.

In the text we are interpreting, Husserl then goes relatively substantially into those problems that relate to understanding the written, documenting expressions of language, but his whole analysis is founded in those analyses that have already been made, and they are developed so that they continue into those quite general reflections on history we started out from. In a way, therefore, we are standing at the "gate" into that field where the "traditional" problems of understanding are gathered. Our main purpose in this context, however, is, not to make specific investigations into that field;[29] what we are engaged in, is primarily the opening of a phenomenological perspective onto the historicity of understanding. And then that quite concrete "space" that this paper defines also sets strict limitations on what we might be doing. Therefore, as at least some of the most specific phenomenological aspects of the whole problem have been put forward, we now have to go into the rest in a short, very summary manner – thereby also laying the ground for what will be our own concluding remarks.

In the continuation of his text, Husserl points out that there ". . . is a distinction, then, between passively understanding the expression and making it self-evident by reactivating its meaning".[30] This is, of course, a distinction both quite obvious and of essential, far-reaching theoretical importance, and this brings us to realise the highly relevant possibility of "... the *seduction of language*", which represent a "constant danger" – and one which the scientific community especially has to maintain a standing "strategy"[31] to handle. What now will be the essential problem, is the active understanding that "one structure" possibly common to all, as it occurs in that "persisting existence" written language constitutes, has a "sedimented" meaning-structure allowing the *evidential reactivating* of the original self-evidence it might contain. Realising that in general "... scientific thinking attains new results on the basis of those already attained", when it comes to geometry with its "finally immense proliferation", the decisive question must be raised – "... what has become of the claim and the capacity for reactivating?"[32] According to Husserl, the situation[33] is genuinely problematic, and again then, put into the perspective of science generally, you have got "... the peculiar 'logical' activity which is tied specifically to language, as well as to the ideal cognitive structures that arise specifically within it. To any sentence structures that emerge within a merely passive understanding there belongs essentially a peculiar sort of activity best described by the word 'explication'".[34]

Husserl goes into that whole field of active-passive cognitive processes so related to language, and he explicates the meaning-structure of sentences implying a "genesis" in this way: "... sentences give themselves in consciousness as reproductive transformations of an original meaning produced out of an actual, original activity; that is, in themselves they refer to such a genesis".[35] Finally, he goes specifically back to geometry again, explicating what, in this context, is his main concern regarding that science. On the one hand, geometry, so to speak, gives the decisive "proof" of the *possibility* of meaning formation existing as *ideal*[36] meaning formations; this it does without, however, itself realising either the structural field or the power of consciousness constituting such a "proof" correlative to what the concrete field itself lays open historically (so it is not only Galilei who is "naive" in that respect). And, on the other hand, he explicates his concrete criticism specifically against the (actual) practices[37] of that science, as this now has got its phenomenological grounding in that broadly, concretely reflected analysis of what then might be called the "historicity of ideality". Finally, he then, ends up saying that "the whole pregiven deductive sciences ... is first only a claim which can be justified ... only through the actual capacity for reactivation", and so he asks if "so-called 'epistemological grounding'"[38] might constitute an adequate alternative, but all this is rejected. The only possible way seems to be through the application of that *actual* capacity for *reactivation*, that, of course, is "specific" to those sciences as demonstrated in the above genetic analysis of the historicity of (geometry's) ideality, but that at the same time implies *genuine understanding* of the *total complex of history* the analysis then goes into. There we also will be facing a profound criticism on the "ordinary" way of understanding history,[39] then being intimately integrated into the criticism of geometry. What is missing, is that genuine perspective of the profound historicity constituting the precondition for the adequate understanding of them both. And that again situates the whole problem into the total complex of the historical life-world and the universal a priori of history.

This is where our analysis started out from, and in a way we have therefore completed the "circle" of our reflected interpretation, trying to reflect the concept of 'understanding' more explicitly in this very complex field of the reflected historicity the text lays open to us.

6. CONCLUDING REMARKS – TRYING TO "REALISE" THE "HISTORICITY OF UNDERSTANDING"

Bringing the concept of understanding into an interpretation of this text, might, of course, be "understandable" in quite a trivial way. My reflecting reason, however, focuses primarily on the texts strongly emphasising the capacity of *reactivating* what one is conceiving along with the evidence that might be contained in it. In the text this is related specifically to geometry, but one of the main points even then, is that this implies the prescientific; as Husserl puts it: ". . . without the 'what' and the 'how' of its prescientific materials, geometry would be a tradition empty of meaning".[40] Then you have that overall perspective of history and tradition explicating the whole structural field emerging with its evidences into this "middle-between" history and geometry; Husserl's "real" problem, then, seems to be constituting the preconditions for realising (the possibility of) ideality quite generally, in that way really setting the whole of reflected historicity at the heart of his own transcendental phenomenology. To us this might give interesting "material clues" for finding our way into that phenomenology, as a similar capacity for reactivating "with evidence" will be as essential as it is difficult in facing the challenges of understanding[41] that philosophy.

All taken together, grounding the concept of 'understanding' upon that evidential capacity for reactivation, thus realising its "core", situated in the historical environment that gives material for such evidences, understanding might itself become a *genuinely constituting* capacity, as it already is highly communicable in the ongoing total historical process that both the sciences and life itself constitute together. This might even fit into that "vital movement of the coexistence and interweaving of original formations and sedimentations of meaning"[42] – as this constitutes Husserl's general conception of history, the one cited at the start of our interpretation. Let us then, finally, take a look into that – trying to keep in mind the horizon of historicity we have been explicating.

Speaking of "original formations and sedimentation of meaning", "coexisting and interweaving", this takes place *only* in correlation with and through the human capacity for cognition and understanding. This capacity, itself then historically constituted, relates to history, but also to the actual situation in the sciences, both historical and non-historical, and, of course, to ourselves as bodily situated persons, situated in actual and potential communalization, through praxes with an open horizon of a so to speak endless variety of states of affairs. To cut the matter short, all this might be loosely gathered into the conception of our historical *life-world*. Regarding Husserl's statement on history, one of the crucial component elements then will be the "original formation" (of meaning); formulated as it is, that expression is, however, genuinely *ambiguous* in a rather pregnant sense. "Original formation" might, of course, refer to the "first" historical formation in the "traditional" sense, but this is *not* (necessarily) the "first" phenomenological original, the formation genuinely constituted in and by understanding, actually constituting it as the *presenting*-presented "unity" of them both. As the actual understanding and the cognate formation constitute each other, phenomenologically speaking, understanding is something the person, or the cognating subject itself, realises with *actual* "evidence" *presented* – as the matter in question itself gives rise to it – and so we (also) have all the (actual) sedimentations of meaning, both general and personal, more or less vividly *appresenting*, coexisting and interweaving in this whole conception. As history discloses itself this way, the historical "reality", in its vital, pulsating movement, does, of course, always appresent; it is always including more than it actually shows: But this appresentation then (also) has *its presentation* – with its potential seated at the heart of that material-matter (for instance given in the appearance of a text), as it motivates – activating our cognitive capacity *as* reactivation and "vitalisation", understanding "in concretio" something with "real" historical significance.

To our capacity for concrete understanding, this conception of history, then, constitutes a core-problem, including both the question of *actual evidence* in correlation to the reactivating "formatting" constituted in this "giving" *and* the question of this *whole* sedimented "store" of meaning, that through this always pulsating correlation might also bring to the surface a deep-seated *teleology* penetrating this total horizon of meaning, then giving genuine, historically grounded, *philosophical* power to our actual understanding. As this correlates with the actual historical situation – motivating our cognitive capacity in this genuine manner, we will ourselves, as living beings, be part of that great vital movement of meaning that history constitutes in this phenomenological sense of historicity.[43] Then, of course, you might have to differentiate the concept of "evidence" since this meaning-potential, the actual total horizon of life and understanding, gives rise to all the different aspects contained in this whole "vital movement" going into our experiencing and understanding the historical world. But there are strategies, praxes of genuine genetic understanding, that concretely "adapt" to this total field of human life and understanding. This is what we have been asking "back" into – then explicating – trying ourselves to reactivate the meaning-genesis, thus disclosing a "hidden" historicity in what we, probably all of us as reflecting, speaking people, both do and understand in the most "obvious" manner. Doing this, then, we might also have been approaching the concrete meaning of "the universal a priori of history",[44] as this meaning can not but emerge from evidence given in and by that totality.

Those statements on history now going into – and "surrounding" – our conception of the "historicity of understanding", obviously calls for the total horizon of the *Crisis* complex, and maybe for a genetically reflected analysis of the total complex of Husserl's phenomenology. Implicitly, specifically the general horizon of the *Crisis* has given much to such an "idea" (of a project) just being explicated. It might then, of course, have been advantageous to have more explicitly presented that total horizon of the *Crisis*[45] itself, to help in the understanding of all the implications of meaning and preconditioning functions emerging out of this exposition. But then we probably also would find ourselves asking for more from the total complex of phenomenology, from its very beginning, and that has to be another story.

University of Bergen
Norway

NOTES

1 Edmund Husserl, *Die Krisis der Europäischen Wissenschaften und die Transzendentale Phänomenologie* (The Hague, 1962). In this context I will be using the English version: *The Crisis of European Sciences and Transcendental Phenomenology*, trans. D. Carr (Evanston: Northwestern University Press, 1970), cited hereafter as *Crisis*.
2 *Crisis*, p. 371.
3 This has to do with Husserl's relation to the Humanities (Historicism, Dilthey etc.), maybe most clearly stated in "Philosophie als strenge Wissenschaft" (1911), but also substantially discussed in that appendix we will be interpreting – towards the end, and it does, of course, fit into the total conception of the *Crisis*.
4 Of course, this text we are going to interpret, has been interpreted before, and perhaps the most famous interpretation is that given by J. Derrida (in his "Introduction à 'L'Origine de la géometrie' de Husserl," 1962), – on the one hand using the text for an analysis opening the total field of Husserl's phenomenology, and on the other opening a criticism that might be "fatal" to the whole project of the same philosophy. But there are others (D. Cairns, 1940; E. Fink, 1939; G. Brand, 1969) both before and after Derrida, that directly interpret or seem to be using the text for grounding a perspective on the whole of Husserl's phenomenology. We will, however, not be going into a discussion of that "tradition" of interpretation.
5 "The Origin of Geometry" does not belong to the main text, but it seems like it was meant for it – in the second part, which deals with the philosophical-scientific tradition. According to the recently published *Ergänzungsband* (*Husserliana*, bd XXIX, 1993) to the *Crisis* (pp. lvi–lvii), Eugon Fink, Husserl's assistant at that time, suggested that the text should be placed before the actual §9 – "Galileo's Mathematization of Nature," as a new §9 called "Die Rückfrage nach dem Ursprungssinn der Geometrie als historisches Problem." This paragraph should then be divided into the following sub-points: a) die Traditionalität der Geometrie (Tradierung überhaupt); b) die Bedingungen des weltlichen Daseins eines Idealgebildes als die Bedingungen seiner Tradierbarkeit (Identität des Gebildes, Sprache, Schrift); c) Tradition und Reaktivierung der Traditionsstiftung (Bedingungen einer vollständigen Reaktivierung); d) die Rückfrage in den Sinnesursprung als Freilegung einer verborgenen Geschichtlichkeit; e) die apriorischen Voraussetzungen aller Historie. Husserl did not however follow this suggestion, and why we don't know, but the text was published – by Fink – in 1939, under the title "Die Frage nach dem Ursprung der Geometrie als intentionalhistorisches Problem" (in the *Revue Internationale de Philosophie*, Brussels, 1. Jahrgang, No. 2). Then we might, of course, speculate into why all this happened that way. One possible explanation might be that Husserl simply realised that the analysis of that text was too "heavy" burdened with very important problems that in a way would have complicated the understanding of those analyses the

main text actual contains on Galilei; bringing it into the text could then have had effects on its understandability contrary that which was the main purpose of the whole project. We don't insist on this explanation, but find it interesting, not least because one might then construct an interesting analogy between the actual analysis of Galilei and the text we are focusing on: The first might be looked at as an analysis of how the modern conception of nature is constituted through historicity in the grounding of our life-world; the other, then analogically, one of how the core "substances" of understanding, as "idealities" given in the relevant "self-evidences," are constituted through historicity, so to speak constituting a field of the "historicity of understanding" analogous to that of "nature," but then going deeper really into the whole matter. Fink's way of structuring the text might then give decisive clues, but this might also have seemed problematic for Husserl's whole project in his *Crisis*. We don't really make this the "hypothesis" of our interpretation, but find it interesting enough to mention.

6 *Crisis*, p. 52.

7 *Crisis*, p. 354.

8 *Crisis*, pp. 354–355.

9 The three last quotations from *Crisis*, p. 355.

10 As already stated at the outset, and now made understandable on the above consideration, geometry is, thus, a tradition. Husserl also points out that we – as we are understanding "our geometry, available to us through tradition (we have learned it, and so have our teachers)" – know it "to be a total acquisition of spiritual accomplishments which grows through the continued work of new spiritual acts into new acquisitions. We know of its handed-down, earlier forms, as those from which it has arisen; but with every form the reference to an earlier one is repeated" (*Crisis*, p. 355). Then you have got a "new" dimension explicitly brought forward; it is not really separated from the endless variety of particularities in history (that you grasp only through "subsumption") – nevertheless it has another potentiality for presenting the "elements" contained, as that of "total acquisition" in geometry (as we understand it handed-down) constitutes an "internal link" where all the accomplishments are so to speak repeated and repeat*able*, in exactly that spiritual "space" (our capacity for) understanding might be constituting.

11 *Crisis*, p. 355. But what exactly is true of *every* science? Husserl is saying that geometry "necessarily has this mobility and has a horizon of the geometrical future in precisely this style; this is its meaning for every geometer who has the consciousness (the constant implicit knowledge) of existing within a forward development understood as the progress of knowledge being built into the horizon". What then is it that, according to Husserl, seems to be true for every science? He probably does not mean that every science has the specific style, mobility and future as the science of geometry; but they all have a "rational" style, mobility and a horizon of their future – defined by *their* style constituted through their meaning – as there also exist scientists with a consciousness and a constant implicit knowledge of existing within a forward development understood as the progress of knowledge being built into that horizon. This might be true for every science – that they have their horizon of their future, with scientists having both their explicit and implicit knowledge, where the basic meaning is constituted through that horizon as both progress and style, always pulsatingly, through working praxis, melt into the horizon, in that way also always revitalising the basic meaning, that has its "origin" in "history" precisely as it is emerging out of that *historicity* constituted this way. This might even constitute a *universal* meaning going into the horizon of *every science as such.*

And doesn't *this* consideration, then, constitute a *horizon* for *understanding* science *teleologically*, quite generally? Husserl continues his regressive inquiry, by going back into the quite obvious: ". . . every science is related to an open chain of the generations of those who work for and with one another, researchers either known or unknown to one another who are the accomplishing subjectivity of the whole living science. Science, and in particular geometry, with its ontic meaning, must have had a historical beginning; this meaning itself must have an origin in an accomplishment: first as a project and then in successful execution" (*Crisis*, p. 356). Then we have been brought to the beginning – as it might be seen quite "rationally," considered in the perspective of "common," natural understanding – since it also obviously includes the aspect of history and the moment of (a) beginning.

12 As geometry given in its actual appearance, contains a total (to be sure, an uncompleted total) meaning, it seems obvious that this meaning, as Husserl then points out, ". . . could not have been present as a project and then as mobile fulfilment at the beginning. A more primitive formation of meaning necessarily went before . . . that it appeared for the first time in the self-evidence of successful realization. But this way of expressing it is actually overblown," (*Crisis*, p. 356).

13 *Crisis*, p. 356.

14 The problem might well be compared and be considered similar to that problem Roman Ingarden formulated in relation to the *Cartesianische Meditationen*, pp. 205–206, and characterised as "the problem of the beginning".

15 *Crisis*, p. 356.

16 What Husserl then is investigating, whether it is the problem of the ("first") objectivity of geometry *or* that of the historicity of cognition, might then be neither easy nor so important to decide. The main point seems to be that they essentially belong together, and – while geometry poses the problem of both the "first" self-evidence and objectivity upon us – in this phenomenological perspective, they can't be separated. But here we have to add one important precondition; this becomes clear as one realises that the objectivity of geometry really implies and constitutes the problem of ideality quite *generally*, so that what we really are facing is the problem of the historicity of ideality "as such". And, of course, *then* you might realise that the science of geometry constitutes a well-fitted "example" – as it undoubtedly might lead us into asking for evidential grounding, also giving substantial sense to the quest for the beginning, at the same time making it possible to realise the distinction between questing for a historical origin and a genuine origin constituted in this philosophical sense of historicity.

17 *Crisis*, pp. 356–357.

18 *Crisis*, p. 357.

19 *Crisis*, pp. 357–358.

20 *Crisis*, p. 358. What Husserl is actually saying about that relation is, of course, laden with difficult problems, but now he is describing the whole matter as if it is quite obvious, in this manner: ". . . living wakefully in the world we are constantly conscious of the world, whether we pay attention . . . the horizon of our life . . . of things . . . interests and activities. Always standing out against the world-horizon is the horizon of fellow men, whether there are of them present or not. Before even taking notice of it at all we are conscious of the open horizon of our fellow men with its limited nucleus of our neighbours. . . . We are thereby conscious of the men on our external horizon in

each case as "others"; in each case "I" am conscious of them as "my" others, as those with whom I can enter into actual . . . relations of empathy. ". . . It is precisely to this horizon of civilization that common language belongs. One is conscious of civilization from the start as an immediate and mediate linguistic community. . . . far-reaching documentations, as possible communications . . . horizon of civilization . . . open and endless one, . . ."

[21] *Crisis*, p. 359.

[22] *Crisis*, p. 359.

[23] *Crisis*, p. 359; The relevant continuation is: "Vivid self-evidence passes – though in such a way that the activity immediately turns into the passivity of the flowingly fading consciousness of what-has-just-now-been. Finally this 'retention' disappears, but the 'disappeared' passing and being past has not become nothing for the subject in question: it can be reawakened".

[24] "To the passivity of what is at first obscurely awakened and what perhaps emerges with greater and greater clarity there belongs the possible activity of a recollection in which the past experiencing is lived through in a quasi-new and quasi-active way," *Crisis*, p. 360.

[25] *Crisis*, p. 360.

[26] *Crisis*, p, 360; The following leads to the just quoted: "In this full understanding of what is produced by the other, as in the case of recollection, a present coaccomplishment on one's own part of the presentified activity necessarily takes place; but at the same time there is also the self-evident consciousness of the identity of the mental structure in the productions of both the receiver of the communication and the communicator; and this occurs reciprocally. The productions can reproduce their likenesses from person to person, and in the chain of the understanding of these repetitions what is self-evident turns up as the same in the consciousness of the other,"

[27] *Crisis*, p. 360.

[28] *Crisis*, pp. 360–361; The whole relevant sequence is: "The important function of written, documenting linguistic expression is that it makes communications possible without immediate or mediate personal address; it is, so to speak, communication become virtual. Through this, the communalization of man is lifted to a new level. Written signs are, when considered from a purely corporeal point of view, straightforwardly, sensibly experienceable; and it is always possible that they be intersubjectively experienceable in common. But as linguistic signs they awaken, as do linguistic sounds, their familiar significations. The awakening is something passive; the awakened signification is thus given passively, similarly to the way in which any other activity which has sunk into obscurity, once associatively awakened, emerges at first *passively* as a more or less clear memory. In the passivity in question here, as in the case of memory, what is passively awakened can be transformed back, so to speak, into the corresponding activity: this is the capacity for reactivating that belongs originally to every human being as a speaking being. Accordingly, then, the writing-down effects a transformation of the original mode of being of the meaning-structure, [e.g.] within the geometrical sphere of self-evidence, of the geometrical structure which is put into words. It becomes sedimented, so to speak. But the reader can make it self-evident again, can reactivate the self-evidence" pp. 360–361.

[29] This will, of course, constitute an extensive and very important field of investigation in the continuation of the inquiry this paper conducts.

[30] *Crisis*, p. 361; The continuation is as follows: "But there also exist possibilities of a kind of activity, a thinking in terms of things that have been taken up merely receptively, passively, which deals with significations only passively understood and taken over, without any of the self-evidence of original activity. Passivity in general is the realm of things that are bound together and melt into one another associatively, where all meaning that arises is put together passively. . . . apparently . . . possible reactivation . . . instead of being fulfilled, comes to nothing; . . ."

[31] *Crisis*, p. 362; The problem to "tackle" has this structure: ". . . free play of associative constructions. In view of the unavoidable sedimentation of mental products . . . persisting linguistic acquisitions . . . can be taken over by anyone . . . constant danger. . . . avoided . . . if its capacity to be reactivated and enduringly maintained . . . in scientific community . . . of knowledge living in the unity of a common responsibility . . . 'once and for all' . . . 'stands fast', forever identically repeatable with self-evidence and useable for further theoretical or practical ends – as indubitable reactivatable with the identity of its actual meaning".

[32] *Crisis*, p. 363; Husserl sees a possible difference here between the deductive and descriptive sciences.

[33] At *Crisis*, p. 366 he asks if the genuine meaning could really be "cashed in," and the answer is negative: "Unfortunately, . . . this is our situation, and that of the whole modern age."

[34] *Crisis*, p. 364. This concept of "explication" (*Verdeutlichung*) that would have been a core-concept, going into the problem of understanding more substantially at this level, is explicated by Husserl in the following way: "A passively emerging sentence . . . memory . . . , or one heard and passively understood, is at first merely received with a passive ego-participation, taken up as valid; and in this form it is already our meaning. From this we distinguish the particular and important activity of explicating our meaning. Where as in its first form it was a straightforwardly valid meaning, taken up as a unitary and undifferentiated – concretely speaking, a straightforwardly valid declarative sentence – now what in itself is vague and undifferentiated is actively explicated, . . . But there is something special . . . to have the intention explicated, to engage in the activity which articulates what has been read . . . , extracting one by one, in separation from what has been vaguely, passively received as a unity, the elements of meaning, thus bringing the total validity to active performance in a new way on the basis of the individual validities. What was a passive meaning-pattern has now become one constructed through active production. This activity, then, is a peculiar sort of self-evidence; the structure arising out of it is in the mode of having been originally produced. And in connection with this self-evidence, too, there is communalization. The explicated judgment becomes an ideal object capable of being passed on . . . logical . . . thus the *domain of logic* is universally designated; this is universally the sphere of being to which logic pertains insofar as it is the theory of the sentences (propositions) in general."

[35] *Crisis*, p. 365.

[36] The "proof" might be indicated by the following: Husserl asks, ". . . what about the possibility of complete and genuine reactivation in full or originality, through going back to the primal self-evidences . . . based on the fundamental law (of logic) . . . with unconditionally general self-evidence, is: if the premises can actually be reactivated back to the most original self-evidence, then their self-evident consequences can be also.

. . . chain of logical inferences . . . no matter how long it is. . . . However, if we consider the obvious finitude of the individual and even the social capacity to transform the logical chains of centuries, truly in the unity of one accomplishment, into originally genuine chains of self-evidence, we notice that the (above) law contains within itself an idealization: namely, the removal of limits from our capacity, in a certain sense its infinitization. . . . later" *Crisis*, p. 365.

[37] This is what is taking place in the *Crisis*, pp. 366–369; Husserl then goes into his general analyses on history.

[38] *Crisis*, p. 368.

[39] "The ruling dogma of the separation in principle between epistemological elucidation and historical, even humanistic-psychological explanation, between epistemological and genetic origin, is fundamentally mistaken, unless one inadmissably limits, in the usual way, the concepts of 'history', 'historical explanation', and 'genesis'," (*Crisis*, p. 370). This is then developed through the rest of the text and is the perspective ruling it all – from the very beginning.

[40] *Crisis*, p. 366.

[41] To Husserl himself the concept of understanding "as such" was of no prominence, and he might have had his reasons for that. To us, however, trying to understand his phenomenology, it will, of course, be of decisive importance, and then who would be the better guide and "supplier" of the intellectual equipment we need, then Husserl himself?

[42] *Crisis*, p. 371.

[43] As might be realised from the exposition of this paper, Husserl does not give any "clear-cut" definition of this expression, having that total "concreteness" of cognition and understanding in mind, and he has different ways of explaining, correlating, genuinely emerging out of his analysis; here we just give one relevant one from a footnote: "The historical world is, to be sure, first pregiven as a social-historical world. But it is historical only through the inner historicity of the individuals, who are individuals in their inner historicity, together with that of other communalized persons. Recall what was said in a few meager beginning expositions about memories and the constant historicity to be found in them." (*Crisis*, p. 372, footnote).

[44] What we just have been saying, might, of course, cause severe problems – not only for understanding that complex line of thought as it is expressed by us, but also for accepting such concepts as 'the universal a priori of history'. To many people this would appear a plain "contradiction in terms," and, if it is "resulting necessarily" from "self-evidences" in these investigations Husserl is conducing, then that leads us to realise that there has to be something quite wrong with his whole project looking into the "origin" of Geometry considered to be an "intentional-historical problem" – as the German text puts it. As this obviously is meant to "constitute" the "transcendental" founding elements of that science, then you have got the rather odd situation where geometry, which from the very beginning of science has been considered to be the science operating with truths that "stand fast," with universal validity, so to speak transcending history, is now made into a historical problem – as a science also taking in profoundly elements of the subjective, so that the universality might be questioned of a grounding that seems to be threatening that very same universality. At least this might raise a quite natural question. To the phenomenologist, especially having the general perspective of the *Crisis* in mind, this putting together of elements that might appear unreconcilable, will, however, not come

as an surprise; the project of the appendix is well integrated into that total perspective of transcendental phenomenology – such as it had been developed at that time of the *Crisis*. It even might be directly related to the beginning of phenomenology itself – where Husserl analyses the founding elements of mathematical science giving them an intentional-subjective foundation.

[45] Of course, "The Origin of Geometry," now having given the "material" for our reflection on the general problem of history, is first and foremost connected to that analysis of the origin of mathematical-natural science in the *Crisis*, as it describes Galilei – in the Renaissance – so to speak transforming history, bringing forward elements from the ancient tradition, transforming them into a new praxis and in that way laying the foundation for the modern age. As it then comes to that quite general characterisation of history – also including its "universal a priori," this might, of course, be seen as a genuine component of the total *Crisis*-perspective. It fits quite harmoniously into that teleological grounding of philosophy, and relates not least to what Husserl in the first part of the main text calls "the greater historical phenomena" and then characterises as "humanity struggling to understand itself" (*Crisis*, p. 14). In the *Crisis*, history becomes a substantial part of transcendental phenomenology, which thereby – as historicity – includes the actual historical situation, the sciences, the philosophical-scientific tradition and, of course, the life-world as that which gives the concrete horizon for even transcendentally grounding such a philosophy in the most profound teleological manner. In that way this phenomenology seeks to establish itself as maybe the key part of that "greater phenomena" – going into the core of that "struggle" penetrating the whole of history, where humanity has been and very much still is struggling to understand itself, clarifying the grounding preconditions of such an understanding. Then this understanding of history as that "vital movement of the coexistence and the interweaving of original formations and sedimentations of meaning," will be of the most decisive importance, as it genuinely will be constituting what we in this context call 'the historicity of understanding', profoundly situated in "the transcendental problem," as the paths into phenomenological transcendental philosophy are constituted by "inquiring back from the pregiven lifeworld" and "from psychology" just as the main part of the *Crisis* (III A&B) explicates this whole matter.

GIAN MICHELE TORTOLONE†

THE FRAME OF EVENTS

The passage from the stream of consciousness to the stream of existence, and their mutual intrinsic connection, constitutes one of the knots of phenomenology. My thesis is that the problematic core of the whole process is to be found in the corporeity and complexity of phenomenological dynamics, and in particular in the ambiguous notion of *event*. I shall try to highlight the permanence of the problem from Husserl's original outlines to Merleau-Ponty's developments; I shall eventually suggest a possible outcome in a new determination of event frame.

Let's begin from the definition of *Erlebnis* in *Ideen I.*[1] *Erlebnis* implies "all that is to be found in the stream of *Erlebnisse*", of which the *Erlebnis* represents the *real moment*. Every act of consciousness, in each of its elements, is realized in unities.[2] However the notion of "real moment" immediately broadens the field of *Erlebnisse*, including for instance the data of sense, and at once arranges a twofold polarity which refers to existence: *cogito* and sensitivity, related to the unity of the self.[3] Quite uncertain is the process by which every *Erlebnis* is included, as an act, in the lived experiences it is related to. The determination of temporality related to the constitution of the self and to the way the self's world yields itself is linked to such a process. These are the essential aspects of the research into the constitution of the self, which Husserl will try to develop in *Ideen II*. Husserl's research, however, seem to encounter difficulties on the very definition of event frame. In *Ideen II* the point of departure is therefore the notion of the corporeal structure which represents the main sphere of connection between psyche and materiality. The flux of *Erlebnisse* is connected with the unity of the self, which as self-man is linked in its turn to the reality of one's own body. In the empirical subject, as Husserl points out, "the psyche (*die Seele*) is constituted as a reality connected or intertwined with the reality of the body (*Leib*)"[4] But the very weft of this entanglement is not examined.

† Gian Michele Tortolone died unexpectedly on May 21st, 1998 due to a tragic motorcycle accident. His professor – Dr. Ciancio (University of Vercelli) – helped to correct the article.

A-T. Tymieniecka (ed.), Analecta Husserliana LIX, 423–431.

In *Ideen*, the body provides individual perception with objectivity, by the immediate position of intersubjectivity. Objectivity, just like the space-time unity, is provided by the body as a center of the world dynamics. Everything is thus solved in the evidence: "It is evident that the apprehension of one's own body [causes the apprehension of objects] . . . as things . . . in a unique objective world".[5] But more evident are the numerous implicit assumptions of Husserl's thesis, which will be criticized and corrected in *Krisis*. (§54, *Husserliana* VI, 187ff.).

The *unsolved paradox* of the phenomenological base of the "(natural) reality-man" is that Husserl arrives at the definition of psychophysical unity, maintaining the stream of *Erlebnisse* and the constitutive dynamics of the body to be entirely distinct from one another – which is the theoretical reason for the difficulties in the general framing of existential events. It is not possible to come out of the "fatal" polarity. "In this way the body is apprehended. . . . For through this layer [. . .] the material body is intertwined with the psyche . . . This self, that is the psyche, "has" its own body; . . . The psyche and the psychic self "have" their own body. . . ."[6] The *Leib* too is characterized by an internal/external polarity. From the inside the body appears to be an organ of the subject, nucleus and substratum of sensations, which forms however with the psyche a concrete unity (by means of the body?). From the outside the body is a "material thing" (*materielles Ding*), and at the same time it represents the "overturning point" (*Umschlagspunkt*) which transforms causal relations into subject's conditional relations. It is generally assumed that there are two attitudes present at the same time, though their different modalities are not specified.[7] The unity of the flowing psychic existence is connected with the stream of corporeal existence, which should be thus interpreted as the stream of corporeal events, even if such homogeneity is not put into topic.[8] *The body is the interface of reality*, which maintains the ambiguity of being at the same time "thing" and constitutive process. Here is the prospect of a necessary correlation between *Erlebnis* and *Ereignis*, in corporeity, but such prospect is not put into a frame: "what we should oppose to material nature as a second kind of reality is not the psyche, but the concrete unity of body and psyche, the human subject,"[9] – which is exactly the possibility given by the determination of event constitution, which proceeds from the redefinition of corporeity as well as of correlated space-time. The corporeity, as an implicit notion, changes into a problematic dimension as far as the correlation between the self and its world is concerned, putting into evidence the *pre-comprehension* underlying the phenomenological analysis. The circle by which any act or moment of the subject is always a real event cannot be avoided. The perceived event is connected with lived experiences. Husserl will arrive at stating that the psyche and the body are constituted together, however, even in the final considerations in *Ideen III*, Husserl will not overcome the obstacle of the twofold internal/external, psyche/materiality polarization, etc., the implicit dualism representing its horizon of meaning.[10] The following research joined in *Krisis* seem to confirm these results.

The *Lebenswelt* develops as flux and horizon of intentional acts and experiences, forming a living horizon, in its turn constantly flowing [11] – all this in the subjective evolution of the modes in which the consciousness of one's presence in the world take place.[12] The world as a horizon, by means of *epoché*, is defined as the result of the universal correlation of world and consciousness of the world, the latter "being the very existence of subjectivity consciousness which produces the validity of the world, subjectivity which in its continuous attenuations has always its own world and is always actively creative."[13] Moreover, the world is defined as the correlation between being and subjectivity. The paradox of subjectivity, and the connection between consciousness and intentend world,[14] should have led to consider the question of the constitutive frame of the horizon-world, and therefore of events. Husserl mentions this task, but only to solve it in a non-ontological sphere, under the topic of presence, the certainty of existing in the world. In fact, an ontology of the *Lebenswelt* could allow a connection and a passage from perceived to lived experiences.[15] On the contrary, the main questions remain unsolved. What does subjectivity as constitutive flux in the real world mean? What is the relation between subject and the world about? How are the horizons of facticity (*Faktizität*) elaborated in the self-world relation? What happens if the transcendental quest is found out to be part of the process of one's world constitution? By explaining the *un-said*, the horizon which Husserl refers to, we find this: it is the event (intentional) frame constitutive of reality, which realizes the certainty of existing in the world. What could then such a frame consist of?

This is the background of Merleau-Ponty's research; he will try to define the link between temporality, acts of consciousness and constitution of the world about. His point of departure, the *ambiguity of presence*, is immediately connected with corporeity and time: ". . . the ambiguity of the existence in the world manifests itself together with the ambiguity of the body, and the later is related to the ambiguity of time."[16] A synthesis of *Erlebnis* and *Ereignis* appears at the once as inevitable. In the living present it is important that we can understand how the synthesis of subject, living thought and corporeity endowed with plasticity is realized, and in what way the environment is arranged in relation to the polarity of our body. Analyzing the symbolic function, in the definition of world meaning, Merleau-Ponty argues that consciousness (for instance, in movement) *has no* power, *but is* that power; exactly as the subject has no body, but is that specific living body.[17] Thus, *consciousness* does not only imply the variety of psychic facts, but also *the whole constitutive process of existing in the world, exactly as corporeity* in the way it is realized: two synonyms reflecting the same reality from different points of view. Consciousness projects into the body and vice versa, in every constitutive act of existence: that is to say in every event. If we consider a complex event, for example, a gesture, we can define the movement as original intentionality. The body catalyzes the absolute consciousness of "here" and "now", and in the gesture I am not in space or time; but I am related to them, my body applies to and embraces them. In the instant of the movement I can seize its totality: I realize the field, the horizon, in which consciousness concretes in the act of being related to (*être à*) the world (the thing) through the body. *The act is mixed up with the structure of the world*: ". . . the psychophysical event can no longer be conceived [. . .] as the contiguity of a process in itself and of a *cogitatio*. The unity of soul and body [. . .] is accomplished at every instant in the movement of existence."[18] An event is the act of the constitution of reality for myself. What will its frame then consist of? *Everything is related to time*. The act is realized in time; and the perceptive synthesis is a temporal synthesis. Consciousness perceives time, whereas the subject perceives a world. It can be thus argued that time is the event frame, as the event reciprocally frames time: in so far as the subject as a body makes and interprets time, while as consciousness it unfolds it.[19] The problem however is not totally solved yet. If an event is the frame of the present, consciousness, as movement of temporality, unfolds time in the void (*creux*) of the future.[20] And if void were the event frame?

In *Le visible et l'invisible* the world is considered again as an open question.[21] However the event dimension together with the consciousness we have of it (as we shall discuss afterwards, Merleau-Ponty here avoids and put into question the word "perception") represents at once the central thesis: the world secret is our being in contact with it.[22] The contact lies in the corporeal dimension, the act by which I put myself in the world and perceive it. The present, which surrounds me, I see and perceive, is given by the instant of presence.[23] In the thickness of the body, events catalyze in a world. But this thickness, instead of being a "fullness", has different characteristics, such as reference, latency, absence.[24] The constitutive paradox, which Merleau-Ponty succeeds in touching, is that an event is framed as a reference between an absence and a subsequent absence. The depth of the event reveals the invisible as the structure of existence and presence.[25] And that is exactly what we find in consciousness,[26] pure ideality living on carnality as well as on the ambiguous horizon of present: the *Erlebnis* is framed in the world, as both *lived experiences*. It is Thus revealed to be the *un-said* of Husserl's analysis: the concepts deriving both from psychological and transcendental reflection are often correlates or counterparts of the objective world. As Merleau-Ponty explains: "It is necessary to abandon [. . .] such notions as 'acts of consciousness', 'matter', 'form', and also 'image' and 'perception'."[27] The event falling in its constitution makes possible the interpretation of existing in the world.

In Merleau-Ponty's work notes we find the essential point. In May 1959 Merleau-Ponty wrote: "Étudier exactment l'*Erfüllung* du présent: danger de cette métaphore: croire qu'il a un *certain vide* qui a ses dimensions et qui est rempli par une quantité définie de présent."[28]

For that reason Merleau-Ponty had always preferred not to use the word "void", which was linked in his opinion to Sartre's theses. In fact Merleau-Ponty identifies void with "nothingness", and tended to consider it in a nihilistic perspective. The previous passage ends thus with the following statement: "Bref: le néant (ou plutôt le non être) est creux et non pas trou." We find here, in counterlight, the definition of event as a refilling of the present, solved, according to the coincidence void-nothingness, *nichtiges Nichts*, in he breadth of horizon. In any case, even at the level of the instant, an event is a process, which is constituted in – and constitutes – the double polarity of the psychic and somatic. However, a few months later, in February 1960, referring to Husserl's *Erwirken* Merleau-Ponty corrects: "L'invisible est un creux dans le visible . . . Mais ce qui est beau c'est I'idée de prendre à la lettre l'*Erwirken* de la pensée: c'est vraiment du *vide*, de l'*invisible* . . . Il n'y a pas à chercher des choses spirituelles, il n'y a que des structures du vide.

Simplement je veux planter ce vide dans l'Être visible, montrer qu'il en est l'*envers*. . . ."[29]

The text is composed of at least six passages, which should be briefly examined. Thought seizes existence in its development, but I am also conscious of producing my thought, as a development of existence. Consciousness is not yielded in he invisible, but as it is the invisible. The accomplishment is related to the unaccomplished. As such, the accomplishment is related to a void which is not nothingness, but the hollow which precedes and makes it possible what has to be. The invisible is thus the chasm persisting at the bottom of the visible, it is a sinking in the very act of constitution. The concretization of existence, as it appears in that act of existence which thought is, is really invisible, the void characterizing every event. The spiritual frame is realized as flux in the clefts of existence: thus, the only framework is the void, for what is void is recognized as the opposite of the visible.

If all this is true, we have to change our vision of the frames of reality, starting with the definition of event. In fact Merleau-Ponty arrives at a working out of the concept of the "flesh of the world". This concept implies in its turn a metamorphosis in the consideration of time, which Merleau-Ponty however did not succeed in concluding.[30] In any case, Merleau-Ponty has suggested a research line which goes from body to temporality, from thought to the original void, towards a comprehensive frame of reality.

In fact, if we really realize the radical *epoché* which Husserl himself mentions in *Krisis, we no longer find the self*, but *the invisible of the self*, its constitutive elements and their factual dynamics.

In this perspective, an event becomes the neutral frame on which the double tension of reality is organized. The psychophysical *continuum*, so much investigated by Merleau-Ponty, has to be organized in an original dimension, constitutive of both reality (*Ereignis*) and consciousness (*Erlebnis*). When Merleau-Ponty states that there is no room for events in the objective world, because events would be cut out by a supposed external observer in the space-time continuum, in reality he is talking about *instants* (identical to events according to the general meaning). But while instants flow within consciousness, the event yields within the frame of presence. *Totality* is the enigma of the event, the flesh of time. And now we know that this flesh has the thickness of the void. For this reason the future can open before a present which is absorbed in the past. Therefore, if we want to try a new determination of an event we can define it as the *concretized processing element of the net of individual lived experiences in intentionality as well as of reality in general*. The existential net is the connection of events, constituent/constituted of and by a certain subjective polarity. Thus we are not actors on a stage, but indifferently and altogether actors, stage, audience and theatre in the play of life. Every event yields itself together with its own reference outline, and can do this thanks to the frame it is made of – thus generating the background, the site, the horizon of its own temporality, as well as the necessity of dissolution. Every event appears, in the beginning, indifferently psychic and material. An event is the yielding of reality according to its own codes of concretization.

If a void is the original background of an event, as a nucleus of reality, the event has to be constituted of the void as well as of the tension generated in the void. What has to be found out and defined, is the phenomenon energetically emerging from and in the void, and the way it emerges. Let's start from the latter. If the void expends, nothing happens. And if it contracts? If the void implodes on itself, what could happen? The tension imaginable in the contracting void is the frame of that original nucleus of existence, which in its casualness can be defined a *trial*. Now, if the trial is the dynamic unity and the energetic form emerging from the constitutive void, an event can be defined as a link in the harvest of infinite trials that realizes existence, in any of its forms. *For this reason every event is implosive, not explosive and centrifugal, and it is consumed in itself*. An events is formed by the psycho-material net within which the trials constituting the granulation of existence are ordered and concretized. This net is realized and embodied in each corporeity. Resulting from trials, an event is a neutral frame, i.e., anterior to any psychic or materials qualification, put in continuity with the void it is made of. The body is the interface of reality, for it catalyzes in its existing in the world a continuous series of events, that we, and just us, interpret in our horizon as internal and external. As a constitutive synthesis of existence on the polarity of the subject, an event, in which is synthesized the whole universe, is an implosive process of indefinite micro-events, of infinite trials, synaptically interconnected: which by imploding are deposited on the frames of corporeity.

A further question is the *determination of event extension*. On the one hand, an event coincides with the subject's world about. On the other hand, it gets over it, being in correlation with the universal harvest of trials. In the environment the extension of an event is determined by intentionality. Outside, it is the synaptical frame of trials that can determine the co-extensibility of events. An event is thus a whole life, in any of its yielding and unfolding.

In order to visualize the intentional net constitutive of the event, we can symbolize (intentional) egological rays converging on a virtual pole in which the determination of the self is synthesized in process. The event wraps us as a sphere with an indefinite (though complete) ray, of which we are at the same time center and periphery.

By saying that the subject's life is all in its own making, we mean that there cannot be an existential *surplus*, but only the globality of the event. For this reason I exist as my present, because it is made in the metabolizing of all the events constitutive of myself, with all the possible links. In this *existential metabolism* corporeity and temporality coincide.

As a synthesis of trials an event embodies time. On the basic frame of the trials yielding as well as of their confluence in events, which can follow one another through their emerging and their imploding in the void, there can be explained the insight according to which time coincides with event, and is the unfolding of life. In the same way, it is possible to interpret consciousness as movement of temporality, or flux of events. Time is originally the constitution of event in a single "movement", an act which as it emerges implodes. Temporality is the paradoxical dimension of the present. However in the structure of a void, it is no longer necessary for the event to look for the interpretation of the present in retention and projection between the past and the future. The present appears as *the result of a peristaltic movement* of trials yielding and solving. In the implosion of events, time is always virtual. It is related to the invisible, to that overturn of existence in which it is embodied. Time is rhythm, the oscillation of trials. For this reason the time of life is irreversibly consumed. An event is immediately and globally one's own time. Time is the present in which the event occurs. The occurring is together time and event.

The circle closes.

University of Turin
Italy

NOTES

[1] Husserl's texts are quoted in the current Italian translation, with reference to the German work *Husserliana, Edmund Husserl Gesammelte Werke*, and in particular: Vols. III–V, *Ideen zu einer reinen Phänomenologie und phänomenologischen Philosophie* (The Hague: M. Nijhoff, 1950–1952), (cited here as *Ideen I–III*; Ital. ed. [Turin: Einaudi, 1965–1976, 3rd); and Vol. VI, *Die Krisis der europäischen Wissenschaften und die transzendentale, Phänomenologie* (The Hague: M. Nijhoff, 1954), (cited here as *Krisis*; Ital. ed. Milan: il Saggiatore, 1961, 1975, 5th). The German works will be followed by the page directions.

[2] *Ideen I*, pp. 80, 75.

[3] On the question of *cogito* as an act and the issues it raises cf. *Ideen I*, p. 76.

[4] *Ideen II*, pp. 92–93 (pp. 488–489). The quotation is at p. 93.

[5] Cf. the whole passage in *Ideen II*, p. 81.

[6] Cf. § 40 in *Ideen II*, pp. 153ff.

[7] Cf. § 42 in *Ideen II*, p. 161.

[8] *Ideen II*, p. 139.

[9] *Ideen II*, p. 139.

[10] *Ideen III*, p. 117 ff.; in particular p. 124.

[11] *Krisis*, p. 152.

[12] *Krisis*, p. 149.

[13] *Krisis*, pp. 154ff.

[14] On the paradox of subjectivity cf. *Krisis*, pp. 182, 265–266ff. On the question of connection cf. among the others *Krisis*, p. 257.

[15] The mention of the ontological necessity is explained *Krisis*, § 37, p. 145.

[16] Cf. *Phénoménologie de la perception* (hereinafter *PhP*) (Paris: Gallimard, 1945), p. 101.

[17] *PhP*, pp. 140ff; in particular pp. 150–151.

[18] *PhP*, p. 105.

[19] Cf. in particular *PhP*, p. 277; and pp. 479ff.

[20] *PhP*, p. 473

[21] *Le visible et l'invisible* (hereinafter *VI*), edited by C. Lefort (Paris: Gallimard, 1964). We shall follow the page numeration of the 1983 edition.

[22] *VI*, p. 53.

[23] *VI*, pp. 152ff, pp. 173ff

[24] *VI*, p. 179.

[25] *VI*, p. 195.

[26] *VI*, p. 200.

[27] *VI*, p. 209.

[28] *VI*, p. 249.

[29] *VI*, p. 289.

[30] *VI*, p. 296.

JOSEF CIBULKA†

DIE ZIRKELSTRUKTUR UND DER ZEITASPEKT DER ERFAHRUNG

Über die Gegenseitige Angewiesenheit der Husserl'schen und der Heidegger'schen Phänomenologie

Jan Patočka erzählte, dass Eugen Fink anfangs der dreissiger Jahren meisterhaft die Heidegger'sche Phänomenologie vom Standpunkt Husserl's aus, und zugleich die Husserl'sche Phänomenologie vom Standpunkt Heidegger's zu kritisieren wusste. "Der Kontrast beider arbeitete in Fink".[1] Patočka erzählte es als eine bibliographische Besonderheit, aber Paul Ricoeur[2] schreibt sogar in seinen philosophischen Analysen der Systembeziehungen zwischen dem Husserl'schen und dem Heidegger'schen Strome der Phänomenologie, dass es "Anleihen" des einen Stromes von der Systemoptik des anderen Stromes gibt.

So stellt sich von selbst eine Hypothese heraus, ob es nicht fruchtbar wäre vorauszusetzen, dass beide Strömungen gegenseitig angewiesen sind, dass es die Optik des einen ist, die immanente und einzigartige Zirkelstruktur durchzuschauen, in die jede von beiden durch eine für sie spezifische Weise verzaubert ist, deren Zirkelhaftigkeit aber nur mit Hilfe der gegenseitigen Strömung durchzuschauen ist.

1. ÜBER DIE ZIRKELSTRUKTUR DER HUSSERL'SCHEN FUNDIERUNGSIDEE

Es ist eine grundlegende Vision von Husserl, dass das fungierende Leben der Subjektivität vor unseren Augen die zugehörigen Geltungen fundiert.[3] Aber das Erlebnis, das eine bestimmte Bedeutung fundieren soll, ist noch zuvor konzeptuell "zugeschnitten", damit daraus eben eine solche und nicht eine andere Bedeutung resultiere.[4] Das Bewusstsein schaut seine Gegenstände als phänomenologisch sichtbar und deskriptiv verstehbar, erst nachdem es seine Teilnahme an der Transzendenz der materiellen Natur mittels der Form der Leiblichkeit schon verwirklicht hatte. Die Leistung der Phänomenologie stellt ein Nachvollziehen desjenigen Weges

† Josef Cibulka died unexpectedly on 27th March, 1998.

A-T. Tymieniecka (ed.), Analecta Husserliana LIX, 433–439.

dar, den das Bewusstsein schon zuvor verwirklicht hatte.[5] Die retrospektive Konstitution interpretiert das, was die historische intersubjektive Tätigkeit der Generationen schon geleistet hatte, in ein non-zeitliches ideales Universum des Bewusstseins.[6]

Ähnlich, wie der Arzt in die inneren Leibesorgane einen Kontrast-Stoff im Interesse ihrer Untersuchung einleitet, kann man auch die einzelnen Momente des Bewusstseins nur durch Vermittlung ihres Gegenpoles phänomenologisch sichtbar machen.[7]

2. DER ZEITASPEKT DER ERFAHRUNG IN DER PHÄNOMENOLOGIE

Durch seine Theorie der Apperzeption verwirklichte Husserl seine Auffassung der Formierung einer aktuellen intentionalen Erfahrung auf Grund der Gewesenheit. Meine einzelne Empfindung ist, nach Husserl, durch sein implizit historisches A priori "beseelt". Indem ich ein mir bisher unbekanntes Tier erkenne, erkenne ich gleichzeitig auch seine Gattungsfähigkeiten. Sobald ich später ein anderes Tierexemplar dieser Gattung erblicke, erblicke ich bei ihm dieselben Gattungseigenschaften[8] Die Determinierung einer einzelnen Wahrnehmung durch die ganze Historie meines Wahrnehmens ist auf diese Weise in eine "Apperzeption" projiziert, in ein ideal apriorisches Moment meiner wahrnehmenden Tätigkeit.

Wie steht es aber, bei Husserl, mit einer analogen Formierung einer einzelnen Wahrnehmung durch die Zukunft, durch die Erwartung einer, die Zukunft antizipierenden und vorgreifenden Zielstrebigkeit?

Husserl bezeichnet – im Allgemeinen – die Dimension der Fiktion als ein Lebenselement der Phänomenologie, wie aller eidetischen Wissenschaften[9] Die Möglichkeiten sollen in keine Kopien des Bestehenden uminterpretiert werden: ". . . die Wirklichkeiten müssen behandelt werden als Möglichkeiten unter anderen Möglichkeiten. Dies geschieht nur dann, wenn jede Bindung an vorgegebener Wirklichkeit auf das Sorgsamste ausgeschlossen ist".[10] Auf diese Weise verbindet aber Husserl die Dimension des Möglichen einseitig mit dem Aspekt einer Befreiung des eidetisch Allgemeinen von der Faktizität, von der Zufälligkeit.[11] "Sie beruht auf der Abwandlung einer erfahrenen oder phantasierten Gegenständlichkeit zum beliebigen Exempel, das zugleich den Charakter eines leitenden 'Vorbildes' erhält, des Ausgangsgliedes für die Erzeugung einer offen endlosen Möglichkeit von Varianten . . . wir lassen uns vom Faktum als Vorbild für seine Umgestaltung in eine Phantasie leiten . . .".[12]

Diese Husserl'sche Umwandlung einer Faktizität zum Exempel soll also den Weg in die Dimension der Möglichkeiten eröffnen, die eidetische Sphäre von der Hegemonie der Faktizität befreien. Was legitimiert aber ein einzelness faktisches Reales, als ein Vorbild, als eine Norm zu gelten?

Kommt auf eine solche Weise nicht dazu, dass im Gegenteil jener Horizont, der sich durch meine bisherigen Erfahrungen etabliert hat, auch meine imaginativen Variationen vorzeichnet?

Bernhard Waldenfels problematisiert diesen Husserl'schen Zugang, eine schlichte Gegebenheit unvermittelt durch universelle Strukturen in ihrem Gegebensein gelten zu lassen. Er fordert eine variable Auffassung von Strukturen selber, eine Möglichkeit der Restrukturalisierung der schon etablierten Sinnstruktur, eine Möglichkeit einer konkurrierenden Sinnbildung.[13]

Im Laufe einer konkreten phänomenologischen Beschreibung taucht oft das Moment einer Erwartung, einer Antizipation auf. Zum Beispiel Aron Gurwitsch schreibt: "Höre ich aus einem Nebenzimmer die Stimme meines Freundes, antizipiere ich, dass ich in einer Weile meinen Freund auch erblicke."[14] Durch Husserl's allgemeine phänomenologische Interpretation wird aber dieses Moment einer intensiven Antizipation, bei welcher mein Bestreben so bald als möglich den Augenblick des Zusammentreffens zu kürzen, involviert wäre, nicht als eine aktive Antizipation betroffen.

Paul Ricoeur schreibt mit Recht, dass der bisherige Erfahrungsraum nicht hinreicht, den Erwartungshorizont vorzuzeichnen.[15]

Diese allgemein philosophische Sachlage ist vorhanden, trotzdem es in der Wirklichkeit eben die Rückwirkung der Erwartungen ist, die die aktuelle Erfahrung antizipitaw gestaltet.

Die Heidegger'sche Auffassung der Erschlossenheit betrifft das Moment des antizipativen Entwerfens.

3. ÜBER DIE ZIRKELSTRUKTUR DER HEIDEGGER'SCHEN ONTOLOGIE DES VERSTEHENS

Man interpretiert es als eine befreiende Tat, dass Heidegger die Gegebenheit der Gegenstände übersprang, dass er als das primäre Begegnen des Menschen mit der Wirklichkeit das praktische Bewenden verstand.[16] Gegen das aristotelische Ableiten der Möglichkeiten von den gegebenen, schon existierenden Gegenständen geht Heidegger von den Möglichkeiten aus, und stellte das Postulat auf, die Möglichkeit als Möglichkeit zu begreifen, sie als Möglichkeit zu bewahren.[17] Ontologisch noch ursprünglicher als das gegenwärtige Vorliegen ist, bei Heidegger, die zeilhafte Erschlossenheit, die Zeit als der ursprüngliche Horizont, von woher das Dasein das # in der Mehrafalt seines Sinns versteht.[18]

Die Dynamik, die diese Ontologie der Erschlossenheit von sich aus entwickelt, entbehrt aber, in ihrer inneren Verfassung, der Negativität, obwohl schon der frühe Heidegger behauptete, dass seine Hermenautik ihre Aufgabe nur auf dem Wege der Destruktion erfüllt.[19]

Wenn wir sagen, dass Heidegger's Auffassung der Entschlossenheit der Negativität entbehrt, können wir es annähern, wenn wir Gadamers Auffassung der Erfahrung erwähnen.

Gadamer betrachtet als eine "wirkliche" Erfahrung diejenige Erfahrung, die permanent detypisiert, was vor dieser Erfahrung als normal, als erwartet galt. Wenn ich – Gadamer zufolge – eine Erfahrung mit irgendeinem Gegenstande mache, bedeutet das, dass ich bisher diesen Gegenstand nicht richtig gesehen habe. Die Dynamik der Erfahrung besteht darin, dass wir diese Erfahrung fortwährend überschreiten. Ein erfahrener Mensch ist derjenige, der für die neuen Gegenstände der Erfahrung maximal geöffnet ist.[20]

Im Vergleich mit dieser Gadamer'scher Auffassung der Erfahrung reduziert sich die Eigendynamik der Erschlossenheit auf ein durchaus positives Entfalten. Auf Grund der Erschlossenheit bleiben wir Apologeten derjenigen Richtung unserer Erschlossenheit, die für sie authentisch ist. Walter Biemel illustriert es durch Ibsen's Drama *Die wilden Enten* – jeder Versuch, sein eigenes Sein zu negieren, scheitert.[21]

Hans Lipps schreibt sogar in seiner "hermeneutischen Logik", dass es uns bei einem Gespräch nicht so sehr darum geht, sich über etwas zu verständigen, dass wir eher bestrebt sind, nur die Linie unserer Begegnung mit den anderen Leuten zu festigen.[22]

Manfred Frank bezichtigt Heidegger, dass er die Frage nach dem Sein in eine Frage nach der Seinsweise des Seins verwandelt, das diese Frage stellt.[23]

Das Moment der Antizipation und der Negativität wird auf diese Weise unterdrückt.

Die Zirkelstruktur der Heidegger'schen Sicht enteert – in diesem erwähnten Sinne – die Offenheit der Gegenstandssphäre gegenüber, die einem schöpferischen Geist früher eigen war.

Im Lebensgefühl von Goethe waren es die gegenständlichen Inhalte, die die ganze strömende Subjektivität vorwärtstrieben. Diese strömende Subjektivität stieg durch ihr Selbsttranszendieren empor. Die gegenständlichen Schöpfungen sind ursprünglich keineswegs dem Erleben gegenüber entfremdet. Erst Sartre kritisiert Husserl's Intentionalität, dass sie uns in die Entfremdung stürzt.[24]

Wie es möglich ist, den gegenständlichen Aspekt der Husserl'schen Phänomenologie fruchtbar zu entwickeln, zeight die ästhetische Konzeption von Roman Ingarden. Der Leser eines literarischen Werkes muss sich, Ingarden zufolge, an der Vision des Autoren beteiligen, die in die Realisierung des Werkes verzaubert ist. Er muss, mit dem Autoren, seine ganze frische Kraft zur derjenigen Spannung mobilisieren, die es ihm ermöglicht, gemeinsam mit dem Autoren die ästhetischen Gegenstände des Werkes zu konstituieren.[25]

Nun geht es darum, diese gegenstandsorientierte divinatorische Spannung der künstlerischen Imagination von den einseitigen philosophischen Interpretationen zu befreien. Diese Befreiung ist als eine zweifache Befreiung zu betrachten: 1) Als eine Befreiung von der Annahme, dass eine perspektivierte und situierte allgemeine Sicht gegeben sei, die das Individuelle allgemein erklärbar macht[26] – also als eine Befreiung von der eidetischen Sicht im Sinne der Husserl'schen Phänomenologie. 2) Als eine Rehabilitierung der eben erwähnten gegenstandsorientierten Spannung.

4. DIE BERÜHRUNGSPUNKTE DER HUSSERL'SCHEN UND DER HEIDEGGER'SCHEN PHÄNOMENOLOGIE

Wenn wir die Berührungspunkte der Husserl'schen und der Heidegger'schen Phänomenologie suchen, können wir auf der Seite des Husserlianismus Eugen Fink, auf der Seite der Heideggerinterpreten Günther Figal nennen.

Fink's Studie *Vergegenwärtigen und Bild. Beiträge zur Phänomenologie der Unwirklichkeit*, aus dem Jahre 1927, ist gleichursprünglich mit Heidegger's *Sinn und Zeit*. Sie Fasst die intentionale Formung der Erlebnisse als eine Oberfläche einer Tiefenschicht der Prozesse der Depräsentation und der ursprünglichen Zeitlichkeit. Eben diese, dem Anschein nach destruierenden Impulse sind es, bei Fink, die einen Untergrund der Positivität der Gegenstände darstellen, die die "Bedingungen der Möglichkeit" für jede Gegenständlichkeit konstituieren. In der Auffassung Fink's die Reflexion, die die "schlafenden Intentionen" weckt, vermag aber ihre temporale Ferne und Dunkelheit nicht zu tilgen, belässt sie in ihrer Dunkelheit und Unbestimmtheit.[27]

Auch in anderen seinen Texten integrierte Fink die Kritik Heidegger's an Husserl.

Günther Figal interpretiert Heidegger als einen Philosophen des "möglichen Seins". Das praktische Bewenden mit den Werkzeugen bildet, der Figal'schen Interpretation nach, eine Gelegenheit für eine Variabilität der entstehenden Möglichkeiten.[28] Wenn wir aber praktisch handeln, sind wir in eine deformierte, nicht-authentische Gestalt des Seins inkarniert. Wir fixieren die ursprünglichen Antworten auf verschiedene Möglichkeiten unserer Lage in fixierte Projekte und destruieren so die ursprünglichen Möglichkeiten.

Um den Charakter der Möglichkeiten als Möglichkeiten wiederherzustellen, empfiehlt Figal unsere Projekte "zurückzunehmen", sie in ihre ursprüngliche Dimension der Antworten auf die Möglichkeiten zurückzukehren.

In der Optik der Figal'schen Interpretation ist die Geschichte ein Prozess der historischen Möglichkeiten, die auch durchaus anders realisiert werden konnten, als sie sich praktisch realisierten. Die Unbestimmtheit des entsteheden Seins fordert solche Antworten auf die authentischen Fragen, die uns davon befreien, Apologeten unserer Praxis zu sein.

Fink und Figal konkurrieren die Linie der einen und der gegenseitigen Strömung der Phänomenologie auf eine solche Weise, die die Sicht der Gegenseite integriert.

c/o Ing. Dušan Cibulka
Astrova 50
82101 Bratislava
Slovenia

ANMERKUNGEN

1 Zumr, J., "Mit Jan Patočka über die Philosophie und über die Philosophen," in: *Filozofický časopis* (Prag, 1967/5), S. 589.

[2] Ricoeur, P., *Zeit und Erzählung*. Bd. III. *Die erzählte Zeit* (München, 1991), T. 39.
[3] Husserl, E., *Formale und Transzendentale Logik* (Halle, 1929), S. 216.
[4] Grünewald, B., *Der phänomenologische Ursprung des Logischen* (Kastellaun, 1977).
[5] Schuhmann, K., *Die Fundamentalbetrachtung der Phänomenologie* (Haag, 1971), S. 120.
[6] Tugendhal, E., *Der Wahrheitsbegriff bei Husserl und Heidegger* (Berlin, 1967), S. 253.
[7] Merleau-Ponty, M., *Phänomenologie der Wahrnehmung* (Berlin, 1966), S. 10.
[8] Husserl, E., Manuscript A VII, S. 33.
[9] Husserl, E., *Ideen zu einer reinen Phänomenologie und phänomenologischen Philosophie*. In: Husserliana III, S. 129.
[10] Husserl, E., *Erfahrung und Urteil* (Hamburg, 1972), S. 423–424.
[11] ". . . das sich abhebende allgemeine vor allem von seinem Charakter der Zufälligkeit befreit werden muss . . ." (Husserl: *Erfahrung und Urteil*), S. 410). Vlg. Ströker, E.: "Die Phänomenologie Husserl's. Vorbedingungen, Ausarbeitung, Wirkungen," in: E. Ströker and P. Jansson, *Phänomenologische Philosophie* (Freiburg-Müchen: K. Alber, 1989), S. 85.
[12] Husserl, E., *Erfahrung und Urteil*, S. 410–411.
[13] Waldenfels, B., "Die Abgründigkeit des Sinnes. Kritik and Husserl's Idee der Grundlegung," in: Waldenfels: *In den Netzen der Lebenswelt* (Frankfurt a/M.: Suhrkamp, 1985), S. 21–22.
[14] Gurwitsch, A., *Bewusstseinsfeld* (Berlin, 1975), S. 228.
[15] Ricoeur, P., *Zeit und Erzählung*, III. *Die erzählte Zeit* (München, 1941), S. 336–337.
[16] Herrmann, F.-W. von, *Subjekt und Dasein*, 2. Ausg. (Frankfurt a/M., 1985), S. 56.
[17] Heidegger, M., "Prolegomen zur Geschichte des Zeitbegriffs," in: GA 20, S. 104.
[18] Herrmann, *op. cit.*, siehe die Bemerkung Nr. 1, S. 87–88.
[19] Heidegger, M., "Phänomenologische Interpretationen zu Aristoteles. Anzeige der hermeneutischen Situation," in: *Dilthey-Jahrbuch*, 1989/6, S. 238.
[20] Gadamer, H.-G., *Wahrheit und Methode* (Tühinyen, 1965), T. 335–338.
[21] Biemel, W., *Martin Heidegger in Selbstzeugnissen* (Hamburg. Rowohlt, 1973), S. 60.
[22] Lipps, H., *Untersuchungen zu einer hermeneutischen Logik* (Frankfurt a/M., 1959), S. 31.
[23] Frank, M., *Das individuelle Allgemeine* (Frankfurt a/M., 1977), S. 15.
[24] Sartre, J.-P., "Une ideé fondamentale de la phénoménologie de Husserl," in: *La Nouvelle Revue francaise*, 1939, S. 130.
[25] Ingarden, R., *Das literarische Kunstwerk* (Halle, 1931), S. 160 Vgl. Frank, M.: *Das Sagbare und das Unsagbare* (Frankfurt a/M., 1990), S. 31–32.
[26] Frank, M., siehe die Bemerkung Nr. 9.
[27] Fink, E., "Vergegenwärtigung und Bild," in: *Studien zur Phänomenologie* (Haag, 1966), S. 24.
[28] Figal, G., *Martin Heidegger – Phänomenologie der Freiheit* (Frankfurt a/M., 1988), S. 99.

JESÚS CONILL

CRÍTICA DE LA RAZÓN IMPURA: ENTRE NIETZSCHE Y ZUBIRI

Es conocida la reviatlización del pensamiento de X. Zurbiri a través de las aportaciones de José Luis Aranguren, Pedro Laín e Ignacio Ellacuría. Pero, a pesar de su fructífera influencia, sigue siendo verdad que, para comprender a Zubiri, se requiere una adecuada contextualización de su pensamiento, porque más de uno se ha desanimado y ha abandonado la ardua tarea de leer sus libros, al no contar con los puntos de referencia precisos, o bien al haber tenido la sensación de enfrentarse a una filosofía encerrada en sí misma.

Ahora bien, hoy en día es ya un logro bien reconocido, por ejemplo, la contextualización fenomenológica de Zubiri, gracias a la meritoria labor de Antonio Pintor-Ramos.[1] Sin embargo aun cuando haya que proseguir en esta línea (a todas luces imprescindible), convendría ampliarla en un sentido no exclusivamente historiográfico; porque lo importante no es sólo *de dónde* viene una filosofía como la de Zubiri, sino *a dónde* va. Se necesita ampliar la contextualización de su pensamiento, reinterpretándolo no únicamente desde el *texto* sino desde la *realidad vital*, por tanto, repensarlo – recrearlo – con el propósito de descubrir sus virtualidades para responder a los envites del mundo en que vivimos.

En este sentido creo que, para contextualizar hoy en día a Zurbiri, es muy relevante su relación con Nietzsche. Y, aunque algunos la consideren extravagante, debo aclarar que esta conexión Nietzsche-Zubiri se estableció por sí sola al ir leyendo a ambos autores por separado: en cada uno de ellos me resonaba el otro. No fue algo premeditado o una hipótesis de trabajo ficticia, sino una asociación espontánea que surgía continuamente en el curso del pensamiento.

Pero esta conexión entre Nietzsche y Zurbiri empezó a tomar cuerpo de modo más sistemático cuando afronté el problema fundamental que planteaba la crítica a la metafísica, proveniente de la genealogía nietzscheana. La metafísica contemporánea tenía que responder al radical reto nietzscheano: si realmente en una época reconocida públicamente como "postmetafísica" el saber metafísico ha quedado definitivamente arrumbado o si, por el contrario, su permanencia es insoslayable.[2]

A-T. Tymieniecka (ed.), Analecta Husserliana LIX, 441–455.

La experiencia actual de abismo (*Abgrund*) apunta a la sospecha de que todo en el fondo es contingente y caótico, de que sólo contamos con un orden funcional y convencional, de modo que escepticismo y nihilismo resultan inevitables. Pero, precisamente por eso, tal vez recupera su sentido aquel esfuerzo intelectual que surgió para orientarse en el mundo, para saber estar en la realidad, al que se denominó metafísica; y me parece que fue lúcido O. Külpe al anunciar a comienzos de siglo que en el umbral de la filosofía del futuro se encontraba el *problema de la realidad*.

Nietzsche recogió con tono trágico el problema de la realidad, que se venía arrastrando desde la teoría del conocimiento de Kant y que subsiste como tema de su tiempo en gran parte de la filosofía del XIX y comienzos del XX.

Curiosamente el problema de la realidad es el motor de la preocupación filosófica en el primer escrito de M. Heidegger[3] – otro de los maestros de Zubiri, junto a Ortega y Husserl – donde se lamenta de la trayectoria postkantiana de la filosofía, debido a que ésta se ha alejado cada vez más del problema de la realidad. Heidegger se acoge en este momento a las posibilidades que ofrece el realismo crítico de Külpe, de enorme importancia para la transformación realista del criticismo transcendental kantiano en el Racionalismo crítico de K. Popper y H. Albert.[4]

A este problema latente en toda la filosofía moderna y contemporánea, agigantado y llevado al extremo por Nietzsche, es decir, al de la aparente difuminación de la realidad, responde la filosofía de Zubiri.

A partir de Zubiri puede surgir una metafísica postnietzscheana, capaz de contribuir a que nos vayamos recuperando de la merecida crisis de la razón moderna, ya que la crítica de Nietzsche a la metafísica no alcanza a la metafísica zubiriana o, al menos, habría que reformularla de un modo preciso; pero en tal caso se descubrirían *aspectos comunes* entre ambas filosofías, que provocan la sospecha de si Zubiri, con intención o sin ella, de un modo más o menos consciente, no pudo estar movido por el interés de ofrecer un camino filosófico a la altura del reto nietzscheano.

Por lo menos, en lo concerniente al grave problema de la realidad, las referencias a Külpe y Heidegger abogarían en favor de esta hipótesis. Pero hay otros aspectos de las filosofías de Nietzsche y Zubiri que revelan una posible conexión muy fructífera, como por ejemplo, el modo de entender la razón, el lenguaje y lo moral.

Comenzaré con una reflexión sobre el horizonte filosófico en que se sitúan Nietzsche y Zubiri, porque sus obras les han parecido a muchos "intempestivas" o "extemporáneas". Y es que no es tan fácil entenderlos rebasando las presentaciones panfletarias – como ha sido habitual en la divulgación de Nietzsche – o los deformados estereotipos, en el caso de Zubiri.

1. HORIZONTE DEL PENSAMIENTO: ¿SUPERACION DEL NIHILISMO?

Un aspecto común a Nietzsche y Zubiri fue sin duda la admiración que sintieron por los griegos, especialmente los presocráticos. Pero a pesar de sentirse cerca de su pensamiento, inmerso en el ámbito de la *physis*, y recurrir a él en muchas ocasiones, no es ese el horizonte filosófico en que se encuentran.

Tampoco se sitúan ya sin más en el horizonte del criticismo moderno, que todavía confía en la *conciencia* y en la *lógica*, sino que, más bien, intentarán superar el subjetivismo en que quodó atrapado y el idealismo en que desembocó.

Antes bien, Nietzsche y Zubiri coinciden en atribuir una importancia epocal a la *desfundamentación* y al *ateísmo*, así como al *desarraigo* característico de nuestra situación intelectual.

Por lo que respecta al ateísmo, Zubiri llega a afirmar que "hay un ateísmo de la historia"; y añade a continuación: "el tiempo actual es tiempo de ateísmo, es una época soberbia de su propio éxito. El ateísmo afecta hoy, *primo et per se*, a nuestro tiempo y a nuestro mundo" (*NHD*, pp. 394–395).

No resulta difícil comparar este ateísmo epocal con el sentido nietzscheano del anuncio de la "muerte de Dios". Porque se trata en ambos casos de un acontecimiento histórico experienciado, de una verdad histórica vivida, que caracteriza el mundo en que estamos inmersos.

Por otra parte, Zubiri se percata también del estado de desfundamentación en que vivimos: "Nuestra época es rica en ese tipo de vidas (. . .) ante las cuales surge siempre un último reparo: 'Bueno, ¿y qué? . . .' (. . .) Como época, nuestra época es época de desligación y de desfundamentación" (*Ibid.*).

Además, esta desligación y desfundamentación produce un desarraigo vital, que tiene importantes repercusiones para la configuración del sujeto humano: "no podemos olvidar que es también la época de la crisis de la intimidad" (*Ibid.*). Este "desarraigo", según Zubiri, es "el problema más hondo de la existencia". Así que "sólo lo que vuelva a hacer arraigar nuevamente la existencia en su primigenia raíz puede restablecer con plenitud el noble ejercicio de la vida intelectual" (*NHD* 25).

Nietzsche y Zubiri coincidirán en detectar que el ámbito en que se viven y resuelven (en su caso) este tipo de problemas no es el de los razonamientos y las estructuras lógicas, sino en un nivel más profundo: el de la *experiencia*. Sólo el camino de la experiencia nos sitúa donde se vive el sentido del ateísmo, según Zubiri (en la posibilidad de sentirse religado o desligado), y de la "muerte de Dios", según Nietzsche.

De ahí que Zubiri escribiera lo siguiente: "es necesario *apurar aún más la experiencia*".[5] Lo mismo que Nietzsche con el nihilismo, apuesta Zubiri por ir al fondo de la experiencia humana, aunque sean diferentes los sabores que al final nos ofrezcan.

Asi, Zubiri confía en que, apurando la experiencia, "llegará seguramente la hora en que el hombre, en su íntimo y radical fracaso, despierte como de un sueño encontrándose en Dios y cayendo en la cuenta de que en su ateísmo no ha hecho sino estar en Dios. Entonces se encontrará religado a Él, no precisamente para huir del mundo, de los demás y de sí mismo, sino al revés, para poder aguanatar y sostenerse en el ser" (*NHD* 395–396).

Sin embargo, Nietzsche sólo pone su esperanza en la superación del nihilismo reactivo (pasivo), si somos capaces de crear un nuevo "sentido de la tierra" a través de lo que se expresa mediante la metáfora del "superhombre".[6]

Pero ninguno de ellos consideraría el nihilismo como una propuesta teorética, sino como un acontecimiento histórico expriencial y vital, un "destino", que cambia la orientación del pensamiento y de la vida humana. Tanto Nietzsche como Zubiri se debaten dentro de un horizonte que ya no es ni antiguo (premoderno) ni moderno, sino posgriego y poscristiano, y que Antonio Pintor ha intentado sintetizar con la fórmula "*horizonte de la intramundanidad*".[7]

Por mi parte, añadiría que la intramundanidad, que ya hizo su aparición como tal en el enfoque moderno, tiene ahora rasgos propios, como la desfundamentación y el desarraigo; y sobre todo adquiere un nuevo carácter experiencial, cuyo resultado configura otro modo de estar en el mundo, que cada vez confía menos en los esquemas racionalizadores – presuntamente liberadores – propios de la conciencia moderna.

Así pues, conscientes del fracaso del idealismo de la razón moderna, se sitúan ambos en un horizonte de *pensamiento experiencial*, al que cabe considerar, por tanto, de condición posmoderna, postconceptista y postidealista; y en el caso de Zubiri, a su vez, postnietzscheana,[8] ya que sólo desde el *abismo* y el *enigma* – que nos abre la experiencia vital – será "quizás" posible superar el nihilismo.

2. MÁS ALLÁ DEL CRITICISMO E IDEALISMO MODERNOS

Nietzsche y Zubiri pretenden ofrecer una alternativa al criticismo moderno, basado en la autoconciencia psicológica o lógica del sujeto, rebasando el enfoque epistemológico,[9] puesto que los conceptos y las ideas producidos por las diversas modalidades de filosofía de la conciencia ni cuentan con ningún punto de apoyo frime, ni se han percatado del proceso y trasfondo del que realmente surgen.

Ni la lógica, ni la epistemología han llegado al fondo de la cuestión; tal vez una "genealogía" (Nietzsche) o una "noología" (Zubiri) alcancen las *infraestructuras de la razón*. Por eso la búsqueda nietzscheana y zubiriana de un ámbito previo a lógica, a la conciencia, al lenguaje gramaticalizado, a los conceptos y a la relación sujeto/objeto, constituye una común preocupación, que tiene por objeto superar el "ideísmo" moderno, en terminología de Zubiri, y las deformaciones (ocultaciones e incluso "mentiras", según Nietzsche), producidas por la conciencia moderna.

Así, mediante la genalogía Nietzsche descubre, por ejemplo, que la conciencia lógica "no es más ni menos que el comentario fantástico de un texto desconocido, quizá incognoscible, pero *sentido*" (*Aurora*, pp. 74 y 75). Algo parecido podría decirse del ulterior desarrollo que las reactualizaciones del logos y de la razón llevan a cabo, según Zubiri, a partir de *lo sentido* en la "aprehensión primordial". El logos y la razón aportarían asimismo el comentario libre de lo que de modo inefable estamos sintiendo.

Y esto vale igualmente para el lenguaje, porque no debe olvidarse que, para Nietzsche, conciencia y lenguaje van de la mano (*La gaya ciencia*, prgf. 354); de ahí que resulte ridículo oír tantas veces en nuestros días – sin más reflexión autocrítica – que mediante el presunto cambio de paradigma desde la conciencia al lenguaje se va a lograr un pensamiento postmetafísico.

Por tanto, lo que hace falta para superar realmente la filosofía moderna es un nuevo hilo conductor, que no nos aleje de la realidad por odio a los sentidos, ni construya mundos ideales como los de la epistemología y la ontología moral.[10] Nietzsche lo encuentra en el *cuerpo* y Zubiri en el análisis del *sentir*; en ambo casos se desemboca en una razón experiencial: corporal como la "gran razón" en Nietzsche y "razón sentiente" en Zubiri. De esta manera se espera superar la "logificación de la inteligencia", denunciada por ambos, y desbaratar así todo dualismo (antiguo o moderno), a partir del "instinto de realidad" (en Nietzsche) o de la "impresión de realidad" (en Zubiri).

Esto supone un cambio profundo en el modo de entender la realidad, la razón, el lenguaje y la vida moral. Pero a estos cambios subyace un nuevo método, el único que es capaz de superar el idealismo y que podríamos denominar "*physikós*". Este método nos ofrece una nueva vía de acceso originario – físico – a lo real a través del cuerpo y del sentir, por tanto, sin dejar de lado ni menospreciar la animalidad humana, sino incorporándola en el proceso intelectivo y en todos los ámitos de la vida.

Nietzsche y Zubiri se oponen a la tendencia idealista a fabricar el ser a la medida del logos.[11] Con lo cual, el posible "fundamento" (*Grund*) tendrá aquí un carácter abierto a la eperiencia, de la que tampoco está fuera el abismo (*Ab-grund*) y el caos, puesto que la filosofía no tiene ya como matriz el orden lógico. Antes bien, el valor de la lógica se determina desde instancias no-lógicas, previas a la conciencia y al lenguaje gramaticalizado, donde *sentimos* la fuerza de las impresiones y *vivimos* el poder de lo real.

3. LA REALIDAD COMO PODER

La primera novedad que introduce la noción zubiriana de realidad, y que la asemeja a la de Nietzsche, consiste en que no alude a la realidad "allende" toda aprehensión sensible, ni a la realidad "en sí" de una cosa real en un mundo independiente de mi percepción. Pues no hace falta salir de la aprehensión para sentir la realidad; antes bien, "realidad" designa una relación o, en terminología zubiriana, una "*formaldad* en impresión" (el "de suyo" de lo sentido), desde donde es preciso rectificar toda la historia de la metafísica, ya que ésta se ha dedicado a entificar y objetualizar la realidad.

Con esta nueva idea de la realidad va unido inseparablemente un nuevo modo de entender la inteligencia, que permite sequir descubriendo semejanzas entre ambos filósofos. Porque se trata de una inteligencia, cuya primera función es estrictamente bilológica, de manera que la realidad tendrá que tomarse también primordialmente como formalidad, pero no de carácter lógico, gramatical o lingüístico, sino dependiente de una *habitud* o modo de habérselas con las cosas, que está en relación con las estructuras del cerebro, entendido como "órgano de hiperformalización".

Estas ideas zubirianas confluyen con la radical crítica nietzscheana de los dualismos idealistas, antiguos y modernos, que se basan en la escisión entre dos mundos, en la separación entre lo fáctico y lo ideal normativo, en el recurso a "trasmundanons" por desprecio del mundo real y en la necesidad de fingir un mundo "en sí".

Y algo parecido ocurre con la idea de la "verdad real" zubiriana, que se asemeja a lo que algunos[12] vislumbran como "verdad originaria" o "modo de habérnosla" con el mundo, y que consiste en una verdad antepredicativa, prejudicativa, en un contacto originario con lo real, todavía no deformado por las instancias que se dedican a ordenar lógica y moralmente el caos de las impresiones originarias.

Otra importante novedad de la noción zubiriana de realidad, que converge con las posiciones nietzscheanas, es su tratamiento de la realidad como *poder*. Zubiri ha prestado cada vez más atención al problema de la experiencia del poder de lo real y su filosofía de la readidad se ha ido perfilando como filosofá del poder.

El análisis zubiriano – del que no estaría lejos el nietzscheano – detecta que en el fondo de la experiencia de la realidad el hombre hace probación física de lo que es el poder de lo real, experienciando que éste se apodera de nosotros; y como este apoderamiento acontece ligándonos al poder de lo real, al "hecho" de este apoderamiento le da Zubiri el nombre de "religación".

Además, en esta experiencia de la religación se manifiesta el carácter enigmático del poder de lo real y el problematismo de nuestra vida. Y, como lo enigmático nos inquieta, la experiencia del enigma del poder de lo real nos mueve hasta que nos sentimos lanzados a buscar el fundamento. "La voluntad de verdad se radicaliza hasta convertirse en voluntad de fundamentalidad".[13]

En cambio, Nietzsche abandona la exigencia de fundamentación, porque no quiere perpetuar el "error del fundamento", requerido por los esquemas lingüístico-gramaticales de la razón lógica y la conciencia. El dinamismo vital y la metaforicidad del lenguaje exigen renunciar al persamiento fundamentador. "El valor para la vida decide en último término".[14] Y ¿qué es la vida? Vida es voluntad de poder, contrapuesta a la voluntad de verdad.

Se impone la comparación con Zubiri, para quien el dinamismo dirigido hacia el fundamento no surge de la razón concipiente y logificada, sino de la fuerza misma con que la realidad se actualiza en la "razón sentiente". Entonces, ¿quién tiene más "razón" en la cuestión del fundamento, Nietzsche o Zubiri? ¿No tiene sentido abrirse al fundamento desde una razón sentiente, en cuyo ámbito experienciamos el enigma y el abismo del poder de lo real, como en Nietzsche?

A mi juicio, en la perspectiva zubiriana no habría ruptura entre volunted de poder y voluntad de verdad, porque ambas brotan de una *voluntad de realidad*, de donde surge, a su vez, la voluntad de razón, estructurada por la libertad.

Por eso, aunque, según Zubiri, la experiencia del poder de lo real nos abre el ámbito de la fundamentalidad, el hombre puede optar libremente por suspender tal fundamentalidad, reduciendo la realidad-fundamento a una realidad-objeto. Aquí lo decisivo es la respuesta (tras probación e interpretación) que se dé al enigma de la experiencia de la realidad como poder: *factualidad o fundamentalidad* se nos presentan como dos posibles experiencias del poder de lo real y su carácter enigmático.

El análisis zubiriano de la religación y el nietzscheano del poder apuntan a un fenómeno común, que se resuelve de modo diferente, tanto intelectiva como opcionalmente, porque se "hacen" dos experiencias de realidad diferentes. Una conduce al problema de Dios; la otra, al ateísmo y al nihilismo: no hay fundamento del poder de lo real. Basta su facticidad.

Pero debe quedar claro que no es por exigencia lógica y meros conceptos, sino por *experiencia* (probación física) del poder de lo real y su carácter enigmático, como llega Nietzsche a una interpretación y opción favorable a la facticidad del poder de lo real. Como Nietzsche, Zubiri ha seguido también el camino de la *experiencia* y ha asumido el reto que supone la opción por la *factualidad* del poder de lo real en el horizonte de la intramundanidad, una experiencia decisiva en la vida contemporánea, marcada por el nihilismo.

A mi juicio, esta renuncia contemporánera al fundamento nos sumerge en una facticidad horadada de nada, de la que vivimos en nuestro horizonte cultural, y cuyo resultado es el imperante pragmatismo nihilista; dicho de otro modo: somos fenoménicamente pragmáticos y nouménicamente nihilistas. Por eso, constituye un mérito de Zubiri haber intentado mostrar que la marcha (el dinamismo) de la razón experiencial no conduce necesariamente al nihilismo, sino que como razón sentiente se abre a la *fundamentalidad* desde su estructura libre, a partir de la primigenia impresión de realidad.[15]

Esta concepción de la realidad como poder no sólo es la clave para entender el fondo de la experiencia (como hemos visto), sino también el punto de conexión entre la realidad y el sentido. Según Zubiri, no podemos abrirnos al sentido (y, en su caso, al valor), sin pasar por una consideración del poder. "Porque lo que es el sentido de todos los sentidos, el constructo de todos los constructos y la condición de todas las condiciones, el poder de todos los poderes es justamente el poder de lo real en tanto que real" (*EDR* 235).

Aunque Zubiri considera que el carácter primario de las cosas con las que el hombre tiene que habérselas es la realidad, el sentido de esa realidad aparece, cuando se la mira desde el punto de vista de lo que el hombre quiere hacer (*SH* 327), por tanto, en la medida en que abre *posibilidades* de las que pueda apropiarse para acondicionar su vida.

De un modo análogo, también para Nietzsche, el sentido depende de la realidad, puesto que "la voluntad de poder es el *factum* último a que descendemos" [40 (61)] y "la voluntad de poder *interpreta*" (VIII 1, 137). El proceso en que consiste este interpretar pone en marcha un dinamismo de apropiación de posibilidades: "en realidad *la interpretación es ella misma un medio para enseñorearse de algo*" (VIII 1, 137).

La interpretación es una actividad simplificadora a partir de una originaria "fuerza *creadora*", que genera "nuestras *posibilidades*", nuestras formas de vida y actividades, por consiguiente, los más diversos sentidos y valoraciones en dependencia de unos impulsos que son reductibles a la *voluntad de poder*.

Así pues, también la voluntad de poder funciona en la línea de las posibilidades, estableciéndose una relación entre la realidad como poder y el sentido a través de las posibilidades y apropiaciones, por las que acondicionamos nuestra vida. Tanto en Zubiri como en Nietzsche se ponen en juego dinamismos de posibilitación, apropiación y acondicionamiento, de lo que es buen ejemplo el fenómeno de la técnica.[16]

Ahora bien, la *distancia entre sentido y realidad*, propuesta por Zubiri,[17] permite una *crítica* del sentido a la luz de la realidad: evaluar en toda *apropiación* la solidez del sentido, si nos enriquece o empobrece, desde el punto de vista de la realidad. Porque los "recursos" pueden convertirse en obstáculos y las posibilidades, en amenazas. Hay que estar en guardia frente a tales abusos: por ejemplo, ante el ambivalente progreso técnico; o bien, según Nietzsche, ante la degeneración de los instintos a través de determinadas configuraciones de sentido, como las del judaísmo y del mundo moderno.

4. LA "GRAN RAZON" DEL CUERPO Y LA INTELIGENCIA SENTIENTE

Frente a la hegemonía moderna de la conciencia Nietzsche aporta un nuevo comienzo (¿equivalente a otro "discurso del método"?): "partir del *cuerpo* y utilizarlo como hilo conductor", dando como resultado, a mi juicio, una filosofía del cuerpo,[18] una alternativa a las – hasta ahora – prevalentes filosofías del *ser*, de la *conciencia* y del *lenguaje*.

En definitiva, somos cuerpo y en él encontramos la auténtica razón, ya que "el cuerpo es una gran razón":[19] la razón en el cuerpo (*Vernunft im Leibe*) y en la vida (*Vernunft im Leben*); todo lo demás son "instrumentos" del cuerpo".

Tanto este seguimiento del hilo conductor del cuerpo en Nietzsche como el análisis de la intelección sentiente en Zubiri constituyen una *crítica de la razón lógica* como vía de acceso a la realidad.

Según Nietzsche, no hay ninguna exigencia objetiva para pensar lógicamente, dado que las regularidades y el orden de las cosas proviene de una coerción subjetiva práctica. Las categorías lógicas no tienen valor cognoscitivo; sólo consisten en esquemas ficticios o en imperativos prácticos para arreglarnos un mundo a nuestro servicio.

La lógica es útil, pero es una falsificación de la realidad, una "mentira", que es inaceptable como "medida de la vida". La lógica no es lo originario y primordial, sino que tiene profundas raíces en la vida orgánica. Es un invento de "animales inteligentes" en su lucha por la existencia. Pero hubo animales inteligentes antes de que hubiera razón lógica.

Por consiguiente, tal vez sólo una *formalidad biológica* al estilo zubiriano podría hacer frente a esta crítica nietzscheana de la razón presuntamente autónoma y pura, basada en la formalidad lógica. Porque en la base de lo lógico encontramos un acto nológico: "Antes de que se haya 'pensado' (*gedacht*), se tiene que haber 'inventado' (*gedichtet*), el sentido formador es más originario que el 'pensante'" (VII 3, 377).

La fuerza que mueve a logificar (por ejemplo, a igualar) es la "voluntad de poder". Pero no podría haber igualación judicativa, si primero dentro de las sensaciones no hubiera actuado una cierta igualación: tiene que haberse realizado un "proceso de *asimilación*", "una actividad intelectual" originaria, que todavía no cae dentro de la conciencia (VII 3, 366 s.).

Estas consideraciones nietzscheanas de la lógica son comparables con los aspectos biológico y pragmático del "inteleccionismo" zubiriano, del que debería resaltarse con más rotundidad su "sensismo"[20] práctico. Al igual que hemos visto que, para Nietzsche y Zubiri, no todo lo real es (de estructura) racional, tampoco la razón misma es primariamente algo lógico, sino que la razón intelige la realidad con la *fuerza coercitiva* propia del cuerpo y del sentir. Así lo muestran sus respectivas genealogía y noología de la *experiencia*.[21]

La "genealogía de la experiencia" de Nietzsche[22] nos hace retroceder al mundo primitivo de las impresiones, intuiciones y metáforas, al que se contrapone otro mundo construído de conceptos, como instancia reguladora que impone un orden, y bajo el que queda sepultado el fogoso torrente primordial, compuesto por una masa de imágenes, "que surgen de la capacidad originaria de la fantasía humana", ya que el hombre es un "animal fantástico".

También Zubiri da enorme importancia al "pensar fantástico": "Es absolutamente imposible eliminar los esquemas de pensamiento que yo llamaría fantásticos". "Junto a él [pensar fantástico] está el otro pensar raciocinante en forma de lógica, del que se cree que piensa nada más que con conceptos abstractos, cuando en realidad el pensar fantástico y el pensar raciocinante son pura y simplemente dos vertientes de una sola realidad que es el pensamiento humano".[23]

En último término, la genealogía nietzscheana descubre que el origen de la experiencia está en las *impresiones* y en el *impulso hacia la construcción de metáforas*, es decir, en vivencias originarias, que son el trasfondo de la libre creación. De ahí que los "estados de conciencia" sean estados "de quinto orden comparados con el valor de los instintos".

Zubiri parece conciliar en su noología lo que en la genealogía nietzscheana se presenta en clave de conflicto. Armoniza intuición y razón, de cuya contraposición se sirve Nietzsche para distinguir dos tipos humanos, el hombre intuitivo y el racional. Tampoco hay oposición entre intuición y concepto, sino que simplemente son dos formas de intelección; pues, antes de ser dos fuentes de conocimiento, son modos de actualización de lo real, en la que lo decisivo no es el contenido, sino la unidad proporcionada por la formalidad de realidad (proveniente de la impresión de realidad). Asimismo, para Zubiri, hay continuidad entre impresiones y conceptos, entre el mundo sensible y el inteligible, basándose en la impresión de realidad. Y, si Nietzsche optó por seguir el hilo conductor del cuerpo para regenerar la razón, también Zubiri descubre la radicación de la razón en el orden sentiente, que constituye el origen estructural de la intelección racional, capacitada para desarrollarse como razón poética,[24] donde las metáforas son un modo de *dar razón*, porque tampoco hay ruptura alguna entre impresiones, emtáforas y razón. Las metáforas tienen, según Zubiri, tanto de racional como los conceptos, pues los unos y las otras se mueven en el espacio de la libre creación experiencial de la realidad.

Así pues, la noología zubiriana puede considerarse en bastantes aspectos como una armonización de elementos de la experiencia humana, que en la genealogía de Nietzsche se presentaban dislocados y trágicamente enfrentados.

5. GENEALOGIA Y NOOLOGÍA DEL LENGUAJE

La *genealogía* nietzscheana conduce hasta las raíces metalógicas y metalingüísticas: a un instinto fundamental del hombre, al impulso constructor de metáforas. Pero esta concepción genealógica se comprende mejor, si se la resitúa en el contexto de la tradición neokantiana, tanto gnoseológico-fisiológica (F. A. Lange), como crítico-lingüística (especialmente G. Gerber), aun cuando esta última haya pasado más desapercibida.[25]

Porque es precisamente a partir de esa tradición de donde recibe la influencia en los aspectos más destacables; ante todo, el carácter metafórico del lenguaje, que explica, según Nietzsche, el fracaso de todas pretensiones de alcanzar la verdad, entendida como expresión adecuada de la realidad. Pero incorpora también que la función del lenguaje no es reproducir la realidad en sí, sino que constituye una forma de arte, que surge de instinto artístico del hombre y que goza en su proceso de producción de un cierto margen de libertad, ya que las palabras son tropos y las proposiciones, figuras. De ahí la defensa de la retórica (como prosecución de los recursos artísticos ínsitos en el lenguaje).

Avanzando en esa línea, Nietzsche critica la gramática,[26] porque nos fuerza a pensar bajo la forma del juicio y nos somete a la estructura sujeto-predicado; critica la fe en los "esquemas" lingüístico-gramaticales, porque no garantizan la relación del lenguaje con la realidad, ni la verdad. Antes bien, con la fijación gramaticalizada se pierde la vida y la experiencia primordial del lenguaje en su carácter energético (Humboldt).

Nietzsche quiere ir "más allá de la gramática", penetrando en el dinamismo de la vida y de los sentidos. También en este punto cabe una comparación, por ejemplo, con la crítica zubiriana del esquema lógico-lingüístico de la predicación y con su defensa del estado contructo a partir del logos sentiente (vía *physikós*).

Ahora bien, el problema de la relación entre el pensamiento y el lenguaje se complica enormemento en Nietzsche. Porque hace sospechar que la determinación lingüística del pensamiento desfigura la realidad e incapacita para lograr la verdad. Este es el problema que se plantea en el fondo, si se radicaliza de modo relativizador (negativo) la consideración del lenguaje como acepción del mundo (*Weltansicht*). El lenguaje ya no es mera imagen o figura (*Bild*) del mundo sino imagen engañosa (*Trugbild*), es decir, una quimera lingüística. Porque, debido a la coerción lingüística, el hombre se ve sometido a la necesidad de desfigurar la realidad y no tiene más remedio que engendrar una esfera de engaño, fraude y mentira.[27] Y la razón de Nietzsche es conocida: "los medios de expresión del lenguaje no se prestan para expresar el devenir" (VIII 11 (73)).

Lo que ofrecen los medios de expresión es, a lo sumo, "una relación": "las relaciones de las cosas con los hombres" (*WL* 5). Así, por ejemplo, la dureza de una piedra nos es conocida sólo en forma de contacto: se *nota* dura. El predicado "duro" expresa sólo una relación. Y esto vale para cualquier medio de expresión: en todos ellos se expresa sólo cómo *es* algo *para* el que hace tal experiencia.

Así que el lenguaje actualiza las cosas, no como son, sino en una perspectiva extraña a ellas. Pero ¿no pasa Nietzsche demasiado rápidamente del carácter metafórico del lenguaje a su crítica (hiperbólica) como "ilusión", "engaño" y "mentira"? ¿No haría juego con la posición nietzscheana alguna otra noción de realidad, como la que ofrece Zubiri?

Por ejemplo, la noología de Zubiri nos muestra la *realidad* de las cualidades sensibles (*IS*, 171 ss. [Apéndice 5]). Para Zubiri las cualidades sensibles son reales y no meramente subjetivas, dado que su noción de realidad incorpora la "relación" a que hacía referencia Nietzsche y, por tanto, el "*de suyo*" zubiriano tiene un carácter relacional: no es un "en sí", sino una peculiar "realidad para mí".

Y en lo concerniente al lenguaje, Zubiri está más cerca de la tradición en que se sitúa Nietzsche que de los habituales formalismos. Ya en su etapa fenomenológica fue consciente de los límites del formalismo lingüístico, defendiendo una radicación experiencial del lenguaje en el mundo vital y destacando su carácter instrumental y pragmático.[28] Pero el lenguaje no se encuentra en el orden de la aprehensión primordial de realidad, a la que considera "inefable", sino en el proceso intelectivo del logos y de la razón, caracterizado por la libre creación. El lenguaje es resultado de la capacidad para crear nuevas formas de actualizar lo real. La expresividad lingüística, ganada en la distancia y en la libertad, tiene la fuerza creativa y transformdora de las metáforas, desarrollando un fondo potencial que enriquece nuestra experiencia de lo real.

En esa línea cabe entender firmaciones zubirianas como las siguientes: "El lenguaje mismo no es, para los efectos de la intelección, algo meramente representativo [. . . , sino que] tiene además una función diferente. [. . .] no se trata de *constatar* la realización de representaciones, sino de *experienciar* una dirección" (*IR* 215). Lejos de toda logificación lingüística de la intelección, aquí se destaca el carácter *experienciador* del lenguaje y se muestra su capacidad para ampliar creativamente la expriencia de la realidad.

El aprecio zubiriano del carácter metafórico del lenguaje en la razón sentiente conecta con la posición de Nietzsche, para quien el lenguaje es originariamente metafórico y está ligado a un impulso corporal a la creación de ritmos, formas y figuras ("*Erfinden von Gestalten*"). Ahora bien, desde la perspectiva zubiriana tampoco habría ruptura entre lingüísticidad y realidad, sino que el lenguaje serviría para potenciar la creatividad.

Universidad de Valencia

NOTAS

1 A. Pintor-Ramos, "Zubiri y la fenomenología," *Realitas* III–IV (1979), pp. 389–565.
2 J. Conill, *El crepúsculo de la metafísica* (Barcelona: Anthropos, 1988).
3 "Das Ralitätsproblem in der modernen Philosophie" de 1912.
4 cfr. J. Conill, *El crepúsculo de la metafísica* (Barcelona: Anthropos, 1988), cap. 3 y *El enigma del animal fantástico* (Madrid: Tecnos), cap. 3.
5 Subrayado nuestro.
6 J. Conill, *El enigma del animal fantáastico* (Madrid: Tecnos, 1991), cap. 8.
7 A. Pintor, *Realidad y sentido* (Salamanca: Universidad Pontificia, 1993), p. 303.
8 D. Gracia, *Voluntad de verdad* (Barcelona: Labor, 1986), p. 15.

[9] "Cartesianismo epistemológico," en terminología de H.-G. Gadamer, *Verdad y método*.
[10] J. Conill, *El crepúsculo de la metafísica*, cap. 6.
[11] J. Ortega y Gasset, "La idea de principio en Leibniz."
[12] Granier, Stegmaier, Vermal.
[13] Diego Gracia, *Voluntad de verdad*, p. 250.
[14] J. Conill, *El poder de la mentira*, Tecnos, Madrid, 1997.
[15] Cfr. J. Conill, *El enigma del animal fantástico*, pp. 193–197.
[16] J. Ortega, J. Conill, *El enigma del animal fantástico*.
[17] Crf. A. Pintor-Ramos.
[18] Para una filosofía del cuerpo en línea zubiriana, cfr. P. Laín, *El cuerpo humano* (Madrid: Espasa-Calpe, 1989); *Cuerpo y alma* (Madrid: Espasa-Calpe, 1991).
[19] F. Nietzsche, *Así,* Así habló Zaratustra, p. 60.
[20] A. Pintor-Ramos, *Realidad y sentido*.
[21] Cfr. J. Conill, *El enigma del animal fantástico*, pp. 189 ss.
[22] Cfr. *Verdad y mentira en sentido extramoral* (= WL).
[23] *El problema filosófico de la historia de las religiones*, p. 128.
[24] G. Marquínez, "Las ideas estéticas de Zubiri y el realismo fantástico latinoamericano," *Estudios filosóficos* 105 (1988), pp. 297–316.
[25] Meijers, *art. cit.*
[26] Cfr. J. Simon, "Grammatik und Wahrheit"; "sprache und Sprachkritik bei Nietzsche."
[27] J. Conill, *El poder de la mentira*, Tecnos, Madrid, 1997.
[28] A. Pintor-Ramos, *op. cit.*

CARLA DANANI

LIFE, EXPERIENCE AND UNDERSTANDING IN DILTHEY'S THOUGHT

THE TASK OF PHILOSOPHY

In an essay dated 1907,[1] which deals with the "essence of philosophy", Dilthey states that the youth of every thinker is full of the struggle with the mystery of life and of the world.[2] Philosophy is about life and our living in the world. In Dilthey's thought this concept is developed in various ways.

A precise grasp of the essence of philosophy is to be gained through an analysis of the way in which philosophy has been practiced in the different historical periods.[3] The subject matter of philosophy is essentially the mystery of the world and life; if this makes philosophy different from those sciences which focus on specific aspects of the world, it is its attempt to give a solution which is universally valid that distinguishes it from art and religion.[4] Philosophy feeds on the cultural background of time, place, human relationships and personality itself.[5] Moreover, whatever man does stems from his spiritual life (*Seelenleben*) and from his relationship with the external world.[6] According to Dilthey philosophy is "an undertaking (*Leistung*) that derives from the need felt by the individual spirit to reflect on its activity; to rework and consolidate from within its activity; to give a firm shape to its relations with the whole of human society."[7] How works philosophy? Dilthey considers that the criterion should be that "life must be explained on the basis of life itself".

Heidegger said that speaking about "philosophy of life" was like saying "botany of plants";[8] however, the history of philosophy taught Dilthey the opposite.[9]

But this was not all. In the phrase "philosophy of life", "life" is both subject and object: life is meant as the source of philosophy and also as its theme. For Dilthey an individual who carries out philosophical enquiries is not unrelated to the object of his studies; the point of view from which his reflection starts cannot be outside his theme, life.[10]

For Dilthey this means reconnecting the thinking to the living process.[11] But the fact that thinking lies in life has an important conse-

A-T. Tymieniecka (ed.), Analecta Husserliana LIX, 457–472.

quence: it cannot know what lies behind it.[12] To explain thinking we must start with life; the reversal cannot be done: life cannot be fully understood with our thoughts.[13] He points out that "whatever we know has its roots in something that we will never be able to understand in depth. But we can describe it."[14] When he says that life cannot be known, he means that it cannot be analysed, that no formulae or explanations can tell us what it is – as there are no concepts that come before it and to which it can be reconnected.

Reason is not the primary creating factor in the world.

We can, following Dilthey's thought, do nothing more than try to describe this life, in its various forms, elements and functions.

NATURE AND SPIRIT: DISTINGUISHING AND SEPARATING

Let us develop some of the themes we have sketched out above. Dilthey says that life is given to us directly, to each of us, as our own life, and that it is only starting from our own life that we understand life around us, the other forms of human and animal life.[15] Therefore, the one who wonders about the mystery of life cannot do anything but begin his investigation from here.

Starting from here means starting from lived experience. "We call lived experience", Dilthey says in a work he wrote before 1880, "that process in man's consciousness through which something real [*ein Wirkliches*] is disclosed to consciousness itself. This essential reality [*Wirkliches*] can be an external thing, an external process or a fact of psychic life."[16] The subject of *Erlebnis* is a psycho-physical unity; the whole of human nature cannot be divided, except as an abstraction. This unity must be taken into account even by a science that is interested only in historical and social events: it cannot deal only with the spiritual side of man. It is here that the distinction between the natural sciences and the human sciences encounters a problem which the process of justifying that distinction cannot disregard.

Dilthey recognizes the connection between spiritual and physical reality. He says that sciences create and expound their theories, their concepts, always with reference to mankind; and this process involves the living connection between what we later try to distinguish as physical and psychic. "We ourselves are nature", Dilthey states, "and nature works inside us, unconsciously, with obscure impulses; our states of consciousness are always expressing themselves in sounds, facial gestures, words and are objectified in institutions, states, churches, scientific academies: it is through these connections that history proceeds."[17] History too regards the connection between what is physical and what is psychic.

While this does not prevent us from making a distinction, we must recognize that the abstractions which derive from this distinction do not represent entities, full realities.[18] The ground of the distinction between the human sciences and the natural sciences cannot be found in facts taken by themselves. On the contrary, these different elements, objects, are made, and obviously, "the object of human studies is made differently from the object of the natural sciences."[19]

The concept of object, Dilthey says, is grounded in the relations between sensorial impressions and something which is different from the self, that is a connection between these sensorial impressions in a whole which stands therefore independently in front of the self. Thus, there are two different situations. If there is not any reason why the self or something analogous to it should be projected onto some of these objects, Dilthey specifies, they are said to be "nature". On the other hand, if we are forced to project a psychic event onto animal and human beings what we find is a spiritual reality. Dilthey realizes that we are not able to determine the point from which we can project that psychic event onto living beings; he concludes his reasoning by stating the principle that "the existence of psychic life is revealed by the presence of voluntary movements."[20] A further criterion of discrimination is later established on the basis of the regularity, coordination and familiarity of psychic life as it appears in human beings compared with animals.

Throughout his work, Dilthey tries to give the framework of the human sciences a configuration independent of the natural sciences. This appears to be possible, but Dilthey's conclusions seem to be reached by passing without adequate justifications from distinction to separation of the two fields, and by ignoring the original premises *Erleben* is reduced to the field of spiritual reality.[21] Dilthey accepts the theory of the intellectuality of perception, according to which perceptions already contain the formal categories of thought. On this basis the intellectual faculty can be seen as a special development of perception and conversely experience is understood as something complex and rich – which comes before the distinction between sensibility and intellectuality. On the other hand, however, Dilthey reduces lived experience to the "psychic" dimension. This passage does not seem to be adequately justified.

Let us try to clarify our preoccupation: for example, we can reflect upon the investigation into the legitimacy of a belief in an "external world". The legitimacy of this belief can be found in the experience and in the related feelings we have when something resists our will. For example I am not able to remove visual impressions, and this shows that they are not subject to my will, but they dominate it, and through bodily motions, touch and our experience of something resistant I experience them as reality.[22] Therefore it can be said that there is something "external" to the self; the borderline Dilthey says, coincides with the boundary of my body: it is the body, in fact, that is organized to express movements and feel resistance.[23] The "other" is what is completely beyond our bodily life. Obviously this does not mean, according to Dilthey, that the psycho-physical unity is known in its complexity from birth: the awareness that it is constituted of different parts is acquired through experience, through which our sense of life extends to every stable expansion of our own person, which takes part in our voluntary movements and gives us sensations of pressure and resistance. So a child, for example, becomes aware that his toes belong to him only through the pain he feels when he plays with them.[24]

Let us pose this question: is the immediate experience of a human being the experience of a psycho-physical unity or is it a psychological, spiritual experience that has nothing to do with the body? Let us consider the possibility of matching the terms "external reality" and "physical nature", "internal reality" and psychic, spiritual dimension".

What is "other" than the self, what is in this sense "external" is said to be so, as it is "located beyond the area of our own bodily life", beyond our skin.[25] Dilthey speaks also about the phenomenality of the corporeality of one's own self,[26] as we can speak of the phenomenality of apparitions, of space, etc. Now, if one's own corporeality marks the limit which defines something as "external", it is evident that this corporeality is immediately felt to be of a different order from the objects which are called "external".

We have to make it clear that the principle of phenomenality states that everything that exists for a subject must be a fact of his consciousness.[27] Moreover, the consciousness we are speaking about cannot be defined through the relationship between subject and object, through the double transcendance of subjectivity and reality: it is a consciousness which always refers to something, because it feels, wills and conceptualizes. This is what occurs in *Erlebnis* and this, according to Dilthey, cannot be disregarded. The *ego*, the external world, are concepts that have been found out through a splitting up or an analysis of *Erlebnis*, of what it contains. It is therefore impossible to found knowledge on the self-evident-truth of the transcendental *ego cogito*: the structure of knowledge turns out to be circular. This is expressed in the formula which says that knowledge cannot go behind life.[28]

What can be called "external" does not coincide with what can be called a "natural element or object", we find that also other *egos* are external because they are outside us. We should not forget that without a natural ground there can be no exteriorization, no expression of psychic life, no culture: from a factual point of view, natural ground and cultural superstructures are closely interwoven and inseparable.[29]

We have said that for Dilthey all science is experiential science, "analysis of experience"; he calls experience the process that takes place in one's consciousness and in which consciousness is given a real [*wirklich*] character". Dilthey states that: "external experience originates when consciousness is confronted with something real (*wirklich*), different from itself", whereas "internal experience learns the real (*wirklich*) character of consciousness itself."[30] However, he does not conclude from this that there is identity between external and natural or between internal and psychic. Nevertheless, Dilthey adds that psychological insight and external understanding never coincide in the same act, and from this he concludes the consequence that "the reality of spiritual life is never given us at the same time as the reality of our bodies."[31] It should be noted that a few lines above Dilthey states that it is only through abstraction that the spiritual life of a human being can be isolated from the living psycho-physical whole. He observes that two paths have been taken to grasp that connection between natural and spiritual events which is represented by the psycho-physical unity: on the one hand, there is transcendental philosophy, according to which the laws of nature depend on states of consciousness; on the other hand, there are the natural sciences, which have developed a general picture in which spiritual reality is conditioned by the material world. The opposition derives, therefore, from the contrast between their respective points of departure: the former from internal experience, the latter from external experience.[32]

Dilthey is therefore well aware of the problem which, he believes, "we will be able to consider solved only after we have settled the initial contrast between the transcendental [. . .] and the empirical point of view, the latter being directed to the object [. . .]".[33] Dilthey's aim to conceive of and develop philosophical research as an analysis of life seems to be an excellent premise from which to develop a solution, but Dilthey does not pursue both sides of the question to the same extent. In the end the terms inner, spiritual, immediate are used interchangeably without justification.

"Nature is foreign to us, since for us it is only something external, not an inner reality":[34] this is his conviction.

It seems here as if nature is foreign and dumb because it lies "outside", but we must note that other people are external to us, yet they are not complete strangers. Moreover, in both cases "knowledge of that which is independent of consciousness and transcends it is not gained immediately".[35] So we can hypothesize that the assumption (which Dilthey does not discuss) on which all these terms are based is the negation *in actu exercito* of the psycho-physical unity of the human being. It seems to follow that the natural sciences and the human sciences, rather than reflecting on the same experience from different points of view, are concerned with different fields, and one of these fields is mute and foreign.[36] The legitimacy of the distinction between the natural sciences and the human sciences does not seem to require an affirmation of the fact that nature and spirit are mutually strangers, independent and separate because nature too is in some way familiar, in *Erlebnis* of the psycho-physical unity.

Setting ambiguities aside, in Dilthey's theory of knowledge there is a caesura between man as a living unity and dumb nature. As regards human beings the situation is different, however. "I myself, that live and know myself from within, am a part of this social body, and all the other parts are similar to me, and therefore for me they can be understood from within."[37]

For a philosophy that tries to clarify life through life, the path through man's "own life" that comes out in Dilthey's work ends in an investigation of man as a historical, social being: the concept of life is reduced to socio-historical connections.[38]

LIFE AND KNOWLEDGE

Thought – Dilthey states – is something which comes out in the living process.[39] This means that "knowledge exists and is connected to *Erleben* without our being aware of it, and we cannot find its origin or foundation outside *Erleben*."[40] In every conscious process, analytical thought is forced to select something different from the process itself, something which cannot come from any relationship. Regarding its origin, Dilthey states that he has no ready answer, but nevertheless it exists, it is clearly different from intellectual acts and appears as a separate entity only through an analytical process. However, "we always deal with realities which are ordered and interpreted through an intellectual process."[41] This is the difference between form and content in our knowledge, which Dilthey takes up, criticizing Kant, Locke, Schopenhauer, whilst he notes with approval that Schleiermacher's *Dialectics* shows how the two dimensions are mutually linked. However, we cannot have a "pure" experience of a fact: because everything that happens in my soul contains something well-ordered, or singled out, or connected, or learned through an intellectual process.[42]

In order to understand what knowledge is we have to start from man as a whole, as a global entity, in his vitality and fullness.

In lived experience there is a connection between feeling, intellect and will. According to Dilthey, there are different levels of the thinking experience, through which from implicit thought one passes to discursive thought, something which, according to Kant, was not possible. The different levels are seen as different moments in which thought is variously driven, structured, organized; implicit thought is conceived as a form of thinking which is immanent in our attitudes. The connection between implicit and explicit thought seems to affirm the existence of continuity between the dimension of sensibility and that of the intellect, which Dilthey also expresses in his theory of the intellectuality of perception.[43]

We could say that science is a later form of knowledge, which derives from a channelling and organization of what we experience in accordance with a particular logic. It does not seem to be correct, then, to say that for Dilthey the lived experience is something irrational: this can be said only by those who use reasoning as the instrument by which to conceive life, which is the whole from which reasoning itself derives.[44] We can instead detect an expansion of the concept of knowledge: it encompasses both the human sciences and the natural sciences. Furthermore, such an expansion highlights the derivative nature of the sciences, which build up their objects in an approach which claims to be purely conceptual: they are so specialized that they are unaware of the acts of abstraction and composition on which they rely.

In short, our effort to attain knowledge needs to be directed in two opposing courses: whilst it looks ahead to its object, it has also to look back toward its origin. This means that knowledge also has a critical dimension.[45]

The unity of life is a link: this can be grasped instantly. The unity of life means that there is no simple unity which pre-exists the single parts, no discrete multiplicity, and no whole coming before the single parts. For this reason Dilthey can say that "melody tells us a lot more about life than all definitions of it."[46] Intellect, instead, can only distinguish, separate, identify, connect, relate. The core of life itself is, instead, in its being an organized reality, a link. This living link (it does not matter whether it refers to the object or to the subject, because they are "very similar") is expressed in a variety of categories. As Kant established and clarified, a category is "a concept that in its turn expresses or produces a link."[47] It is generally thought, however, that the link contained in a category is grounded on reason, on the intellect's ability to make a synthesis.

Dilthey identifies, instead, two groups of categories: one group consists of formal categories rooted in reason (i.e., the concepts of identity, equality and difference); the other consists of real categories rooted in the living link itself. Formal categories indicate relations through which thought clarifies to itself what is real (*wirklich*); they are clear and unambiguous. Real categories are indefinable in terms of number, order, nature: their content is inexhaustible because they express the living connection itself. The category of self-sameness, for example, expresses that unity which can only be lived and which keeps all the differences and modifications together in the same living unity. However, Dilthey considers it impossible to express and understand that category through intellect: this confirms that such a category is not a product of intellect.[48] The fact that intellect aspires to reach by its own means that which lies beyond the living connection which is expressed in the category of sameness leads to some contradictions.

"In explanation we rely on processes which are purely intellectual, but in understanding we rely on co-operation between various spiritual forces in a process of apprehension, moving from the general to grasp the specific."[49] It is impossible to elevate lived experience to a conceptual state: the antinomies derive from the work that human thought does on experiences, which cannot be completely reduced to concepts. We have therefore to recognize that antinomies are immanent in our field of knowledge of what is real and acting in experience.[50] Intellect, which claims to reduce experience to full logical clarity, does not realize that our experience of the world and also our self-awareness have derived from the vitality of our self, which is "something more than ratio."[51] So *Er-leben* comes before *Er-kennen*.[52] It is this that is the main point of what Dilthey calls tragedy: thought connects different parts; this conflicts with the concept of "life in the whole". We can, however, conceive of life only in this way.[53]

To sum up, Dilthey thinks that what intellect does not grasp is the link, the whole. It remains to investigate whether he considers that another form of proper knowledge is possible. Relevant in this respect are his writings on aesthetics, where he states that aesthetic experience has a cognitive value. He says that the whole configuration of the historical human world can be comprehended first and foremost through poetry rather than through science.

According to Dilthey, "art tries to express what life is."[54] We must not forget that he states that there is a mutual inner relationship between representation, feeling and will and he thinks that every form of experience is a special expression of this unity. He believes that if every image stands for something which cannot be expressed completely, the artist must be recognized as the one who possesses the most sincere language because he speaks only through images and symbols.[55]

THE SENSE OF TRAGEDY

We said that the living psycho-physical unity, which is ultimately seen as a link, as a psychic structure, is taken by Dilthey as a means to investigate the mystery of life and of the world. Through an analysis of *Erlebnis* he comes to the conclusion that the self is constructed through its relations with the other and with its own milieu. The self does not seem to be therefore, "life itself".

Dilthey states that the certainty of *Erlebnis* needs no mediation, and therefore can be defined as immediately conscious.[56] However, he also recognizes that all our psychic conditions have a beginning, change and disappear in time;[57] "I possess the particular link of my own life only in the context of time, as I remember its course."[58] He clarifies that present *Erlebnis* is constituted by the past, but the past belongs to our *Erleben*[59] as something that transcends the consciousness that lives in the present *Erleben*.[60] Therefore we can call the immediacy of the present

Erlebnis "the immediacy of transcendence". Dilthey says that "is" does not even stand for the shortest period of time: "the briefest moment in the progress of time encloses within itself yet another course of time;"[61] "the present never exists."[62]

The conclusion is that we cannot grasp the essence of life instantly: in fact, "even when our consciousness of the flowing of time is intensified, every moment that is observed is a moment remembered and no longer the flowing itself, because it is held by our attention, which fixes on that which is flowing."[63] There is no prompt access to life, even in the sense of the life of the self.

Moreover, what we know of our life (of which we are aware through *Erleben*) is always something singular; it is in every case a form of knowledge which is relative to something unique.[64] Dilthey states therefore that we need to leave the subjectivity of *Erleben*, not least because, as he points out, to say that man comes before history and society is an explanatory fiction adopted by geneticists.

It is in the act of understanding other people that "the obscurity of *Erleben* is clarified"; the mistakes which derive from the limited understanding of the subject are corrected, *Erlebnis* itself is amplified and completed.[65]

If, on the one hand, Dilthey can say that life is "the inner relationship between psychic functions in the links of each person,"[66] on the other hand he states that it is also "the link between the mutual relations between people, which take place in the conditions of the external world. [. . .] it consists in the reciprocal activity of living unities."[67] "Life itself", therefore, transcends the psycho-physical unity, the psychical link, the individual.

But Dilthey affirms that it is the same spirit that lies in the ego, in the you, in every individual, in a community, in every cultural system and finally in the whole of the spirit and of universal history. The other is not different but similar. He gives understanding the task of reassuring us that there is a companionship between men. This companionship, he says, is expressed in the fact that men share the same reason, have a capacity for sympathy and are bound together through reciprocal duties and rights.[68] The transcendence of our "own lives" leads, therefore, to a reassuring socio-historical world.

Here the finite nature of man seems to find both foundation and protection. In this way the risk seems to have been avoided of having to admit that there is tragedy in life itself, as we have had to recognize that it exists in thought. Moreover, the objectivations of life – that is, communities and works, which are the "fulfilment of the spirit in the sensible world" extend the field of homogeneity, of that which we can share: we can say therefore that they play a reassuring role in the context of Dilthey's thought.

On the whole, however, Dilthey's analysis does not hide ambivalences. It points out the eternal contradiction which exists between the creative spirit and historical consciousness: if man gives certainty to his own life by extracting that which he creates, as something durable, from the flow of time, each individual creation is nevertheless characterized by relativity and transcience.[69] The external expression of the human spirit is not at all reassuring; through our critical awareness we can see it as a mirror that multiplies the image of its own incapacity of founding itself. There is no subject without a world, but there is not even a meaningful world without the continuing contribution of the various subjects; at the same time synthesis in the 'absolute spirit' is rejected. The person does not come to an end in the objective spirit, and this goes beyond the subject, as well.[70]

In the fragment dated 1895, *Dichter als Seher der Menschheit*, after emphasizing that there is no statement that contains the truth, Dilthey says: "every religious dogma and every scientific formula about the link of the world is a symbol of something that no myth, no dogma and no concept can fully comprehend or express as it is in itself. Life is like the Yggdrasil tree, the ash tree in the middle of the world, on the most remote branches of which this individual [lives]. How could he embrace the whole with his own eyes?."[71] In *Ideen über eine beschreibende und zergliedernde Psychologie* Dilthey says that the psychic conncection is the subsoil of cognitive acts;[72] however, this subsoil is not a foundation in its turn.

Sometimes he uses the expression "life itself", even if not as a central and well-defined term; on the other occasions, as in *Leben und Erkennen*[73] he speaks about the living connection in the foreign thing and person and also about the "general living connection that comprehends everything". We can find here that *Leben* overcomes individuality.

It seems as if this overcoming can only in part coincide with the human world. Even if it is the philosopher himself who clarifies that he uses the word "life in the human sciences to refer only to the human world", he adds that "here it is defined by the context in which it is used."[74]

We can speak, therefore, not only of the antinomies of logical knowledge, but also of the antinomies of life; moreover, Dilthey never asks what life is, but only how it can be understood. In *Die Typen der Weltanschauung*[75] he focuses on life as a mystery. "When an attempt is made to grasp it in its totality", its face emerges from the various experiences of life "as contradictory", as "vitality and, at the same time as law, reason and arbitrary act, a face that shows ever changing features, perhaps clear in its details, but completely mysterious in its whole.

Dilthey even uses the phrase "estrangement of life". This "increases as man experiences continuous struggle in society and in nature". He detects contradictions that he calls "strange": between universal transience and our own will to find something stable, between the power of nature and the autonomy of our will, between the narrowness of everything in space and time and our faculty to overcome every limitation.

This is not a question of antinomies as they concern thought, but rather as they are lived and experienced. "Tragedy" does not exist only in thought; it is not an issue which concerns only thought, therefore. Tragedy emerges from experience; it concerns one's own life, connected to but different from "life itself".

"This is the deepest and most tragic point in individual life: to own eternity", Dilthey admits to his friend Paul Yorck von Wartenburg.[76] In man's experience life has many different facets and always lacks something.[77] Even if we do not want to consider the irreducibility of natural reality (but this fact is not of secondary importance, as Dilthey could speak even of a tree as of a "living unity"), for Dilthey death remains a key question, for which not only *Erkennen* and *Erklären* are inadequate, but also *Verstehen*. Life is afraid of death as something that concerns life.[78]

Dilthey acknowledged that he had no solution to the enigma of life, rather he wanted to share with his interlocutors his attitude towards life. This attitude came to him through his reflections on the consequences of historical consciousness.[79]

What philosophy can do, through awareness (self-reflection) and analysis of the all-embracing socio-historical connections of life, is to prepare the ground for an understanding of history and for control over reality.[80]

We can, it seems, detect an important suggestion, not beyond but emerging through and thanks to the ambivalences that his reflections do not claim to minimise: we are not to conceive of life as an object, *Gegenstand*, because it is never simply before us, fully comprehensible, but rather that which is before us is a symbol, an echo, a memory of it. Dilthey states that "there is no further supposition concerning some single agent in history, whether immanent agent or real condition that can be considered in the philosophy of history as providence or immanent purpose or force that drives history onwards".[81] We cannot disregard this. Nevertheless, something beyond the metaphysical view that tried to find an absolute foundation as object comes to mind: namely, that "in the face of relativity the continuity of a creating force stands out as the essential historical element".[82]

Università degli Studi di Macerata
Italy

NOTES

[1] W. Dilthey, *Das Wesen der Philosophie* (1907), *Gesammelte Schriften* V, p. 364 (to be referred to in subsequent notes as *GS*) 19 vols. (Stuttgart: B.G. Teubner) (Göttingen: Vandenhoeck & Ruprecht), 1914–1982. All translations from Dilthey in this essay are mine. As regards notes, I preferred to maintain quotations in their original languages.
[2] *Ibid.*, pp. 346, 365.
[3] *Ibid.*, p. 340.
[4] *Ibid.*, p. 365.
[5] *Ibid.*
[6] W. Dilthey, *Plan der Fortsetzung zum Aufbau der geschichtlichen Welt in den Geisteswissenschaften, GS* V, p. 372.
[7] *Das Wesen der Philosophie*, p. 413.
[8] M. Heidegger, *Sein und Zeit*, in *Gesamtausgabe*, II (Frankfurt a M.: V. Klostermann, 1977), p. 62.
[9] See G. Misch, *Lebensphilosophie und Phänomenologie* (Bonn: F. Cohen, 1930), p. 6.
[10] F. Donandio, "Ragione storica e ragione ermeneutica: Dilthey e il conte Yorck," in *Wilhelm Dilthey. Critica della metafisica e ragione storica*, ed. G. Cacciatore, G. Cantillo (Bologna: Il Mulino, 1985), affirms that: "il ritorno alla vita come fondamento cercato e sempre perduto è l'acquisizione teorica più rilevante della riflessione di Dilthey." See W. Dilthey, *Frühe Entwürfe zur Erkenntnistheorie und Logik der Geisteswissenschaften*, (before 1880), *GS* XIX, p. 13.
[11] *Leben und Erkennen. Ein Entwurf zur erkenntnistheorethischen Logik des 19. Jahrhunderts* (1892–1893), *GS* XIX, p. 344: "Wir ziehen diese Formel der vor, dass auf den psychologischen Zusammenhang, in dem Denken auftritt, zurückzugehen sei. Denn willkürliche Abgrenzung und Abstraktion auf das Innere wird so vermieden."
[12] *Ibid.*, p. 347; *Plan der Fortsetzung . . .* , *GS* VII, pp. 261–262: "Leben ist nun die Grundtatsache, die den Ausgangspunkt der Philosophie bilden muss. Es ist das von innen Bekannte, es ist dasjenige, hinter welches nicht zurückgegangen werden kann. Leben kann nicht vor den Richterstuhl der Vernunft gebracht werden."

[13] *Leben und Erkennen . . .* , *GS* XIX, p. 347: "Das Leben bleibt für das Denken unergründlich, als das Datum, an welchem es selber auftritt, hinter welches es daher nicht zurückgehen kann."
[14] *Ibid.*, p. 346.
[15] *Ibid.*, p. 304: "Der Ausdruck Leben drückt hier zunächst das aus, was jedem das Bekannteste, Intimste ist. Was Leben sei, ist so in der Erfahrung gegeben. Wir erleben es, und es ist uns doch ein Rätsel. Aber wir wissen, wie es auftritt und sich darstellt. Es ist, wo eine Struktur besteht, welche von Reiz zu Bewegung geht. Dieser Fortgang von Reiz zu Bewegung ist überall an die Erscheinung eines Organischen gebunden. In dieser Struktur, welche vom Reiz zur Bewegung geht, pulsiert gleichsam das Geheimnis des Lebens. Die Lebenseinheit ist immer in dem Zusammenhang dieser Struktur. Wir kennen diese Struktur von innen nur an uns selber."
[16] *Frühe Entwürfe . . .* , *GS* XIX, p. 23.
[17] W. Dilthey, *Der Aufbau der geschichtlichen Welt in den Geisteswissenschaften* (1910), *GS* VII, p. 80.
[18] *Ibid.*, p. 81.
[19] *Ibid.*, p. 86.
[20] See W. Dilthey, *[Über vergleichende Psychologie] Beiträge zum Studium der Individualität* (1895–1896), *GS* V, p. 249.
[21] *Plan der Fortsetzung . . .* , *GS* VII, p. 196: "Auf dem Boden des Physischen tritt das geistige Leben auf; es ist der Evolution als deren höchste Stufe auf der Erde eingeordnet. Die Bedingungen, unter denen es auftritt, entwickelt die Naturwissenschaft, indem sie in den physischen Phänomenen eine Ordnung nach Gesetzen entdeckt. Unter den phänomenal gegebenen Körpern findet sich der menschliche, und mit him ist hier in einer nicht weiter angebbaren Weise das Erleben verbunden. Mit dem Erleben aber treten wir aus der Welt der physischen Phänomene in das Reich der geistigen Wirklichkeit. Es ist der Gegenstand der Geisteswissenschaften und die Besinnung über diesen und ihr Erkenntniswert ist ganz unabhängig vom Studium ihrer physischen Bedingungen."
[22] *Frühe Entwürfe . . .* , *GS* XIX, p. 22.
[23] *Ibid.*; see W. Dilthey, *Beiträge zur Lösung der Frage vom Ursprung unseres Glaubens an der Realität der Aussenwelt und seinem Recht* (1890), *GS* V, pp. 96, 105.
[24] *Ibid.*, pp. 107–108.
[25] *Ibid.*, p. 107.
[26] *Ibid.*, p. 137.
[27] *Ibid.*, p. 90.
[28] See M. Riedel, *Verstehen oder Erklären? Zur Theorie und Geschichte der hermeneutischen Wissenschaften* (Stuttgart: Klett-Cotta, 1978).
[29] A. Marini, *Alle origini della filosofia contemporanea. Wilhelm Dilthey*, (Florence: La Nuova Italia, 1984), p. 241: "La teoria della conoscenza, secondo Dilthey, deve farsi carico anche di questa relazione, deve chiarirla e non può dimenticarla; questa è la differenza, la divergenza fondamentale tra Dilthey e Yorck, per il quale la teoria della conoscenza non deve impicciarsi minimamente del problema naturale."
[30] *Frühe Entwürfe. . .* , *GS* XIX, p. 53.
[31] W. Dilthey, *Einleitung in die Geisteswissenschaften* (1883), *GS* I, p. 14 (to be referred to in subsequent notes as *EGW*).
[32] *Ibid.*, pp. 18–21, Dilthey speaks about "uns umgebende Naturganzen."

[33] *Ibid.*, p. 21; see pp. 34–35.

[34] *Ibid.*, p. 36. Something different is said in *Leben und erkennen . . .*, *GS* XIX, p. 356: "Ein Baum hat ihn. Derselbe bildet ein Ganzes. Wir fühlen in ihm die bildende Kraft nach, die uns mit seinen Früchten oder seinem Schatten oder seinem Blütenduft und seiner Blütenpracht beschenkt. Seine Blätter, Äste, Stamm, Wurzeln bilden einen Zusammenhang funktionierender Teile. Auch er hat eine Struktur, vermittelst deren er lebt und zu seinem Milieu in lebendigen Verhältnissen steht. Die Früchte gehören zu ihm, sie sind dem Stamm und den Ästen gleichsam eigen. Er selbst ist überall in allen seinen Zeilen und Kräften."

[35] *Beiträge zur Lösung . . .*, *GS* V, p. 134: "Dass dem Ich ein Du oder ein Es gegenübersteht, das heisst nichts anderes, als dass so von meinen Willen ein ihm gegenüber Unabhängiges erfahren wird." See also *Frühe Entwürfe . . .*, *GS* XIX, p. 55: "[. . .] ist der Ton als ein von dem Selbst verschiedener Bestandteil des Bewusstseins gegeben, das Hören selbst aber als mein Zustand. Das so in dem Bewusstsein ihm gegebene Objekt ist Aussenwelt, der Akt des Bewusstseins, in welchem dieses Objekt gegeben ist, ist ein Bestandteil meines Selbst."

[36] *Frühe Entwürfe . . .*, *GS* XIX, p. 25.

[37] *Einleitung in die . . .*, *GS* I, p. 41.

[38] N. Bausch, *Der Begriff Leben im Werk Friedrich Nietzsches im Vergleich zu den Objektivationen des Lebens bei Wilhelm Dilthey* (Berlin, 1974), p. 218: "Die generelle Differenz zu Nietzsches Lebensbegriff ist damit schon angedeutet, denn Nietzsche suchte den Begriff Leben als begriffssprengende kosmische Metapher für Sein schlechtin zu erweitern."

[39] See *Leben und Erkennen . . .*, *GS* V, p. 344.

[40] W. Dilthey, *Studien zur Grundlegung der Geisteswissenschaften* (1905–1910), *GS* VII, p. 18.

[41] *Leben und erkennen . . .*, *GS* XIX, p. 336.

[42] *Ibid.*, pp. 335–336.

[43] *Ibid.*; W. Dilthey, *Ideen über eine beschreibende und zergliedernde Psychologie* (1894), *GS* V, pp. 150–151.

[44] A. Marini, *Alle origini della filosofia contemporanea: Wilhelm Dilthey* (Florence: La Nuova Italia, 1984), p. 252: "[. . .] lungi dal cercare di individuare la struttura della volontà e del sentimento proiettando su di esse quella del ragionamento deduttivo, o di diversi tipi di ragionamento dobbiamo, secondo Dilthey, compiere l'operazione inversa e concepire la ragione intellettuale, i suoi strumenti, i suoi procedimenti, metodi, ecc. come specializzazioni, impoverimenti (anche se dal punto di vista tecnico sono rafforzamenti), di una ragione assai più potente."

[45] See *ibid.*, pp. 253–273.

[46] *Leben und erkennen . . .*, *GS* XIX, p. 353.

[47] *Ibid.*, p. 360.

[48] See *ibid.*, pp. 362–369. Here Dilthey criticizes Kant.

[49] *Ideen über eine beschreibende . . .*, *GS* V, p. 172.

[50] *Ibid.*, p. 175.

[51] *Ibid.*, p. 196.; see also [*Über vergleichende Psychologie*] . . ., *GS* V, p. 278.

[52] N. Bausch, *Der Begriff Leben . . .*, p. 238: "Damit ist der Übergang zur Hermeneutik gebeben."

[53] W. Dilthey, *Das geschichtliche Bewusstsein und die Weltanschauungen, GS* VIII, p. 71.
[54] [*Über vergleichende Psychologie*] . . . , *GS* V, p. 280.
[55] W. Dilthey, *Dichter als Seher der Menschheit*, in W. Dilthey, *Die grosse Phantasiedichtung*, ed. H. Nohl (Göttingen, 1954), p. 3.
[56] *Studien zur Grundlegung* . . . , *GS* VII, p. 26.
[57] *Ibid.*, p. 23.
[58] *Ibid.*, pp. 73–74.
[59] *Ibid.*, p. 72.
[60] *Ibid.*, p. 29.
[61] *Plan der Fortsetzung* . . . , *GS* VII, p. 194.
[62] *Studien zur Grundlegung* . . . , *GS* VII, p. 73.
[63] *Plan der Fortsetzung* . . . , *GS* VII, p. 195.
[64] *Der Aufbau* . . . , *GS* VII, p. 141.
[65] *Ibid.*, p. 145. Dilthey thinks that *Erleben* and understanding affect each other.
[66] *Das Wesen der* . . . , *GS* V, p. 408.
[67] *Plan der Fortsetzung* . . . , *GS* VII, p. 229.
[68] *Der Aufbau* . . . , *GS* VII, p. 141.
[69] *Das Wesen der* . . . , *GS* V, p. 364.
[70] See R. Bodei, "Dialettica, ermeneutica e costruzione della personalità," in *Dialettica ed ermeneutica*, ed. T. F. Geraets (Naples: Quattroventi, 1993), p. 17.
[71] *Dichter als Seher* . . . , p. 3.
[72] *Ideen über eine beschreibende* . . . , *GS* V, p. 151.
[73] *Leben und erkennen* . . . , *GS* XIX, p. 354.
[74] W. Dilthey, *Fragmente zur Poetik* (1907–1908), *GS* VI, p. 314.
[75] W. Dilthey, *Die Typen der Weltanschauung und ihre Ausbildung in den metaphysischen Systemen*, GS VIII, pp. 78–79.
[76] *Briefwechsel zwischen Wilhelm Dilthey und Graf Paul Yorck von Wartenburg*, ed. Sigrid v.d. Schulenberg (Halle: 1923), letter n. 40 (Summer 1884).
[77] W. Dilthey, *Handschriftliche Zusätze und Ergänzungen der Abhandlung über die Typen der Weltanschauung, GS* VIII, p. 147.
[78] *Die Typen der* . . . , *GS* VIII, p. 81; *Handschriftliche Zusätze* . . . , *GS* VIII, pp. 140, 143: "Das furchtbarste und fruchtbarste dieser Rätsel ist, dass der lebendige den Tod erblickt, ohne ihn verstehen zu können, da er immer dem Leben ein Unfassliches Fremdartiges ist. Dies ist so wirksam, dass jeder sinnigere Mensch sich erinnert, wie Gedanken über Tod und Unsterblichkeit sein philosophisches oder religiöses Nachdenken zuerst beschäftigten."
[79] "Ich habe keine Lösung des Lebensrätsel, aber die Lebensstimmung, die aus dem Sinnen über die Konsequenzen des historischen Bewusstseins mir erwachsen ist, diese wollte ich ihnen mitteilen": W. Dilthey, *Traum, GS* VIII, p. 220.
[80] W. Dilthey, *Zur Weltanschauungslehre, GS* VIII, p. 188.
[81] *Plan der Fortsetzung* . . . , *GS* VII, p. 287.
[82] *Ibid.*, p. 291.

PAOLO VOLONTÉ

IMAGINATION: RESCUING WHAT IS GOING TO BE CANCELLED

1. INTRODUCTION

What does "imagination" mean? Strawson noticed that "the uses, and applications, of the terms 'image', 'imagine', 'imagination', and so forth make up a very diverse and scattered family".[1] Edward Casey is also right when he adds that "the very multiplicity of such accounts suggests that there are as many kinds of imagination as there are explanations or descriptions of it".[2] In this paper I will argue that the phenomenological analysis of imaginative phenomena worked out by Edmund Husserl in the course of his philosophical career on the one hand gives us a clue to an understanding of these "scattered" phenomena in terms of a unified field and on the other provides us with grounds for the claim that, prevailing Western philosophical attitudes notwithstanding, the cognitive performance of imaginative acts cannot be ignored.

The theory of intentionality is one of the most familiar themes in Husserlian phenomenology. Of all the various aspects of Husserl's idea of intentionality I here intend to discuss his so-called "confutation of the image-theory" (*Widerlegung der Bildtheorie*), i.e., the confutation of all those theories of knowledge which assume the existence of some thing interposed between the apprehended object and the subject's apprehension of it. In Empiricism as well as in Kantianism and the post-Hegelian theory of knowledge, there was a tendency to identify this third element with an image or a schema of the productive imagination, or sometimes (as is the case with von Helmholtz) with a sign. The concept of intentionality, on the other hand, allows for an account of the way perception works that need not assign any task whatsoever to imagination. Perceiving is one kind of thing, imagining a quite different kind of thing. In both cases intention is directed towards the object, not to any representative (*Repräsentant*) of the object in consciousness.

According to Sartre, the confutation of the image-theory on the basis of the concept of intentionality was Husserl's most innovative contribution to the contemporary theory of imagination.[3] This is beyond question, but it is not the whole story. It is also a gateway to a new

A-T. Tymieniecka (ed.), Analecta Husserliana LIX, 473–492.

domain of phenomenological reflection, that envisaged by Tugendhat as the true "core of intentional psychology",[4] namely, the domain of the complete differentiation of all possible kinds of intentionality. Since image-representations are intentional acts no less than the perceptions to which they correspond, and yet are typically different from the latter, it will be necessary to inquire where the essential difference lies between these two different types of intentionality.

We have, then, to single out that "unitary, and essentially unitary, kind of apprehension"[5] which is usually called imagination. We must set image-representation over against any other kind of representation, and in particular over against perception. According to Husserl, perception means the "bodily" presence of the thing itself. In general it is a "positional presentation (*Gegenwärtigung*)". It is a "presentation" because it makes its object present "bodily" instead of reproducing a mere simulacrum of the object. It is "positional" because it posits its object as existing now, that is as a real entity. All image-representations therefore share an opposition to positional presentation. If they did not, they would not be essentially distinct from the perception. Rather, they would belong to the same kind of apprehension as perception itself. That is, non-perceptual representations share the circumstance of either not being positional or not being a presentation at all. Husserl would say they are either neutralizations ("non-positional") or presentifications ("non-presentations"), or both at once.

In this way Husserl's phenomenological description and eidetic analysis provide us with some essential criteria for defining the various kinds of intentional acts; but they do not make it possible for us to arrive at a unified concept of imagination. I mentioned that non-perceptual acts must be neutralizations, presentifications or neutralized presentifications. None of these covers the whole of imagination. I will argue in the following pages that both neutralized presentifications and neutralizations of perceptual acts (neutralized presentations) can legitimately claim to qualify as imaginative acts. Positing presentifications, which fall outside the scope of this paper, are manifold. Even if they belong *prima facie* to so-called memory, in such a way that they cannot be assimilated by imagination, a few of them – namely, those presentifications of the present that Husserl at times also calls figurations (*Ausmalungen*) – share a number of characteristics with what is usually called imagination.

Among the various kinds of act that, in the framework of Husserl's phenomenology, can be defined as imaginative acts, my discussion will be confined to the two main kinds: image-consciousness and phantasy. The former is the consciousness of something arrived at by means of something else which is called an "image" (e.g., the picture of an acquaintance). The latter is the consciousness of something as some thing that is now non-existent (e.g., the imaginative intuition of a centaur). An analysis of both kinds of intentional acts will make it possible to single out a characteristic which is common to both these acts and which, in my opinion, is shared by all acts of an imaginative kind. I would like to suggest here that *all imaginative acts are rescue operations of that which is otherwise bound to be cancelled.* The meaning of this statement will become clear in the course of the discussion which follows, but at this point I can perhaps outline the main argument. In both patterns of imaginative act – image-consciousness and phantasy – non-real hyletic data superimpose themselves on the original perceptual field. This give rise to a conflict between two incompatible noematic senses: one is consistent with the perceptual field, while the other is inconsistent with it. According to Husserl, every occasion on which we become conscious of conflict ends with the cancellation of that element which is most weakly motivated. Consistency with the perceptual field is a very strong source of motivation, and hence, as a rule, the consistent noematic sense prevails over the inconsistent sense. Accordingly, the latter is usually cancelled. A very special pattern of consciousness, the imaginative act, is what is required in order to make it possible to "rescue" the noematic sense that would otherwise be cancelled. It enables us, by means of a special device, to conserve the weakly motivated noemata in the conscious world.

2. IMAGE-CONSCIOUSNESS

According to Husserl, image-consciousness takes place where something is not represented directly in and of itself, but rather through an image which acts as its representative. For example, a man is perceived not "bodily" but by means of a picture. In this case one object is present instead of another and represents it. It is perceived not in and of itself, but rather as a mere representative of the other object, which is intended to at the same time. Husserl states: "And though we already have the one in the field of intuition, we are not directed towards it, but through the medium of a secondary apprehension are directed towards the other, that which is copied or indicated".[6] That is to say, this kind of lived experience bases itself on a *double objectivity*. Image-consciousness is marked by a constitution of two objects at the same time, by a coexistence of two apprehensions founded on each other.

What makes this double objectivity possible? According to Husserl, it depends on three points.[7] Image-representation represents *something* by means of an image. The object represented is a reproduced object, namely, that object which the image reproduces. If the image represents a man, the reproduced object is the man who is intended by means of the image. Husserl calls this object the *picture subject* (*Bildsujet*). In addition, however, the image itself is twofold. On the one hand, it is an *image as a physical thing* (*physisches Bild*), i.e., the real object that is apprehended in terms of perception or memory or as a phantasy representation of an image: the painted canvas, the printed paper, the piece of marble, and so on; but this physical thing as such, i.e., the canvas or marble, is not a reproduction of the object intended. Over and above the physical thing, a *picture-object* (*Bildobjekt*), i.e., the mental image of the represented thing, is required. The mark of the image as a physical thing consists of a few aspects that do not relate to its character as an image. There are undeniable differences between a piece of marble and a piece of plaster, even if they represent the very same image of the selfsame subject. The picture-object, on the other hand, is this very image itself, the proper representative of the picture subject. Indeed it is a constituted object, and yet it is not just perceived in itself, but at one and the same time as the image of another object.

Husserl gives a synthesis of this as follows: "There are three objects: 1) the physical image, i.e., the thing made of canvas, marble etc.; 2) the object representing or reproducing something; 3) the represented or reproduced object, which I will call simply the *picture subject*".[8] In order to clarify these distinctions I can use an example, for instance, the black-and-white picture of Peter. Thus I have: 1) a printed sheet of paper; 2) an image of Peter, i.e., something which reproduces Peter even if it is not the correct size and the colours differ (and hence it cannot be identified as Peter himself); 3) Peter himself.

This is a well-known and oft-quoted theory.[9] Husserl himself frequently has recourse to it. However, I am going to question it. Is it phenomenologically correct and carried through in a truly rigorous fashion? Careful analysis shows that it is not really a phenomenological description at all, as the use of such terms as "real object" would seem to suggest. If consciousness is the consciousness of a picture-object, from the phenomenological point of view there is no physical image, and even the image subject is not a real object at all, but simply a characteristic of the noematic sense of the picture-object. I presume Husserl used this account simply in order to define his own terminology and to take advantage of the clear distinctions it gave rise to. At the same time he also afforded us with the tools to develop a real phenomenological description of the same phenomenon. Let us now turn to this new kind of description he offered.[10]

Although conflict is often given less importance than it is due by scholars of Husserl, for transcendental phenomenology it is an important clue for our understanding of the way human knowledge works. What is usually given in normal experience is simply perceived most of the time as a real thing, i.e., its reality is not questioned.[11] In my relationship with the world it is a matter of course for me that the world really does exist around me, so that the problem of its existence does not even arise. Yet this uniformity of experience can also break down: the object of experience breaks down and as a result a state of conflict ensues.[12] At this point many different forms of modalization may arise, such as doubt, supposition, negation, and so on and so forth. All these forms of modalization somehow result from a conflict between two different apprehensions of the same hyletic data.[13] The example of a model in a wax-works – an example that is often found in Husserl's writings – is enlightening in this respect. What I see sometimes appears to be a person, and sometimes appears to be a model. The apprehension "person" enters into conflict with the apprehension "wax model". Let us now suppose that the first apprehension becomes more strongly motivated than the second, even if the grounds for supposing this apprehension to be correct are not strong enough to entirely cancel out the possibility of the second. At this point I have the perception of what is *probably* a person. Let us now suppose that, on a different occasion, both apprehensions are just as strongly motivated as one another. At this point I cannot decide which one to adopt, and hence I am in *doubt*. Finally, one of the apprehensions becomes so strongly motivated that it cancels out the other. I discover it was a wax model after all, and when I thought I was perceiving a person I was wrong; it was simply an illusion.

Husserl stresses the universal power of conflict in the cognitive process when he notes that "wherever one perceptive apprehension enters into conflict with another – and this presupposes that they are both wholly or partly based on the same hyletic data – what is felt to be actually present is determined by the apprehension that is able to consistently fit in with the unity of the present perceptual field as a whole, in such a way as to constitute a comprehensive global perception which shares in the strength of the mutually founded intentions of belief. The other apprehension that takes possession of the sensible content [. . .] and give place to an appearance, on the other hand, constitutes a mere figment, an illusory object".[14] In this passage Husserl states a general principle: wherever I meet a situation in which apprehensions enter into conflict with one another, whenever I perceive an inconsistency in the world around me, I tend to eliminate the apprehension that disturbs my perceptual field by confining it to a world of hallucinations and perceptual deceptions. This *principle of consistency* carries a great deal of weight in motivational terms in the way the world of experience is constituted. The hyletic constellation "person or wax model", for example, is going to be perceived as a person when it appears in the context of a lively street, and as a wax model if it appears in a shop-window or a wax-works. However, if what we have just apprehended as a model in a wax-works begins to speak to us, the apprehension "person" suddenly re-acquires a great deal of weight in motivational terms, even if the context itself has remained unaltered. This means that even an apprehension which initially appeared to be a disturbing factor in the perceptual field may reveal itself as actually real or, more precisely (and in phenomenological terms), it may acquire a motivational basis which prevails over other apprehensions of the selfsame phenomenon. In cases such as these we are forced to alter our previous perception of the phenomenon: the illusory object revealed itself as real, and the real object as illusory. However, the standard of judgement which is employed remains the same: the principle of consistency.[15] Resolving the element of conflict endemic in a given experience necessarily requires some changes in the way the subject interprets that experience, sometimes in favour of one "pole", and sometimes in favour of the other. A landscape in the middle of a wall gives rise to a consciousness in which conflict is involved. It will therefore be experienced as an image, as a kind of "window looking out onto an imaginary world" in the middle of the perceptual field; if the landscape should for some reason assert itself as being actually present, i.e., if the motivation that upholds its apprehension as a real perceived landscape grows stronger, the conflict in the subject's lived experience between wall-perception and landscape-perception revives. This conflict once again demands to be resolved. We at once search for an explanation that can resolve the conflict by attributing a decisive motivation to one of the conflicting possibilities ("what seemed to be a picture is actually a window", or maybe "the wall I thought I perceived is actually a hallucination", etc.). How far this principle of consistency is justified and why we are unable to bear conflict and contradiction – all this is a profound philosophical enigma that even Husserl himself did not solve.[16]

In image-consciousness, then, there is a conflict between opposing apprehensions just as in the case of the wax model (illusion).[17] For example, the same hyletic constellation may be apprehended either as a series of areas of paint of various colours on a canvas or else as the picture of a woman. The real existence of the picture or of the areas of paint cannot become a subject for phenomenological investigation. The conflict between our apprehension of an image and our apprehension of a series of areas of paint, however, is a datum that belongs entirely to a purely phenomenological description.

What kind of conflict is it? Husserl states that opposing apprehensions in a state of conflict-consciousness are as a rule perceptual apprehensions,[18] and that the picture-object presents itself as a perceptual apprehension.[19] However, a common perception presents us with a thing instead of an imaginative relationship. For image-consciousness, imaginative modification of the perceptual apprehension is required. According to Husserl, a perception becomes imaginative in character when it involves a special kind of conflict – not the conflict that ends with a modalization but a conflict that ends with the definitive identification of one of the conflicting apprehensions as an image-representation: "The appearance of the picture-object differs from that of a normal perception on an essential point that makes it impossible to envisage it as such: it has the character of *non-reality*, that is the character of a *conflict with the now-present*".[20] This new "conflict with the present" gives rise to a new kind of relation between the opposing apprehensions. I shall call it by the more general Husserlian term of "superimposition".[21]

Allow me to now explain the meaning of this superimposition. As I have already pointed out, image-consciousness is the "appearance now of a now-now".[22] What *appears* without reference to its not-being-now, appears as present, i.e., as an object of perception. On the other hand, what appears as *not-being-now* and at the same time as a non-perceptual object, appears as a remembered or phantastic phenomenon, not as one present in terms of image. So the *essence* of image-consciousness involves the concept that what appears, on the one hand, belongs to the perceptual field, and, on the other hand, differs from the rest of the perceptual field insofar as it does not share with it the characteristics of "now-existence". A field of presentification is superimposed on the perceptual field (field of presentation). Since what is presentified does not become a consistent part of the perceptual field, this superimposition involves a conflict with the now-present. A mountain landscape does not easily fit into my usual perceptual field, in which the unanimous concordance of experiential data dictates the appearance of a flat in town. As soon as I perceive a mountain landscape on the wall of my flat in town, I become aware of a certain conflict between the appearance of the landscape and its context. As a rule a conflict such as this would be sufficient reason to cancel out the appearance of the mountain landscape – now seen in terms of mere illusion – in favour of an opposing apprehension of the same hyletic data, for example the apprehension of "an area on the wall". However, the same conflict also allows for a different solution, namely, a solution that is made possible by a non-reality modification. Our awareness of the fact that the perceived landscape, instead of claiming to be really present, has about it the character of "non-reality", puts us in a position to insert it in a harmonious, uniform fashion into the rest of the perceptual field. It is in this way that the landscape does not claim to be bodily present, but rather "in terms of an image", that is, through something that in its turn belongs consistently to the present perceptual field. Since it is present-in-terms-of-an-image, the landscape itself falls within the perceptual field without giving rise to a sense of conflict. In this way it avoids being turned into mere illusion. *In brief, it is discordance within the perceptual field which yields imagination, which, in its turn, should be understood as a means of overcoming the discordance itself.*

What is this resolutive character of "non-reality"? This is the crux of Husserl's argument. There are two points which are of fundamental importance in his account. On the one hand, the appearance of the picture-object is of a perceptual kind, and this is "because it is based on a sensation that is subjected to an apprehensions".[23] On the other hand, it is not entirely a perceptual apprehension, since it enters into conflict with the surrounding perceptual field. This conflict is not the direct source of image-consciousness. Rather, thanks to this conflict, the appearances of the picture-object "do not involve the consciousness of being or consciousness of presence. On the contrary, they are for us – as they appear – *non-being*".[24] So it is non-being-consciousness that yields image-consciousness.

These considerations could be formulated into a universal principle: *the appearance of a picture-object is not a perceptual appearance insofar as it lacks "belief", that is, lacks the character of reality.*[25] In other words, we might say that the appearance of the picture-object is *neutralized*. On this point, Husserl is very clear in the first book of his *Ideas*: "This *depicting picture-object* stands before us *neither as being nor as non-being*, nor in any *other positional modality*; or rather, we are aware of it as having its being, though only a quasi-being, in the neutrality-modification of Being".[26] The picture-object is present before us, yet despite this it is not posited as being present. Although the subject of a presentation, in terms of presentation it is necessarily given as being absent. *According to a strict phenomenological description, image-consciousness is a neutralized presentation*: "An example for the neutrality-modification of the perception".[27]

As a consequence, the lived experience which is called image-consciousness is essentially different from perception, and especially from that particular kind of perception which is illusion.[28] As a neutralized presentation it can be rightfully considered an imaginative phenomenon. No less than in the case of illusion, image-consciousness is founded on real conflict. What is it, then, that prevents it from simply being cancelled out as an apprehension which gives rise to conflict, as might be the case with an illusion or any other form of perception? According to Husserl, the distinction between illusion and image-consciousness is that "the genuine figment makes its appearance within the unity of reality itself (the wax-model) whereas an image does not properly 'appear' in it, but rather in a space of its own",[29] i.e., in the space of neutrality. This is not sufficient explanation, however. In itself, "appearing in a space of its own" is a good reason for cancelling out the imaginative apprehension, not for accepting it. The imaginative apprehension can impose itself only insofar as the neutrality-modification can resolve the conflict between imaginative space and perceptual space. From a functional point of view, image-consciousness as neutrality-modification acts as a guarantee of the uniformity of experience. The picture-object, with its own space, would stand in the way of this uniformity. Only image-consciousness as neutralized presentation is able to restore the uniformity of experience without causing the image-object to lose all meaning by being transformed into one of those apprehensions which are illusory and hence, in Husserl's words, cancelled out. Only image-consciousness allows us to constitute a neomatic sense that would otherwise be cancelled from the field of consciousness.

3. PHANTASY

I would now like to turn to Husserl's account of phantasy lived-experiences. I will argue that here, too, descriptive results can be obtained which are very similar to those obtained with regard to image-consciousness.

On this point, let me say straight away that initially, i.e., approximately up to the lectures of 1905 on phantasy and image-consciousness, Husserl thought he could explain phantasy lived-experiences in the same way as image-consciousness lived-experiences. He thought that an image as representative of the object intended to also acts in phantasy representation. Yet the 1905 lectures themselves are witness to how he gradually became aware that such an account does not agree with the descriptive datum of phantasy representations, and, what is more, clashes with the confutation of the image-theory, because it restores to intentional lived-experience the intermediation of something different from what is intended to.[30] In a phantasy lived-experience we intend *straight* to an object and not through an image of it, just as in perceiving we intend to an object without the intermediation of anything else.[31] In other words, the double objectivity that is characteristic of image-consciousness is no longer present in phantasy: it is Julius Caesar himself who is the subject of a phantasy, not his picture. That is why it is not based on any kind of image.

Thus we need to establish the true character of phantasy appearances without resorting to the example offered by image-consciousness. The latter is a neutrality-modification that arises as a consequence of an inconsistency in th perceptual field. The consciousness of non-reality that marks every single image-representation is the indispensable neutralization of that lived-experience which is most weakly motivated in order to prevent its noematic sense from being cancelled out. If the phantasy apprehension were to take place within the perceptual field, it, too, would at once reveal itself as consciousness of non-reality. However, the failure of this attempt to explain phantasy representation as a mode of image-consciousness implies that phantasy representation does not appear inside the perceptual field. On the other hand, Husserl states that it has "a field of its own, wholly separated from that of perception".[32] According to him, perceptual appearance and phantasy appearance cannot belong to the same world of experience. A thesis such as this is not self-evident. As Husserl himself admits,[33] it is possible, for example, to phantasise a drawing on a white sheet of paper which I now perceive before me. Yet he clearly thinks that the insertion into the perceptual field of something like this that is merely phantasised would be apparent. In fact, in this kind of experience the awareness that what is phantasised does not really belong to the given perceptual field always remains. Husserl is right when he points out that the "inserted picture never shows itself as a perceptual appearance, as a part of the perceptual field".[34] This is because if something apparently belongs to the perceptual field in every respect, it presents itself not as the subject of a phantasy representation but as the subject of a perception, which may be true perception, illusion or image-consciousness. By phantasising a drawing on the white sheet of paper which I can now perceive in front of me I may be aware that the drawing is merely phantasied. In this case I superimpose the phantasy field on the perceptual field. This act of superimposition rests on the fact that the phantasy field is wholly identical with the perceptual one, with the only exception of what pertains to the object that is apprehended as a phantasised one. On the other hand, maybe I am not aware that the drawing is merely phantasised. In this case I apprehend the object of experience as if it were real – I "believe" I perceive what is actually merely phantasised; or else, to use more precisely phenomenological terms, the object of experience *is now in every respect a perceived one.*[35]

The transition from phantasy world to the world of perception and vice versa is therefore of a discontinuous kind. It is a question of two separate and incompatible worlds conflicting with each other and being superimposed on one another. A new form of conflict therefor results, as Husserl himself points out: "Here, too, we have a case of conflict-consciousness, but it is a very different kind of conflict from that which occurs in the visual field. What is in conflict here, without any interpretation, is the whole perceptual field and the whole phantasy field".[36] The phantasy object thus belongs to a world that places itself alongside and opposite the perceptual field, without the two fields intermingling in any way. The perceptual world has an "absolute" unity of its own which is founded on a binding temporal form: every object in perceptual world is characterized by a unique and definite temporal situation; it occupies a fixed place in the unitary and single flux of the one objective time. This is not the case with the phantasy world, because phantasy objects have no absolute temporal position.[37] Furthermore, since perception is the "appearance of the now-present", *that which "conflicts with perception" is accordingly non-present.*[38] So what appears in phantasy is a non-present object, and phantasy is a non-presentation.

The description outlined above raises the same problem that I have already discussed with regard to image-consciousness: what allows us to apprehend some superimposed noematic senses as phantasies? Husserl's answer in this case is the same as the one he gave with regard to image-consciousness: "Let us go back to the case of an inhibited apperception: we would never go beyond a nullity-consciousness [. . .] and we would never get over the stage of inhibited or confounded reality, if it were not possible to assume a new standpoint that prevents a conflict-producing experience from arising, a type of experience which is, as a consequence of its conflict-producing character, also inhibitory in nature. Altering our standpoint in this way implies a transition from experience (or negation of experience) to *phantasy*, that is, to that special kind of consciousness which is a consciousness of the as-if".[39] What here is called "inhibited apperception" is tantamount – in the superimposition-conflict – to the apperception of phhantasy. It is labelled as "inhibited" since in the dynamic of the conflict itself it is supplanted by an apprehension of the same hyletic data which is more strongly motivated. On the other hand, the "conflict-producing experience" is the opposite apperception, or that apprehension that is in accordance with the world of perception. Thus, the passage by Husserl quoted above means nothing more than that *phantasy – no less than image-consciousness acts as a lived-experience preserving noematic senses that otherwise would be destroyed.* Only the transition to an apprehension of a phantasic kind allows us to prevent noematic senses that are "in themselves" not adequately motivated from being cancelled. Only the intentional lived-experience called phantasy allows us to retain a multiplicity of senses and worlds that the dynamics of the conflict would otherwise cancel out.

"Perception" means all those acts that constitute the now-present. As a consequence, phantasy means those acts that enter into conflict with the now-present by breaking down its homogenous continuity. That which is in conflict with the present is necessarily non-present. All those lived-experiences that do not "present" (*gegenwärtigen*) their sense content must therefore be classified as "phantasy". Everything that is given to the consciousness and is not present can only be given as something *presentified*. Although belonging to presence (*Gegenwart*), it is not given "as present" (*anwesend*).

In the context of Husserl's phenomenological eidetics of objectivating acts, phantasy occupies the place of a presentification.[40] In this way Husserl settles an essential difference between perception and phantasy, a difference that does not consist "in the objectivation – which is common to both – where the 'appearance' of the object takes place, but in the characteristic that makes up the difference between the present and the presentified".[41]

The lectures of 1905 conclude with the definition of phantasy as a presentification. This is an important issue, since it allows Husserl to make an essential distinction between image-consciousness and perception. Yet the definition of phantasy as a simple presentification still leaves the problem of its relation with the memory open, and memory shares in the character of the pressentification in a way which is paradigmatic. In fact it is not by chance that Husserl always talks of memory and phantasy as if they were part of the same kind of acts in the lectures and manuscripts of the same period.[42] In his *Einleitung* to Vol. XXIII of "Husserliana", E. Marbach expounds on Husserl's various attempts to explain memory and phantasy as two related instances of the same mode of givenness.[43] I shall not go into Husserl's accounts in any detail here, but clearly the results he achieved in 1909–1910 (in the manuscripts) and published in 1913 (in *Ideas I*) are of great importance for our discussion.

From this point of view the manuscript that was published as no. 13 in Volume XXIII of "Husserliana" is particularly important. In this manuscript Husserl asks, with regard to intuitive acts, whether we should not admit "that, indeed, we cannot get off with *just one* modification".[44] He thus becomes aware that memory and phantasy are modes of intentionality which are as different as memory and perception or perception and phantasy. Memory is not "an ordinary characteristic of consciousness" availing itself of the matter of "phantasy-appearance",[45] just as phantasy is not a modification of memory. What is remembered, no less than what is phantasised, are *non-originary* entities. As such they set themselves up in opposition to the object of perception. So they are presentified entities. Yet the meaning of the expression "non-originary" is different in both cases. Although what is remembered is actually given in a non-originary manner, it is apprehended as something existing in the past. Its being is therefore posited. It becomes a consistent part of the perceptual world: the absent object. On the other hand, what is phantasised enters into conflict with the world of perception. It is not just absent from the here and now; it also always and necessarily belongs to a quite different world. Its being is neutralized, its non-originarity points not just to an absence of presence, but also to an absence of certainty (or uncertainty): "The quality 'non-originarity' means in the one case *actual* presentification [. . .], and in the other case *inactuality*: mere representation".[46]

In this way Husserl finds a solution for the problem of the organization of the objectivating acts – a solution he will never again change. If we concentrate our attention on the carrying through of an eidetics of imaginative acts such as this, we must first of all bear in mind that all acts that are usually considered phantasies share the characteristics of *not presenting* their objects, since, for example, a phantasised centaur is not experienced as existing here and now. In the second place, however, neither are these acts simple neutralizations of present lived-experiences, since they would otherwise be image-representations of their objects, as when for instance a centaur is experienced by means of a corresponding picture-object. Finally, they are not even mere presentifications, since, even if a positional presentification does not posit its object in the here andnow, it does however posit its object in some spatial and temporal location that is regarded as real. On the other hand, the phantasised centaur is not just absent now; *it has never been experienced in my perception* as a "bodily" present centaur, since it would otherwise be a remembered one. Phantasy is thus necessarily a *neutralized presentification*.[47]

Why do we call the one experience world "perception" and the other "phantasy", and not vice versa? Or, as Husserl says, "why do we not consider sometimes the one and sometimes the other as now-present?"[48] Moreover, what forces us to set a world of continuity against a world of discontinuity and to regard the first as the world of what is present and the other as the world of what is not present? And even supposing that there are two worlds, and that one of them disturbs the continuity of the other, "why is the world which disturbs continuity considered as phantasy?"[49]

Questions such as these bring us to a very important problem. It is better to start from descriptively certain matters of fact: among all the objects of experience there is something which presents itself as part of a perceived world, so that it is posited as really existing; on the other hand, there is something which enters into conflict with this world and is therefore apprehended as part of one or many alternative worlds of phantasy. Experience, then, consists of two conflicting kinds of world. They are not entirely unrelated to one another, but instead intersect at certain points where the differences between them temporarily disappear, only to reappear later. I refer here to very special kinds of experience, such as falling asleep, waking up, falling into a trance and so on – generally speaking, every time the strict consistency of the real world temporarily fails, such that the parallel courses of perceptual world and phantasy world intersect and perhaps even intermingle with one another (as, for instance, when we dream).

The question we have to answer refers directly to this situation, but it can be formulated in two distinctly different ways. The formulation given above results from an *ontological* standpoint: Why does the principle of consistency work? There are two worlds. One is consistent, the other inconsistent. We call the consistent one the world of perception, by which we mean that it is the actual (*wirklich*) world and even – from the natural standpoint – the real (*real*) world. Yet why should the consistent world be real too? Why do we attach importance to consistency instead of inconsistency? I have already pointed out that Husserl does not answer this question. I imagine that the reason for this could be formulated as follows: Husserl is not interested in the mere ontological formulation of the problem because his theory assumes a phenomenological standpoint.

From a phenomenological point of view, the ontological question does not, in fact, even arise. The phenomenological question is *always a question about sense*, so that in phenomenology the question of being *is nothing more than one part of the question of sense*. Hence there is no reason to ask why the principle of consequence ought to work. The phenomenological question is more radical: What is it that makes us define one kind of givenness as the world of perception, and the other kind of givenness as the world of phantasy? Furthermore, why do we ascribe the belief of reality only to the first, and not to the second? A phenomenological kind of answer must refer to the descriptive experience-datum. Now, the descriptive content of a single experience may be more or less consequent on the context of the experience; but its sense-for-me is going to vary considerably according to its degree of consistency with the whole of the present content of experience. The noematic sense of that which becomes a consistent part of the perceptual process is essentially different from the noematic sense of that which enters into conflict with the consistent world, at the very least because only the first can be foreseen and interact with its surroundings, such that it becomes useful and meaningful for my existence. *The difference between objects of phantasy and objects of perception lies in the essential difference of their sense-for-me*. The meaning of the noematic sense "presence of a wolf" is of a certain kind when it is consequent on the perceptual context; it is of a very different kind when it enters into conflict with the same perceptual context. Only in the first case, in fact, does it instill fear in me, because only in the first case do the effects of its posited existence actually come true. The wolf whose existence is inconsistent with the perceptual world can do no more than devour me in an inconsistent fashion.[50]

In this way phenomenology completely upturns the natural standpoint. We are used to thinking that a perceived wolf scares us because it really exists (and so can really devour us). Phenomenology teaches us to realise that the intentional object "wolf" becomes the intentional object "really existent wolf" only by its assuming a certain sense for the knowing subject, namely, a sense which involves, among other things, fear – yet not fear by chance, but fear caused by the consistency that the perception of the wolf has with the world of perceptual reality.

4. CONCLUSION

Phantasy and image-consciousness, then, are neutrality-modifications. As such, they are non-reality modifications. What is represented through an image, as well as what is phantasised, is constituted into a non-reality world where the question of being is neutralized. As far as image-consciousness is concerned, it would appear that the sum of both the character of non-reality and the character of reproductive imagination is the condition for preventing a sense-apprehension from being lost, i.e., for preserving a sense-apprehension that would otherwise be cancelled out because it is in conflict with the surrounding perceptual field. Likewise, phantasy would also appear to be consciousness of non-reality, an as-if consciousness which is able to preserve or to uncover – thanks to the non-reality modification – a contested and inhibited apperception. In both cases we reach the same conclusion: *phantasy and image-consciousness are in a sense rescue operations of neomatic senses that otherwise would be lost.*

To make this conclusion more general: imagination is that special modification in cognitive apprehension that makes it possible to "rescue" those apprehensions of hyletic data which, because of insufficient motivation, would otherwise inevitably be cancelled out by other superimposed (perceptual) apprehensions. Instead, they are preserved through this modification – and preserved not merely as an empty representation, but with that strength of determination which belongs to intuition.

As a consequence, we should ascribe a *fundamental constitutive function* to representations of an imaginative kind. Image-representations are nothing less than acts which constitute sense, no more and no less than perception constitutes sense. However, they constitute sense that perception will not accept, as the sense they constitute is incapable of offering sufficient motivational grounds to be accepted as perceptual appearance. *Imagination is a constitution of such "entities" that are disallowed as far as perception is concerned.* They are, it is true, entities which are lacking in that noematic characteristic which we call "being"; but they nonetheless have a sense of their own and are thus a part of my sense of reality as a whole. What kind of world would we live in if all those senses that are "rescued" by phantasy were to be cancelled? Would it be a more "meaningful" world? Would its sense be, so to speak, "deeper", "better" or more "beautiful"? Surely not. It would be a world without fables and dreams, without images, fact-representations or phantasy-modifications of "reality"; and it would be a world without historical planning, without inventive power, without the gift of prophecy. It would be a poorer world by far, a world without complexity.

Università Cattolica di Milano
Italy

NOTES

[1] P. F. Strawson, "Imagination and Perception," in *Experience and Theory*, ed. L. Foster and J. W. Swanson (London: Duckworth, 1970), p. 31.

[2] E. S. Casey, *Imagining. A Phenomenological Study* (Bloomington/London: Indiana University Press, 1976), p. 5.

[3] Cf. J.-P. Sartre, *L'imagination* (Paris: Presses Universitaires de France, 1965[6]), pp. 146–148.

[4] E. Tugendhat, *Der Wahrheitsbegriff bei Husserl und Heidegger* (Berlin: de Gruyter, 1970²), p. 27.

[5] E. Husserl, *Phantasie, Bildbewußtsein, Erinnerung. Zur Phänomenologie der anschaulichen Vergegenwärtigungen. Texte aus dem Nachlaß (1898–1925)*, ed. E. Marbach (The Hague: Nijhoff, 1980), "Husserliana" XXIII, p. 6.

[6] E. Husserl, *Ideen zu einer reinen Phänomenologie und phänomenologischen Philosophie*, First Book, *Allgemeine Einführung in die reine Phänomenologie*, Part One, ed. K. Schuhmann (The Hague: Nijhoff, 1976), "Husserliana" III/1, p. 90.

[7] On the following cf. *ibid*, p. 252; E. Husserl, *Erste Philosophie (1923/24)*, second part, *Theorie der phänomenologischen Reduktion*, ed. R. Boehm (The Hague: Nijhoff, 1959), "Husserliana" VIII, pp. 111–112; Husserl, *Phantasie, Bildbewußtsein Erinnerung, op. cit.*, pp. 18–20, 29.

[8] Husserl, *Phantasie, Bildbewußtsein, Erinnerung*, *op. cit.*, p. 19.

[9] Cf. R. Bernet, I. Kern, and E. Marbach, *Edmund Husserl. Darstellung seines Denkens* (Hamburg: Meiner, 1989), p. 140; M. M. Saraiva, *L'imagination selon Husserl* (The Hague: Nijoff, 1970), pp. 93–95.

[10] The thread of this inquiry is to be found in the analysis that Husserl developed in the lectures of 1905 on "Phantasy and Image-Consciousness," now edited in *Phantasie, Bildbewußtsein, Erinnerung*, *op. cit.*, pp. 1–108.

[11] Cf. E. Husserl, *Erfahrung und Urteil. Untersuchungen zur Genealogie der Logik*, ed. L. Landgrebe (Hamburg: Meiner, 1985⁶), pp. 359–360.

[12] Cf. E. Husserl, *Formale und transzendentale Logik. Versuch einer Kritik der logischen Vernunft*, ed. P. Janssen (The Hague: Nijhoff, 1974), "Husserliana" XVII, p. 242.

[13] Cf. E. Husserl, *Analysen zur passiven Synthesis. Aus Vorlesungs- und Forschungsmanuskripten. 1918–1926*, ed. M. Fleischer (The Hague: Nijhoff, 1966), "Husserliana" XI, p. 34: "The very same stock of hyletic data is the common base of two superimposed apprehensions. [. . .] They are in conflict one with the other, each of them in it own way is strongly motivated and as it were 'claimed' by the perceptual situation up to now and its intentional content."

[14] Husserl, *Phantasie, Bildbewußtsein, Erinnerung*, *op. cit.*, p. 48.

[15] In order to understand more fully how this principle of consistency works, it may be worthwhile to refer to Hume's argument with regard to the nature of miracles: a miracle can in no way exist, since we accept believing in a "wonder" only when not to believe in it would mean to admit something even more wonderful (Cfr. D. Hume, *Enquiry Concerning Human Understanding*, Vol. IV of D. Hume, *The Philosophical Works*, ed. Th. Green and Th. Grose, [Aalen: Scientia, 1964], Sect. X).

[16] Cf. Husserl, *Phantasie, Bildewußtsein, Erinnerung*, *op. cit.*, p. 77: the question "Why do we set value on continuity instead of discontinuity?" remains unanswered.

[17] Cf. *ibid.*, p. 40: "The oft-cited example of a wax-works shows that if the imaginative function is put to one side, an imaginative phenomenon transforms itself and gives rise instead to a common perceptual apprehension, and perhaps even into a perception in the strict sense of the term, with a basis in normal belief. We might at first mistake the wax model for a person. In this case we have a normal perception, though it reveals itself afterwards as a mistake. When we suddenly become aware of the deception, a consciousness of an imaginative kind occurs."

[18] Cf. *ibid.*

[19] Cf. *ibid.*, p. 47.
[20] *Ibid.*
[21] Cf. *ibid.*, pp. 482–485.
[22] *Ibid.*, p. 47.
[23] *Ibid.*, p. 489.
[24] *Ibid.*, p. 66.
[25] Cf. Husserl, *Ideen I, op. cit.*, pp. 247–248; Husserl, *Phantasie, Bildbewußtsein, Erinnerung, op. cit.*, p. 508.
[26] Husserl, *Ideen I, op. cit.*, p. 252.
[27] *Ibid.*
[28] Cf. Husserl, *Phantasie, Bildbewußtsein, Erinnerung, op. cit.*, pp. 479–480, 486–491.
[29] *Ibid.*, p. 480.
[30] Cf. Husserl, *Ideen I, op. cit.*, p. 90: "In my lectures delivered at Göttingen (since the summer term of 1904, in fact) I have substituted the inadequate exposition which I gave in the *Logical Investigations* with an improved version of the same."
[31] Cf., for instance, Husserl, *Phantasie, Bildbewußtsein, Erinnerung, op. cit.*, pp. 83, 150, 161.
[32] *Ibid.*, p. 66.
[33] Cf. *Ibid.*, p. 67, note 1.
[34] *Ibid.*, p. 76.
[35] Husserl's conclusion leaves us in no doubt: "Of course there is no real insertion of phantasy in perception as such, as if a mixture could really originate from it. [. . .] Synthesis leads the corresponding parts of both fields to a synthetical unity, [. . .] but this unity in the intellective consciousness of transition is not the unity of appearance" (*ibid.*, cf. also pp. 67–68, 74, 86).
[36] *Ibid.*, p. 67.
[37] Cf. Husserl, *Erfahrung und Urteil, op. cit.*, §§ 36–39, and the manuscript these paragraphs are based on (MS. M III 3 VII: "Wirklichkeit und Phantasie. Untersuchungen zum Problem der Individuation"). M. M. Saraiva's synthesis is also particularly well-done. (cf. *L'imagination selon Husserl, op. cit.*, p. 223).
[38] Cf. *Husserl, Phantasie, Bildbewußtsein, Erinnerung, op. cit.*, p. 67.
[39] *Ibid.*, p. 513.
[40] Cf. *ibid.*, pp. 85–87; Saraiva, *L'imagination selon Husserl, op. cit.*, pp. 130–133.
[41] Husserl, *Phantasie, Bildbewußtsein, Erinnerung, op. cit.*, p. 100.
[42] Cf. for instance *ibid.*, p. 191: "Memory can be used just as well as imagination."
[43] Cf. Marbach, *Einleitung, op. cit.*, pp. LXIII–LXV
[44] Husserl, *Phantasie, Bildbewußtsein, Erinnerung, op. cit.*, p. 294.
[45] *Ibid.*, p. 296.
[46] *Ibid.*, p. 298. Here a fleeting terminological innacuracy leads Husserl to use the terms "actuality" and "inactuality" instead of "positionality" and "neutrality." Cf. E. Husserl, *Zur Phänomenlogie des inneren Zeitbewußtseins (1893–1917)*, ed. R. Boehm (The Hague: Nijhoff, 1966), "Husserliana" X, p. 102: "'Actuality' and 'inactuality' here have the same meaning as 'potentiality' and 'neutrality' in the sense of the *Ideen*." On these topics cf. also G. Piana, *Elementi di una dottrina dell'esperienza* (Milan: Il Saggiatore, 1979), p. 114.
[47] Cf. Husserl, *Ideen I, op. cit.*, p. 250; Husserl, *Erste Philosophie, op. cit.*, pp. 112–113;

E. Husserl, *Phantasie, Bildewußtsein, Erinnerung, op. cit.*, p. 322; Bernet, Kern, and Marbach, *Edmund Husserl. Darstellung seines Denkens, op. cit.*, pp. 137–138; Saraiva, *L'imagination selon Husserl, op. cit.*, pp. 204–216. On the way this idea matured in the course of time, see Husserl, *Phantasie, Bildbewußtsein, Erinnerung, op. cit.*, pp. 217, 245–248, 253–255, 277, 281–282. To define phantasy as a neutralized presentification is to put it in the same frame of reference as image-consciousness, which is a neutralized presentation. According to this argument, Belussi's distinction between a-positionality (image-consciousness) and quasi-positionality (phantasy) is unjustifiable (cf. F. Belussi, *Die modaltheoretischen Grundlagen der husserlschen Phänomenologie* [Freiburg/Munich: Alber, 1990], pp. 251–260).

[48] Husserl, *Phantasie, Bildbewußtsein, Erinnerung, op. cit.*, p. 77.

[49] *Ibid.*

[50] Cf. Husserl himself: "Genetically speaking a discredit [*scil.* neutralization] of this nature is very important. A phantasy-will, namely a discredited one, yields no action, a phantasy-judgement yields no willing, etc." (Husserl, *Phantasie, Bildbewußtsein, Erinnerung, op. cit.*, p. 104). A similar argument – but in a different context – is put forward by Jean-Paul Sartre in *L'imaginaire. Psychologie phénoménologique de l'imagination* (Paris: Gallimard, 1940), p. 158.

NIKOLAY MILKOV

WHAT IS ANALYTICAL PHENOMENOLOGY?

AN INTRODUCTORY REMARK ON THE PROGRESS IN PHILOSOPHY

In this paper we shall try to show how small, in extent, is the progress of philosophy from a certain – analyzed – point of view. It is more like progress in some natural language, than in a science.[1] From this point of view it becomes clear that Husserl and Wittgenstein moved (with different success) to virtually one and the same idea, and independent of each other. This idea can be described as that of *analytical phenomenology*. The aim of the paper is to make this idea clear. Because of its general character, our method of exposition will be dogmatic.

1. THE *SUBJECT* OF (PROFESSIONAL) PHILOSOPHY IS PHENOMENOLOGY

Phenomenology investigates the ***given***. The two most important characteristics of the latter are that it appears as series of *indefinables*; and that we learn them *intuitively*. Indefinables are what *exists*.

1.1. *The Phenomenology of Plato-Aristotle*

The characteristics of the given were well outlined already in Plato's and Aristotle's phenomenology. A central concept in Aristotelian phenomenology is that of substance (on this see also §2.1).[2] From a *logical* point of view, a substance is "that which is not predicated of a substratum, but of which all else is predicated" (*Met.*, 1029a7). Plato before that gave an *ontological* description of substances: they are both forms, and mathematical objects – but also "the substances of sensible bodies" (*Met.*, 1028b19–20). For Aristotle, ontological substances are: essences, universals, genera, and substrata. Just like Plato, he accepted that "some of the sensible substances are generally admitted to be substances" (*Met.*, 1029a34–35). Apparently, for both Plato and Aristotle, substances are not only universals, but also sense-data. Despite the

A-T. Tymieniecka (ed.), Analecta Husserliana LIX, 493–500.

apophantic (sentential) theory of truth assumed by both, they accepted also the *existence* (i.e., the giveness) of sensible essences (not only of universals) which are indefinable, are given, and are known intuitively.

1.2. *The Phenomenology of Husserl-Wilson*

The modern form of phenomenology – Husserlian phenomenology, which today is for many phenomenology proper – was anticipated, among others, by Cook Wilson, who recommended we "try to find out what a given activity of thought presupposes as implicit or explicit in our consciousness, without allowing precious results to be interfered with by any preconceived opinion."[3]

Husserl specified that this investigation of the phenomena of "pure" thought can be seen as eidetic intuition (*Wessensanschauung*). The essences, or phenomena, are known intuitively, and are indefinables.

It is important to notice here that we contemplate the stream of phenomena, by which one phenomenon replaces another, etc., in the same way in which we contemplate the stream of Hume's "impressions." It can be said "that the intuiting of essences is nothing but an 'experience' in the sense of perceiving, remembering, or some other act of the same type."[4] That is why the paradigm of essences, of phenomena, are the sense-data.[5]

Both kinds of phenomena – sense impressions, and essences – are to be expressed only in descriptions. They cannot be said, but only described.[6] And since philosophy studies the indefinables, it is a descriptive discipline.

In connection with this conclusion, we are reminded that also according to Wittgenstein (from 1913) philosophy "is purely descriptive" (*NB* 106), describing the only essences for him at the time – objects and logical forms. The latter essences are grasped in a moment of intellectual intuition (as shown in *NB* 109, in the *T* 6.122, etc.).

In the 30s and the 40s Wittgenstein described new essences – different language games. They are again intuited – not statically, however, in an act of contemplation, but dynamically. They are namely *invented* in a moment of a creative intuitive insight. That is why Wittgenstein now defined philosophy (just as did Kant before him) as a synthetic *a priori* (*RFM* III 39).

1.3. *Husserl's Fallacy*

When compared with the phenomenology of Plato-Aristotle, and with the first presentation of Wittgenstein's phenomenology given in §1.2, it becomes clear that Husserl committed a fallacy. He realized that "essences" must have no characteristics, and no causal relations, that they are unchangeable, etc. They are both synthetic and necessary. Thus, they have no alternatives.[7] At the same time, he continued to insist that phenomenology is concerned with mental phenomena only. In order to save the phenomena chosen as "psychic," Husserl defined them as "purely psychic."

A consequence of this fallacy is the assumption that the realm of the (purely) psychic is totally different from the realm of the objective. In the objective world, in contrast to the psychic one, there is no difference between appearance and reality. A symptom that on this point Husserl's phenomenology commits a conceptual mistake, which is not made in Wittgenstein's phenomenology, is that it is essentially redundant. Husserl's phenomenology has a *Doppelgänger* in empirical psychology. The phenomenology is supposed to be necessary, and the empirical psychology – contingent.

1.4. *The Phenomenology of Moore-Russell*

Manifestly anti-duplicationist, in contrast, was the phenomenology of the first "analytical" philosophers. From the very beginning, they accepted that there are different kinds of data, or giveness. The question of their quality – material or psychic – was recognized as irrelevant. In this way the old materialism vs. idealism problem was eliminated (*ausgeklammert*).

At the beginning, Moore introduced (in *Principia Ethica*) the general indefinables of "good" and "bad," "yellow", and other colors. In *The Principles of Mathematics* Russell assumed five other indefinables of logic. In 1909 Moore introduced the term "sense-data". In 1914, under the influence of Wittgenstein, Russell accepted as basic indefinables the logical forms also.[8]

It is very instructive that both Russell and Moore were explicit that the new philosophy (later called "analytical") must deal with the "discussion of indefinables." After 1909 the first thing they tried to do was to construct on these indefinables philosophical systems – Moore in *Some Main Problems in Philosophy*, Russell in *Theory of Knowledge*.

1.5. *Wittgenstein's Phenomenology*

The movement towards introducing new indefinables in philosophy continued with Wittgenstein. In the *Tractatus*, besides the undefined objects (*Gegenstände*), also accepted as indefinables were aesthetic, and mystical objects – in a word, all things seen as being either beyond the time or *sub specie aeternitatis*; but also accepted were the ultimately analyzed logical forms. In the 30s and the 40s, the idea of what we already have called "dynamic" indefinables (*a priori*) was developed: mathematical proofs, new physiognomies, new language-games, and pragmatics in general.

In order to make this dogmatic statement clearer, two additional explanatory remarks are needed. 1) Wittgenstein's differentiation of December 1929 between physics and phenomenology is well known. According to it, "the world we live in [exist] is the world of sense-data; but the world we talk about is the world of physical objects" (*CL* I 82). Physics is concerned with establishing, constructing, laws (games!); phenomenology with *static* finding (in reality), or *dynamic* inventing (in thought experiments) of possibilities (with "contemplating possibilities", by Hegel and John Wisdom). Thus, 2) the given of Wittgenstein's phenomenology is of two types: a) static sense-data; and b) dynamic mental-data. The latter are the data of our "mental eye" – the possibilities we contemplate when we are active mentally (as distinct from sensually). The proof that the latter are data is the fact that they are indefinables.

What we observe, and describe in phenomenology, are slices of the construction (the game) which the laws of science suggest, or chose (*Waismann*, 161). Otherwise, their choice is contingent: "the form is optional" (*T* 6.341). A necessary condition for such a game being successful is: to connect different slices in a coherent construction. These slices are both empirical (synthetic), but also necessary (*a priori*). What is contingent are the – choice of – different kinds of physical hypotheses.

2. THE *METHOD* OF PHILOSOPHY IS ANALYTICAL

2.1. *Radical Indefinables*

That phenomenology is analytical was assumed also by Husserl and stated in his insistence that it is *Wesens**analyse***. In the early 30s, Wittgenstein saw the intrinsic connection between phenomenology and analysis, assuming that "phenomenology would be the grammar of the description of those facts on which physics builds its theories" (*PR* 51). In what can be called Wittgenstein's "later phenomenology"[9] (from *Remarks on Colour*, 1950–51) he continued to believe that being a contemplation of possibilities (of the possible connections of indefinables), phenomenological investigation is in fact a logical (or grammatical) investigation. He is only insistent that what he is doing is not a phenomenology – in the sense of a special *theory*, which lies between science and logic – but a discussion of *phenomenological problems* (*RC* I 53, III 248). The latter are also called *conceptual problems*: "Phenomenological analysis is analysis of concepts" (*RC* II 16). More precisely, its task is to clear up "which are the simple [indefinable] colour concepts" (*RC* III 69), the "sameness of colour" (*RC* III 265), etc. In other words, it is a search for the indefinables of color.

They are indefinables, however, not only in the sense that we cannot give them a verbal definition. We also cannot define them ostensively. We cannot pick them out, or point at them – otherwise, they could be defined. They are "radically indefinable." We only see them, intuit them. They manifest themselves.

Besides, we can describe them, and this – only pointing to the rules according to which they "behave," to the possibilities in which they can occur (and which we later can contemplate), etc. The task of the phenomenological (= logical) study of indefinables is to set up the rules of the games in which they are, or can be, involved. (Hence, the importance of rules in Wittgenstein's philosophy.) Unanalyzed, they are quite unclear, and "as a rule" – misleading. These considerations explain why phenomenology can only be analytical.[10]

The importance of the indefinables comes from their being principles – the items giving the ultimate explanations of the complex constructions. That is why Leibniz called them "genuine units," "primary forces," or "metaphysical points", which are to be differentiated from the mathematical points. See *Die philosophischen Schriften von G. W. Leibniz*, Hrsg. C. J. Gerhardt, 7 Bände (Hildesheim: Olms), Vol. IV, p. 482. Aristotle called them *entelechies*.

Since the terms of logic are the substances, or phenomena, which are given, logic means a *basic* calculation. Science is a Basic calculation too, but a secondary one. While phenomenology investigates (contemplates) the possibilities of the given,[11] science investigates the possibilities of these possibilities. That is why the intrinsic procedure of philosophy is that of deliberating, making speculations, of *peirastikê*.

Here it is even more apparent that the dichotomy between psychological, and material, of which Husserl was so strongly convinced, is a sham dichotomy. The advantage of Wittgenstein's phenomenology comes from the clear recognition of this. The quasi-dichotomy retained is between calculation and indefinables.[12] It is "quasi" since it is between mental- and sense-data. Both, however, are phenomena, being ineffable and known by intuition.

Many authors have recognized this "fusion or duality of the phenomenal and the objective" by Wittgenstein.[13] What is important is just that both are *a priori*, intuitive, and ineffable. An earlier formulation of this fusion was the otherwise perplexing assumption of Moore and Russell that sense-data are objective; and the assumption of Plato and Aristotle before them that they are substances. Another fusion is that between calculation and contemplation. The realization of the grounds of this fusion can also resolve the apparent controversy between the East Coast Wittgensteinians (the Hintikkas, N. Malcolm), who insist that the objects of the *Tractatus* are sense-data, and the British Wittgensteinians (B. McGuinness, H. Ishiguro among them) who believe that they are *a priori*.

2.2. *The Variety of Logic*

The varieties of phenomenology give rise to different types of logical calculi: 1) mereology; 2) the logic of states of affairs; 3) apophantic logic; 4) truth-functional logic; 5) sense-data anlayses; 6) colors for philosophers; 7) particulars-universals; 8) necessary-contingent; 9) substance-accident; 10) topics. Calculations closely connected with these are a) arithmetic; b) algebra; c) geometry; d) theoretical physics; e) grammar; f) stylistic; g) chess, etc. Wittgeinstein demonstrated the principle unity between them very well by often mixing mathematics, grammar, logic, and chess. He spoke of "the logic of colour concepts" (*RC* I 22), or a "geometry of colour" (*RC* I 66, III 86), of "mathematics of colour." These all are different kinds of logics, which can be developed with different levels of precision.

A typical example of discovering a new variant of logic is, what can be called, "Fregean phenomenology." In 1891 Frege introduced two indefinables – truth and falsehood – which were accepted as the only possible meanings of propositions. In contrast, Wittgenstein's (later) phenomenology is concerned with sense, not with the truth/falsehood (*Waismann* 63).[14]

Of special interest is the Moorean-Russellian phenomenology of particulars and universals (from Moore's paper "Necessity," and Russell's "On the Relations of Particulars and Universals"). F. P. Ramsey developed some of its points in "On Universals." Strawson's phenomneology in *Individuals* (re)introduced some additional types of calculation: those of monads, sounds, persons and the missing, but easy to "presuppose" items of this scheme: smells, tastes, kinematics. J. L. Austin's "linguistic phenomenology" of concept was developed partly in opposition to Wittgenstein's phenomenology in both its entitetic/apophantic (Tractarian) and pragmatic (language-game) form, etc.

Being basic, these different logics are immediate, and because of that most economical (in the sense in which J. L. Austin spoke of "language economy"). That is why the varieties of phenomenology can be used for *training* our thinking: they all contain the ultimate "minimal conceptual scheme" which is repeated in all logic games. This can be done since the apparatus of each type of phenomenology is, in its own way, basic (ultimate). This means only that their "material" elements (i.e., the elements which are *now* not seen as indefinables, but which can be the next indefinables) are "put in brackets." Otherwise, any attempt to limit philosophy with only one of these types of phenomenological analyses leads to a "one-sided diet" (*PI* §593).

University of Bielefeld
Germany

ABBREVIATIONS FOR WITTGENSTEIN'S WORKS

CL I	*Cambridge Lectures. 1930–1932.*
NB	*Norebooks 1914–1916.*
PI	*Philosophical Investigations.*
PR	*Philosophical Remarks.*
RC	*Remarks on Colour.*
RFM	*Remarks on the Foundations of Mathematics.*
T	*Tractatus logico-philosophicus.*
Waismann	F. Waismann, *Wittgenstein and the Vienna Circle.*

NOTES

[1] Or of the streets of an old twon. Indeed, exactly like our language, philosophy too "can be seen as an ancient city: a maze of little streets and squares, of old and new houses, and of houses with additions from various periods" (*PI* §2).

[2] See G. E. M. Anscombe and P. T. Geach, *Three Philosophers* (Oxford: Blackwell, 1973), p. 5.

[3] J. Cook Wilson, *Statement and Inference*, Vol. I (Oxford: Clarendon Press, 1927), p. 328.

[4] E. Husserl, *Philosophie als strenge Wissenschaft* (Frankfurt: Klosterman, 1967), p. 46.

[5] The fact that by the development of his new study of phenomenology Husserl started from Hume is well known to the historians of philosophy. See, e.g., R. T. Murphy, "Husserl and Hume: Overcoming Scepticism?," *Journal of the British Society for Phenomenology* 22, No. 2 (May 1991), pp. 30–44.

[6] Cf. with Wittgenstein: "Objects can only be *named*. Signs are their representatives. I can only speak about them: I cannot *put them into words*. Propositions can only say *how* things are, not *what* they are" (*T* 3.221).

[7] Here too, it can be seen how essential and unavoidable Kant's synthetic *a priori* truths are for philosophy *per se*.

[8] Now it is clear why we assumed (in §1.1) Aristotle's logical subjects and Plato's Forms and mathematical objects to be indefinables.

[9] There is a relatively large amount (a dozen of) papers, and one book, on Wittgenstein and phenomenology (and Husserl). *All* of them, however, discuss Wittgenstein's phenomenology of the early 30s. [See, e.g., the most recent papers of J. Hintikka, "Wittgenstein and the Problem of Phenomenology," in: I. Niniluoto *et al.*, eds., *Language, Knowledge, and Intentionality* (Helsinki: The Academic Bookstore, 1990), pp. 15–46; and Garth Hallett, "The Genesis of Wittgenstein's Later Philosophy in His Failure as a Phenomenologist", *Philosophy & Theology* 5, No. 4 (Summer 1991), pp. 297–312.] In fact Wittgenstein remained a phenomenologist in the 50s.

[10] On this Husserl and Kant agree.

[11] "The possibility of its occurring in a state of affairs is the form of an object [i.e., of an indefinable]" (*T* 2.0141).

[12] See on this, H.-J. Schneider, *Phantasie und Kalkül* (Frankfurt a.M.: Suhrkamp, 1992).

[13] See, e.g., M. and J. Hintikka, *Investigation Wittgenstein* (Oxford: Blackwell, 1986), p. 51.

[14] In March 1930, however, Wittgenstein states that a law in physics is possible, i.e., convenient. It predicts. What is true/false, are phenomenological statements (*Waismann* 101).

STANLEY RIUKAS

THE PROBLEM OF THE TRANSCENDENTAL EGO IN HUSSERL

The concept of the transcendental ego is the fundamental concept of Husserl's phenomenology. it is the foundation on which, in Husserl's own view, the entire edifice of transcendental phenomenology and scientific philosophy rests and which is thus of crucial importance for the understanding of his doctrine.[1] It is to be expected that a concept of so great an importance would be of considerable complexity and rather difficult to understand and that Husserl would warn us, on many occasions, against various possible misunderstandings of it. The purpose of this paper is, therefore, to critically examine the concept of transcendental ego in order to find out whether frequent misunderstandings of it on the part of many philosophers and other thinkers are due to their intellectual inability to think profoundly or perhaps to the serious, if not fatal, difficulties involved in the logical structure of this concept itself.

Accordingly, this paper will inquire, first, into the nature of the transcendental ego; secondly, into the assumption it makes; and thirdly, into the function it is supposed to perform.

I

Since the transcendental ego is described by Husserl as the agent that performs innumerable activities of various kinds and thus serves as the ultimate foundation not only for transcendental philosophy and psychology but for all the natural and social sciences,[2] one is tempted to infer from this that this ego is essentially different and perhaps even separate from the natural ego. So it comes as a bit of surprise when Husserl explicitly denies that the transcendental ego is separate or even different from the human ego in its natural attitudes.[3]

The human ego or mind is, according to Husserl, in the natural attitude when it naively believes that it can attain to a genuine knowledge of things and of the world which they make up as they are in themselves rather than merely of their phenomena which constitute the internal world of the transcendental ego.[4] The transcendental ego comes into existence when the human ego abandons its natural attitude or a naive belief in

A-T. Tymieniecka (ed.), Analecta Husserliana LIX, 501–510.

its ability to know things as they are in themselves and comes to the all-important insight that it can know only their phenomena, and know them reliably and with certainty.[5] This abandoning of the natural attitude is achieved through what Husserl calls, by a technical term, *epoche*, which is a universal suspension of our belief in the existence of the entire external world including even our bodies and our minds in their natural attitudes.[6] We do not deny that the external world exists, but we do deny to ourselves belief in its existence.

The process through which the transcendental ego arises is consistently called by Husserl phenomenological, transcendental reduction rather than formation or production in order to avoid the slightest suggestion that it is something newly created. The truth is that there is only one ego or mind, which, in Husserl's own words, alternates between the natural and transcendental attitudes. One could perhaps express the same idea more simply by speaking of two distinct and essentially different function of one and the same human ego or mind – the natural and the transcendental.

In Husserl's view, the transcendental epoche is not his own invention. It was already employed by Descartes as the point of departure for his philosophy,[7] as it is evident from his explicit resolve to treat all of his past beliefs about this-worldly and even other-worldly realities to be not only doubtful but outright false[8] – at least for a certain time. In agreement with Gassendi's charge that Descartes, by deciding to treat all of his beliefs as false is only replacing his previous prejudice of indiscriminate belief with a prejudice of indiscriminate unbelief,[9] Husserl points out that, by his radical epoche, Descartes, instead of enabling himself to explore the world of phenomena and develop a truly scientific philosophy, winds up by burning the bridges that connect the world of philosophy and science with the everyday world. The correctness of the above assessment of Descartes' project can be seen from the fact that the three indubitable truths that are alleged to constitute the axiomatic base of his philosophy are, in essence, nothing more than an expansion of the notorious ontological argument. For it is hardly debatable, at least after Kant, that from the idea of a Supremely Perfect Being nothing more than the *idea* of the existence of such a being can be inferred and *not* a real *existence* itself;[10] and that, likewise, from the idea of infinite holiness or truthfulness nothing more than the *idea* of the existence of the external world, but *not* the *existence* itself of that world, can be inferred. The consequence of this line of reasoning is that, logically,

Descartes cannot retrieve even his own body from the abyss of universal doubt except at the cost of relinquishing his project.

In contrast to Descartes' epoche, Husserl's epoche refuses to break the link between the transcendental world and that of everyday realities or to declare beliefs about these realities false, and thus it seems at first to avoid the disastrous consequences of the Cartesian predicament. In reality, however, Husserl's epoche or suspension, and with it his entire project of a scientific philosophy, is caught in essentially the same predicament. To be sure Husserl, since he has not burned the bridge linking the transcendental and the natural egos, can alternate between them at will, i.e., suspend his beliefs about this world and lift the suspension and thus return to it again. But since the transcendental function of mind is distinct and even separate from the function of the ego or mind in its natural attitude, whatever the ego accomplishes by way of achieving superior or scientific knowledge while functioning transcendentally, it has no way of transferring or communicating it to the same ego while it is functioning within the boundaries of the natural attitude. The gulf between the two functions is apparently unbridgeable. Otherwise, the two functions would cease to be two distinct functions and would become one, which would make the distinction between the transcendental ego and the ego in its natural attitude vacuous of any sense. One could hardly object effectively to his conclusion by pointing out that the natural ego would be greatly enriched by sharing the wealth of knowledge etc. attained by the transcendental ego in its own sphere. For, by sharing in that alleged wealth, the natural ego would have to function as the transcendental ego, and, to that extent, would cease to function as itself. As a result, Husserl's epoche is just as successful as Descartes' in imprisoning him in the coils of the notorious ontological argument, from which there is no escape.

The above considerations may not be fatal to Husserl's assumption of the transcendental ego, but they seem to make that assumption at least highly improbable, if not self-contradictory.

II

Let us assume, for the sake of argument, that the concept of the transcendental ego survives the above critical scrutiny of its nature, that it functions in the way postulated by Husserl, and that it thus provides the indubitable foundation for transcendental phenomenology and scientific philosophy as well as for all the sciences and mathematics. And let us further assume that all the innumerable activities of crucial importance attributed by him to the ego or mind both in its transcendental and natural capacity logically imply that the ego or mind is an agent which carries out all these activities. These assumptions with regard to the Husserlian ego would be nothing new since we had to make them already with regard to the Cartesian ego and its *cogito*, a fact which Husserl seems to be concerned to emphasize by consistently making the *ego* explicit whenever quoting the famous *cogito*, a usage not required at all by idiomatic Latin.

Let us therefore examine Descartes' claim that the ego is an agent which carried out the activity of "cogito" in the broadcast or Cartesian sense of that term. Descartes explicitly states that, far from proving the reality or existence of the ego or "I" by an argument, he infers this existence of "I," – "a thing that thinks," "the soul," or "spirit" – from the analysis of "thinking," which he assumes from the beginning to be the *activity* of the "I," an assumption for which he gives no support. But if thinking is only an impersonal process – a possibility which can hardly be disproved for all we know about thinking – or a cluster or processes, the only thing Descartes is entitled to trivially infer is another process, or a process from process. Moreover, since the inference process fails to produce the required "I" and as a result there is no "I," no "thinking thing," no "soul or mind," there can be no thinking activity carried out by this non-existent "I."[11] In other words, Descartes' argument or analysis is undeniably circular since its antecedent "*I* think" already assumes the existence of an "I" as an agent that is engaged in an activity, whose existence is then supposed to be "inferred" in the consequent.

Unlike Descartes, Husserl spares himself the unrewarding trouble of attempting to explicitly infer the reality of the ego – the transcendental and the natural – from an analysis of its any specific activity. He clearly regards the existence of the ego as the necessary consequence and the best explanation of the bulk of activities, which he unhesitatingly ascribes to the problematic ego, viewing it as an agent from which they emanate. And thus inevitably he winds up in the same predicament as Descartes by treating the ego as the agent and cause of all psychological or mental activities or events and by treating the latter as effects of the ego viewed as an agent. It may be readily granted to Husserl that such a relationship between the alleged agent and its alleged activities is self-evident by a common man's criteria, but it must be denied that it is self-evident to critically thinking philosophers or psychologists.

One must, then, perhaps reluctantly, conclude that Husserl's valiant efforts have failed to extricate him from the perennial dualistic predicament, which, as is well known, postulates in one form or another "the ghost in the machine," due to the notorious "category mistaken."[12]

III

Suppose now we set aside all of the above difficulties and grant for the sake of argument that the transcendental ego is a true agent exactly in the sense postulated and described by Husserl and proceed to subject to a critical analysis its specific functions as conceived by him.

As is well know, much of Husserl's transcendental phenomenology hinges on his familiar distinction between the everyday or external world which we perceive through our senses and the internal world which consists of phenomena, including thoughts, ideas, judgments, values, etc., and which we perceive through reflection (a term familiar since Locke), and the systematic exploration of which is appropriately called transcendental phenomenology or scientific philosophy.

Some serious difficulties are involved in this proposal of the internal or phenomenological world. First, the argument for the existence of the internal world is based here on an analogy with the everyday world, and such arguments are notoriously weak since the differences between the analog and the analogate may be greater than the similarities. For while the external world consists of at least relatively stable objects, events, and various structures, which, at least in principle, are accessible to public and critical analysis, scrutiny, comparison, exploration, synthesis, etc. of various kinds, there are no objects whatever making up the phenomenological world, which is only metaphorically called a world. If it is asserted that the phenomena are the objects making up this world, it will have to be conceded that they are such only in a metaphorical sense, and that in reality they are only processes. But what kind of processes? Are they perhaps physico-chemico-electrical brain or neuronal processes? Hardly. Husserl's metaphysical scheme is plainly incompatible with this kind of materialist interpretation. But even if these processes are mental in a sense compatible with his scheme, they still cannot be regard as constituting a world of their own since, as processes, as forms of mind in flux, they lack of necessity any other stability except that of a perennial Heraclitean flux. The same must be said if the internal world is described, as Husserl seems to prefer, in terms of acts rather than processes.

Since the proximate task of the transcendental ego is alleged to be the *description* of phenomena before comparing, analysing, and classifying them for the purpose of a theory about them, it must be inquired into how such a description of instantaneous or punctiform occurrences is possible.[13] After all, we are able to engage in description of objects making up the external world only because they possess temporal duration and in description of events only because they are inextricably connected with objects. Our inspection can go back to the object repeatedly in an attempt to make sure we correctly and adequately describe its qualities whereas we cannot do this at all with mental or psychological occurrences that hardly have any temporal duration.

If it is, however, claimed that these occurrences do possess temporal duration, that acts or occurrences of cognition, volition, emotion, or valuation, for instance, are apt to perdure indefinitely, it must be pointed out that, to be correct, it is not these acts themselves that perdure but only their natural consequences, memory traces, and perhaps increased dispositions to perform them again. It may be suggested that we can bring back the same mental or psychological act through memory. A moment's reflection shows that a remembered occurrence is by definition no longer the original one. Thus we are precluded from a repeated return to the original act for purposes of a more adequate description, as we can do in case of everyday objects which are always there.

Thus we must conclude that the transcendental ego may be an impossibility since there seem to be no appropriate objects or events which it is supposed to describe.

But surely our analysis must be mistaken. For after all we have recognized at least an instantaneous or punctiform existence of mental events; and so the transcendental ego could very well exert itself in describing and otherwise exploring the realm of these events. Let us therefore look into this possibility.

The acts of reflection through which the transcendental ego is supposed by Husserl to carry out observation, scrutiny, and description of mental or psychological events have, generally speaking, only two categories of objects to describe: mental acts or phenomena actually occurring and the memories of them. To begin with the latter category, what assurance do we have that memories correctly reflect the phenomena of which they are memories when it is notorious that even the written memory is far from being completely reliable? If it is claimed that memories are self-certifying since transcendental reflection can compare different stages of a memory and correct possible distortions due to its being a "decaying sense," it must be pointed out that such a comparison already presupposes the correctness of memory. Apparently the transcendental ego, in spite of the claim that mental phenomena are self-intimating, cannot achieve certainty and thus provide a highly desirable foundation for philosophy and the sciences.

Suppose, however, that the primary task of transcendental phenomenology is not the description of mental phenomena, that is, not of memories, but rather the description of mental acts or events actually occurring. Suppose further that the transcendental ego focuses its attention, through an act of reflection, upon some particular mental act such as a mathematical calculation or a solution of some scientific or moral problem in an attempt to inspect and to describe it. The activity of calculation or solution of a difficult problem may very well demand the maximum of concentration or attention. The act of transcendental reflection, description etc., is apt also to demand the maximum of attention. A truly complete or total engagement of mental resources, of total available energy in the activity of calculation logically excludes any additional, let alone total, commitment of the same resources to another activity, that of reflection. If we say that we engage in these two activities with only partial output of our energy, we would be charged with half-hearted efforts, lack of seriousness, and the value of our project would be discounted.[14]

It may be objected that we are capable of performing quite successfully several fairly complex activities at the same time such as driving a car, talking to a friend, listening to a radio music, watching the traffic etc., thus giving to each one of them only divided attention. This may be quite true. But the question still remains whether we are giving our attention to all of them at the same time or rather shifting it back and forth among them. Also notoriously such activities are routine, carried out largely by force of habit and with a minimum of attention. By contrast, our attention would be total if the President's assassination were announced on the radio or if we became involved in a serious accident.

The hypothetical case of a serious accident points to the category of phenomena which by their very nature demand our total attention such as fright, fear, anger, madness, lust, and other violent or paralyzing passions. It is quite unlikely that a clam, transcendental reflection on such experiences is even possible, let alone that the resulting description of them is of any value. When our total attention is completely focused, as in a fit of anger, on the object of our anger, we may surely try to subjugate our anger by turning it from its object to the act of anger itself. But by then the act or event of anger will be gone, and we will be focusing only on a more or less reliable memory of it.[15] It is clear thus that transcendental reflection cannot provide us with a certain and indubitable foundation of philosophy and knowledge in general, as Husserl was hoping. The same must be said about another often discussed category of phenomena, that of pains whose perception is supposed to be incorrigible. For it is well known that, in certain cases, pains are indistinguishable from pleasures. So they cannot contribute to an indubitable foundation of knowledge either.

The preceding reflections on the function of the transcendental ego tend to suggest that it does not possess any capacity to scrutinize and describe mental phenomena so as to provide information that could be used in developing an indubitable foundation of all certain knowledge.

In conclusion, then, I would like to say that the Husserlian project of phenomenology with its transcendental ego represents one of the most sustained attempts to vindicate metaphysical dualism, an attempt which must be regarded, in the light of our remarks, as no more successful than any previous ones.

NOTES

1 Edmund Husserl, *The Paris Lectures*, trans. Peter Koestenbaum (The Hague: Martinus Nijhoff, 1967), Introduction, p. lv: ". . . Husserl feels justified in saying that the transcendnetal Ego is the source and basis of all knowledge," a feeling to which Husserl gives a repeated expression in his writings.

2 Edmund Husserl, "Phenomenology," Edmund Husserl's Article for the *Encyclopaedia Britannica* (1927) in *Husserl Shorter Works*, ed. Peter McCormick and Frederick A. Elliston (Notre Dame: University of Notre Dame Press; Brighton, U.K.: The Harvester Press, 1981), p. 33: "If we take the universe of all possible empirical sciences whatever and demand a *radical* grounding that will be free from all 'foundation crisis,' then we are led to the all-embracing a priori of the radical and that is (and must be) *phenomenological* grounding."

3 Edmund Husserl, Husserl's Inaugural Lecture at Freiburg im Breisgau (1917). trans. R. N. Jordan in *Husserl Shorter Works*, op. cit., "On the other hand, he can convert his natural attentional focus into the phenomenologically reflective one; he can make the currently flowing consciousness and, thus, the infinitely multiform world of phenomena

at large the theme of his fixating observations, descriptions, theoretical investigatins . . . , which, for short, we call "phenomenlogical."; ". . . pure reflection excludes everything that is given in the natural attitude and excludes therefore all of Nature."

4 Edmund Husserl, *The Paris Lectures*, *op. cit.*, p. 14: "Before the *epoché*, I was a man with the natural attitude and I lived immersed naively in the world. I accepted the experienced as such, and on the basis of it developed my subsequent positions . . . but I did not focus on the experiencing of my life, on the act of being interested, on the act of taking a position. . . . I was a transcendental ego even while in the living natural attitude, but knew nothing about it."

5 *Ibid.*, p. 15: "In order to become aware of my true being I needed to execute the phenomenological *epoché* . . . the phenomenological attitude, with its *epoché*, consists in that *I reach the ultimate experiential and cognitive perspective thinkable. In it I become the distinterested spectator of my natural and worldly ego and its life*."

6 *Ibid.*, p. 15: "The phenomenological reduction . . . tends to split the ego. The transcendental spectator places himself above himself, watches himself, and sees himself as the previously world-immersed ego . . . The natural man (in whom the ego, in the last analysis, is a transcendental one, but of which he knows nothing) possesses a world and a science which he naively takes to be absolute. However, the transcendental spectator who has reached the consciousness of a transcendental ego conceives the world merely as a *phenomenon*. . . ."

7 *Ibid.*, p. 3; ". . . phenomenology might almost be called a new, a twentieth century, Cartesianism."

8 René Descartes, Meditation I, in *Philosophical Works of Descartes*, Vol. I, trans. E. S. Haldance and G. R. T. Ross (New York: Dover publications, 1955), p. 148: ". . . I consider that I shall not be acting amiss, if, taking of set purpose a contrary belief, I allow myself to be deceived, and for a certain time pretend that all these opinions are entirely false and imaginary. . . ."

9 René Descartes, The Fifth Set of Objections by P. Gassendi, in *Philosophical Works of Descartes*, trans. Haldene and Ross (Cambridge: The University Press, 1967), p. 136: ". . . why you should not have preferred to indicate simply and with few words that what you previously knew was uncertain . . . rather than by regarding everything as false, not so much to dismiss an old prejudice, as to take up with a new one."

10 Immanuel Kant, The Impossibility of an Ontological Proof of the Existence of God, *Critique of Pure Reason*, trans. Norman Kemp Smith (New York: St. Martin's Press, 1965), p. 504; ". . . but in that case . . . we have presupposed an existence as belonging to the realm of the possible, and have then, on that pretext, inferred its existence from its internal possibility – which is nothing but a miserable tautology."

11 It may be objected that the failure of the inference process to produce the required ego or "I" does not mean that it does not exist. Quite true. But then an ego which is unknown and may not even exist surely cannot serve as a fundamental concept in developing phenomenology.

12 Gilbert Ryle, "Descartes' Myth," *The Concept of Mind* (New York: Barnes & Noble, 1971), pp. 11–24: It would be "ridiculous to construct the disjunction 'She came home either in a flood of tears or else in a sedan-chair.' Now the dogma of the Ghost in the Machine does just this. It maintains that there exist both bodies and minds; that there occur physical processes and mental processes; . . . ."

[13] Husserl, *The Paris Lectures*, *op. cit.*, p. 19: "Phenomenology always explains meanings, that is intentionality, by producing these sense-ful-filling syntheses. The tremendous task placed upon description is to expound the universal structure of transcendental consciousness in reference to and creation of its meanings."

[14] Gilber, Ryle, *op. cit.*, p. 165: ". . . even if it is claimed that in introspeating we are attending twice at once, it will be allowed that there is some limit to the number of possible synchronous acts of attention, and from this it follows that there are or must be some mental processes which are unintrospectable . . . But if this knowledge [knowledge of one's own mental processes] does not always rest on introspection, it is open to question whether it ever does."

[15] *Ibid.*, p. 166: "No one could introspectively scrutinise the state of panic or fury, since the dispassionateness excercised in scientific observation is, by the definition of 'panic' and 'fury,' not the state of mind of the victim of those turbulences."

YOUNES MERROUCH

POUR UNE PHÉNOMENOLOGIE DE LA CHAIR

Le désir ne procède pas du *manque* – c'est la *plénitude* qui est mise en évidence tout le long d'une phénoménologie de la chair, à condition que l'*époché* soit maintenue, aussi radicale que possible. Pour dévoiler la chair en tant que vécu immédiat, il faut réduire phénoménologiquement sa teneur "humaine" (psycho-physiologique, anthropologique, théologique, etc.). Il apparaît alors, entre autres, que *désirer* consiste à *percevoir*. La *hylè*, aussi engagée soit-elle dans le processus intentionnel, est une plénitude jaillissante. Mais la visée signitive – à vide – serait-elle justement le "manque" qui viserait la pléntude hylétique du remplissement intuitif-sensible?[1]

Examinons donc un acte signitif visant un "objet" charnel; en l'occurrence, et pour partir d'un exemple clair et précis: la visée à vide "sein". Les *data* hylétiques visuels et tactiles sont absents; mais cette visée est bien un acte complet, une noèse avec son corrélat noématique, et il serait difficile, voire aberrant d'affirmer que le vécu en question est dépourvu de composante hylétique. Or la phénoménalité est polymorphe et la visée signitive "baigne" dans un certain nombre de phénomènes plus ou moins "obscurs" qui ne la concernent pas directement et dont les corrélats intentionnels lui sont hétérogènes. Cependant, si l'on écarte tout ce qui implique l'intentionnalité pour retenir la seule visée "sein", il subsistera encore un vécu irréductible: *la coenesthésie*. Celle-ci, à moins qu'elle ne se vise elle-même, n'est pas un vécu intentionnel; néanmoins sa phénoménalisatin est liée à la visée: il est évident que les fluctuations du vécu charnel non spécifique et diffus sont connectées au dèroulement de la pensée avec ses noèses successives. Or, si la coenesthésie n'est pas animée en une "noèse coenesthésique", elle est: soit un vécu hylétique libre, délié, soit la composante hylétique d'un vécu intentionnel. Les deux cas sont envisageables; le deuxième nous permet de formuler la proposition suivante: *La coenesthésie est la composante hylétique de la visée signitive*. Il serait préférable d'éviter le terme "à vide" puisque la visée signitive est d'emblée remplie coenesthésiquement. La coenesthésie n'est pas un vide, ni un creux: c'est un plein hylétique qui peut s'animer signitivement. Le pouvoir constituant,

A-T. Tymieniecka (ed.), Analecta Husserliana LIX, 511–520.

quelque noématique qu'il soit, procède de la plénitude et non du manque.

La visée signitive "sein" a sa propre *hylè*; celle-ci lui assure un remplissement initial, partiel. Or le remplissement sensible n'est jamais définitif, il s'opère en une série d'apparitions, par esquisses successives dont les possibilités sont illimitées. Nous dirons que la visée en question est une perception initiale qui peut être orientée vers le percevoir-plus à travers un remplissement hylétique progressif. Il ne s'agit pas d'un manque qui attend d'être rempli; la visée "désirante" est un plein tendant vers plus de plénitude. A partir du noème, cette "tention vers" peut se déployer immédiatement dans l'imagination. Un "sein" apparaît aussitôt; mais il est évident que les *data* constituant l'image se distinguent nettement de ceux qui constituent l'apparition perceptive "en chair" – il y aurait une sorte d'abîme phénoménologique infranchissable entre l'imagination et la perception. Le désir serait-il donc l'imagination en tant que manque *de* perception, *de* "donné en chair"?

Il n'y a aucune rupture entre la visée signitive et la première esquisse imaginative, celle-ci découle de celle-là. La coenesthésie, tendue noétiquement vers le "sein", prend forme en une série de noèses-esquisses convergeant vers une image. – L'image serait constituée de *hylé* coenesthésique. Car l'on ne résout pas le problème en considérant que le *datum* imaginatif est "image *du datum* hylétique originaire".[2] La couleur et la lumière sont *visées* coenesthésiquement. "Image *de*" est un corrélat noématique éventuel; quant à la *hylè* de l'image elle-même, elle ne peut être que la coenesthésie qui se trouve ainsi engagée dans une sorte de "projection" noétique. La phénoménalisation de l'image – en particulier sa pléntiude – variera selon le mode et l'intensité de l'investissement coenesthésique. La constitution de l'image "sein" est liée, dans une sorte de "tension vers", à la visée signitive, les deux actes ayant la même composante hylétique (la coenesthésie) – aussi le noème "sein en chair" n'est-il en aucun cas le corrélat d'un manque. L'imagination n'est pas un "manque de" perception puisque le noème "percevoir" est lui-même constitué par une pléntitude coenesthésique. Il faut rappeler ici que l'imagination intervient au coeur même du processus perceptif, dans le déroulement protentionnel des noèses constituant les esquisses successives de l'object. La rupture imagination/perception se situe au niveau de la *hylè*: la *hylè* coenesthésique ne peut se transformer en *data* hylétiques de couleur et de lumière. La "couleur" de "l'image-sein" n'est pas un *datum*, elle est le corrélat intentionnel du *datum* coenesthésique; néanmoins elle est visée signitivement en tant que *datum*: constituée de telle façon qu'elle apparaisse comme donnée. Quant au "manque", il n'est que le corrélat noématique d'une visée signitive dont la noèse consiste à *nier* sa propre *hylè*: la coenesthésie projetant son auto-négation – c'est une possibilité intentionnelle. Mais la "matière" de toute intentionnalité n'est autre que la plénitude du jaillissement hylétique.

L'imagination est désir dans la mesure où elle est un investissement hylétique pour la constitution du "sein", une "perception coenesthésique" s'intensifiant, tendue vers le percevoir-plus, un état d'accroissement, de "débordement". Quand le "sein", visé signitivement et constituté imaginativement, apparaît "en chair", la phénoménalisation ne correspond pas au remplissement d'un vide: au pôle intentionnel d'une coenesthésie noétiquement tendue vers le "sein" qu'elle constitue, de nouvelles *data* jaillissent: la *hylè* de lumière et de couleur fait surgir en chair l'objet charnel en *illuminant* sa constitution. Ces *data* jaillissent à la pointe de la tension coenesthésique; la "protubérance" du "sein en chair" adhère à la plénitude coenesthésique imaginative. La coenesthésie "épouse" la *hylè* lumineuse. L'apparition en chair ne comble pas un manque, elle ne prend pas la place – ou le vide – du vécu signitif-imageant: elle est un *plus*, un remplissement supérieur.

Examinons à présent le cas d'une *Apparition* du "sein": apparition *subite*: sans protention, sans aucune visée signitive ou imaginative préalable. La constitution de l'apparition ne peut s'accomplir en tant que *vécu charnel* sans l'intervention de la coenesthésie. Car le jaillissement hyléique de couleur et de lumière ne prend forme en tant que "sein", "chair", qu'en s'unissant, dans un remplissement mutuel, avec un jaillissement charnel, en l'occurrence tout le flux coenesthésique qui se projette ainsi soudainement dans la lumière de l'apparition. La totalité du vécu hylétique charnel se trouve engagée brusquement, avec les *data* hylétiques visuels, dans une noèse unique, subite, d'une intensité excessive: il s'agit là d'une sorte de projection "explosive" – les noèmes "chair" et "sein" eux-mêmes se "dissolvent" dans l'ampleur du jaillissement hylétique. Vu l'intensité excessive immanente à la phénoménalité de l'*Apparition*, il serait phénoménologiquement opérationnel de considérer que la *hylè* est une force et que *l'apparition de la chair est un jaillissement excessif de forces.*[3] Qu'est-ce donc que le "désir", dans cette plénitude explosivement débordante? – *Désirer c'est percevoir, le désir est un jaillir.*

Le désir est la plénitude du jaillissement hylétique en tant que *jaillir plus*, il est le *devenir* des forces – qui ne sont ni les mêmes ni autres (l'altérité, envers de l'identité, est un concept qui doit être mis hors-circuit). Nous ne nous référerons pas à la Volonté de Puissance parce que nous proposons la *force* comme concept *phénoménologique* qui s'applique au jaillissement hylétique et qu'il ne faut pas confondre avec la force au sens nietzschéen. Cependant, nous pouvons considérer, dans les strictes limites de l'analogie, que l'*Apparition* de la chair est *tragique* dans la perspective précise de l'esthétique de Nietzsche. Les *data* hylétiques de couleur et de lumière constitueraient l'élément apollinien; la *hylè* coenesthésique (obscure) serait l'élément dionysien. Les force dionysiaques se projetteraient ainsi dans une illumination apollinienne d'une soudaineté explosive.

La perception de la chair a cela de particulier qu'elle est une mise en lumière de la coenesthésie. Car le jaillissement conenesthésique est précisément la *chair*: vécu informe, obscur, diffus, im-médiat, irréductible; l'apparition est le jaillissement de cette *chair* vers le jaillissement de la lumière. En ce sens la *hylè* coenesthésique *apparaît*: la *chair s'*apparaît. C'est la *hylé* elle-même qui est visée: le contenu du complexe noématique n'est autre que la pléntiude hylétique. La constitution du noyau noématique "chair" (et des divers éléments possibles du complexe noématique: "tendre", "clair", "protuberant", "galbé", etc.) est bien une composante de la perception de la chair, mais elle est une constitution *par* la chair *pour* la chair: en jaillissant, en constituant l'apparition et le sens, la *hylé* s'apparaît, s'intensifie, jaillit-plus. Nous appellerons *jouissance* la perception de la chair à partir du moment où celle-ci consiste en une *hylétisation* du vécu. Une fois que, à partir des *data* hylétiques, l'objet charnel est constitué à travers les premières esquisses noétiques, le noème est acquis, "rempli", "plein" de la plénitude hylétique qui le constitue et dont il est en même temps le corrélat intentionnel. La constitution même de l'apparition et de son noème provoque alors, en retour, une intensification du jaillissement hylétique que nous appellerons *noèse inversée*. Car, l'objet intentionnel étant déjà pleinement constitué, le processus d'intensification ne saurait tendre vers une "sur-constitution", ni vers une "hyper-intentionnalité". Le noème tendrait plutôt à "éclater", libérant ainsi les forces (la *hylé* coenesthésique) qui le constituent. La *hylè* se "gonfle" alors à l'image même de l'apparition charnelle constituée qui jaillit justement à la manière d'un "gonflement": protubérance, saillie. La perception de la chair persiste, mais *hylétisée*, concentrée sur les seuls *data* purement hylétiques. La *jouissance* ou *noèse inversée* atteint *la hylè constituante au-delà de l'objet constitué*. Les forces, en jaillissant, constituent l'objet charnel pour le "trans-tituer" en le hylétisant, faisant ainsi *jaillir-plus-de-forces*. Il s'agit là d'un *plus* paroxystique sans être "explosif" (il faut distinguer nettement jouissance et Apparition "tragique"). La jouissance est une sorte d'"affirmation" qui ne peut avoir lieu qu'au-delà du sens: au-delà du noème, la *hylè* charnelle adhère à son propre jaillissement; les forces échappent à la structuration noématique et débordent, déliées, ne constituant que leur propre plénitude jaillissante.[4]

Les différents modes d'apparition de la chair se ramènent à deux composantes principales que nous introduirons d'emblée à partir de la dichotomie corrélationnelle "horizontalité/verticalité".[5] La corrélation sensible constituant l'apparition est fondée sur les rapports de contraste et de similarité: l'une des deux composantes prédomine et constitue une *homogénéité* (jaillissement hylétique similaire) contrastant avec l'autre composante et qui ne peut apparaître que sur le mode du contraste.[6]

L'apparition du type "horizontal" se constitue en *s'étalant* tout en jaillissant "tendue"; mais elle ne se tend que pour mieux "s'étaler" en "profondeur", en "rondeur": "gonflement", "protubérance", "mollesse" (jaillissant avec "fermeté"), "tendreté" etc. Tous ces éléments se ramènent à une double *plénitude*, hylétique *et* intentionnelle: la plénitude du jaillissement hyléique "coïncide" avec son corrélat noématique dans la mesure où l'apparition constituée est précisément une plénitude (charnelle) apparaissante, illuminée. L'homogénéité du jaillissement hylétique ne peut "s'étaler" noétiquement sans les *data* de lumière: la "clarté" charnelle (avec la gamme illimitée des nuances jaillissant en *data* de couleur) illumine la totalité de l'apparition jusqu'à ses composantes obscures: le flux coenesthésique, les cénesthèses (que nous examinerons plus loin) et les *data* tactiles sont connectés à la *hylé* lumineuse. La *clarté* et la *plénitude* fusionnent dans une homogénéité commune; la lumière est charnéisée, "coenesthésiée". Les *data* de lumière jaillissent en constituant une "profondeur" éclatante (et non pas l'éclat d'une "surface"): luminosité protubérante, débordante. Si la "forme" charnelle apparaît nettement tout en étant "voilée", les *data* de clarté restent en quelque sorte "implicites", jaillissant dans un "horizon obscur", l'intentionnalité constituante se concentre alors sur une homogénéite-plénitude illuminée imaginativement. – Un jaillissement ponctuel et étroit de “nudité” lumineuse peut ainsi “éclairer” la totalité de l’apparition voilée.

Quand la verticalité prédomine, l’apparition *se dresse* alors (noétiquement) et se *tend* en “hauteur”, plénitude “*ascendante*”, “dureté”, “fermeté”: jaillissement OSSEUX. Les corrélats intentionnels de l’apparition du type “vertical”, que nous appellerons également apparition *ithyphallique*, se ramènent aux noèmes que nous désignerons par “os” et “phallus”. Il s’agit là d’une phénoménalisation de la chair *se constituant en tant qu’elle contraste avec la charnéité même*. Le jaillissement ithyphallique est une éclatante et inflexible “montée” osseuse. Les *data* hylétiques osseux, jaillissement de forces, constituent une apparition dont le noyau noématique est “*la force*”; ils jaillissent avec les *data* charnels constrastants qu’ils “dominent” en les “soumettant” au déploiement d’une luminosité “verticale”. Le *phallus* n’est autre que le corrélat intentionnel de la visée charnelle constituant (“désirant”) une apparition *osseuse* – c’est une forme spécifique d’intentionnalité charnelle ayant cela de particulier qu’elle est en quelque sorte une projection-illumination du *jaillir* des forces: le jaillir hylétique polymorphe prend forme, s’*illumine* dans la régularité et la rectitude d’une “élévation” inébranlable; en ce sens le *phallus* est exactement le contraire de “l’organe” (“sexuellement”) fonctionnel. C’est la composante (hylétique) ithyphallique-osseuse qui donne forme et consistance à l’informe profusion de la charnéité.

Il faut préciser que sur le plan hylétique, le jaillissement coenesthésique, quelque diffus qu’il soit, est variable même s’il n’est pas engagé dans une projection noétique constituante. La coenesthésie en tant que telle (obscure) peut être “tendre” ou “dure”, “molle” ou “tendue”; autrement dit: elle peut jaillir *charnéité charnelle* ou *chair osseuse*. L’auto-aperception des “os” est bien une composante du jaillissement coenesthésique mais elle peut prédominer, et c’est tout le vécu charnel qui se tend alors à la mainière d’un “*os*” unique et uni.

Le désir, en tant que perception, jaillirait selon les fluctuations corrélationnelles de contraste et de similarité sensibles; à partir de celles-ci, la charnéité charnelle et la chair osseuse auraient pour corrélat intentionnel l’apparition ithyphallique ou l’apparition charnelle; et, vu que l’apparition de la chair est un jaillissement de forces s’apparaissant, le facteur de similarité serait probablement prépondérant. – Celui-ci consiste dans le fait que la *hylé* coenesthésique osseuse ne peut se projeter visuellement qu’en *data* osseux, et la *hylé* charnelle en *data* charnels. Si la chair osseuse constitue (désire) une apparition charnelle et vice versa, c'est grâce à une projection en chisme qui procède d'une inversion des rapports hylétiques: en se projetant, la composante faible s'amplifie et se déploie alors que la composante prédominante se réduit et se concentre. C'est ainsi que la chair constituante peut contraster avec la chair constituée, et c'est cela qui rend possible la polarité du désir (os-pallus/chair; "masculin"/"féminin"). L'accentuation du contraste intensifie le vécu charnel polarisé. Un jaillissement ithyphallique est encore plus osseux quand il est noétiquement tendu vers une apparition charnelle; une charnéité est d'autant plus pleinement charnelle qu'elle perçoit une apparition ithyphallique.

Cependant les possibilités de la perception de la chair ne se réduisent pas à quelques cas-limites de contraste et de polarité. La phénoménalité charnelle est *polymorphe*: elle ne cesse de fluctuer entre l'osseux et le charnel, entre l'imagination et la perception, la coenesthésie, l'esthésie et la kinesthésie; les différentes composantes de l'apparition de la chair jaillissent et se combinent d'une manière illimitée. Le flux coenesthésique peut d'ailleurs se spécifier, outre les projections imageantes et perceptives, en *data* charnels "locaux" constituant l'*auto*-aperception du "doigt", de "l'estomac", de "la cuisse" etc. Nous appellerons *cénesthèse* ce vécu obscur (bien qu'il puisse être associée à des *data* visuels constituant ainsi la "chair-corps-propre") qui jaillit variablement selon une sorte d' "intentionnalité viscérale", passant, par exemple, de "la-pointe-du-petit-doigt" à "la-plénitude-du-ventre". Les cénesthèses sont connectées à l'apparition de la chair à travers les projections coenesthésiques, et, d'une manière plus directe, grâce aux *data* tactiles. Le tact pourrait être considéré comme un "affleurement" cénesthésique; mais il y a un seuil sensible entre la cénesthèse (sentir-"mon"-doigt) et le toucher (sentir-"mon"-doigt-sentant-un-visage). Hylétiquement, ce seuil tient à une différence au niveau du mode et de l'intensité du jaillissement. Les forces constituant la cénesthèse jaillissent à la manière d'un flux "circulaire" qui se rejoint sans cesse quelle que soit son intensité (cette description n'implique aucune spatialisation: "circulaire" est une métaphore). Le passage du cénesthésique au tactile consiste en une sorte de "petite explosion locale" (qui ne saurait être comparée à celle de l'Apparition et se situe plutôt dans la perspective de la jouissance) qui *romp* le cercle: les forces "giclent" alors vers la constitution de l'apparition tactile. Au-delà de ce seuil les *data* cénesthésiques et tactiles se fondent en une plénitude croissante; car les composantes de la phénoménalité charnelle ne s'excluent pas, elles se mêlent, se cumulent et s'intensifient mutuellement. Dans certains cas paroxystiques la polymorphie de la chair se réduit à l'hégémonie d'*une* homogénéité – mais celle-ci ne peut jaillir qu'en une pluralité de forces: un *datum* unique, ponctuel, ne peut former une homogénéité.

Les désir est un jaillissement polymorphe; la plénitude hylétique, multiplicité de forces, jaillit de manières multiples. Examinons donc le cas éventuellement problématique où le désir semble se fixer sur un "objet" unique (vécu désigné communément, dans le contexte "humain", par "passion amoureuse"). Cela commence avec la perception d'une chair-corps à travers une série d'apparitions de la chair. A partir de cette constitution (désir) polymorphe, la chair constituante (percevante) opère une analogie entre "sa"-chair-corps et la chair-corps constituée, ce qui lui permet, en inversant les pôles, de *se* projeter coenesthésiquement en tant que chair-corps *percue* "à distance". La chair *s'imagine* constituée chair-corps par la chair-corps qu'elle constitue. Le vécu en question est donc composé essentiellement de: 1° chair (hylétique) constituante; 2° chair-corps percue; 3° chair-corps imaginée; 4° coenesthésie-constituante imaginée; 5° chair-corps-propre s'imaginant "percue à distance". Les forces jaillissent d'une maière polymorphe, mais selon un schème unique s'auto-reproduisant sans cesse – jusqu'à ce qu'une nouvelle perception se constitue en tant que *dé*-ception mettant ainsi fin à la circularité du déploiement imaginatif. Le jaillissement des forces a pour corrélat intentionnel leur propre re-jaillissement; elles tentent ainsi de se fixer en constituant une temporalité dèsirante circulaire. Mais *une force ne peut se fixer*: l'identité est une fiction, un vécu imaginatif, un mode particulier du *jaillir*. La "passion-amouruse" est le corrélat intentionnellement unifié d'un jaillir charnel polymorphe ("aimer" consiste essentiellement à imaginer) – "l'autre" n'est que l'*hypothétique* corrélat intentionnel d'une visée signitive associée à un aspect particulier de la polymorphie charnelle imaginative. Cet aspect consiste précisément à *imaginer*, à travers l'apparition de la chair, une charnéité, une "sentance" charnelle, une coenesthésie, des cénesthèses, une perception de la chair, etc. Ce sont là des actes imaginatifs charnellement *pleins*; et la coenesthésie est "libre" de constituer signitivement les corrélats qu'elle "veut": "altérite", "manque": ce ne seraient que des noèmes qui n'*alter*eraient en rien la *plénitude* hylétique.

La "sexualité" est un champ intentionnel qui, quelque large qu'il soit, délimite le vécu charnel en lui assignant signitivement des "moyens" et des "fins", réduisant ainsi la chair aux "organes" et autres "zones érogènes", aux "pulsions" et leur "satisfaction" etc., c'est-à-dire à une *fonction* "humaine". La visée "sexuelle" rate la constitution (l'apparition) de la chair puisqu'elle coïncide avec "le *projet* sexuel": l'acte noématiquement fonctionnel l'emporte sur la visée purement charnelle. Là où "l'organe" apparaît, la chair disparaît: la signification "sexuelle" *masque* la chair; elle atténue, voire annihile l'intensité phénoménale (l'excès) du vécu charnel en le fonctionnalisant. Les forces sont "prises" dans le complexe noématique qu'elles constituent: le jaillissement hylétique-charnel se résout dans ses noèmes "fonctionnels". Aussi toute perspective esthétique est-elle éliminée et le péril de l'*Apparition* est écarté: la "sexualité" supprime le possible "tragique" (au sens nietzschéen), "sublime" (au sens kantien) de la chair. C'est ainsi que celle-ci "s'humanise".

Si la phénoménalité charnelle devait se réduire à la "sexualité", alors le *désir* serait effectivement un manque que la jouissance "sexuelle" ne saurait combler puisqu'il ne pourrait procéder dans ce cas que du "phantasme", de "l'inconscient" (freudisme), ou de l'irréductible "écart", "manque à être" etc. (philosophies de "l'altérité"). Mais il n'en est rien: "manque", "négativité", "altérité" et "sexualité" sont – comme *tout corrélat intentionnel* – constituées *par* la *plénitude hylétique*.[7] – La phénoménologie de la chair nous montre la double plénitude du désir: *PLENITUDE HYLETIQUEMENT CHARNELLE constituant une PLENITUDE noétiquement CHARNELLE.*[8]

5 Rue Mansour
Tanger
Maroc

NOTES

[1] Autrement dit: la visée à vide est-elle une noèse sans *hylè*? (Nous envisagerons plus loin le cas inverse.) *Cf.* le § 85 des *Ideen I* (trad. Paul Ricoeur): "Quant aux deux possibilités laissées en suspens ci-dessus, on pourrait les intituler: *matières sans forme* et *formes sans matière*." (E. Husserl, *Idées directrices pour une phénoménologie*, Gallimard, 1985, coll. Tel, p. 290.) Husserl écrit plus loin (p. 294) à propos de la dissociation phénoménologie hyletique/phénoménologie noétique: "Les analyses de loin les plus importantes et les plus fructueuses sont du côté noétique" – quoiqu'il affirme par ailleurs le caractère primordial de la *hylè*, en l'occurrence le fait qu'elle est constituante et in-constituée: *cf.* le § 36 des *Leçons pour une phénoménologie de la conscience intime*

du temps, trad. Dussort, PUF, 1964, p. 99). La phénoménologie de la chair nous amènera à nous intéresser davantage au "côté" hylétique, et ce dans la perspective d'une tentative de réponse à la question que posent les "deux possibilitiés".

[2] Cf. Husserl, *Idées directrices pour une phénoménologie*, *op. cit.*, pp. 374–375.

[3] Nous ne pouvons exposer ici plus en détail cette thèse (*hylè* = force); nous l'avons fait dans "*Epochè* et force," *Analecta Husserliana* LVII, pp. 75–81, 1998, Kluwer Academic Publishers. Le concept phénoménologique de *force* surgit d'une *épochè* radicalement radicale: définitive, irréversible et interdisant le recours à l'*ego* quelque transcendantal qu'il soit – c'est précisément à ce niveau-là que se situe ce que l'on pourrait appeler *une phénoménologie hylétique*.

[4] Husserl, à propos de la jouissance: "Mais de même qu'il y a, pour ainsi dire, un mode de représentation à distance, une visée représentative vide hors de l'être présent soi-même, de même il y a a un mode du sentir qui se rapporte à l'objet à vide; et de même que le premier se remplit dans la représentation intuitive, de même le sentir à vide est rempli par la jouissance." (*Ideen II, Recherches phénoménologiques pour la constitution*, trad. E. Escoubas, PUF, 1982, p. 33.) Il y a là une séparation entre le "sentir" et la visée, entre la jouissance et le remplissement hylétique. Pour nous, "sentir" n'est autre que percevoir (jaillir hylétique noétiquement constituant); et la jouissance (hylétisation par noèse inversée) n'est autre que la plénitude paroxystique du jaillissement hylétique: la teneur "affective" du vécu n'est pas une qualité qui l'affecte ou qui lui est associée, elle est son jaillissement même.

[5] Nous décrivons ici la *morphè*, sachant qu'elle dépend immédiatement de la *hylé* – Husserl l'a bien noté: "tout changement dans le statut hylétique de la perception, s'il ne supprime pas franchement la conscience de perception doit avoir au moins pour résultat que l'objet qui apparaît devienne objectivement "autre", soit en lui-même, soit dans le mode d'orientation lié à son apparaître, etc." (*Idée directrices . . .* , *op. cit.*, p. 338). Il n'est donc absolument pas question pour nous d'une "spatialisation": l'horizontalité et la verticalité (comme l'étalement, la protubérance, l'élévation, la profondeur etc.) sont les formes intentionnelles de deux modes différents du jaillir hylétique. D'ailleurs "flux" et "jaillissement" sont eux-mêmes des termes "métaphoriques". Il n'y a pas de termes exactement adéquats pour décrire la pure phénoménalité. – "*Les forces jaillissent*" est la proposition qui nous semble, phénoménologiquement, la plus adéquate (la "pureté" phénoménologique d'un énoncé est proportionnelle à la radicalisation de l'*époché* dont il émane).

[6] Sur le contraste et la similarité sensibles *cf.* Husserl cité par Marc Richir dans "Synthèse passive et temporalisation/spatialisation" in *Husserl*, dir. Richir et E. Escoubas, Grenoble, Ed. Jérome Millon, 1989, p. 20.

[7] Tout acte signitif est une constitution dont la composante hylétique-constituante est le jaillissement coenesthésique. Il en serait de même pour la perception catégoriale.

[8] Les propositions "sentir = percevoir" et "désirer = percevoir" sont en corrélation immédiate avec la thèse "*hylè* = force". La qualité et l'intensité Phénoménologiques du vécu ne sont autres que *le comment du jaillir des forces*.

ANDRZEJ ZUCZKOWSKI

SPEECH ACTS AND EMOTIONAL CAUSALITY IN EVERYDAY LIFE

According to J. L. Austin's (1962) theory of speech acts, a Speaker (S) by means of locutionary and illocutionary acts produces some consequential effects on a Listener's (L) thoughts, feelings, actions, such as, for instance, conviction, dissuasion, surprise, offence, humiliation, etc.[1] Such effects can be intentional or unintentional.[2] If S produces them intentionally, i.e., with the intention of producing them, then in this case, and only in this case, he performs a perlocutionary act. The sufficient and necessary conditions for performing a perlocutionary act seem then to be mainly the following three:

(i) S *intends* (i.e., he has the intention) to produce an effect X on L (for instance, S wants L to close the door);

(ii) S *says something* (i.e., he performs locutionary and illocutionary acts) in order to produce X ("Close the door!"; "Please, close the door!"; "Would you mind to close the door?", etc.);

(iii) the effect X, which corresponds to S's intention, occurs in L (he closes the door).

If we consider Austin's distinction between *attempt* at and *achievement* of a speech act (i.e., between "the act of attempting to do X" and "the act of doing, performing, achieving X")[3], we can reformulate the last two conditions in the following way:

(ii) S *attempts* to produce X by means of locutionary and illocutionary acts;

(iii) S *performs* a perlocutionary act (if and) only if he *achieves* (the production of) X.

If S doesn't achieve (if he fails) to produce X, he performs no perlocutionary act. Thus, the performance of a perlocutionary act corresponds to the achievement of the effect X and vice versa. Performance and achievement occur when intentions (i) and effects (iii) correspond to each other, meet each other. This means that S performs a perlocutionary act if and only if the effects he produces on L by saying something (i.e., by performing locutionary and illocutionary acts) are intentional. And when are such effects intentional? When they

A-T. Tymieniecka (ed.), Analecta Husserliana LIX, 521–532.

correspond to S's intentions. The performance (i.e., the achievement) of a perlocutionary act seems then to be made up of:

(i) S's "perlocutionary intentions" to produce X;
(ii) S's locutionary and illocutionary acts, which are means (-to-an-end) that S chooses in order to *act* on L's thoughts, feelings, actions;
(iii) "perlocutionary effects", which occur in L as a result of S's locutionary and illocutionary acts and which meet S's intentions.

In the present paper I intend to deal only with perlocutionary acts which produce effects on feelings and emotions.

If, for example, S performs the locutionary act of telling L "You are stupid!" or "You did a very good job!", in saying what he says he also performs the illocutionary act of *insulting* L or of *praising* him. Let us suppose that S's perlocutionary intentions are to *offend* L in the first case and to *flatter* him in the second one. Let us also suppose that L correctly understand the locutionary *meaning* (i.e., sense + reference) and the illocutionary *force* (it is an insult or praise) of S's utterances and that L feels offended after the first utterance and flattered after the second one. In this case, according to Austin, S would perform the perlocutionary acts of offending or flattering L. In fact, in order to *describe* from a perlocutionary viewpoint what happened between S and L, we can say "S offended/flattered L" or "S made L feel offended/flattered" or something along those lines.

These two linguistic structures are in fact the main ones we normally use in everyday life in order to *describe* perlocutionary acts, performed by S, which produce effects on feelings and emotions: we use verbs and verbal expressions of the following two types:

(i) "She *amused* me"; "You're *boring* me"; "I'll *astonish* him";
(ii) You *make me feel* angry/sad/glad/happy . . .".

Though such verbs and verbal expressions have a different syntactic surface structure, yet they share a *causal* semantic deep structure which could be schematized in the following way: somebody causes, caused or will cause a certain feeling or emotion in somebody else: anger, sadness, joy, boredom, astonishment, happiness, amusement . . .

We are dealing here with S's "causation", "production", "generation", "creation" of something new in L: a feeling, an emotion. For that reason, this kind of causation could be called *feeling causation* or *emotional causality*.

Focus on S

Verbs and verbal expressions of type (i) and (ii) solve in a causal way the problem concerning the relation between S, what he says and L: it is S who causes, by saying what he says, certain feelings in L; it is S who offends, flatters, amuses, bores . . . L.

Inside the communicative structure which is made up of S, his words and L, the whole focus is put on S and on what he says: L's effects are caused by S's words and then, after all, by S who utters them. The main role is given to S and to his words; he makes all the job: it is he who is the only one who acts, is active, performs an action, while L seems to be passive and dependent on what S says. Utterances (i) and (ii) do not acknowledge any autonomy to L but envision a whole dependence of L on S: L is wholly dependent on S.

AN IMPLICIT THEORY OF INTERPERSONAL RELATIONS (THE COMMON SENSE VIEWPOINT)

Utterances (i) and (ii) *are not simply ways of speaking*; they also *show what people believe* (think, are convinced) happens in linguistic communication, i.e. they convey an implicit theory of interpersonal relations, common sense, confirmed and reinforced every day by the language itself.[4]

As further example, let us suppose that after you have said something to me, I feel angry and I tell you: "You made me feel angry"; by this utterance I mean it is you who aroused my anger; because of what you said to me, I got angry. I also mean that if you had not said to me what you did or if you had said something different, I would not have got angry.

In this way, I *attribute* some effects which are inside me, internal to me (i.e., my feeling), to some causes which are outside me, external to me (i.e., your words and you). Such an attribution implies my firm belief that after your words I had no other way to react except that of feeling angry, i.e., my belief that your words *necessarily* caused (i.e., it is so and cannot be otherwise) my anger.

In some contexts, besides my thinking of you as the one who caused those effects, I can also attribute to you the intention of and the responsibility for having caused these effects. My "attributive reasoning" can then be paraphrased in the following way: 1) I feel angry (the effect) because 2) *you* made me feel angry by saying what you said (the causal link between your words and my anger), therefore, 3) you *intended* to make me feel angry (intention) and you succeeded in doing so (responsibility).

But let us even suppose that you actually intended to make me feel angry: in that case the effects which your words produce on my feelings meet your intentions. That would be – so to speak – the highest degree of emotional causality: you intended to cause a certain feeling in me and you performed a speech act which caused that feeling.

But even in this case, in which the effects in me meet your intentions perfectly, is it proper to describe what happened between you and me by saying "You made me feel angry" and to maintain that you and your words are the cause of my anger?

Similarly, in the first example it is quite proper to use illocutionary verbs to describe what happened between S and L and to say "S insulted/praised L", because to insult and to praise are two illocutionary acts and the performance of illocutionary acts depends on S. Is it proper as well to use perlocutionary verbs (or verbal expressions) to describe what happened between S and L and to say "S offended/flattered L" or "S made L feel offended/flattered"? In other words, just when the perlocutionary effects in L perfectly meet S's intentions (and then the perlocutionary act succeeds perfectly), is it proper to state that S and his words cause L's feelings?

PARADOXICAL CONSEQUENCES OF EMOTIONAL CAUSALITY

If such were the case, i.e., if my feelings were caused by what others said to me, then I grant myself no power over my feelings and so take upon myself the responsibility for them; I am not in charge of them.

I attribute power and responsibility to others: it is you who make me feel the way I feel. Vice versa, I can cause other people's feelings and have power over them: I can control them and take charge of them.

If such were the case, I would be the cause of and the one responsible for your feelings but not my own; you would be the cause of and the one responsible for my feelings but not your own. Each of us would be in the other's thrall. We both would be totally dependent on other people; we could have great power over others' feelings but no power over our own.

If I have such a belief, when I experience "unpleasant" feelings (sadness, anger, fear . . .), I can think of myself as a "victim" of others and so can accuse and blame them for that reason ("It's their fault if I'm sad/angry/scared . . ."); vice versa, when I experience "pleasant" feelings (joy, happiness, gladness . . .), I can reward others for that reason ("Thanks to him/her, I feel happy . . .").

Moreover, I feel I bear responsibility (i.e., blame or credit) for the feelings I arouse in other people. In any case, we have always to do with an emotional dependence: my feelings depend on other people (it is they who can make me feel bad or good) and other people's feelings depend on me (it is I who can make them feel good or bad).

TWO ALTERNATIVE THEORETICAL VIEWPOINTS

Focus on L

In the psychological field some theories of psychotherapy (such as F. Perls's Gestalt therapy, E. Berne's Transactional Analysis, R. and M. Goulding's Redecision therapy, R. Bandler and J. Grinder's Neurolinguistic Programming) maintain a viewpoint which is centered on L's emotional autonomy and independence and which is then antithetical to the any that would take S as its focus. According to them, no one is responsible for other people's actions, thoughts and feelings; each person is responsible not only for his own actions but also for his own thoughts and feelings (but not for those of other people); he has enough power and capability to be master of his own life. Personal responsibility, power and capability are often denied and externalized with different reasons being given, for example, the utterances (i) and (ii), so that people consider out of their control feelings which are their own responsibility.

> Sentences of this type, in fact, identify situations in which one person does some act and a second person *responds* by feeling a certain way. The point here is that, although the two events occur one after another, there is no necessary connection between the act of one person and the response of the other. Therefore, sentences of this type identify a model in which the client assigns responsibility for his emotions to people or forces outside his control. The act itself does not cause the emotion; rather, the emotion is a response generated from a model in which the client takes no responsibility for experiences which he *could* control. (Bandler and Grinder, 1975, pp. 51–52).

According to this viewpoint, in our examples, it is not S who, by saying what he says, offends/flatters L or makes him angry, but on the contrary it is L who offends/flatters himself or makes himself angry. In these expressions perlocutionary verbs are used in a reflexive way: the grammatical subject and direct object are no longer two different persons (as in the case of "*S* makes *me* feel angry"), they are the same and only one person ("*I* make *myself* feel angry").

For that reason, in a psychotherapy session, addressing a client who says "S's talk make me feel angry", R. and M. Goulding do not ask him questions such as "What did S tell you to make you feel angry?", because such questions would confirm client's belief that his anger is caused by S; on the contrary, they ask him: "When you are listening to S, what do you tell yourself in your head to make yourself angry?". By this kind of questioning they shift the focus away from S (where it lays in the client's description) to the client himself and they repropose to him his anger as a feeling which is totally his, which is dependent on him and not on S, as a feeling, then, which he can begin to feel himself the master of and be responsible for.

Thus, instead of using sentences with focused-on-S causal expressions such as "S's talk makes me feel angry", it would be more proper to use such *correlative sentences* as "S talks such and such talk *and* I feel angry/I make myself angry" or "*When* S talks such and such talk, I feel angry/I make myself angry", i.e., *co-ordinate* or *subordinate* sentences, in which causative verbs are used reflexively.

According to T. Kahler (1978, p. 277), in fact, the existence of a high correlation between two events (What *you* say and the anger *I* feel) and the probability of predicting exactly that if the first event occurs (i.e., if you say such and such), then the second one will occur too (i.e., I will feel angry), do not necessarily imply that the former causes the latter. Correlation and probability may have nothing to do, as in the present case, with cause and effect. I can make myself angry whenever you tell me such and such, but that does not mean that you make me feel angry; it only means that, by saying what you say, *you invite me* to feel angry. And invitations may be either accepted or refused . . .

Focusing on the Relation Between S and L

In the field of perception, Gestalt psychologist W. M. Metzger (1954, pp. 64–65) distinguishes three categories of global qualities of the objects we perceive: shape qualities ("round", "linear", . . .), material qualities ("smooth", "transparent", . . .), expressive qualities ("cheerful", "sad", . . .). To these he adds a fourth category of qualities ("attractive", "pleasant", "repugnant", "amusing", "boring", "interesting", . . .) which, unlike shape, material and expressive qualities, are not *object* qualities, i.e., they do not appertain to objects as objects: these concern the relation between the perceived object and the perceiving subject, and more precisely the particular *effect* of the relation on the perceiving subject. We can call them *effect qualities*.

Metzger's viewpoint, applied to linguistic communication, is not solely focused either on S or on L, but is focused on both, or – more precisely – on the *relation* between S and L. According to language and common sense, perlocutionary effects on L are the result of S's linguistic *action* (locutionary and illocutionary acts) on L. According to Metzger, effect qualities are the global outcome, which is experienced with phenomenal immediacy, of the *interaction* between S and L. Since they are global qualities, i.e., Gestalt qualities, they reflect some phenomenal *conditions* which are *structural* inasmuch as they concern some features of S, some of L, some of what S says and some of the relation between S and L, i.e., features of the particular and wider Gestalt that they all together constitute.[5]

According to this viewpoint, S and L are on the same level; the term "conditions" recognizes the contribution both of them make to the effect: the effect of a linguistic action by which S "causes" a feeling in L depends not only on the features of the one who "performs the action" (i.e., S), but also on the features of the one who "undergoes the action" (i.e., L). In contrast, causal language tends to consider L's features to be nonexistent and to exalt the features of S and his words.

In fact, though effect qualities are properties not of the perceived object or person but of the perceiving subject's feelings and result from the subject's relation with the object or person, everyday language attributes such qualities not to the perceiving subject but to the perceived object or person.

When I say "That movie is amusing", "This exercise is boring" or "Mary is depressing", I use the adjectives "amusing", "boring" and "depressing" as if they were object qualities, in particular as if they were expressive qualities of the movie, of the exercise, of Mary, as when I say "That movie is cheerful" or "Mary is sad". Expressive qualities such as "cheerful" and "sad", which are the movie's and Mary's properties, have to be distinguished from relational effect qualities such as "amusing" and "depressing", which concern the effect that my relation with that movie or with Mary has on *my* feelings.

A movie may be cheerful and not amuse me, just as my interaction with a sad person may not depress me. Cheerfulness and sadness are the movie's and Mary's properties, they do not depend on my feelings in contrast, amusement and depression are feelings which *I* experience over the movie and Mary. A person other than I could experience different feelings. But, instead of saying, as would be more correct, "I feel depressed, when I'm with Mary" (because, for example, she is sad), language allows me to say "Mary is depressing me" or even simply "Mary is depressing".

In this way qualities, which are *relative to the perceiving subject*, are presented by language as *object's absolute qualities* and the relation between object and subject is presented as a cause-effect relation: I – who am the (experiential or phenomenal) subject who is depressing, amusing or boring himself – become a (linguistic or grammatical) object, i.e., a direct object of the action of something else (movie, exercise) or someone else (Mary) which in its (or her) turn becomes the subject of the sentence: "It's the movie that amuses *me*"; "It's the exercise that bores *me*"; "It's Mary who depresses *me*". Here again appears a causal structure focused on the other-than-I, whether object or person. But, if the feelings I experience over objects and persons depend on me too, how is it possible to maintain that it is that-other-than-I which causes my feelings and to say "That movie is amusing me", "This exercise is boring me", "Mary is depressing me", i.e., how is it possible to use a linguistic structure according to which my feelings depend on objects and on other persons?

We could then ask ourselves: though there are other theoretical viewpoints and other possibilities in linguistic expression, why in everyday usage is causal language focused on S? How did a syntactic-semantic and conceptual organization focused on S come to prevail over the other ones possible?

WHY IS CAUSAL LANGUAGE FOCUSED ON S? THREE POSSIBLE ANSWERS

The common sense answer takes the following line: we talk the way we do (i.e., we use causal expressions (i) and (ii)), because emotional causality *really exists*; *it is true* that others make us feel a certain way, and just as true that we make others feel a certain way.

According to Gestalt theory, such an answer is to be thought of as "naive", because it implies that language refers to *transphenomenal reality*: here the existence of emotional causality in transphenomenal reality accounts for the existence of emotional causality in language.

In contrast, as far as the relations between language and the non-linguistic reality which language refers to are concerned, according to the "critical" or "less naive" viewpoint of Gestalt theory, language refers to *phenomenal reality*, i.e., not to the world but to our experience of the world. Thus, the answer to our question has to be looked for, first of all, in the scope of the relations between language and phenomenal reality.

Metzger (1954) thinks of phenomenal reality as a *continuum* in which it is possible to distinguish *perceived phenomenal reality* (here and now I perceive [inside or outside of me] something) from *represented phenomenal reality* (here and now I think/believe/imagine/remember . . . something).

If in our consideration of the relations between language and phenomenal reality we take up the thesis that linguistic structures determine native speakers' conceptualization of experience,[6] then this answer becomes possible: we believe, we think, we are convinced that other people cause our feelings because our language gives us a particular semantic structure which conveys precisely that causal logic. In short, we think of emotional causality the way we do because we talk the way we do.

This answer, compared with the common sense answer, has the merit of considering not transphenomenal reality but thought (i.e., our conceptualization of experience, i.e., represented phenomenal reality) to be the referent of language.

But I think that such an answer is only a pseudo-explanation, for it does not answer our question: Indeed, we think that way because we talk that way, but *why do we talk that way*? Where does that way of talking come from? Why did just that way of talking (and not one of other possible ways) prevail in the past and why does it still prevail every day?

A Gestalt-inspired answer comes from Albert Michotte's (1954) experimental phenomenology of the perception of mechanical causality. His experiments show that 1) mechanical causality is a phenomenal datum, 2) it is a perceived phenomenal datum before becoming a represented phenomenal datum, and 3) it is a perceived phenomenal datum without a transphenomenal correlate.

This means that mechanical causality is an immediate perceptual datum which strictly depends on a well-defined system of stimuli, i.e., on well-defined structural conditions of a spatial, temporal and kinetic nature. These conditions make the causal impression coercive; but it is sufficient to change them only a little in order that a causal impression disappear. Michotte's experiments show that a causal impression is not a question of "interpretation" due to acquired knowledge or thought; the causal meaning of an event is *intrinsic*, *immanent* in the event itself and independent of past experience or thought. In other terms, causality – as well as objects' shapes, movements etc. – is a global property, a Gestalt quality which imposes itself in a coercive way on our perception without any mediation of thought or past experience.[7] In short, causality could be thought of as an a priori of perception.

Thus, the answer to our question could be found by seeing the perceptual experience which language refers to as governing: then the correlate of the language of emotional causality would be seen to lie first of all on the perceptual phenomenal level.

If I am convinced (i.e., do not doubt it) that other people cause my feelings, it is because I find the highest degree of consistency between what I *experience* daily in the course of my communication with others and the meaning that these particular causal expressions offer me in conceptualizing (or representing) my *experience*. The word "experience" here has to be understood not as represented but as perceived phenomenal reality for the causal link between language and feelings is first of all not a representation, a thought, but a perception, and only then does the link become a representation (thought, belief, conviction or prejudice), it being an evidence which we perceive with immediacy.

University of Macerata
Italy

NOTES

[1] See Austin 1962, pp. 101, 108, 120.
[2] See Austin 1962, pp. 105, 106, 109.
[3] See Austin 1962, pp. 105, 109, 125.

[4] Whoever has direct experience of group psychotherapy knows how this belief is widespread and deep-rooted and how obvious and natural its content appears to those who share it.

[5] On the relational character of causal events see Lewin 1936, Chapter V. On the prescientificality of the notion of "cause" and on the scientificality of the notion of "conditions", see Metzger 1954, Chapter VII.

[6] See, for example, Whorf 1956.

[7] See Michotte 1950b, 1955a, 1955b, 1959.

[8] On the regarding the emotions as functional connections, see Michotte 1950a. Among the types of causality Michotte investigated, the one closest to emotional causality in language and verbal communication seems to be that of instrumental causality (see Michotte 1951): locutionary and illocutionary acts are S's *means* of achieving perlocutionary effects on L. For a wider discussion on perlocutionary acts and the relations between language and emotional causality, see Zuczkowski 1995, Chapter VII.

REFERENCES

Austin, J. L., *How to Do Things with Words* (Oxford: Oxford University Press, 1962).

Bandler, R. and Grinder, J., *The Structure of Magic* (Palo Alto: Science and Behavior Books, 1975).

Berne, E., *Transactional Analysis in Psychotherapy* (New York: Grove Press, 1961).

Goudling, M. M. and Goulding, R. L., *The Power is in the Patient* (San Franciso: TA Press, 1978).

Goudling, M. M. and Goulding, R. L., *Changing Lives through Redecision Therapy* (New York: Brunner and Mazel, 1979).

Kahler, T., *Transactional Analysis Revisited* (Little Rock, Arkansas: Human Development Publications, 1978).

Lewin, K., *Principles of Topological Psychology* (New York: McGraw-Hill, 1936).

Metzger, W., *Psychologie* (Darmstadt: Steinkopff, 1954).

Michotte, A., "The Emotions Regarded as Functional Connections," in M. L. Reymert (ed.), *Feelings and Emotions: The Mooseheart Symposium* (New York: McGraw-Hill, 1950a), pp. 114–126.

Michotte, A., "La préfiguration dans les données sensorielles de notre conception spontanée du monde physique," *Proceedings and Papers of the XIIth International Congress of Psychology* (Edinburg: Oliver and Boyd, 1950b), pp. 20–22.

Michotte, A., "La perception de la fonction outil", in *Essays in Psychology, Dedicated to David Katz* (Uppsala: Almquist and Wiksells, 1951), pp. 193–213.

Michotte, A., *La perception de la causalité* (Louvain: Publications Universitaires, 1954).

Michotte, A., "Perception et cognition," *Acta Psychologica* 11 (1955a), pp. 70–91.

Michotte, A., L'influence de l'expériene sur la structuration des données sensorielles dans la perception," in A. Michotte, H. Piéron and J. Piaget (eds.), *La perception: Deuxième Symposium de l'Association de Psychologie de Langue Française* (Louvain, 1953) (Paris: Presses Universitaires de France, 1955b), pp. 31–45.

Michotte, A., "Réflexions sur le rôle du language dans l'analyse des organisations perceptives", *Acta Psychologica* 15 (1959), pp. 70–91.

Perls, F. S., Hefferline, R. F. and Goodman, P., *Gestalt Therapy* (New York: The Julian Press, 1951).
Whorf, B. L., *Language, Thought and Reality* (Cambridge, Massachusetts: MIT Press, 1956).
Zuczkowski, A., *Strutture dell'esperienza e strutture del linguaggio* (Bologna: CLUEB, 1995).

INDEX OF NAMES

Analecta Husserliana

The Yearbook of Phenomenological Research

Editor-in-Chief
Anna-Teresa Tymieniecka
The World Institute for Advanced Phenomenological Research and Learning, Belmont, Massachusetts, U.S.A.

1. Tymieniecka, A-T. (ed.), *Volume 1 of Analecta Husserliana.* 1971 ISBN 90-277-0171-7
2. Tymieniecka, A-T. (ed.), *The Later Husserl and the Idea of Phenomenology.* Idealism – Realism, Historicity and Nature. 1972 ISBN 90-277-0223-3
3. Tymieniecka, A-T. (ed.), *The Phenomenological Realism of the Possible Worlds.* The "A Priori', Activity and Passivity of Consciousness, Phenomenology and Nature. 1974 ISBN 90-277-0426-0
4. Tymieniecka, A-T. (ed.), *Ingardeniana.* A Spectrum of Specialised Studies Establishing the Field of Research. 1976 ISBN 90-277-0628-X
5. Tymieniecka, A-T. (ed.), *The Crisis of Culture.* Steps to Reopen the Phenomenological Investigation of Man. 1976 ISBN 90-277-0632-8
6. Tymieniecka, A-T. (ed.), *The Self and the Other.* The Irreducible Element in Man, Part I. 1977 ISBN 90-277-0759-6
7. Tymieniecka, A-T. (ed.), *The Human Being in Action.* The Irreducible Element in Man, Part II. 1978 ISBN 90-277-0884-3
8. Nitta, Y. and Hirotaka Tatematsu (eds.), *Japanese Phenomenology.* Phenomenology as the Trans-cultural Philosophical Approach. 1979 ISBN 90-277-0924-6
9. Tymieniecka, A-T. (ed.), *The Teleologies in Husserlian Phenomenology.* The Irreducible Element in Man, Part III. 1979 ISBN 90-277-0981-5
10. Wojtyła, K., *The Acting Person.* Translated from Polish by A. Potocki. 1979 ISBN Hb 90-277-0969-6; Pb 90-277-0985-8
11. Ales Bello, A. (ed.), *The Great Chain of Being* and *Italian Phenomenology.* 1981 ISBN 90-277-1071-6
12. Tymieniecka, A-T. (ed.), *The Philosophical Reflection of Man in Literature.* Selected Papers from Several Conferences held by the International Society for Phenomenology and Literature in Cambridge, Massachusetts. Includes the essay by A-T. Tymieniecka, *Poetica Nova.* 1982 ISBN 90-277-1312-X
13. Kaelin, E. F., *The Unhappy Consciousness.* The Poetic Plight of Samuel Beckett. An Inquiry at the Intersection of Phenomenology and literature. 1981 ISBN 90-277-1313-8
14. Tymieniecka, A-T. (ed.), *The Phenomenology of Man and of the Human Condition.* Individualisation of Nature and the Human Being. (Part I:) Plotting the Territory for Interdisciplinary Communication. 1983 *Part II* see below under Volume 21. ISBN 90-277-1447-9

Analecta Husserliana

15. Tymieniecka, A-T. and Calvin O. Schrag (eds.), *Foundations of Morality, Human Rights, and the Human Sciences.* Phenomenology in a Foundational Dialogue with Human Sciences. 1983 ISBN 90-277-1453-3
16. Tymieniecka, A-T. (ed.), *Soul and Body in Husserlian Phenomenology.* Man and Nature. 1983 ISBN 90-277-1518-1
17. Tymieniecka, A-T. (ed.), *Phenomenology of Life in a Dialogue Between Chinese and Occidental Philosophy.* 1984 ISBN 90-277-1620-X
18. Tymieniecka, A-T. (ed.), *The Existential Coordinates of the Human Condition: Poetic – Epic – Tragic.* The Literary Genre. 1984 ISBN 90-277-1702-8
19. Tymieniecka, A-T. (ed.), *Poetics of the Elements in the Human Condition.* (Part 1:) The Sea. From Elemental Stirrings to Symbolic Inspiration, Language, and Life-Significance in Literary Interpretation and Theory. 1985
    For Part 2 and 3 *see below* under Volumes 23 and 28. ISBN 90-277-1906-3
20. Tymieniecka, A-T. (ed.), *The Moral Sense in the Communal Significance of Life.* Investigations in Phenomenological Praxeology: Psychiatric Therapeutics, Medical Ethics and Social Praxis within the Life- and Communal World. 1986 ISBN 90-277-2085-1
21. Tymieniecka, A-T. (ed.), *The Phenomenology of Man and of the Human Condition.* Part II: The Meeting Point Between Occidental and Oriental Philosophies. 1986 ISBN 90-277-2185-8
22. Tymieniecka, A-T. (ed.), *Morality within the Life- and Social World.* Interdisciplinary Phenomenology of the Authentic Life in the "Moral Sense'. 1987
    Sequel to Volumes 15 and 20. ISBN 90-277-2411-3
23. Tymieniecka, A-T. (ed.), *Poetics of the Elements in the Human Condition.* Part 2: The Airy Elements in Poetic Imagination. Breath, Breeze, Wind, Tempest, Thunder, Snow, Flame, Fire, Volcano . . . 1988 ISBN 90-277-2569-1
24. Tymieniecka, A-T., *Logos and Life*. Book I: Creative Experience and the Critique of Reason. 1988 ISBN Hb 90-277-2539-X; Pb 90-277-2540-3
25. Tymieniecka, A-T., *Logos and Life*. Book II: The Three Movements of the Soul. 1988 ISBN Hb 90-277-2556-X; Pb 90-277-2557-8
26. Kaelin, E. F. and Calvin O. Schrag (eds.), *American Phenomenology.* Origins and Developments. 1989 ISBN 90-277-2690-6
27. Tymieniecka, A-T. (ed.), *Man within his Life-World.* Contributions to Phenomenology by Scholars from East-Central Europe. 1989 ISBN 90-277-2767-8
28. Tymieniecka, A-T. (ed.), *The Elemental Passions of the Soul.* Poetics of the Elements in the Human Condition, Part 3. 1990 ISBN 0-7923-0180-3
29. Tymieniecka, A-T. (ed.), *Man's Self-Interpretation-in-Existence.* Phenomenology and Philosophy of Life. – Introducing the Spanish Perspective. 1990 ISBN 0-7923-0324-5
30. Rudnick, H. H. (ed.), *Ingardeniana II.* New Studies in the Philosophy of Roman Ingarden. With a New International Ingarden Bibliography. 1990 ISBN 0-7923-0627-9

Analecta Husserliana

31. Tymieniecka, A-T. (ed.), *The Moral Sense and Its Foundational Significance: Self, Person, Historicity, Community.* Phenomenological Praxeology and Psychiatry. 1990 ISBN 0-7923-0678-3
32. Kronegger, M. (ed.), *Phenomenology and Aesthetics.* Approaches to Comparative Literature and Other Arts. Homages to A-T. Tymieniecka. 1991 ISBN 0-7923-0738-0
33. Tymieniecka, A-T. (ed.), *Ingardeniana III.* Roman Ingarden's Aesthetics in a New Key and the Independent Approaches of Others: The Performing Arts, the Fine Arts, and Literature. 1991
Sequel to Volumes 4 and 30 ISBN 0-7923-1014-4
34. Tymieniecka, A-T. (ed.), *The Turning Points of the New Phenomenological Era.* Husserl Research – Drawing upon the Full Extent of His Development. 1991 ISBN 0-7923-1134-5
35. Tymieniecka, A-T. (ed.), *Husserlian Phenomenology in a New Key.* Intersubjectivity, Ethos, the Societal Sphere, Human Encounter, Pathos. 1991 ISBN 0-7923-1146-9
36. Tymieniecka, A-T. (ed.), *Husserl's Legacy in Phenomenological Philosophies.* New Approaches to Reason, Language, Hermeneutics, the Human Condition. 1991 ISBN 0-7923-1178-7
37. Tymieniecka, A-T. (ed.), *New Queries in Aesthetics and Metaphysics.* Time, Historicity, Art, Culture, Metaphysics, the Transnatural. 1991 ISBN 0-7923-1195-7
38. Tymieniecka, A-T. (ed.), *The Elemental Dialectic of Light and Darkness.* The Passions of the Soul in the Onto-Poiesis of Life. 1992 ISBN 0-7923-1601-0
39. Tymieniecka, A-T. (ed.), *Reason, Life, Culture, Part I.* Phenomenology in the Baltics. 1993 ISBN 0-7923-1902-8
40. Tymieniecka, A-T. (ed.), *Manifestations of Reason: Life, Historicity, Culture.* Reason, Life, Culture, Part II. Phenomenology in the Adriatic Countries. 1993 ISBN 0-7923-2215-0
41. Tymieniecka, A-T. (ed.), *Allegory Revisited.* Ideals of Mankind. 1994 ISBN 0-7923-2312-2
42. Kronegger, M. and Tymieniecka, A-T. (eds.), *Allegory Old and New.* In Literature, the Fine Arts, Music and Theatre, and Its Continuity in Culture. 1994 ISBN 0-7923-2348-3
43. Tymieniecka, A-T. (ed.): *From the Sacred to the Divine.* A New Phenomenological Approach. 1994 ISBN 0-7923-2690-3
44. Tymieniecka, A-T. (ed.): *The Elemental Passion for Place in the Ontopoiesis of Life.* Passions of the Soul in the *Imaginatio Creatrix.* 1995 ISBN 0-7923-2749-7
45. Zhai, Z.: *The Radical Choice and Moral Theory.* Through Communicative Argumentation to Phenomenological Subjectivity. 1994 ISBN 0-7923-2891-4
46. Tymieniecka, A-T. (ed.): *The Logic of the Living Present.* Experience, Ordering, Onto-Poiesis of Culture. 1995 ISBN 0-7923-2930-9

Analecta Husserliana

47. Tymieniecka, A-T. (ed.): *Heaven, Earth, and In-Between in the Harmony of Life.* Phenomenology in the Continuing Oriental/Occidental Dialogue. 1995 ISBN 0-7923-3373-X
48. Tymieniecka, A-T. (ed.): *Life. In the Glory of its Radiating Manifestations.* 25th Anniversary Publication. Book I. 1996 ISBN 0-7923-3825-1
49. Kronegger, M. and Tymieniecka, A-T. (eds.): *Life. The Human Quest for an Ideal.* 25th Anniversary Publication. Book II. 1996 ISBN 0-7923-3826-X
50. Tymieniecka, A-T. (ed.): *Life. Phenomenology of Life as the Starting Point of Philosophy.* 25th Anniversary Publication. Book III. 1997 ISBN 0-7923-4126-0
51. Tymieniecka, A-T. (ed.): *Passion for Place. Part II.* Between the Vital Spacing and the Creative Horizons of Fulfilment. 1997 ISBN 0-7923-4146-5
52. Tymieniecka, A-T. (ed.): *Phenomenology of Life and the Human Creative Condition.* Laying Down the Cornerstones of the Field. Book I. 1997 ISBN 0-7923-4445-6
53. Tymieniecka, A-T. (ed.): *The Reincarnating Mind, or the Ontopoietic Outburst in Creative Virtualities.* Harmonisations and Attunement in Cognition, the Fine Arts, Literature. Phenomenology of Life and the Human Creative Condition. Book II. 1997 ISBN 0-7923-4461-8
54. Tymieniecka, A-T. (ed.): *Ontopoietic Expansion in Human Self-Interpretation-in-Existence.* The I and the Other in their Creative Spacing of the Societal Circuits of Life. Phenomenology of Life and the Creative Condition. Book III. 1997 ISBN 0-7923-4462-6
55. Tymieniecka, A-T. (ed.): *Creative Virtualities in Human Self-Interpretation-in-Culture.* Phenomenology of Life and the Human Creative Condition. Book IV. 1997 ISBN 0-7923-4545-2
56. Tymieniecka, A-T. (ed.): *Enjoyment.* From Laughter to Delight in Philosophy, Literature, the Fine Arts and Aesthetics. 1998 ISBN 0-7923-4677-7
57. Kronegger M. and Tymieniecka, A-T. (eds.): *Life. Differentiation and Harmony... Vegetal, Animal, Human.* 1998 ISBN 0-7923-4887-7
58. Tymieniecka, A-T. and Matsuba, S. (eds.): *Immersing in the Concrete.* Maurice Merleau-Ponty in the Japanese Perspective. 1998 ISBN 0-7923-5093-6
59. Tymieniecka, A-T. (ed.): *Life - Scientific Philosophy/Phenomenology of Life and the Sciences of Life.* Ontopoiesis of Life and the Human Creative Condition. Book I. 1999 ISBN 0-7923-5141-X
60. Tymieniecka, A-T. (ed.): *Life - The Outburst of Life in the Human Sphere.* Scientific Philosophy / Phenomenology of Life and the Sciences of Life. Book II. 1999 ISBN 0-7923-5142-8

Kluwer Academic Publishers – Dordrecht / Boston / London